SIGNALS AND SYSTEMS:
Continuous and Discrete

Signals and Systems: Continuous and Discrete

FOURTH EDITION

Rodger E. Ziemer
University of Colorado—Colorado Springs

William H. Tranter
Virginia Polytechnic Institute and State University

D. Ronald Fannin
University of Missouri—Rolla

PRENTICE HALL
Upper Saddle River, NJ 07458

Library of Congress Cataloging-in-Publication Data
Ziemer, Rodger E.
    Signals and systems : continuous and discrete / Rodger E. Ziemer.
William H. Tranter, D. Ronald Fannin. —4th ed.
        p. cm.
    Includes bibliographical references (p.   –   ) and index.
    ISBN 0-13-496456-x
    1. System analysis   2. Discrete-time systems.   3. Digital filters
(Mathematics)   I. Tranter, William H.   II. Fannin, D. Ronald.
III. Title.
    QA402.Z53   1998
621.382′2—dc21   97–44182
                                                    CIP

Publisher: **TOM ROBBINS**
Acquisitions editor: **ALICE DWORKEN**
Production editor: **ROSE KERNAN**
Editor-in-chief: **MARCIA HORTON**
Cover designer: **BRUCE KENSELAAR**
Director of production and manufacturing: **DAVID W. RICCARDI**
Managing editor: **BAYANI MENDOZA DE LEON**
Manufacturing buyer: **JULIA MEEHAN**

The author and publisher of this book have used their best efforts in preparing this book. These
efforts include the development, research, and testing of the theories and programs to determine
their effectiveness. The author and publisher make no warranty of any kind, expressed or
implied, with regard to these programs or the documentation contained in this book. The author
and publisher shall not be liable in any event for incidental or consequential damages in
connection with, or arising out of, the furnishing, performance, or use of these programs.

Printed in the United States of America

10 9 8 7 6 5 4 3 2 1

ISBN 0-13-496456-X

Prentice-Hall International (UK) Limited, *London*
Prentice-Hall of Australia Pty. Limited, *Sydney*
Prentice-Hall Canada Inc., *Toronto*
Prentice-Hall Hispanoamericana, S.A., *Mexico*
Prentice-Hall of India Private Limited, *New Delhi*
Prentice-Hall of Japan, Inc., *Tokyo*
Simon & Schuster Asia Pte. Ltd., *Singapore*
Editora Prentice-Hall do Brasil, Ltda., *Rio de Janeiro*
Prentice-Hall, *Upper Saddle River, New Jersey*

Preface

The philosophy of the previous edition of *Signals and Systems* is retained in the fourth edition by continuing to stress the systems approach so that students are provided the tools and techniques for understanding and analyzing both continuous-time and discrete-time linear systems. While the systems approach is applicable to a very broad class of problems, liberal examples based on traditional circuit theory are included in the book to illustrate the various systems analysis techniques that are introduced.

The most obvious difference between this edition of *Signals and Systems* and previous editions is the inclusion of MATLAB for solving many of the in-chapter worked examples and the addition of problems and exercises at the end of chapters that require the use of computational aids. The selection of MATLAB as a computer tool was an easy choice for us. Over the past several years MATLAB has become widely used within the engineering curricula and has been integrated into a number of courses. MATLAB provides very powerful computational and graphical capabilities. MATLAB code is very concise and as a result it is possible to express complex ideas using very few lines of code. The many toolboxes available provide the support necessary for investigating problems in a variety of areas. As a result, once a student has learned to use MATLAB effectively as a computational and visualization tool, the student can continue to use it throughout his or her career. The availability of a Student Edition that is low-cost, windows-based, and well-documented was considered a very important attribute of MATLAB for this application. All of the exercises included in this book can be worked using the Student Edition, Version 5.

In developing the MATLAB examples, problems, and computer exercises that appear in this edition we kept two considerations in mind. First we strongly believe that computer support can aid learning by allowing students to attack problems in different ways and to consider a wider variety of problems. Students still need to understand basic analysis and synthesis techniques and therefore need to work a large number of problems using traditional (pencil and paper) methods. We therefore chose to use MATLAB as a *supplement* with the in-chapter example problems rather than as a replacement for the traditional analysis techniques. Taking this approach also allows instructors to use the book without committing to the use of MATLAB. In this case, students having interest in the MATLAB applications can pursue the MATLAB applications as individual study. Students are encouraged to use MATLAB for problem solving where appropriate. For the most part, the end-of-chapter problems do not designate the use of MATLAB or any other tool for problem solving. We have, however, placed a section entitled "Computer Exercises" at the end of each chapter. Most of these exercises extend the text material and give the student an opportunity to bring some originality into the problem solving effort.

The second consideration was a very strong desire to keep it simple. MATLAB, as stated previously, is a very compact language. As a result it is possible to combine many operations into a single line of code. Such operations often lead to code that is very difficult to comprehend without considerable effort. We have therefore avoided complicated statements and data structures. In addition we have avoided the use of graphical user interfaces. The code contained in this book, therefore, is not very elegant but we hope that it is easy for the student to understand. The main effort of the student at this point should be to understand the concepts of signal and system theory. Spending considerable time writing code or trying to understand code is not the best use of student time.

The authors have developed a software supplement to *Signals and Systems: Continuous and Discrete,* 4e. This supplement contains copies of all m-files used in this book. This software is available free of charge via file transfer protocol (FTP) from The Mathwork's world wide web site. The files may be found at

```
ftp://ftp.mathworks.com/pub/books/tranter/
```

Another feature of the fourth edition of *Signals and Systems* is that all of the end-of-chapter summaries are written as a point-by-point review of the important topics covered in the chapter. Such a listing can be a valuable study guide for the student. In addition, a number of new problems are contained in the fourth edition.

The organization of the fourth edition is identical to that of the third edition. The first seven chapters deal with continuous-time linear systems in both the time domain and the frequency domain. References to discrete-time signals and systems are made where appropriate so that the students can appreciate the relationship between discrete-time and continuous-time signals and systems. The principal tool developed for time-domain analysis is the convolution integral. Frequency-domain techniques include the Fourier and the Laplace transforms. An introduction to state-variable techniques is also included, along with treatments of both continuous-time and discrete-time state equations. The remainder of the book deals with discrete-time systems, including z-transform analysis techniques, digital filter analysis and synthesis, and the discrete Fourier transform and fast Fourier transform (FFT) algorithms. A new section on computer-aided filter design has been added to Chapter 9.

This organization allows the book to be covered in two three-semester-hour courses, with the first course being devoted to continuous-time signals and systems and the second course to discrete-time signals and systems. Alternatively, the material can be used as a basis for three quarter-length courses. With this format, the first course could cover time- and frequency-domain analysis of continuous-time systems. The second course could cover state variables, sampling, and an introduction to the z-transform and discrete-time systems. The third course could deal with the analysis and synthesis of digital filters and provide an introduction to the discrete analysis and synthesis of digital filters and provide an introduction to the discrete Fourier transform and its applications. Other groupings of topics are possible.

The assumed background of the student is mathematics through differential equations and the usual introductory circuit theory course or courses. Knowledge of the basic concepts of matrix algebra would be helpful but is not essential. An appendix is included to bring together the pertinent matrix relations that are used in Chapters 6 and 7. We feel that in most electrical engineering curricula the material presented in this book is best taught at the junior level.

Plan of the Text

We begin the book by introducing the basic concepts of signal and system models and system classifications. The idea of spectral representations of periodic signals is first introduced in Chapter 1 because we feel that it is important for the student to think in terms of both the time and the frequency domains from the outset.

The convolution integral and its use in fixed, linear system analysis by means of the principle of superposition are treated in Chapter 2. This chapter deals with system modeling and analysis in the time domain. The evaluation of the convolution integral is treated in detailed examples to provide reinforcement of the basic concepts. Calculation of the impulse response and its relation to the step and ramp responses of a system are discussed. The concepts of system modeling and system simulation are also treated in Chapter 2. This chapter concludes with a section devoted to the numerical solution of system equations using SIMULINK.

The Fourier series is introduced in Chapter 3. We have emphasized the elementary approach of approximating a periodic function by means of a trigonometric series and obtaining the expansion coefficients by using the orthogonality of sines and cosines. We do this because this is the first time many students are introduced to the Fourier series. The alternative generalized orthogonal function approach is included as a section at the end of this chapter for those who prefer it. The concept of the transfer function in terms of sinusoidal steady-state response of a system is discussed in relation to signal distortion.

The Fourier transform is the subject of Chapter 4, with its applications to spectral analysis and system analysis in the frequency domain. The concept of an ideal filter, as motivated by the idea of distortionless transmission, is also introduced at this point. The Gibbs phenomenon, window functions, and convergence properties of the Fourier coefficients are also included. In addition, the DFT has been introduced in Chapter 4 to be covered at the option of the instructor.

The Laplace transform and its properties are introduced in Chapter 5. Again, we have tried to keep the treatment as simple as possible because this is assumed to be a first exposure to the material for a majority of students, although a summary of complex variable theory is provided in Appendix C so that additional rigor may be used at the instructor's option. The derivation of Laplace transforms from elementary pairs is illustrated by example, as is the technique of inverse Laplace transformation using partial-fraction expansion. Optional sections on the evaluation of inverse Laplace transforms by means of the complex inversion integral and an introduction to the two-sided Laplace transform are also provided.

The application of the Laplace transform to network analysis is treated in detail in Chapter 6. The technique of writing Laplace-transformed network equations by inspection is covered and used to review the ideas of impedance and admittance matrices, which the student will have learned in earlier circuits courses for resistive networks. The concepts of zero-state response and zero-input response are discussed along with their relationship to transient and steady-state responses. The transfer function is treated in detail, and the Routh test for determining stability is presented. The chapter closes with a treatment of Bode plots and block diagram algebra for fixed, linear systems.

In Chapter 7 the concepts of a state variable and the formation of the state-variable approach to system analysis are developed. The state equations are solved using both time-domain and Laplace transform techniques, and the important properties of the solution are examined. As an example, we show how the state-variable method can be applied to the analysis of circuits. Discrete-time state equations and the concepts of controllability and observability are briefly introduced.

The final three chapters provide coverage of the topics of discrete-time signal and system analysis. Chapter 8 begins with a study of sampling and the representation of discrete-time systems. The sampling operation is covered in considerable detail. This is accomplished in the context of formulating a model for an analog-to-digital (A/D) converter so that the operation of quantizing can be given some physical basis. A brief analysis of the effect of quantizing sample values in the A/D conversion process is included as an introduction to quantizing errors. As a bonus, the student is given a basis upon which to select an appropriate wordlength of an A/D converter. Both ideal and approximate methods for reconstructing a signal from a sequence of samples are treated in detail. The z-transform, difference equations, and discrete-time transfer functions are developed with sufficient rigor to allow for competent problem solving but without the complications of contour integration. The material on the classification of discrete-time systems has been written in a way that parallels the similar material for continuous-time systems.

Chapter 9 allows students to use their knowledge of discrete-time analysis techniques to solve an important class of interesting problems. The idea of system synthesis, as opposed to system analysis, is introduced. Discrete-time integration is covered in considerable detail for several reasons. First, the idea of integration will be a familiar one. Thus students can appreciate the different information gained by a frequency-domain analysis as opposed to a time-domain analysis. In addition, the integrator is a basic building block for many analog systems. Finally, the relationship between trapezoidal integration and the bilinear z-transform is of sufficient importance to warrant a discussion of trapezoidal integration. The synthesis techniques for digital filters covered in this chapter are the standard ones. These are synthesis by time-domain invariance, bilinear z-transform syntheses, and synthesis through Fourier series expansion. Through the application of these techniques, the student is able to gain confidence in the previously developed theory. Since several synthesis techniques depend on knowledge of analog filter prototypes, Appendix E, which discusses several different filter prototypes, is included. Computer-aided digital filter design is also included in this chapter for the first time in this edition.

The discrete Fourier transform (DFT) and its realization through the use of fast Fourier transform (FFT) algorithms is the subject of Chapter 10. Both decimation-in-time and decimation-in-frequency algorithms are discussed. Several examples are provided to give the student practice in performing the FFT operations. We believe that this approach best leads to a good understanding of the FFT algorithms and their function. Basic properties of the DFT are summarized and a comparison of the number of operations required for the FFT as compared to the DFT is made. Several applications of the DFT are summarized and the use of windows in supressing leakage is discussed. This chapter closes with a discussion and illustration of FFT algorithms with arbitrary radixes and the chirp-z transform.

The book also contains a number of optional sections, denoted by an asterisk, that go deeper into specific topics than is often customary for a junior-level course. These topics can be eliminated without loss of continuity.

A complete solutions manual, which contains solutions to all problems and computer exercises, is available from the publisher as an aid to the instructor. The solutions manual for the fourth edition is in a typeset format and makes extensive use of MATLAB for a large number of problems and for solution of computer exercises. Answers to selected problems are provided in Appendix H as an aid to the student.

Acknowledgments

We wish to thank the many people who have contributed, both knowingly and unknowingly, to the development of this textbook. Special thanks go to our many students who have provided a living laboratory within which we could test-teach many sections of this textbook. Their comments and criticisms have been very valuable and are gratefully appreciated. Special thanks are also due to our many faculty colleagues, now too many to mention, who taught from the first and second editions of this book and have offered many suggestions for improvement. Thanks are also due to the reviewers who provided many insightful comments and suggestions for improvement for the fourth edition.

The reviewers for the fourth edition were:

Ronald H. Brown, Marquette University
Malur K. Sundareshan, University of Arizona at Tucson

We also greatly appreciate the assistance of those who reviewed all or parts of previous editions, and we acknowledge their help and contributions to our efforts. These include:

Peter L. Balise, University of Washington; W. A. Blackwell, Virginia Tech; Peter Bofah, Howard University; Gordon E. Carlson, University of Missouri-Rolla; Ralph S. Carson, University of Missouri—Rolla; Hua Ting Chieh, Howard University; Mickey D. Cox, Louisiana Technological University;

F. Dianat, Rochester Institute of Technology; Jimmy L. Dodd, Mississippi State University; John L. Fleming, Texas A&M University; Victor Frost, University of Kansas; Neal Gallagher, Purdue University; Russell A. Hannen, Wright State University; Brian Harms, Kansas State University; Dennis W. Herr, Ohio Northern University; R. Dennis Irwin, Mississippi State University; K. Ross Johnson, Michigan Technological University; Bruce Johansen, Ohio Northern University; Mark Jong, Wichita State University; Fabien Josse, Marquette University; Saleem Kassam, University of Pennsylvania; John M. Liebetreu, University of Colorado—Colorado Springs; John H. Lilly, University of Kentucky; Wu-sheng Lu, University of Victoria; Rushdi Muammar, University of Illinois—Chicago; D. M. Petrela, SUNY Farmingdale; William A. Porter, Louisiana State University; Arifur Rahman, University of South Alabama; Michael Rudko, Union College; John V. Wait, University of Arizona; Mark Wickert, University of Colorado—Colorado Springs; and Chu Huiw Wu, University of Missouri—Rolla.

Last, but certainly not least, we wish to thank our wives, Sandy, Judy, and Sara, and our families for the patience shown as we worked on the fourth edition of this book. They have by now become well accustomed to our endless writing projects.

R. E. Z.
W. H. T.
D. R. F.

Contents

8 Discrete-Time Signals and Systems   350

9 Analysis and Design of Digital Filters   415

CHAPTER 1

Signal and System Modeling Concepts

1-1 Introduction

This book deals with *systems* and the interaction of *signals* in systems. A system, in its most general form, is defined as a combination and interconnection of several components to perform a desired task.[†] Such a task might be the measurement of the acceleration of a rocket or the transmission of a message from New York to Los Angeles. The measurement of the acceleration might make use of visual observation of its position versus time. An equally unsophisticated solution to the message delivery problem might use a horse and rider. Obviously, more complex solutions are possible (and probably better). Note, however, that our definition is sufficiently general to include them all.

We will be concerned primarily with *linear* systems. Such a restriction is reasonable because many systems of engineering interest are closely approximated by linear systems and very powerful techniques exist for analyzing them. We consider several methods for analyzing linear systems in this book. Although each of the methods to be considered is general, not all of them are equally convenient for any particular case. Therefore, we will attempt to point out the usefulness of each.

A *signal* may be considered to be a function of time that represents a physical variable of interest associated with a system. In electrical systems, signals usually represent currents and voltages, whereas in mechanical systems, they might represent forces and velocities (or positions).[‡] In the example mentioned above, one of the signals of interest represents the acceleration, but this could be integrated to yield a signal proportional to velocity. Since electrical voltages and currents are relatively easy to process, the original signal representing acceleration, which is a mechanical signal, would probably be converted to an electrical one before further *signal processing* takes place. Examples illustrating these remarks will be given in the next section.

Just as there are several methods of systems analysis, there are several different ways of representing and analyzing signals. They are not all equally convenient in any particular situation. As we study methods of signal representation and system analysis we will attempt to point out useful applications of the techniques.

So far, the discussion has been rather general. To be more specific and to fix more clearly the ideas we have introduced, we will expand on the acceleration measurement and the message delivery problems already mentioned.

[†]The Institute of Electrical and Electronics Engineers dictionary defines a system as "an integrated whole even though composed of diverse, interacting structures or subjunctions."

[‡]More generally, a signal can be a function of more than one independent variable, such as the pressure on the surface of an airfoil, which is a function of three spatial variables *and* time.

1-2 Examples of Systems

EXAMPLE 1-1

An accelerometer, consisting of a spring-balanced weight on a frictionless slide, is shown schematically in Figure 1-1. It is to be used to measure the longitudinal acceleration of a rocket whose acceleration profile is shown in Figure 1-2a. (a) If the weight used to indicate acceleration is 2 g, determine the spring constant K such that the departure of the weight from its equilibrium position at the maximum acceleration of 40 m/s², achieved at $t = 72$ s from launch, is 1 cm. (b) If a minimum increment of 0.5 mm of movement by the weight can be detected, to what minimum increment of acceleration does this correspond? (c) Derive the profile for the longitudinal velocity of the rocket and sketch it.

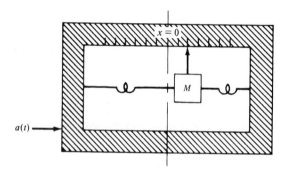

FIGURE 1-1. System for measuring the acceleration of a rocket.

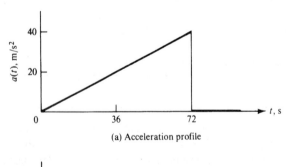

(a) Acceleration profile

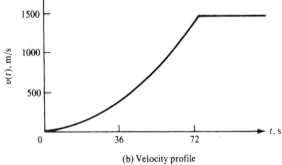

(b) Velocity profile

FIGURE 1-2. Acceleration and velocity profiles for rocket.

Solution:

(a) The force, Ma, on the weight due to acceleration is balanced by the force, Kx, due to the spring tension. Thus

$$Ma = Kx \tag{1-1}$$

where x is the departure from the equilibrium position, taken to be $x = 0$, M is the mass of the weight, K is the spring constant, and a represents acceleration of the rocket. Solving for K, we obtain

$$K = \frac{Ma_{max}}{x_{max}} = \frac{(0.002 \text{ kg})(40 \text{ m/s}^2)}{0.01 \text{ m}} = 8 \text{ kg/s}^2 \tag{1-2}$$

(b) With $\Delta x_{min} = 0.5 \text{ mm} = 0.0005 \text{ m}$, we have

$$\Delta a_{min} = \frac{K \Delta x_{min}}{M} = \frac{(8)(0.0005)}{0.002} = 2 \text{ m/s}^2 \tag{1-3}$$

(c) For $t > 0$, let the acceleration as a function of time be represented as[†]

$$a(t) = \begin{cases} \alpha t, & t \le t_0 \\ 0, & t > t_0 \end{cases} \tag{1-4}$$

where $t_0 = 72$ s and $\alpha = 5/9 \text{ m/s}^3$ for this example. The velocity is obtained by integration of $a(t)$, which yields

$$v_r(t) = \begin{cases} \alpha t^2/2, & t \le t_0 \\ \alpha t_0^2/2, & t > t_0 \end{cases} \tag{1-5}$$

Using the given value for α, the velocity at $t = t_0$ (burnout) is

$$v_r(72) = \frac{5}{9} \frac{(72)^2}{2} = 1{,}440 \text{ m/s} = 4{,}724 \text{ ft/s} = 3{,}221 \text{ mi/hr} \tag{1-6}$$

The velocity as a function of time is shown in Figure 1-2b.

EXAMPLE 1-2 _____

To determine velocity by means of the accelerometer of the previous example, an operational amplifier integrating circuit is used as shown in Figure 1-3. The wiper of the potentiometer at the input is tied to the accelerometer weight and can provide a maximum input voltage of 0.1 V depending on whether the acceleration is positive or negative. The operational amplifier is ideal (infinite input resistance and infinite gain). However, its maximum output is limited to 10 V. Choose RC such that the operational amplifier output will not be overdriven for the maximum velocity expected (assume $R_p << R$).

Solution: Kirchhoff's current law at the node joining R, C, and the amplifier input can be expressed as

$$\frac{v_1 - v_s}{R} + C \frac{d}{dt}(v_1 - v_0) = 0 \tag{1-7}$$

[†]But for a few exceptions, the signals in this chapter are assumed to be zero for $t < 0$.

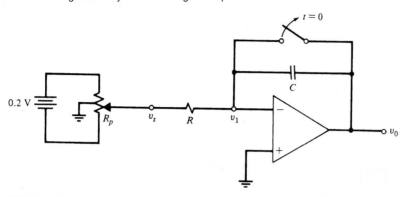

FIGURE 1-3. Integrator circuit used to determine velocity from the accelerometer analyzed in Example 1-1.

However, $v_1 \approx 0$ at the negative input of the operational amplifier since its infinite gain constrains voltages at its inverting and noninverting inputs to be approximately equal, and the positive input is grounded. Thus Kirchhoff's current law equation, when rearranged, simplifies to

$$v_0(t) = -\frac{1}{RC} \int_0^t v_s(\lambda) \, d\lambda \tag{1-8}$$

where the capacitor is assumed initially uncharged. From the previous example, we take the voltage at the potentiometer wiper to be

$$v_s(t) = \beta t, \qquad 0 \le t \le t_0 \tag{1-9}$$

where $v_s(t_0) = 0.1$ V, giving $\beta = 0.1/72 \approx 1.4 \times 10^{-3}$ V/s. Thus, at t_0, the operational amplifier output voltage is

$$v_0(t_0) = -\frac{1}{RC} \frac{\beta t_0^2}{2} \tag{1-10}$$

Setting this equal to -10 V, which is the constraint imposed by the operational amplifier at its output, and solving for RC, we obtain

$$RC = \frac{\beta t_0^2}{20} = \frac{(0.1/72)(72)^2}{20} = 0.36 \text{ s} \tag{1-11}$$

For $R = 50$ kΩ, the required value of C is 7.2 μF.

EXAMPLE 1-3

Another way to carry out the integration operation of the previous example (or any signal processing function, for that matter) is by sampling the voltage analog to the acceleration signal each T seconds, quantizing the samples so that they can be represented numerically, and performing the integration numerically. It will be shown in Chapter 8 that all the information present in a signal can be represented by sample values taken sufficiently often if the signal is *bandlimited* (a term to be precisely defined later). Two simple integration methods that might be used are the rectangular and trapezoidal rules. These are illustrated in Figure 1-4. The former approximates the area under a curve as the sum of a series of "boxcars" or rectangular areas, and can be carried out as the algorithm

$$\hat{v}[(n + 1)T] = \hat{v}(nT) + Ta[(n + 1)T] \tag{1-12}$$

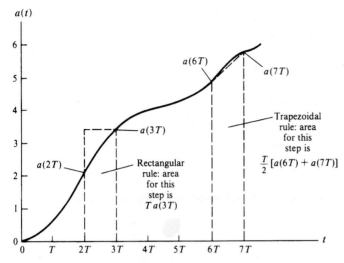

FIGURE 1-4. Illustration of discrete-time integration rules.

where $\hat{v}(nT)$ is the approximation to the velocity at sampling time nT, T is the sampling interval, and $a(nT)$ is the acceleration at sampling time nT. If trapezoidal integration is used, the area under a curve is approximated as a sum of contiguous trapezoids of width T seconds. The algorithm for finding the velocity from the acceleration in this case is given by

$$\hat{v}[(n + 1)T] = \hat{v}(nT) + \{a(nT) + a[(n + 1)T]\}(T/2) \tag{1-13}$$

TABLE 1-1
Numerical Results for Example 1-2

n	nT, s	$v(nT)$, rect.	$v(nT)$, trap.
0	0	0	0
1	4	8.9	4.4
2	8	26.7	17.8
3	12	53.3	40.0
4	16	88.9	71.1
5	20	133.3	111.1
6	24	186.7	160.0
7	28	248.9	217.8
8	32	320.0	284.4
9	36	400.0	360.0
10	40	488.9	444.4
11	44	586.7	537.8
12	48	693.3	640.0
13	52	808.9	751.1
14	56	933.3	871.1
15	60	1066.7	1000.0
16	64	1208.9	1137.8
17	68	1360.0	1284.4
18	72	1520.0	1440.0

Note that these algorithms are *recursive* in that the next approximate value for the velocity is computed from the old approximate value plus the sample value for the acceleration. The recursive structure can be removed by writing the equations for $n = 0, 1, 2, \ldots$ and doing a substitution of the one for $n = 0$ into the one for $n = 1$, then this result into the one for $n = 2$, and so on. In the case of the trapezoidal rule, two sample values for the acceleration are used—the present one and the immediate past one. Note, also, that the T in the argument of the various sampled signals is unnecessary—the index n is the independent variable. More will be said about integration algorithms in Chapter 9. To finish this example, we give a table of values for both algorithms. Note that for the acceleration profile shown in Figure 1-2, the trapezoidal rule gives exact results. Had the acceleration profile not been linear, this would not have been the case.

The starting point of any systems analysis or design problem is a *model* which, no matter how refined, is *always an idealization of a real-world (physical) system.* Hence the result of any systems analysis is an idealization of the true state of affairs. Nevertheless, if the model is sufficiently accurate, the results obtained will portray the operation of the actual system sufficiently accurately to be of use.

The previous examples illustrate the concept of a system and the *design* of systems to accomplish desired tasks. Each example involved the concept of a *signal*. In Example 1-1, the signal was the displacement of the acceleration measuring weight. This was coupled to the operational amplifier system of Example 1-2 to produce *voltage signals*, one proportional to acceleration and the other proportional to velocity.

Another concept demonstrated by the first two examples is that of a subsystem. When taken separately, we can speak of each as a *system,* whereas both taken together also compose a system. To avoid confusion in such cases, we refer to the separate parts (accelerometer and signal-processing integrator in this case) as *subsystems.* The third example illustrated the idea of *digital signal processing,* a concept explored later in the book.

After considering one more example, which is somewhat different from the first three, the remainder of this chapter will be concerned with an introduction to *useful signal models.* In Chapter 2, methods for describing and analyzing systems in the *time domain* will be examined. Throughout the rest of the book we consider other methods of signal and system analysis.

EXAMPLE 1-4 Communications Link

Figure 1-5 is a pictorial representation of a two-way communications link as might exist, for example, between New York and Los Angeles. It might consist entirely of earth-based links such as wire lines and microwave links. Alternatively, a relay satellite could be employed. Regardless of the particular mechanization used, the systems analyst must decide on the most important aspects of the physical system for the application and attempt to represent them by a model.

Let us suppose that the link employs a synchronous-orbit satellite repeater, stationed 35,784 km above the earth's equator. At this altitude the satellite is stationary with respect to the earth's surface since the period of the satellite's orbit is equal to one day. Let us also suppose that the analyst is concerned primarily about echoes in the system due to reflections of the electromagnetic-wave carriers at the satellite and at the receiving ground station. Assume that the transmitter and receiver are equidistant at a slant distance $d = 40{,}000$ km from the satellite, and that the delay (up or down) is

$$\tau = \frac{d}{c} = \frac{(40{,}000 \text{ km})(1000 \text{ m/km})}{3 \times 10^8 \text{ m/s}}$$

$$= 0.13 \text{ s}$$

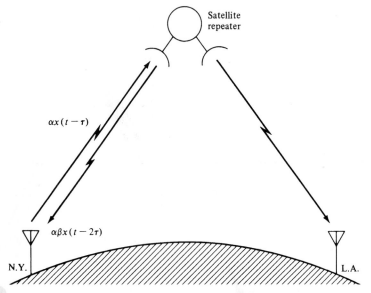

FIGURE 1-5.   Satellite communications link.

where $c = 3 \times 10^8$ m/s is the velocity of electromagnetic propagation in free space. The slant distance d accounts for the satellite being located above the earth's equator midway between the ground stations, each of which is assumed to be at a latitude of 40° north. The signal received at the satellite is

$$s_{sat}(t) = \alpha x(t - \tau), \qquad \tau \le t \le T + \tau \tag{1-14a}$$

where $x(t)$ is the transmitted signal and $[0, T]$ is the interval of time over which the transmission takes place. The parameter α is the attenuation, which accounts for various system characteristics as well as the spreading of the electromagnetic wave as it propagates from the transmitter antenna. A portion of $s_{sat}(t)$ is reflected and returned to the New York ground station terminal to be received by the receiver at that end; we represent it as

$$s_{refl}(t) = \alpha\beta x(t - 2\tau), \qquad 2\tau \le t \le T + 2\tau \tag{1-14b}$$

where β is another attenuation factor. The remainder of $s_{sat}(t)$ is relayed to the Los Angeles ground station after processing by the satellite repeater. A portion of this signal may be reflected, but we will assume its effect to be negligible compared to that of $\alpha\beta x(t - 2\tau)$ when received at the New York station. Thus a speaker at New York will hear

$$s(t) \simeq x(t) + \alpha\beta x(t - 2\tau), \qquad 0 \le t \le T + 2\tau \tag{1-15}$$

That is, the speaker will hear his undelayed speech as well as an attenuated version of his speech delayed by approximately

$$\Delta T = 2\tau = 0.26 \text{ s}$$

Psychologically, the effect of this can be very disconcerting to a speaker if the delayed signal is sufficiently strong; it is virtually impossible for the person to avoid stuttering. Thus a systems analyst would proceed by relating α and β to more fundamental systems parameters and designing for an acceptably small value of $\alpha\beta$.

1-3 Signal Models

Examples of Deterministic Signals

In this book we are concerned with a broad class of signals referred to as *deterministic*. Deterministic signals can be modeled as completely specified functions of time. For example, the signal

$$x(t) = \frac{At^2}{B + t^2}, \qquad -\infty < t < \infty \tag{1-16}$$

where A and B are constants, shown in Figure 1-6a, is a deterministic signal. An example of a deterministic signal that is not a continuous function of time is the unit pulse, denoted as $\Pi(t)$ and defined as

$$\Pi(t) = \begin{cases} 1, & |t| \leq \frac{1}{2} \\ 0, & \text{otherwise} \end{cases} \tag{1-17}$$

It is shown in Figure 1-6b.

A second class of signals, which will not be discussed in this book, are random signals. They are signals taking on random values at any given time instant and must be modeled probabilistically. A random signal is illustrated in Figure 1-6c.

Continuous-Time Versus Discrete-Time Signals

The signals illustrated in Figure 1-6 are examples of *continuous-time signals*. It is important to note that "continuous time" does not imply that a signal is a mathematically continuous function, but rather that it is a function of a *continuous-time variable*.

In some systems the signals are represented only at discrete values of the independent variable (i.e., time). Between these discrete-time instants the value of the signal may be zero, undefined, or of no interest. An example of such a *discrete-time* or *sample-data signal* is shown in Figure 1-7a. Often, the intervals between signal values are the same, but they need not be.

A distinction between discrete-time and quantized signals is necessary. A *quantized signal* is one whose values may assume only a countable[†] number of values, or levels, but the changes from level to level may occur at any time. Figure 1-7b shows an example of a quantized signal. A real-world situation that can be modeled as a quantized signal is the opening and closing of a switch.

Discrete-time signals will be considered in Chapter 8. The following example provides an illustration of two such signals.

EXAMPLE 1-5 _____

Two common discrete-time signals are the unit pulse and unit step signals. The first is defined by the equation*

$$\delta[n] = \begin{cases} 1, & n = 0 \\ 0, & \text{otherwise} \end{cases} \tag{1-18}$$

and the second by the equation

$$u[n] = \begin{cases} 1, & n \geq 0 \\ 0, & n < 0 \end{cases} \tag{1-19}$$

where n takes on only integer values.

[†]A *countable* set is a set of objects whose members can be put into one-to-one correspondence with the positive integers. For example, the sets of all integers and of all rational numbers are countable, but the set of all real numbers is not.

*The brackets are used to indicate a discrete-time signal.

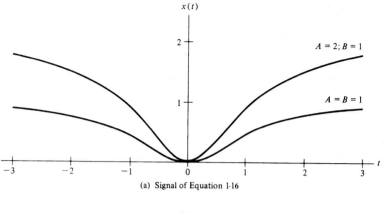

(a) Signal of Equation 1-16

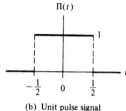

(b) Unit pulse signal

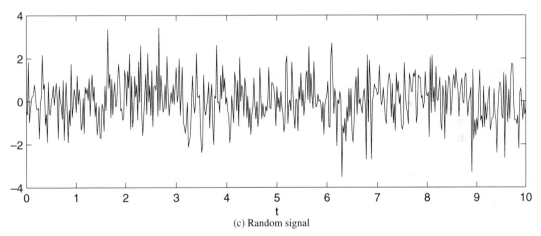

(c) Random signal

FIGURE 1-6.   Graphical representation of two deterministic continuous-time signals and a random signal.

Other discrete-time signals may be built from these elementary signals. For example, a pulse consisting of five 1's in a row starting at $n = 0$ and ending at $n = 4$ and 0's elsewhere may be represented in terms of the discrete-time unit step as

$$p[n] = u[n] - u[n - 5] = \begin{cases} 1, & 0 \leq n \leq 4 \\ 0, & \text{otherwise} \end{cases}$$

The student should verify this by sketching $u[n]$, $u[n - 5]$, and their difference.

$x_s(t)$

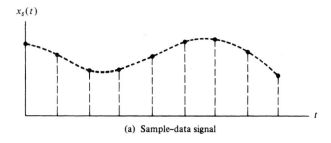

(a) Sample–data signal

$x_d(t)$

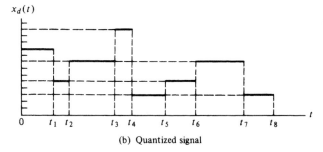

(b) Quantized signal

FIGURE 1-7. Sample-data, or discrete-time, and quantized signals.

Periodic and Aperiodic Signals

A signal $x(t)$ is *periodic* if and only if

$$x(t + T_0) = x(t), \qquad -\infty < t < \infty \tag{1-20}$$

where the constant T_0 is the period. The smallest value of T_0 such that (1-20) is satisfied is referred to as the *fundamental period,* and is hereafter simply referred to as the *period.* Any deterministic signal not satisfying (1-20) is called *aperiodic.*

A familiar example of a periodic signal is a sinusoidal signal that may be expressed as

$$x(t) = A \sin(2\pi f_0 t + \theta), \qquad -\infty < t < \infty \tag{1-21}$$

where $A, f_0,$ and θ are constants referred to as the amplitude, frequency in hertz,[†] and relative phase, respectively. For simplicity, $\omega_0 = 2\pi f_0$ is sometimes used, where ω_0 is the frequency in rad/s. The period of this signal is $T_0 = 2\pi/\omega_0 = 1/f_0$, which can be verified by direct substitution and use of trigonometric identities as follows. We wish to verify that

$$x\left(t + \frac{2\pi}{\omega_0}\right) = x(t), \qquad \text{all } t \tag{1-22}$$

But

$$x\left(t + \frac{2\pi}{\omega_0}\right) = A \sin\left[\omega_0\left(t + \frac{2\pi}{\omega_0}\right) + \theta\right]$$

$$= A \sin(\omega_0 t + 2\pi + \theta)$$

$$= A[\sin(\omega_0 t + \theta) \cos 2\pi + \cos(\omega_0 t + \theta) \sin 2\pi] \tag{1-23}$$

[†]One hertz is one cycle per second.

$x_s(t)$

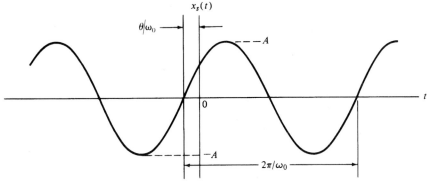

(a) Sinusoidal signal

$x_t(t)$

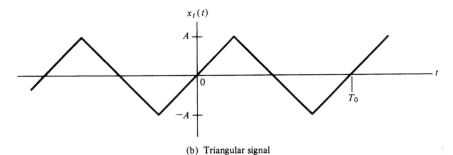

(b) Triangular signal

$x_{st}(t)$

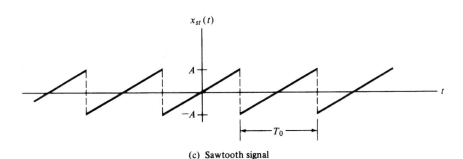

(c) Sawtooth signal

FIGURE 1-8. Various types of periodic signals.

Since $\cos 2\pi = 1$ and $\sin 2\pi = 0$, the required relationship for periodicity is verified. Note that $T_0 = 2\pi/\omega_0$ is the smallest value of T_0 such that $x(t) = x(t + T_0)$. Several types of periodic signals are illustrated in Figure 1-8 together with the sinusoidal signal.

The sum of two or more sinusoids may or may not be periodic, depending on the relationships between their respective periods or frequencies. If the ratio of their periods can be expressed as a rational number, or their frequencies are commensurable,‡ their sum will be a periodic signal.

---

‡Two frequencies, f_1 and f_2, are commensurable if they have a common measure. That is, there is a number f_0 contained in each an integral number of times. Thus, if f_0 is the largest such number,

$$f_1 = n_1 f_0 \quad \text{and} \quad f_2 = n_2 f_0$$

where n_1 and n_2 are integers; f_0 is called the *fundamental frequency*. The periods, T_1 and T_2, corresponding to f_1 and f_2, are therefore related by $T_1/T_2 = n_2/n_1$.

EXAMPLE 1-6 _____

Which of the following signals are periodic? Justify your answers.
(a) $x_1(t) = \sin 10\pi t$
(b) $x_2(t) = \sin 20\pi t$
(c) $x_3(t) = \sin 31t$
(d) $x_4(t) = x_1(t) + x_2(t)$
(e) $x_5(t) = x_1(t) + x_3(t)$

Solution: Only (e) is not periodic. To show that waveforms $x_1(t)$, $x_2(t)$, and $x_3(t)$ are periodic, we note that

$$\sin(\omega_0 t + \theta) = \sin \omega_0\left(t + \frac{\theta}{\omega_0}\right) = \sin \omega_0 t \tag{1-24}$$

if θ is an integer multiple of 2π. Taking the smallest such value of θ, we see that the (fundamental) period of $\sin(\omega_0 t + \theta)$ is

$$T_0 = \frac{2\pi}{\omega_0} \tag{1-25}$$

Thus the periods of $x_1(t)$, $x_2(t)$, and $x_3(t)$ are 1/5, 1/10, and $2\pi/31$ s, respectively.
 The sum of two sinusoids is periodic if the ratio of their respective periods can be expressed as a rational number. Thus $x_4(t)$ is periodic. Indeed, we see that the second term goes through two cycles for each cycle of the first term. The period of $x_4(t)$ is therefore 1/5 s.
 After a little thought, it is apparent that there is no T_0 for which $x_5(t) = x_5(t + T_0)$, because the ratio of the periods of the two separate terms is not a rational number (i.e., their frequencies are incommensurable).

Phasor Signals and Spectra

Although physical systems always interact with real signals, it is often mathematically convenient to represent real signals in terms of complex quantities. An example of this is the use of phasors. The student should recall from circuits courses that the phasor quantity

$$\tilde{X} = Ae^{j\theta} = A\underline{/\theta} \tag{1-26}$$

is a shorthand notation for the real, sinusoidal signal[†]

$$x(t) = \operatorname{Re}(\tilde{X}e^{j\omega_0 t}) = A\cos(\omega_0 t + \theta), \qquad -\infty < t < \infty \tag{1-27}$$

We refer to the complex signal

$$\tilde{x}(t) = Ae^{j(\omega_0 t + \theta)}, \qquad -\infty < t < \infty$$

$$= \tilde{X}e^{j\omega_0 t} \tag{1-28}$$

[†]We could also project the sinusoidal signal onto the imaginary axis. Projection onto the real axis is used throughout this book, however.

as the rotating phasor signal.[†] It is characterized by the three parameters A, the amplitude; θ, the phase; and $\omega_0 > 0$, the radian frequency. Using Euler's theorem, which is $\exp(ju) = \cos u + j \sin u$, we may readily show that $\tilde{x}(t)$ is periodic with period $T_0 = 2\pi/\omega_0$.[‡]

In addition to taking its real part, as in (1-27), to relate it to the real, sinusoidal signal $A \cos(\omega_0 t + \theta)$, we may relate $\tilde{x}(t)$ to its sinusoidal counterpart by writing

$$x(t) = \tfrac{1}{2}\tilde{x}(t) + \tfrac{1}{2}\tilde{x}^*(t) \tag{1-29}$$

$$= \frac{A}{2}\, e^{j(\omega_0 t + \theta)} + \frac{A}{2}\, e^{-j(\omega_0 t + \theta)}, \qquad -\infty < t < \infty$$

which is a representation in terms of conjugate, oppositely rotating phasors. These two procedures, represented mathematically by (1-27) and (1-29), are illustrated schematically in Figure 1-9. These expressions for $x(t) = A \cos(\omega_0 t + \theta)$ are referred to as *time-domain* representations.

Note that (1-29) can be thought of as the sum of positive-frequency and negative-frequency rotating phasors. This mathematical abstraction results from the fact that it is necessary to add complex conjugate quantities to obtain a real quantity. It is emphasized that negative frequencies do not physically exist, but are merely convenient mathematical abstractions that result in nicely symmetric equations and figures such as (1-29) and Figure 1-9b.

[†]A rotating phasor signal is not necessarily associated with any physical quantity that rotates. It is simply a mathematical convenience to represent sinusoids by the $\text{Re}(\cdot)$ operation.

[‡]Complex algebra basics: Many times students will do more work than necessary in complex algebra manipulations. This footnote hopefully will give some pointers to minimize such manipulations. A complex number may be represented in the alternative forms of cartesian and polar:

$$X = a + jb = Re^{j\theta} = R \cos\theta + jR \sin\theta$$

where a is its real part, b its imaginary part, R its magnitude or modulus, and θ its argument or angle. The latter equation follows by virtue of Euler's relationship, or $\exp(j\theta) = \cos(\theta) + j \sin(\theta)$. Matching real and imaginary parts, we see that $\text{Re}(X) = a = R \cos(\theta)$ and $\text{Im}(X) = b = R \sin(\theta)$, where $\text{Re}(\cdot)$ and $\text{Im}(\cdot)$ denote the real and imaginary parts, respectively. The inverse relationships are $R = (a^2 + b^2)^{1/2}$ and $\theta = \tan^{-1}(b/a)$. Most programmable calculators have built-in routines for going back and forth between the cartesian and polar forms. Consider a second complex number, defined by

$$Y = c + jd = Se^{j\phi} = S \cos\phi + jS \sin\phi$$

Addition and subtraction are facilitated with cartesian representation; i.e.,

$$X \pm Y = (a + jb) \pm (c + jd) = (a \pm c) + j(b \pm d)$$

Multiplication and division may be carried in either cartesian or polar form, but usually the polar form is most convenient. For example, multiplication in cartesian form becomes

$$XY = (a + jb)(c + jd) = (ac + j^2bd) + j(bc + ad) = (ac - bd) + j(bc + ad)$$

where $j^2 = -1$ has been used. In polar form, multiplication is more simply performed as

$$XY = (Re^{j\theta})(Se^{j\phi}) = RSe^{j(\theta + \phi)} = RS \cos(\theta + \phi) + jRS \sin(\theta + \phi)$$

Division in cartesian form is facilitated by the process of rationalization as

$$\frac{X}{Y} = \frac{a + jb}{c + jd} = \frac{(a + jb)(c - jd)}{(c + jd)(c - jd)} = \frac{(ac + bd) + j(bc - ad)}{c^2 + d^2} = \frac{ac + bd}{c^2 + d^2} + j\frac{bc - ad}{c^2 + d^2}$$

This fairly lengthy process is shortened considerably by representation of the numbers in polar form:

$$\frac{X}{Y} = \frac{Re^{j\theta}}{Se^{j\phi}} = \frac{R}{S} e^{j(\theta - \phi)} = \frac{R}{S} \cos(\theta - \phi) + j\frac{R}{S} \sin(\theta - \phi)$$

As a numerical example, consider $X = 8 + j6$ and $Y = 3 + j4$. Their sum is $(8 + j6) + (3 + j4) = 11 + j10$ and their difference is $(8 + j6) - (3 + j4) = 5 + j2$. Their product computed in cartesian form is $(8 + j6)(3 + j4) = (24 - 24) + j(18 + 32) = j50$. In polar form it is $\{10 \exp[\tan^{-1}(3/4)]\}\{5 \exp[\tan^{-1}(4/3)]\} = 50 \exp(j\pi/2) = j50$. Their quotient in polar form is $\{10 \exp[\tan^{-1}(3/4)]\}/\{5 \exp[\tan^{-1}(4/3)]\} = 2 \exp\{j[\tan^{-1}(3/4) - \tan^{-1}(4/3)]\} = 2 \exp[-j0.284] = 1.92 - j0.56$, which follows by Euler's relation. If the rationalization procedure is used, the same result in cartesian form is obtained.

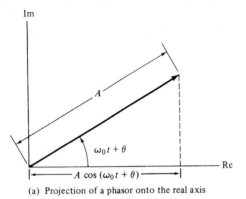

(a) Projection of a phasor onto the real axis

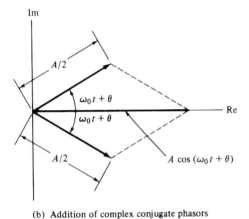

(b) Addition of complex conjugate phasors

FIGURE 1-9. Two ways of relating a phasor signal to a sinusoidal signal.

An alternative representation for $x(t)$ is provided in the *frequency domain*. Since $\tilde{x}(t) = A\exp[j(\omega_0 t + \theta)]$ is completely specified by A and θ for a given value of f_0, this alternative frequency-domain representation can take the form of two plots, one showing the amplitude A as a function of frequency f, and the other showing θ as a function of f. Because $\tilde{x}(t)$ depends only on the single frequency f_0, the resulting plots each consist of a single point or "line" at $f = f_0$, as illustrated in Figure 1-10a. Had we considered a signal that is the sum of two phasors, each plot would have had two points or lines present. The plot of amplitude versus frequency is referred to as the *single-sided amplitude spectrum*. The modifier "single-sided" is used because these spectral plots have points or lines only for positive frequencies.

If a spectral plot corresponding to (1-29) is made, spectral lines will be present at $f = f_0$ *and* at $f = -f_0$, since $x(t)$ is obtained as the sum of oppositely rotating phasors. Spectral plots for the single sinusoidal signal under consideration here are shown in Figure 1-10b. Such plots are referred to as *double-sided* amplitude and phase spectra. Had a sum of sinusoidal components been present, the spectral plots would have consisted of multiple lines. Note that because of the convention of taking the real part [see (1-27)], any signal expressed as a sine function must first be converted to a cosine function before obtaining the spectrum. For this purpose, the identity $\sin(\omega_0 t + \theta) = \cos(\omega_0 t + \theta - \pi/2)$ is useful.

It is important to emphasize two points about double-sided spectra. First, the lines at negative frequencies are present precisely because it is necessary to add complex conjugate phasors to obtain the real sinusoidal signal $A\cos(\omega_0 t + \theta)$. Second, we note that the amplitude spectrum has *even* symmetry about

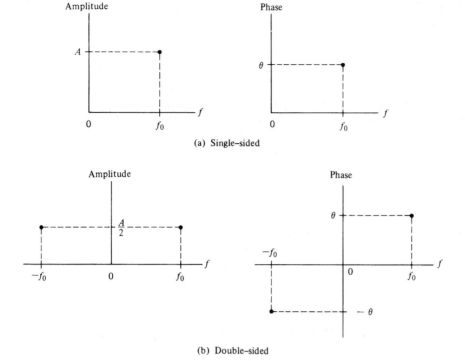

FIGURE 1-10.   Amplitude and phase spectra for the signal $A \cos(\omega_0 t + \theta)$.

the origin, and the phase spectrum has *odd* symmetry. This symmetry is again a consequence of $x(t)$ being a real signal, which implies conjugate rotating phasors must be added to obtain a real quantity.

Figures 1-10a and b serve as equivalent spectral representations for the signal $A \cos(\omega_0 t + \theta)$ if we are careful to identify the spectra plotted as single-sided or double-sided, as the case may be. We will find later that the Fourier series and Fourier transform lead to spectral representations for more complex signals.

EXAMPLE 1-7 _____

We wish to sketch the single-sided and double-sided amplitude and phase spectra of the signal

$$x(t) = 4 \sin\left(20\pi t - \frac{\pi}{6}\right), \qquad -\infty < t < \infty \tag{1-30}$$

To sketch the single-sided spectra, we write $x(t)$ as the real part of a rotating phasor and plot the amplitude and phase of this phasor as a function of frequency for $t = 0$. Noting that $\cos(u - \pi/2) = \sin u$, we find that

$$x(t) = 4 \cos\left(20\pi t - \frac{\pi}{6} - \frac{\pi}{2}\right)$$

$$= 4 \cos\left(20\pi t - \frac{2\pi}{3}\right)$$

$$= \mathrm{Re}\left\{4 \exp\left[j\left(20\pi t - \frac{2\pi}{3}\right)\right]\right\} \tag{1-31}$$

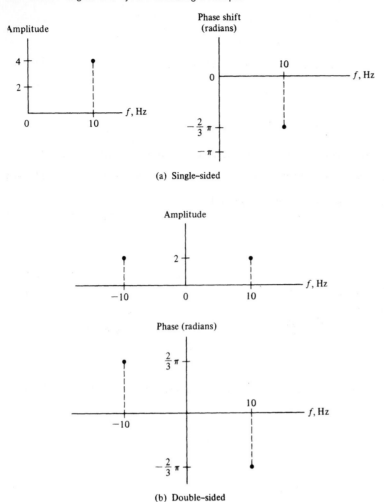

(a) Single-sided

(b) Double-sided

FIGURE 1-11.   Amplitude and phase spectra for Example 1-6.

which results in the amplitude and phase spectral plots shown in Figure 1-11a. To plot the double-sided amplitude and phase spectra, we write $x(t)$ as the sum of complex conjugate rotating phasors. Recalling that $2 \cos u = \exp(ju) + \exp(-ju)$, we obtain

$$x(t) = 2 \exp\left[j\left(20\pi t - \frac{2\pi}{3}\right)\right] + 2 \exp\left[-j\left(20\pi t - \frac{2\pi}{3}\right)\right] \qquad (1\text{-}32)$$

from which the double-sided amplitude and phase spectral plots of Figure 1-11b result.

EXAMPLE 1-8 _____

As an example involving a sum of two sinusoids, consider the spectrum of the signal

$$x(t) = 2 \cos(10\pi t + \pi/4) + 4 \sin(30\pi t - \pi/6) \qquad (1\text{-}33\text{a})$$

Changing the second term to a cosine function gives

$$x(t) = 2\cos(10\pi t + \pi/4) + 4\cos(30\pi t - \pi/6 - \pi/2)$$

$$= 2\cos(10\pi t + \pi/4) + 4\cos(30\pi t - 2\pi/3) \tag{1-33b}$$

This can be written in terms of the real part of the sum of rotating phasors in preparation for plotting the single-sided spectra as

$$x(t) = \text{Re}[2e^{j(10\pi t + \pi/4)} + 4e^{j(30\pi t - 2\pi/3)}] \tag{1-33c}$$

In order to plot the double-sided spectra, we write the signal as a sum of counterrotating phasors as

$$x(t) = e^{j(10\pi t + \pi/4)} + e^{-j(10\pi t + \pi/4)} + 2e^{j(30\pi t - 2\pi/3)} + 2e^{-j(30\pi t - 2\pi/3)} \tag{1-33d}$$

Single-sided and double-sided spectral plots are shown in Figure 1-12.

Singularity Functions

An important subclass of aperiodic signals is the *singularity functions,* which we now discuss. We begin by introducing the unit step and unit ramp functions and using them to represent more complicated signals, such as in the acceleration in Example 1-1.

Consider the *unit step function, $u_{-1}(t)$,* defined as

$$u(t) = u_{-1}(t) = \begin{cases} 0, & t < 0 \\ 1, & t > 0 \end{cases} \tag{1-34}$$

The value of $u(t)$ at $t = 0$ will not be specified at this time except to say that it is finite. Other singularity functions are defined in terms of $u_{-1}(t)$ by the relations

$$u_{i-1}(t) = \int_{-\infty}^{t} u_i(\lambda)\,d\lambda, \qquad i = \ldots, -2, -1, 0, 1, 2, \ldots \tag{1-35a}$$

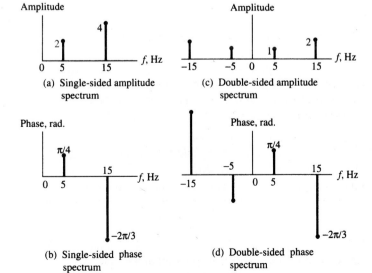

(a) Single-sided amplitude spectrum

(b) Single-sided phase spectrum

(c) Double-sided amplitude spectrum

(d) Double-sided phase spectrum

FIGURE 1-12. Spectra for Example 1-7.

or

$$u_{i+1}(t) = \frac{du_i(t)}{dt} \tag{1-35b}$$

Figure 1-13a shows the unit step. We may graphically integrate it to obtain $u_{-2}(t)$, shown in Figure 1-13b. This singularity function is referred to as the *unit ramp function, r(t)*, and can be expressed algebraically as[†]

$$r(t) = u_{-2}(t) = \begin{cases} t, & t \geq 0 \\ 0, & t < 0 \end{cases} \tag{1-36}$$

Carrying this discussion one step further, we see that a *unit parabolic function* is given by

$$u_{-3}(t) = \begin{cases} t^2/2, & t \geq 0 \\ 0, & t < 0 \end{cases} \tag{1-37}$$

This singularity function is sketched in Figure 1-13c. It is clear that the unit ramp and unit parabolic functions would be convenient for representing the acceleration and velocity of Example 1-1.

We may shift any signal on the time axis simply by replacing t by $t - t_0$. If $t_0 > 0$, the signal is shifted to the right; if $t_0 < 0$, it is shifted to the left. For example, replacing t by $t - \frac{1}{2}$ in the definition of the unit step, (1-34), we obtain

$$u(t - \tfrac{1}{2}) = \begin{cases} 0, & t - \tfrac{1}{2} < 0 \\ 1, & t - \tfrac{1}{2} > 0 \end{cases} \tag{1-38a}$$

or

$$u(t - \tfrac{1}{2}) = \begin{cases} 0, & t < \tfrac{1}{2} \\ 1, & t > \tfrac{1}{2} \end{cases} \tag{1-38b}$$

Similarly, replacing t by $t + \frac{1}{2}$ results in a unit step that "turns on" at $t = -\frac{1}{2}$. Using this technique, we can represent other signals in terms of singularity functions. For example, the *unit pulse function*, defined by (1-17), can be represented as

$$\Pi(t) = u(t + \tfrac{1}{2}) - u(t - \tfrac{1}{2})$$

For the definition of $\Pi(t)$ given by (1-17), it would be consistent to define $u(0) = 1$.

We note, also, that replacing t by $-t$ turns a function around.[‡] Doing this in (1-36), we obtain

$$r(-t) = \begin{cases} -t, & t \leq 0 \\ 0, & t > 0 \end{cases} \tag{1-40}$$

which is a ramp that starts at $t = 0$ and *increases* linearly as t *decreases*.

Also, we can obtain a change of scale on the abscissa simply by multiplying t by a constant, say, β. If $\beta > 1$, the signal is compressed. If $\beta < 1$, the signal is expanded.

To summarize, consider an arbitrary signal represented as $x(\beta t + \alpha)$. We rewrite it as

$$x(\beta t + \alpha) = x[\beta(t + \alpha/\beta)] \tag{1-41}$$

[†]The subscripts are needed on (1-35a) and (1-35b) to identify different members of the class of singularity functions. Since the step and ramp are very commonly used members, they are often denoted by the special symbols $u(t)$ and $r(t)$, respectively.

[‡]The negation of the independent variable is often referred to as *folding*, since the resulting reversal of the signal in time can be viewed as folding a paper on which the signal has been plotted along its ordinate and viewing it from behind the paper.

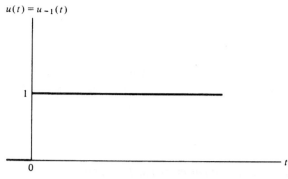

(a) Unit step function

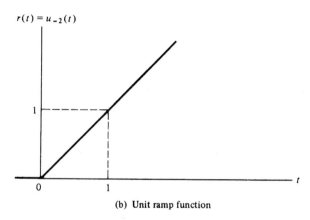

(b) Unit ramp function

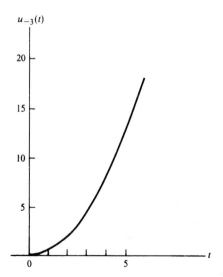

(c) Unit parabolic function

FIGURE 1-13. Examples of three singularity functions.

If $t_0 \triangleq \alpha/\beta$ is positive, $x(t)$ is shifted *left* by α/β; if t_0 is negative, it is shifted *right* by α/β. If $\beta < 0$, $x(t)$ is *reversed* or *reflected* through the origin. If $|\beta| > 1$, $x(t)$ is compressed; if $|\beta| < 1$, $x(t)$ is expanded. An example will illustrate these remarks.

EXAMPLE 1-9

Consider the following signals. Sketch each, discussing how the individual plots are obtained.
(a) $x_1(t) = \Pi(2t + 6)$
(b) $x_2(t) = \cos(20\pi t - 5\pi)$
(c) $x_3(t) = r(-0.5t + 2)$

Solution: For (a), we write

$$x_1(t) = \Pi[2(t + 3)] \tag{1-42}$$

by factoring out the 2 multiplying t. The 3 added to t means that the rectangular pulse function is shifted left by 3 units, and the 2 results in the original 1-unit-wide pulse function being compressed by a factor of 2 to a $\frac{1}{2}$-unit-wide pulse function. The result is shown in Figure 1-14a.

For (b), the 20π multiplying t is factored out to give

$$x_2(t) = \cos[20\pi(t - 0.25)] \tag{1-43}$$

which shows that $x_2(t)$ is a cosine waveform shifted right by 0.25 time unit (assumed to be seconds) with a period of $T_0 = 2\pi/\omega_0 = 2\pi/20\pi = 0.1$ s. This is a shift of 2.5 periods. The result is sketched in Figure 1-14b.

For $x_3(t)$, we write

$$x_3(t) = r[-0.5(t - 4)] \tag{1-44}$$

which says $r(t)$ is reflected about $t = 0$, expanded by a factor of 2 ($1/0.5 = 2$), and delayed or shifted right 4 units. A sketch is shown in Figure 1-14c.

It is sometimes useful to check results by writing out the definition of a singularity function. For part (c), this takes the form

$$r(-0.5t + 2) = \begin{cases} -0.5t + 2, & -0.5t + 2 > 0 \\ 0, & \text{otherwise} \end{cases} \tag{1-45a}$$

$$= \begin{cases} -0.5(t - 4), & t < 4 \\ 0, & \text{otherwise} \end{cases} \tag{1-45b}$$

From this, it is clear that $x_3(t) = 0$ for $t \geq 4$. Checking a few values, we see that $x_3(3) = 0.5$, $x_3(2) = 1$, and $x_3(0) = 2$ so that the result is a reflected or reversed ramp starting at $t = 4$ and increasing 1 unit for each 2-unit decrease in t.

EXAMPLE 1-10

Express the signals shown in Figure 1-15 in terms of singularity functions.

Solution: One possible representation for $x_a(t)$ is

$$x_a(t) = u(t) - r(t - 1) + 2r(t - 2) - r(t - 3) + u(t - 4) - 2u(t - 5) \tag{1-46}$$

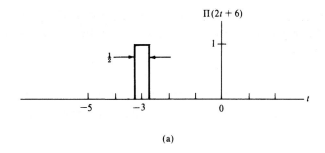

(a)

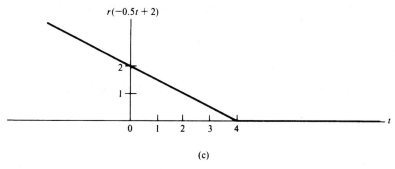

(b)

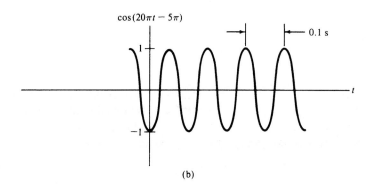

(c)

FIGURE 1-14. Signals relating to Example 1-9.

The step function $u(t)$ is unity for $t > 0$. At $t = 1$, the ramp function $-r(t - 1)$ goes downward with slope -1. The sum of ramps $-r(t - 1) + 2r(t - 2)$ goes upward with slope $+1$ starting at $t = 2$. The ramp $-r(t - 3)$ then flattens out this upward ramp starting at $t = 3$, and the step $u(t - 4)$ increases the plateau of unity height to a value of 2. The final step, $2u(t - 5)$, "shuts off" the preceding sum of singularity functions so that $x(t) = 0$ for $t > 5$.

For $x_b(t)$ we use a product representation. A possible representation is

$$x_b(t) = 2u(t)u(2 - t) + u(t - 3)u(5 - t) \tag{1-47}$$

The product of unit step functions $2u(t)u(2 - t)$ forms the first pulse, and the product $u(t - 3)u(5 - t)$ forms the second pulse. Note that $u(2 - t)$ and $u(5 - t)$ are step functions that are unity from $t = -\infty$ to $t = 2$ and $t = 5$, respectively, and zero thereafter.

The student should attempt to find other possible representations for $x_a(t)$ and $x_b(t)$.

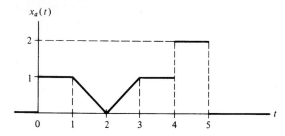

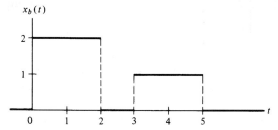

FIGURE 1-15.   Signal to be expressed in terms of singularity functions.

EXERCISE

Represent the acceleration and velocity of the rocket of Example 1-1 in terms of singularity functions.

Answers: $a(t) = 40u_{-2}(t/72)u_{-1}(72 - t)$ m/s^2. $v(t) = \frac{5}{9}u_{-3}(t)u_{-1}(72 - t) + 1440u_{-1}(t - 72)$ m/s.
(Other answers are possible.)

We turn next to the *unit impulse function* or *delta function,*[†] $\delta(t)$, which has the properties

$$\delta(t) = 0, \qquad t \neq 0 \tag{1-48a}$$

and

$$\int_{-\infty}^{\infty} \delta(t) \, dt = 1 \tag{1-48b}$$

Equation (1-48b) states that the area of a unit impulse function is unity, and (1-46a) indicates that this unity area is obtained in an infinitesimal interval on the t-axis.

The motivation for defining a function with these properties stems from the need to represent phenomena that happen in time intervals short compared with the resolution capability of any measuring apparatus used, but which produce an almost instantaneous change in a measured quantity. Examples are the nearly instantaneous increase in voltage across a capacitor placed across the terminals of a battery, or the nearly instantaneous change in velocity of a billiard ball struck by a rapidly moving cue ball. In the first example, the voltage across the capacitor is the result of an extremely large current flowing for a very short time interval, whereas the billiard ball's velocity changes almost instantaneously due to the transfer of momentum from the cue ball at the moment of impact. The current flowing into the capacitor and the force transmitted by the cue ball are examples of quantities that are usefully modeled

---

[†]The two names stem from the use of this function in engineering and physics; the term "delta function" is associated with Dirac, a physicist, who introduced the notation $\delta(t)$ into quantum mechanics. We will refer to $\delta(t)$ simply as a unit impulse function.

by impulse functions. In neither case are we able to measure, nor are we particularly interested in, what happens at exactly the moment the action takes place. Rather we are able to observe only the conditions beforehand and afterward.

Other examples of physical quantities that are modeled by mathematical entities having infinitely small dimensions and infinite "weight" are point masses and point charges.

Since it is impossible for any conventional function to have the properties (1-48), we attempt to picture the unit impulse function by considering the limit of a conventional function as some parameter approaches zero.[†] For example, consider the signal

$$\delta_\epsilon(t) = \frac{1}{2\epsilon} \Pi\left(\frac{t}{2\epsilon}\right) = \begin{cases} \dfrac{1}{2\epsilon}, & |t| \le \epsilon \\ \\ 0, & |t| > \epsilon \end{cases} \tag{1-49}$$

which is shown in Figure 1-16. We see that, no matter how small ϵ is, $\delta_\epsilon(t)$ always has unit area. Furthermore, this area is obtained as the integration is carried out from $t = -\epsilon$ to $t = \epsilon$. As $\epsilon \to 0$ this area is obtained in an infinitesimally small width, thus implying that the height of $\delta_\epsilon(t) \to \infty$ such that *unity area lies between the function and the t-axis*. This limit is shown as a heavy arrow in the figure.

Many other signals exist that provide heuristic visualizations for $\delta(t)$ in the limit as some parameter approaches zero. Examples are

$$\delta_\epsilon(t) = \frac{1}{\epsilon}\left[\frac{\sin(\pi t/\epsilon)}{\pi t/\epsilon}\right]^2 \tag{1-50}$$

which is sketched in Figure 1-17a, and

$$\delta_\epsilon(t) = \begin{cases} (1 - |t|/\epsilon)/\epsilon, & |t| < \epsilon \\ 0, & \text{otherwise} \end{cases} \tag{1-51}$$

[†]The present discussion is aimed at providing an intuitive understanding of the unit impulse function. We will give a more formal mathematical development shortly.

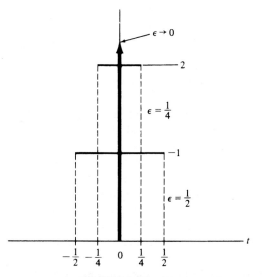

FIGURE 1-16.   Square-pulse approximations for the unit impulse function.

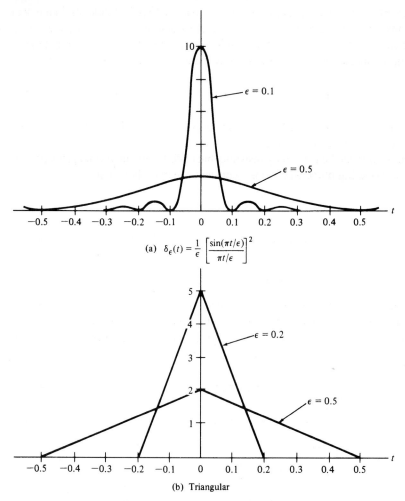

(a) $\delta_\epsilon(t) = \dfrac{1}{\epsilon}\left[\dfrac{\sin(\pi t/\epsilon)}{\pi t/\epsilon}\right]^2$

(b) Triangular

FIGURE 1-17. Two approximations for the unit impulse.

shown in Figure 1-17b. Although the three families of functions illustrated in Figures 1-16 and 1-17 are symmetric about $t = 0$, an equally good example of an approximation for a unit impulse function in terms of the properties (1-48) is $\epsilon^{-1}\exp(-t/\epsilon)u(t)$, which is asymmetric about $t = 0$. These examples illustrate that it is not the actual behavior of $\delta(t)$ at $t = 0$ that is important but rather the integral of $\delta(t)$.

Note that by integrating the square-pulse approximation of Figure 1-16 between $-\infty$ and t we obtain

$$\int_{-\infty}^{t}\delta_\epsilon(\lambda)\,d\lambda = \begin{cases}0, & t < -\epsilon \\ 1, & t > \epsilon\end{cases} \tag{1-52}$$

In the limit as $\epsilon \to 0$, the right side becomes a unit step function. It will be shown later that $u_0(t)$ of (1-35b) and $\delta(t)$ have the same properties as expressed by (1-48).

EXERCISE

Verify that all the families of approximating functions $\delta_\epsilon(t)$ discussed above satisfy (1-48) in the limit as $\epsilon \to 0$.

The considerations above illustrate that the unit impulse function is not really a function in the normal sense. Indeed, it was not until the 1950s that Laurent Schwartz's pioneering work with the theory of distributions provided a firm mathematical basis for the unit impulse function. A more formal mathematical definition of the unit impulse function proceeds by defining it in terms of the *functional*

$$\int_{-\infty}^{\infty} x(t)\delta(t) \, dt = x(0) \tag{1-53}$$

where $x(t)$ is continuous at $t = 0$. That is, $\delta(t)$ is a functional as defined by (1-53), which assigns the value $x(0)$ to any function $x(t)$ continuous at the origin.[‡]

Several useful properties of $\delta(t)$ can be proved by insisting that (1-53) obey the formal properties of integrals. For example, by considering $\delta(at)$, where a is a constant, we may show that

$$\delta(at) = \frac{1}{|a|} \delta(t) \tag{1-54}$$

by changing variables of integration and considering the cases $a > 0$ and $a < 0$ separately. If we let $a = -1$ in (1-54), we have

$$\delta(-t) = \delta(t) \tag{1-55}$$

which agrees with the definition of an even function. Even though $\delta(t)$ is not a function in the conventional sense, we may make use of (1-55) in manipulations involving unit impulse functions as long as we remember that it was a consequence of the defining relation (1-53) and the formal properties of integrals.

A second useful property of the delta function is the *sifting property,* which is

$$\int_{-\infty}^{\infty} x(t)\delta(t - t_0) \, dt = x(t_0) \tag{1-56a}$$

where $x(t)$ is assumed to be continuous at $t = t_0$. This property can be proved from (1-53) by letting $\lambda = t - t_0$ in the integral (1-56a) to obtain

$$\int_{-\infty}^{\infty} x(\lambda + t_0)\delta(\lambda) \, d\lambda = x(t_0) \tag{1-56b}$$

If we define $y(\lambda) = x(\lambda + t_0)$, this integral is the same as (1-53); $y(\lambda)$ is continuous at $\lambda = 0$ as required because $x(t)$ is continuous at $t = t_0$.

An alternative form of (1-53) which will prove useful in Chapters 2 and 3 is

$$\int_{-\infty}^{\infty} x(\lambda)\delta(t - \lambda) \, d\lambda = x(t) \tag{1-57}$$

which is obtained by simple renaming of variables in (1-56a) and by using the "even" property of the delta function expressed by (1-55). The form of the integral in (1-57), known as a *convolution,* is dealt with in more detail in Chapter 2.

Integrals with finite limits can be considered as special cases of (1-56a) by defining $x(t)$ to be zero outside a certain interval, say $t_1 < t < t_2$. Thus (1-56a) becomes

$$\int_{t_1}^{t_2} x(t)\delta(t - t_0) \, dt = \begin{cases} x(t_0), & t_1 < t_0 < t_2 \\ 0, & \text{otherwise} \end{cases} \tag{1-58}$$

[‡]For a further discussion, see A. Papoulis, *Signal Analysis* (New York: McGraw-Hill, 1977), or R. N. Bracewell, *The Fourier Transform and Its Applications,* 2nd ed. (New York: McGraw-Hill, 1978).

A fourth property of $\delta(t)$ is that

$$x(t)\delta(t - t_0) = x(t_0)\delta(t - t_0) \tag{1-59}$$

if $x(t)$ is continuous at $t = t_0$, which follows intuitively from $\delta(t - t_0) = 0$ everywhere except at $t = t_0$.

We may find integrals similar to (1-56a) involving derivatives of $x(t)$. For example, assuming $x(t)$ to have a first derivative that is continuous at $t = t_0$ and using the chain rule for differentiation, we obtain

$$\frac{d}{dt}[x(t)\delta(t - t_0)] = \dot{x}(t)\delta(t - t_0) + x(t)\dot{\delta}(t - t_0) \tag{1-60}$$

where the overdot denotes differentiation with respect to time.[†] Assuming $\dot{x}(t)$ to be continuous at $t = t_0$ and using (1-59), we have

$$\frac{d}{dt}[x(t)\delta(t - t_0)] = \dot{x}(t_0)\delta(t - t_0) + x(t)\dot{\delta}(t - t_0) \tag{1-61}$$

Integrating from $t = t_1$ to $t = t_2$, where $t_1 < t_0 < t_2$, we obtain

$$\int_{t_1}^{t_2} \frac{d}{dt}[x(t)\delta(t - t_0)] \, dt = \int_{t_1}^{t_2} \dot{x}(t_0)\delta(t - t_0) \, dt + \int_{t_1}^{t_2} x(t)\dot{\delta}(t - t_0) \, dt, \qquad t_1 < t_0 < t_2$$

or

$$0 = \dot{x}(t_0) + \int_{t_1}^{t_2} x(t)\dot{\delta}(t - t_0) \, dt, \qquad t_1 < t_0 < t_2 \tag{1-62}$$

where we have used

$$\int_{t_1}^{t_2} \frac{d}{dt}[x(t)\delta(t - t_0)] \, dt = x(t)\delta(t - t_0)\Big|_{t_1}^{t_2} = 0, \qquad t_1 < t_0 < t_2 \tag{1-63}$$

and

$$\int_{t_1}^{t_2} \dot{x}(t)\delta(t - t_0) \, dt = \dot{x}(t_0) \int_{t_1}^{t_2} \delta(t - t_0) \, dt = \dot{x}(t_0), \qquad t_1 < t_0 < t_2 \tag{1-64}$$

which follow because $\delta(t - t_0) = 0$ everywhere except at $t = t_0$ and $\dot{x}(t)$ is continuous at $t = t_0$, so that $\dot{x}(t_0)$ is defined. Rearranging (1-62), we obtain

$$\int_{t_1}^{t_2} x(t)\dot{\delta}(t - t_0) \, dt = -\dot{x}(t_0), \qquad t_1 < t_0 < t_2 \tag{1-65}$$

In general, if the nth derivative of $x(t)$ exists and is continuous at $t = t_0$, it can be shown that

$$\int_{t_1}^{t_2} x(t)\delta^{(n)}(t - t_0) \, dt = (-1)^n x^{(n)}(t_0), \qquad t_1 < t_0 < t_2 \tag{1-66}$$

[†]The derivative of the unit impulse function is sometimes referred to as the *unit doublet,* or simply doublet. It can be visualized by graphically differentiating the triangular approximation for the unit impulse shown in Figure 1-17b. Again, we emphasize that $\dot{\delta}(t)$ is to be interpreted in the context of an integral similar to (1-53). It is not simply two opposite-sign impulses located at the origin.

where

$$x^{(n)}(t_0) \triangleq \left. \frac{d^n x(t)}{dt^n} \right|_{t=t_0} \quad \text{and} \quad \delta^{(n)}(t - t_0) \triangleq \frac{d^n}{dt^n} [\delta(t - t_0)]$$

A unit impulse function and its derivatives may be treated as generalized functions in the sense that if

$$f(t) = a_0\delta(t) + a_1\dot{\delta}(t) + \cdots + a_n\delta^{(n)}(t) \tag{1-67a}$$

and

$$g(t) = b_0\delta(t) + b_1\dot{\delta}(t) + \cdots + b_n\delta^{(n)}(t) \tag{1-67b}$$

then

$$f(t) + g(t) = (a_0 + b_0)\delta(t) + (a_1 + b_1)\dot{\delta}(t) + \cdots \tag{1-68}$$

Equation (1-68) applies also if $a_1, b_1, a_2, b_2, a_3, b_3, \ldots$ are functions that are continuous at $t = 0$.†

EXERCISE

Evaluate the following integrals:

(a) $\displaystyle\int_{-\infty}^{\infty} e^{-\alpha t^2}\delta(t - 10)\, dt$

(b) $\displaystyle\int_{0}^{\infty} e^{-\alpha t^2}\delta(t + 10)\, dt$

(c) $\displaystyle\int_{-\infty}^{\infty} e^{-\alpha t^2}\dot{\delta}(t - 10)\, dt$

(d) $\displaystyle\int_{-\infty}^{\infty} [5\delta(t) + e^{-(t-1)}\dot{\delta}(t) + \cos 5\pi t\, \delta(t) + e^{-t^2}\dot{\delta}(t)]\, dt$

Answers: (a) $e^{-100\alpha}$; (b) 0; (c) $20\alpha e^{-100\alpha}$; (d) $6 + e$.

We now return to the assertion made earlier that

$$u(t) = u_{-1}(t) = \int_{-\infty}^{t} u_0(\lambda)\, d\lambda = \int_{-\infty}^{t} \delta(\lambda)\, d\lambda \tag{1-69}$$

or, equivalently, that

$$\delta(t) = \frac{du(t)}{dt} \tag{1-70}$$

---

†This, again, is formally proved by using (1-53). Another question that may vex the student is, "What about $0 \cdot \delta(t)$?" To show that this is zero, use (1-53) to write

$$\int_{-\infty}^{\infty} 0 \cdot \delta(t)x(t)\, dt = 0 \cdot \int_{-\infty}^{\infty} \delta(t)x(t) = 0 \cdot x(0) = 0$$

where $x(t)$ is continuous at $t = 0$.

which is (1-35b) with $i = -1$. Letting $x(t)$ be continuous at $t = 0$, we consider

$$\int_{-\infty}^{\infty} x(t) \frac{du(t)}{dt}\, dt = x(t)u(t)\Big|_{-\infty}^{\infty} - \int_{-\infty}^{\infty} u(t) \frac{dx(t)}{dt}\, dt$$

$$= x(\infty) - \int_{0}^{\infty} \frac{dx(t)}{dt}\, dt$$

$$= x(\infty) - x(t)\Big|_{0}^{\infty}$$

$$= x(\infty) - x(\infty) + x(0)$$

$$= x(0) \qquad\qquad (1\text{-}71)$$

which was obtained through integration by parts. That is, $du(t)/dt$ has the same property as $\delta(t)$ as expressed by (1-53).

1-4 Energy and Power Signals

Quite often the particular representation used for a signal depends on the type of signal involved. It is therefore convenient to introduce a method for signal classification at this point. It is useful to classify signals as those having finite energy and those having finite average power. Some signals have neither finite average power nor finite energy.

To introduce these signal classes, suppose that $e(t)$ is the voltage across a resistance R producing a current $i(t)$. The instantaneous power per ohm is

$$p(t) = \frac{e(t)i(t)}{R} = i^2(t) \qquad\qquad (1\text{-}72)$$

Integrating over the interval $|t| \le T$, we define the total energy and the average power on a per ohm basis as the limits

$$E = \lim_{T \to \infty} \int_{-T}^{T} i^2(t)\, dt \qquad \text{joules} \qquad\qquad (1\text{-}73)$$

and

$$P = \lim_{T \to \infty} \frac{1}{2T} \int_{-T}^{T} i^2(t)\, dt \qquad \text{watts} \qquad\qquad (1\text{-}74)$$

respectively.

For an arbitrary signal $x(t)$, which may, in general, be complex, the total energy normalized to unit resistance is defined as

$$E \triangleq \lim_{T \to \infty} \int_{-T}^{T} |x(t)|^2\, dt \qquad \text{joules} \qquad\qquad (1\text{-}75)$$

and the average power normalized to unit resistance is defined as

$$P \triangleq \lim_{T \to \infty} \frac{1}{2T} \int_{-T}^{T} |x(t)|^2\, dt \qquad \text{watts} \qquad\qquad (1\text{-}76)$$

Based on the definitions (1-75) and (1-76), the following classes of signals are defined:

1. $x(t)$ is an energy signal if and only if $0 < E < \infty$, so that $P = 0$.
2. $x(t)$ is a power signal if and only if $0 < P < \infty$, thus implying that $E = \infty$.
3. Signals that satisfy neither property are therefore neither energy nor power signals.[†]

EXAMPLE 1-11

Consider the signal

$$x_1(t) = Ae^{-\alpha t}u(t), \qquad \alpha > 0, \tag{1-77}$$

where A and α are constants. Using (1-75), we can verify that $x_1(t)$ has energy

$$E = \frac{A^2}{2\alpha} \tag{1-78}$$

Letting $\alpha \to 0$, we obtain the signal

$$x_2(t) = Au(t) \tag{1-79}$$

which has infinite energy but finite power, as can be verified by applying (1-76):

$$P = \frac{A^2}{2} \tag{1-80}$$

Thus $x_2(t)$ is a power signal.

EXAMPLE 1-12

Consider the periodic, sinusoidal signal

$$x(t) = A\cos(\omega_0 t + \theta) \tag{1-81}$$

The normalized average power of this signal is

$$P = \lim_{T \to \infty} \frac{1}{2T} \int_{-T}^{T} A^2 \cos^2(\omega_0 t + \theta)\, dt \tag{1-82a}$$

Using appropriate trigonometric identities, we may rewrite this integral as

$$P = \lim_{T \to \infty} \frac{A^2}{2T} \int_{-T}^{T} [\tfrac{1}{2} + \tfrac{1}{2}\cos 2(\omega_0 t + \theta)]\, dt = \frac{A^2}{2} \tag{1-82b}$$

which follows because

$$\lim_{T \to \infty} \frac{A^2}{4T} \int_{-T}^{T} \cos 2(\omega_0 t + \theta)\, dt = 0 \tag{1-83}$$

[†]It is easy to contrive examples of signals with infinite energy and zero average power that are nonzero over a finite range of t, but we classify such signals as neither energy nor power. An example is $x(t) = t^{-1/4}, t \geq 1$, and zero otherwise.

Average Power of a Periodic Signal

We note that there is no need to carry out the limiting operation to find P for a periodic signal, since an average carried out over a single period gives the same result as (1-76); that is, for a periodic signal, $x_p(t)$,

$$P = \frac{1}{T_0} \int_{t_0}^{t_0+T_0} |x_p(t)|^2 \, dt \tag{1-84}$$

where T_0 is the period. The proof of (1-84) is left to the problems.

EXAMPLE 1-13 _____

From (1-84), the power of a rotating phasor signal of the form (1-28) is

$$P = \frac{\omega_0}{2\pi} \int_{t_0}^{t_0+2\pi/\omega_0} |Ae^{j(\omega_0 t + \theta)}|^2 \, dt = A^2 \tag{1-85}$$

Using Euler's theorem, we may write

$$Ae^{j(\omega_0 t + \theta)} = A\cos(\omega_0 t + \theta) + jA\sin(\omega_0 t + \theta) \tag{1-86}$$

The power of the real and imaginary components is $A^2/2$. Thus we may associate half the power in a rotating phasor with the real component and half with the imaginary component.

1-5 Energy and Power Spectral Densities

It is useful for some applications to define functions of frequency that when integrated over all frequencies give total energy or total power, depending on whether the signal under consideration is, respectively, an energy signal or a power signal. For an energy signal, a function of frequency when integrated that gives total energy is referred to as an *energy spectral density*. Denoting the energy spectral density of a signal $x(t)$ by $G(f)$, we have, by definition, that

$$E = \int_{-\infty}^{\infty} G(f) \, df \tag{1-87}$$

where E is the signal's total energy. We will give a means for obtaining $G(f)$ for an arbitrary energy signal in Chapter 4.

Denoting the *power spectral density* of a power signal $x(t)$ by $S(f)$, we have, by definition, that

$$P = \int_{-\infty}^{\infty} S(f) \, df \tag{1-88}$$

where P is the average power of the signal. From our consideration of amplitude spectra for periodic signals, we can deduce what their power spectral densities are like. Since, for a single sinusoid of amplitude A and frequency f_0, the two-sided amplitude spectrum has a line at $-f_0$ of amplitude $A/2$ and a line at f_0 of amplitude $A/2$, we associate half of the power of the sinusoid, or $A^2/4$, with the frequency $-f_0$ and the other half, or $A^2/4$, with the frequency f_0. Since the power spectral density is integrated to give total average power, its representation at $f = -f_0$ must be $(A^2/4)\,\delta(f + f_0)$ and that at $f = f_0$ must be $(A^2/4)\,\delta(f - f_0)$. A plot is shown in Figure 1-18a. Generalizing, *for any signal possessing a two-sided line (amplitude) spectrum, we obtain the corresponding power spectral density by taking each line of the amplitude spectrum, squaring its value, and multiplying it by a unit impulse function located*

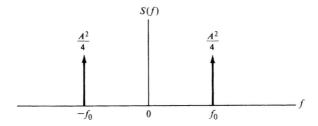

(a) Power spectral density for a single sinusoid

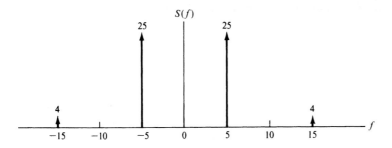

(b) Power spectral density for the sinusoidal sum of Example 1-14

FIGURE 1-18.   Power spectral densities for sinusoidal signals (numbers by impulses show weights).

at that particular frequency to get its power spectral density at that frequency. Note that the power spectral density of any signal is an even function of frequency, and it possesses no phase information about the signal.

EXAMPLE 1-14

Consider the signal

$$x(t) = 10 \cos(10\pi t + \pi/7) + 4 \sin(30\pi t + \pi/8) \tag{1-89}$$

(a) Plot its power spectral density.
(b) Compute the power lying within a frequency band from 10 Hz to 20 Hz.

Solution:

(a) The power spectral density is shown in Figure 1-18b. Analytically,

$$S(f) = 25\delta(f + 5) + 25\delta(f - 5) + 4\delta(f - 15) + 4\delta(f + 15) \tag{1-90}$$

(b) The power in the frequency band from 10 to 20 Hz is obtained by integrating $S(f)$ from $f = -20$ Hz to -10 Hz and from $f = 10$ Hz to 20 Hz since two-sided spectra show equal portions of the signal's spectrum on either side of $f = 0$. The result is 8 W. The total power of the signal is

$$P = \int_{-\infty}^{\infty} S(f)\, df = 50 + 8 = 58 \text{ W} \tag{1-91}$$

1-6 MATLAB in Signal Analysis

It is handy to be able to verify the material presented in this chapter on signal analysis and in future chapters on system analysis with computer programs. This book will utilize MATLAB for this purpose for several reasons, not the least of which is that it includes very good plotting capabilities. MATLAB is an array-based programming language in that every variable is treated as a matrix. Appendix A gives several helpful hints for using MATLAB. The reader is also referred to *The Student Edition of MATLAB*, Version 5 and *MATLAB User's Guide*, Version 5 listed in the references at the end of this chapter.

The purpose of this section is to introduce several MATLAB functions that are used throughout the book for signal and system analysis. These include functions used in MATLAB and user-defined functions to be given here.

One has the option of executing MATLAB statements one at a time from the command window, creating a program and saving it in an M-file for future execution, or creating a function and executing that from the command window. Several useful functions are given below for generating the singularity and other elementary functions defined in this chapter:

```
%       Function for generating a unit step
%
function u = stp_fn(t)
u = 0.5*(sign(t+eps) + 1);          % The constant eps is used to ensure
                                    % against the sign argument being 0,
                                    % where it's value is 0 giving u = 0.5
```

Note that a sign, or signum, function is used to generate this step because it can handle vectors. The same result could have been achieved by using a for loop, but since MATLAB is an interpretive language, this is considerably less efficient than using vector operations. The function stp_fn may be used for generating several other signals defined in this chapter. For example, function programs for generating a unit impulse approximation, a ramp function, and a square pulse signal are given below:

```
%       A function generating a rectangular approximation for
%       the unit impulse function; width is delta
%
function imp = impls_fn(t, delta)
imp = (stp_fn(t+delta/2)-stp_fn(t-delta/2))/delta;
```

```
%       Function for generating a unit ramp
%
function r = rmp_fn(t)
r=0.5*t.*(sign(t)+1);
```

```
%       This function generates a unit-high pulse centered at zero
%       and extending from -1/2 to 1/2
%
function y = pls_fn(t)
y = stp_fn(t+0.5) - stp_fn(t-0.5-eps);
```

EXAMPLE 1-15

To demonstrate the use of these functions we generate and plot the following:

1. A step starting at $t = 2$ and going to the right;
2. A unit impulse occuring at $t = -3.5$;
3. A ramp of slope 2 going backward from $t = 4$.

Note that to give the appearance of continuous-time signals in many cases we must choose the independent variable values sufficiently close together so that the plots appear to be smooth curves. The normal plot routine for MATLAB connects adjacent points with straight lines. In the cases considered here, the signals chosen will appear as they should when plotted because they consist of straight line segments. The MATLAB program listing below is in the form of an M-file. What is given below is exactly what one sees in the command window of *The Student Version of MATLAB* if the "echo on" switch command is executed first (a switch command stays in effect until the off switch command is executed, in this case "echo off" where the quotation marks are not typed but are used here to set the command off from the rest of the text). The results of the program are given in the plots shown in Figure 1-19.

```
EDU»clex15
%     M-file for Example 1-15
%
t=-10:.005:10;
x=stp_fn(t - 2);        % Generate step starting at t=2
y=impls_fn(t+3.5, 0.05); % Generate unit impulse at t=-3.5
z=2*rmp_fn(4 - t); % Generate backwards ramp starting at t = 4
subplot(3,1,1),plot(t,x),axis([-10 10 0 1.5]), xlabel('t'),
ylabel('u(t - 2)')
```

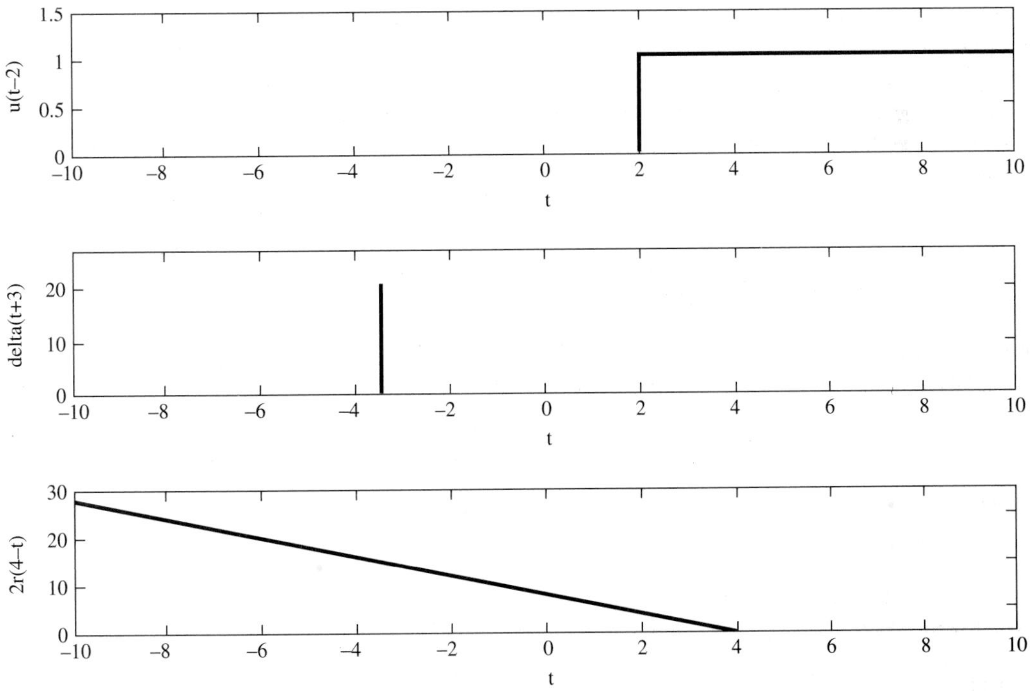

FIGURE 1-19.   Various elementary signals computed and plotted with the aid of MATLAB.

```
subplot(3,1,2),plot(t,y),axis([-10 10 0 25]),xlabel('t'),
ylabel('delta(t + 3)')
subplot(3,1,3),plot(t,z),xlabel('t'),ylabel('2r(4 - t)')
```

Further examples of elementary signal generation and the building up of more complex signals from these are suggested in the computer exercises at the end of the chapter.

EXAMPLE 1-16 _____

As a final example, we plot a sample-data signal using a special plotting feature of MATLAB. The chosen signal is a sinewave of frequency 0.5 hertz sampled at 0.2 second intervals. What appears in the MATLAB command window when the program is executed is given below (assuming "echo on" has been executed first). The output plot, which uses a stem plot function, is shown in Figure 1-20.

```
EDU» clex16
%   M-file for Example 1-16; plots a sample-data sinewave
%
del_t = 0.2;
T0 = 2;
n = 0:10;
x = sin(2*pi*n*del_t/T0);
stem(n*del_t,x),xlabel('n*del_t'), ylabel('x(n*del_t)')
```

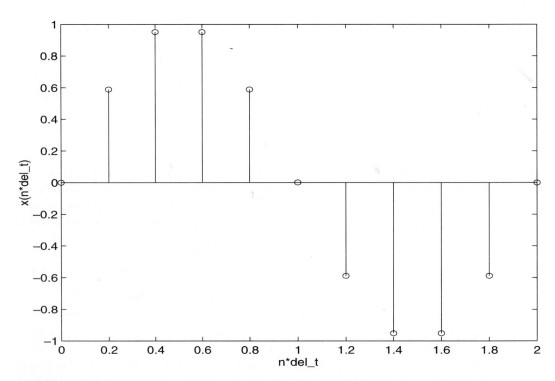

FIGURE 1-20.   A sample-data sinewave plotted with the aid of MATLAB.

Summary

In this chapter, the concepts of a *signal* and a *system* were introduced. Two simple examples of systems were given. The remainder of the chapter then concentrated on models for signals. The following are the main points made in this chapter.

1. A *system* is a combination and interconnection of several components to perform a desired task. Systems may be interconnected with each other to form other systems. In such cases the component systems are referred to as *subsystems.*
2. A *signal* is a function of time that represents a physical variable of interest. Signals may represent voltages, currents, forces, velocities, displacements, etc.
3. *Signal processing* is necessary to convert signals to more convenient forms and to produce desired quantities from measured quantities. An example was provided with the accelerometer, where the displacement of the mass, which was proportional to acceleration, was converted to an electrical voltage proportional to the acceleration, and this voltage was integrated to produce a signal proportional to velocity. A second integration would have produced a signal proportional to distance.
4. Signals can be *deterministic* or *random.* A deterministic signal is modeled as a completely specified function of time, whereas a random signal takes on a random value at any given time instant. This book is concerned only with deterministic signals.
5. Signals can be further categorized as *continuous time* or *discrete time.* The former type of signal is a function of a continuous-time variable, and the latter is a function of an independent variable assuming values from a discrete set. A *quantized* signal is one whose values are taken from a discrete set and should not be confused with a discrete-time signal. All signals discussed in this paragraph may be deterministic or random.
6. Signals may also be categorized as *periodic* or *aperiodic.* A periodic signal is one for which

$$x(t + T_0) = x(t), \qquad -\infty < t < \infty$$

where T_0 is termed the *period.* The smallest value of T_0 for which the above equation holds is called the *fundamental period.* (It should be clear that if T_0 is the fundamental period, then any integer multiple of T_0 is also a period.)
7. A useful periodic signal is the *rotating phasor signal,* defined as

$$x(t) = Ae^{j(\omega_0 t + \theta)}, \qquad -\infty < t < \infty$$

where A is the amplitude, ω_0 is the frequency in radians per second, and θ is the phase angle in radians. The quantity $A \exp(j\theta)$ is called a *phasor.* Rotating phasor signals are convenient in that a cosinusoidal signal can be represented either as the real part of a rotating phasor signal or as one-half the rotating phasor plus one-half its complex conjugate. A signal which is the sum of sinusoids can be represented as the sum of rotating phasor signals, with either the real part or half the rotating phasor sum plus half its complex conjugate taken.
8. Representation of sums of sinusoids in terms of rotating phasor sums allows one to plot signal spectra easily. *Single-sided spectra* are obtained by representing the sum of sinusoids as the real part of the corresponding rotating phasor sum; the phasor amplitudes are plotted versus frequency to obtain the *amplitude spectrum,* while the phasor phase angles are plotted versus frequency to obtain the *phase spectrum.* Since only positive-frequency components are present, the spectra exist only for frequencies greater than zero, which is the reason for the adjective "single-sided." *Double-sided spectra* are obtained by representing the sum of sinusoids as one-half the corresponding rotating phasor sum plus one-half its complex conjugate; the phasor

amplitudes are plotted versus frequency to obtain the *amplitude spectrum,* while the phasor phase angles are plotted versus frequency to obtain the *phase spectrum.* Since both positive- and negative- (due to the conjugation) frequency components are present, the spectra exist for both positive and negative frequencies, which is the reason for the adjective "double-sided."

9. Important examples of *singularity* functions are the *unit impulse,* the *unit step,* and the *unit ramp.* The unit impulse is defined by the property that

$$\int_{-\infty}^{\infty} \delta(t)x(t)\, dt = x(0)$$

where $x(t)$ is any signal continuous at $t = 0$. From this property, it can be deduced that a unit impulse as a "function" can be viewed as occurring at the origin and being infinitesimally narrow and infinitely high such that its area is unity. The unit step is defined to be unity for $t > 0$ and zero for $t < 0$, with the value at zero immaterial as long as it is finite. The unit ramp is a function that is zero for $t < 0$ and that increases linearly with unit slope for $t > 0$. The unit step is the derivative of the unit ramp, and the unit impulse is the derivative of the unit step. Likewise, a "unit doublet" could be defined as the derivative of the unit impulse, and a "unit parabola" could be defined as the integral of the unit ramp. The differentiation process or integration process can be carried out indefinitely to obtain a doubly infinite class of singularity functions. For purposes of the analyses carried out in this book, the unit impulse, step, and ramp functions will suffice. More complex functions can be built up of sums or products of singularity functions.

10. The sifting property of the unit impulse function mentioned above can be generalized to

$$\int_{t_1}^{t_2} x(t)\delta^{(n)}(t - t_0)\, dt = (-1)^n x^{(n)}(t_0), \qquad t_1 < t_0 < t_2$$

where the superscript (n) denotes the nth derivative, and the nth derivative of $x(t)$ is assumed to exist and be continuous at $t = t_0$.

11. Another classification of signals is that of energy, power, or neither. To define this classification, we first define the energy of a signal to be

$$E \triangleq \lim_{T \to \infty} \int_{-T}^{T} |x(t)|^2\, dt$$

and the power of a signal to be

$$P \triangleq \lim_{T \to \infty} \frac{1}{2T} \int_{-T}^{T} |x(t)|^2\, dt$$

The following classes of signals can then be defined:

1. $x(t)$ is an *energy signal* if and only if $0 < E < \infty$, so that $P = 0$.
2. $x(t)$ is a *power signal* if and only if $0 < P < \infty$, thus implying that $E = \infty$.
3. Signals satisfying neither property are neither power nor energy signals.

12. The *energy spectral density* of a signal $G(f)$, is a function which, when integrated over frequency from $-\infty < f < \infty$ (f in hertz), gives the total energy in the signal.

13. The *power spectral density* of a signal, $S(f)$, is a function which, when integrated over frequency from $-\infty < f < \infty$, gives the total power in the signal. For any signal possessing a two-sided line (amplitude) spectrum, we obtain the corresponding power spectral density by taking each line of the amplitude spectrum, squaring its value, and multiplying it by a unit impulse function located at that particular frequency.

Further Reading

Many circuits books have a simplified treatment of some of the material in this chapter. Recent ones are:

L. S. BOBROW, *Elementary Linear Circuit Analysis.* New York: Holt, Rinehart, and Winston, 1987

W. H. HAYT, JR. AND J. E. KEMMERLY, *Engineering Circuit Analysis.* New York: McGraw-Hill, 1993

D. R. CUNNINGHAM AND J. A. STULLER, *Circuit Analysis.* 2nd ed. New York: John Wiley, 1995

L. P. HUELSMAN, *Basic Circuit Theory,* 3rd ed. Upper Saddle River, NJ: Prentice Hall, 1991

The spectrum analysis ideas introduced in this chapter are also treated in various communications texts. One example is:

R. E. ZIEMER AND W. H. TRANTER, *Principles of Communications: Systems, Modulation, and Noise,* 4th ed. New York: Wiley, 1995

There are many other books treating continuous- and discrete-time signal and system theory with the order of topics not necessarily the same as used here. Recently published examples are:

A. V. OPPENHEIM AND A.S. WILLSKY, *Signals and Systems,* 2nd ed. Upper Saddle River, NJ: Prentice Hall, 1997

E. W. KAMEN AND B. S. HECK, *Fundamentals of Signals and Systems Using MATLAB.* Upper Saddle River, NJ: Prentice Hall, 1997

C. L. PHILLIPS AND J. M. PARR, *Signals, Systems, and Transforms.* Upper Saddle River, NJ: Prentice Hall, 1995

Several older books treating signal and system theory should be mentioned to recognize the pioneering efforts in this field. A by-no-means-complete list is:

R. A. GABEL AND R. A. ROBERTS, *Signals and Linear Systems,* 3rd ed. New York: Wiley 1987

W. M. SIEBERT, *Circuits, Signals, and Systems.* New York: McGraw-Hill, 1986

C. D. MCGILLEM AND G. R. COOPER, *Continuous and Discrete Signal and System Analysis.* New York: Holt, Rinehart, and Winston, 1974

R. J. SCHWARZ AND B. FRIEDLAND, *Linear Systems.* New York: McGraw-Hill, 1965

S. J. MASON AND H. J. ZIMMERMAN, *Electronic Circuits, Signals, and Systems.* New York: John Wiley, 1960

The following books provide further reading on the treatment of singularity functions from the standpoint of generalized function theory:

A. PAPOULIS, *Signal Analysis.* New York: McGraw-Hill, 1977

R. N. BRACEWELL, *The Fourier Integral and Its Applications,* 2nd ed. New York: McGraw-Hill, 1978.

For programming help with MATLAB, see the following references:

The MATLAB User's Guide, Version 5, Natick, MA: Mathworks, 1996

The Student Edition of MATLAB, Version 5. Upper Saddle River, NJ: Prentice Hall, 1997

Problems

Section 1-2

1-1. **(a)** Derive an expression for the distance traveled by the rocket of Example 1-1 valid up to the time of burnout.

(b) Suppose a voltage analog to distance is to be provided by another operational amplifier integrator circuit cascaded with the one of Example 1-2. If the same *RC* value is used, design a voltage divider to be placed at the output of the first operational amplifier integrator so that the second one is not overdriven at burnout ($t = 72$ s).

1-2. **(a)** For the rectangular integration rule given in recursive form in Example 1-3, show that the nonrecursive form up through time NT is

$$\hat{v}(NT) = \hat{v}(0) + T \sum_{n=1}^{N} a(nT)$$

Check this using some of the values given in Table 1-1 (page 5).

(b) For the trapezoidal integration rule given in recursive form in Example 1-3, show that the nonrecursive form up through time NT is

$$\hat{v}(NT) = \hat{v}(0) + \left[a(0) + a(NT) + 2 \sum_{n=1}^{N-1} a(nT)\right]\left(\frac{T}{2}\right)$$

Check this, using values from Example 1-3.

1-3. Rework Example 1-1 if the acceleration profile for the rocket is

$$a(t) = 20 \text{ m/s}^2, \qquad 0 \le t \le 50 \text{ s}$$

and zero otherwise. Assume the same parameter values as given in Example 1-1. Find and plot the velocity profile.

1-4. Rework Example 1-1 with the acceleration profile shown. Sketch the resulting velocities.

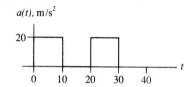

$a(t)$, m/s²

FIGURE P1-4

(a) Assume an initial velocity of zero.
(b) Assume an initial velocity of 1500 m/s.

1-5. In the satellite communications example, Example 1-4, let the transmitted signal be $x(t) = \cos\omega_0 t$.
(a) Show that the received signal can be put into the form

$$y(t) = \sqrt{1 + 2\alpha\beta \cos 2\omega_0 \tau + (\alpha\beta)^2} \, \cos\left(\omega_0 t - \tan^{-1} \frac{\alpha\beta \sin 2\omega_0 \tau}{1 + \alpha\beta \cos 2\omega_0 \tau}\right)$$

(b) Plot the envelope of the cosine in the above equation versus $\omega_0 \tau$ for $\alpha\beta = 0.1$. Note that the received signal will "fade" as τ varies due to satellite motion.

Section 1-3

1-6. (a) A sample-data signal derived by taking samples of a continuous-time signal, $m(t)$, is often represented in terms of an infinite sequence of rectangular pulse signals by multiplication. That is,

$$x_{sd}(t) = m(t) \sum_{n=-\infty}^{\infty} \Pi\left(\frac{t - nt_0}{\tau}\right)$$

If $m(t) = \exp(-t/10)u(t)$, $\tau = 0.5$ s, and $t_0 = 2$ s, sketch $x_{sd}(t)$ for $-1 \le t \le 10$ s.
(b) Flat-top sample representation of a signal is sometimes preferred over the above scheme. In this case, the sample-data representation of a continuous-time signal is given by

$$x_{ft}(t) = \sum_{n=-\infty}^{\infty} m(nt_0) \Pi\left(\frac{t - nt_0}{\tau}\right)$$

Sketch this sample-data representation for the same signal and parameters as given in part (a). Discuss the major difference between the two representations.

1-7. Ideal sampling is represented in two ways depending on whether the signal is considered to be continuous-time or discrete-time.

(a) The continuous-time signal $\cos(2\pi t)$ is to be represented in terms of samples by multiplying by a train of impulses spaced by 0.1 seconds. Let the impulse train be written as

$$\sum_{n=-\infty}^{\infty} \delta(t - 0.1n)$$

and show that the ideal impulse-sampled waveform is given by

$$\sum_{n=-\infty}^{\infty} \cos(0.2\pi n)\delta(t - 0.1n)$$

by using appropriate properties of the unit impulse.

(b) Show that the discrete-time representation is the same, except that the continuous-time unit impulse is replaced by a discrete-time unit pulse function, $\delta[n]$, appropriately shifted. Sketch for the signal $\cos(2\pi t)$ sampled each 0.1 seconds

1-8. Sketch the following signals:

(a) $\Pi(0.1t)$
(b) $\Pi(10t)$
(c) $\Pi(t - 1/2)$
(d) $\Pi[(t - 2)/5]$
(e) $\Pi[(t - 1)/2] + \Pi(t - 1)$

1-9. What are the fundamental periods of the signals given below? (Assume units of seconds for the t-variable.)

(a) $\sin 50\pi t$
(b) $\cos 60\pi t$
(c) $\cos 70\pi t$
(d) $\sin 50\pi t + \cos 60\pi t$
(e) $\sin 50\pi t + \cos 70\pi t$

1-10. Given the two complex numbers $A = 3 + j3$ and $B = 10 \exp(j\pi/3)$.

(a) Put A into polar form and B into cartesian form. What is the magnitude of A? The argument of A? The real part of B? The imaginary part of B?
(b) Compute their sum. Show as a vector in the complex plane along with A and B.
(c) Compute their difference. Show as a vector in the complex plane.
(d) Compute their product in two ways: by multiplying both numbers in cartesian form and by multiplying in polar form. Show that both answers are equivalent.
(e) Compute the quotient A/B in two ways: by dividing with both numbers expressed in cartesian form and by dividing with both numbers expressed in polar form. Show that both answers are equivalent.

1-11. Find the periods and fundamental frequencies of the following signals:

(a) $x_a(t) = 2 \cos(10\pi t + \pi/6)$
(b) $x_b(t) = 5 \cos(17\pi t - \pi/4)$
(c) $x_c(t) = 3 \sin(19\pi t - \pi/3)$
(d) $x_d(t) = x_a(t) + x_b(t)$
(e) $x_e(t) = x_a(t) + x_c(t)$
(f) $x_f(t) = x_b(t) + x_c(t)$

1-12. (a) Write the signals of Problem 1-11 as the real part of the sum of rotating phasors.
   (b) Write the signals of Problem 1-11 as the sum of counterrotating phasors.
   (c) Plot the single-sided amplitude and phase spectra for these signals.
   (d) Plot the double-sided amplitude and phase spectra for these signals.

1-13. (a) Write the signals given in Problem 1-9 as the real parts of rotating phasors.
   (b) Write each of the signals given in Problem 1-9 as one-half the sum of a rotating phasor and its complex conjugate.
   (c) Sketch the single-sided amplitude and phase spectra of the signals given in Problem 1-9.
   (d) Sketch the double-sided amplitude and phase spectra of the signals given in Problem 1-9.

1-14. (a) Express the signal given below in terms of step functions. Sketch it first.

$$x_a(t) = \Pi[(t - 3)/6] + \Pi[(t - 4)/2]$$

   (b) Express the derivative of the signal given above in terms of unit impulses.

1-15. Suppose that instead of writing a sinusoid as the real part of a rotating phasor, we agree to use the convention

$$\sin(\omega_0 t + \theta) = \text{Im exp}[j(\omega_0 t + \theta)]\}$$

or

$$\sin(\omega_0 t + \theta) = \exp[j(\omega_0 t + \theta)]/2j - \exp[-j(\omega_0 t + \theta)]/2j$$

   (a) What change, if any, will there be to the two-sided amplitude spectrum of a signal from the case where the real-part convention is used?
   (b) What change, if any, will there be to the two-sided phase spectrum of a signal from the case where the real-part convention is used?

1-16. Sketch the following signals:
   (a) $u[(t - 2)/4]$ **(d)** $\Pi(-3t + 1)$
   (b) $r[(t + 1)/3]$ **(e)** $\Pi[(t - 3)/2]$
   (c) $r(-2t + 3)$

1-17. Derive expressions for singularity functions $u_i(t)$ for $i = -4$ and $i = -5$. Generalize to arbitrary negative values of i.

1-18. Plot accurately the following signals defined in terms of singularity functions:
   (a) $x_a(t) = r(t)u(2 - t)$
   (b) $x_b(t) = r(t) - r(t - 1) - r(t - 2) + r(t - 3)$
   (c) $x_c(t) = \sum_{n=0}^{\infty} x_a(t - 2n)$ (Plot for $0 \le t \le 8$, and use three dots to indicate its semi-infinite extent.)
   (d) $x_d(t) = \sum_{n=0}^{\infty} x_b(t - 3n)$ (Plot for $0 \le t \le 9$ and use three dots to indicate its semi-infinite extent.)

1-19. (a) Sketch the signal $y(t) = \sum_{n=0}^{\infty} u(t - 2n)u(1 + 2n - t)$.
   (b) Is it periodic? If so, what is its period? If not, why not?
   (c) Repeat parts (a) and (b) for the signal $y(t) = \sum_{n=-\infty}^{\infty} u(t - 2n)u(1 + 2n - t)$.

1-20. Express the signals shown in terms of singularity functions (they are all zero for $t < 0$):

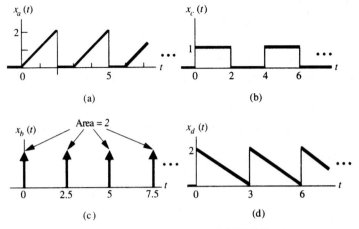

(a)

(b)

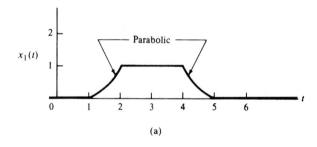

(c)

(d)

FIGURE P1-20

1-21. Represent the signals shown in terms of singularity functions.

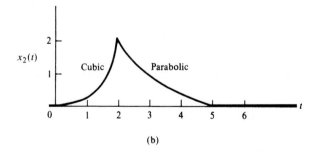

(a)

(b)

FIGURE P1-21

1-22. Write the signals shown in Figure P1-22 in terms of singularity functions.

1-23. **(a)** Show that

$$\delta_\epsilon(t) = \frac{e^{-t/\epsilon}}{\epsilon} u(t)$$

has the properties of a delta function in the limit as $\epsilon \to 0$.

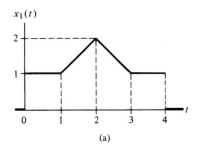

(a)

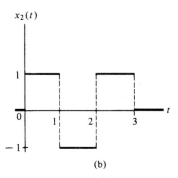

(b)

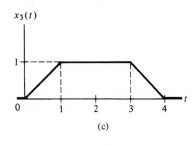

(c)

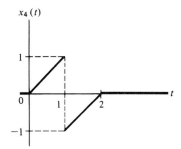

FIGURE P1-22

(b) Show that $\exp[-t^2/2\sigma^2]/\sqrt{2\pi\sigma^2}$ has the properties of a unit impulse function as $\sigma \rightarrow 0$. (*Hint:* Look up the integral of $\exp(-\alpha t^2)$ in a table of definite integrals.)

1-24. (a) By plotting the derivative of the function given in Problem 1-23(b) for $\sigma = 0.2$ and $\sigma = 0.05$, deduce what a unit doublet must "look like."

(b) By plotting the second derivative of the function given in Problem 1-23(b) for the same values of σ as given in part (a), deduce what a unit triplet must "look like."

1-25. In taking derivatives of product functions, one of which is a singularity function, one must exercise care. Consider the second derivative of

$$h(t) = e^{-\alpha t}u(t)$$

Blindly carrying out the derivative twice yields

$$\frac{d^2h}{dt^2} = \alpha^2 e^{-\alpha t}u(t) - 2\alpha e^{-\alpha t}\delta(t) + e^{-\alpha t}\dot{\delta}(t)$$

Use the fact that

$$\alpha e^{-\alpha t}\delta(t) = \alpha e^{-\alpha 0}\delta(t) = \alpha\delta(t)$$

to obtain the correct result.

1-26. Evaluate the following integrals.

(a) $\displaystyle\int_5^{10} \cos 2\pi t \, \delta(t-2) \, dt$

(b) $\displaystyle\int_0^5 \cos 2\pi t \, \delta(t-2) \, dt$

(c) $\displaystyle\int_0^5 \cos 2\pi t \, \delta(t-0.5) \, dt$

(d) $\displaystyle\int_{-\infty}^{\infty} (t-2)^2 \delta(t-2) \, dt$

(e) $\displaystyle\int_{-\infty}^{\infty} t^2 \delta(t-2) \, dt$

1-27. Evaluate the following integrals (dots over a symbol denote time derivative).

(a) $\displaystyle\int_{-\infty}^{\infty} e^{3t} \ddot\delta(t-2) \, dt$

(b) $\displaystyle\int_0^{10} \cos(2\pi t) \dddot\delta(t-0.5) \, dt$

(c) $\displaystyle\int_{-\infty}^{\infty} [e^{-3t} + \cos(2\pi t)] \dot\delta(t) \, dt$

1-28. Find the unspecified constants, denoted as $C_1, C_2, \ldots$, in the following expressions.

(a) $10\delta(t) + C_1\dot\delta(t) + (2 + C_2)\ddot\delta(t) = (3 + C_3)\delta(t) + 5\dot\delta(t) + 6\ddot\delta(t)$

(b) $(3 + C_1)\delta^{(4)}(t) + C_2\ddot\delta(t) + C_3\dot\delta(t) = C_4\delta^{(3)}(t) + C_5\delta(t)$

1-29. (a) Sketch the following signals:

(1) $x_1(t) = r(t+2) - 2r(t) + r(t-2)$

(2) $x_2(t) = u(t)u(10-t)$

(3) $x_3(t) = 2u(t) + \delta(t-2)$

(4) $x_4(t) = 2u(t)\delta(t-2)$

(b) For the signal shown, write an equation in terms of singularity functions.

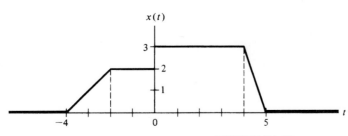

$x(t)$

FIGURE P1-29

1-30. For the signal shown, write an equation in terms of singularity functions.

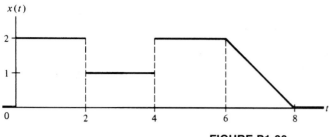

FIGURE P1-30

1-31. Evaluate the integrals given below.

(a) $\displaystyle\int_{-\infty}^{\infty} t^3 \delta(t-3)\, dt$

(b) $\displaystyle\int_{-\infty}^{\infty} (3t + \cos 2\pi t)\dot{\delta}(t-5)\, d$

(c) $\displaystyle\int_{-\infty}^{\infty} (1 + t^2)\dot{\delta}(t-1.5)\, dt$

1-32. Write the signals in Figure P1-32 in terms of singularity functions.

Section 1-4

1-33. Sketch the following signals and calculate their energies.
(a) $e^{-10t}u(t)$
(b) $u(t) - u(t-15)$
(c) $\cos 10\pi t\, u(t)u(2-t)$
(d) $r(t) - 2r(t-1) + r(t-2)$

1-34. Obtain the energies of the signals in Problem 1-22.

1-35. Which of the signals given in Problem 1-18 are energy signals? Justify your answers.

1-36. Obtain the average powers of the signals given in Problem 1-9.

1-37. Obtain the average powers of the signals given in Problem 1-11.

1-38. Which of the following signals are power signals and which are energy signals? Which are neither? Justify your answers.
(a) $u(t) + 5u(t-1) - 2u(t-2)$
(b) $u(t) + 5u(t-1) - 6u(t-2)$
(c) $e^{-5t}u(t)$
(d) $(e^{-5t} + 1)u(t)$
(e) $(1 - e^{-5t})u(t)$
(f) $r(t)$
(g) $r(t) - r(t-1)$
(h) $t^{-1/4}u(t-3)$

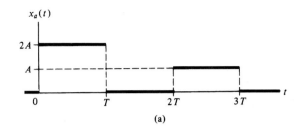

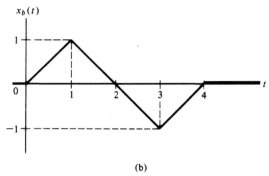

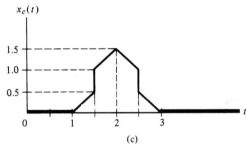

FIGURE P1-32

1-39. Given the signal

$$x(t) = 2\cos(6\pi t - \pi/3) + 4\sin(10\pi t)$$

(a) Is it periodic? If so, find its period.
(b) Sketch its single-sided amplitude and phase spectra.
(c) Write it as the sum of rotating phasors plus their complex conjugates.
(d) Sketch its two-sided amplitude and phase spectra.
(e) Show that it is a power signal.

1-40. Which of the following signals are energy signals? Find the energies of those that are. Sketch each signal.
(a) $u(t) - u(t-1)$
(b) $r(t) - r(t-1) - r(t-2) + r(t-3)$
(c) $t\exp(-2t)u(t)$
(d) $r(t) - r(t-2)$
(e) $u(t) - \frac{1}{3}u(t-10)$

1-41. Given the following signals:

(1) $\cos 5\pi t + \sin 6\pi t$

(2) $\sin 2t + \cos \pi t$

(3) $e^{-10t}u(t)$

(4) $e^{2t}u(t)$

(a) Which are periodic? Give their periods.

(b) Which are power signals? Compute their average powers.

(c) Which are energy signals? Compute their energies.

1-42. Prove Equation (1-84) by starting with (1-76).

1-43. Given the signal

$$x(t) = \sin^2(7\pi t - \pi/6) + \cos(3\pi t - \pi/3)$$

(a) Sketch its single-sided amplitude and phase spectra.

(b) Sketch its double-sided amplitude and phase spectra after writing it as the sum of complex conjugate rotating phasors.

Section 1-5

1-44. Plot the power spectral density of the signal given in Problem 1-43.

1-45. Given the signal

$$x(t) = 16 \cos(20\pi t + \pi/4) + 6 \cos(30\pi t + \pi/6) + 4 \cos(40\pi t + \pi/3)$$

(a) Find and plot its power spectral density.

(b) Compute the power contained in the frequency interval 12 Hz to 22 Hz.

Computer Exercises

1-1. Use the functions given in Section 1-6 for the step and ramp signals to plot the signal shown in Problem 1-30 using MATLAB.

1-2. Use the functions given in Section 1-6 for the step and ramp signals to plot the signals shown in Problem 1-32 using MATLAB.

1-3. Use the elementary function programs given in Section 1-6 to compute and plot the following:

(a) A step of height 3 starting at $t = 3$ and going backwards to $t = -\infty$;

(b) A signal that starts at $t = 1$, increases linearly to a value of 2 at $t = 2$, and is constant thereafter;

(c) A stairstep signal that is 0 for $t < 0$, jumps to a value of 1 at $t = 0$, a value of 2 at $t = 1$, a value of 3 at $t = 2$, a value of 4 at $t = 3$, and stays at 4 thereafter;

(d) A ramp starting at $t = 2$ and going downward with a slope of -3.

1-4. (a) Generate a cosine burst of frequency 2 Hz, lasting for 5 seconds; (b) generate a sine burst of frequency 2 Hz and lasting for five seconds; (c) combine the results of (a) and (b) to produce a sinusoidal burst of frequency 2 Hz, 5 seconds long, and with starting phase at $t = 0$ of $\pi/4$ radians. [Hint: recall the trigonometric identity $\sin(x + y) = 0.5(\sin x \cos y - \cos x \sin y)$. Set $y = \pi/4$.]

1-5. (a) Write a MATLAB function to generate a sequence of impulses spaced by an arbitrary amount `t_rep`, lasting for `t_width` and centered on $t = 0$. Call it `cmb_fn(t,t_rep, t_width,delta)`
(b) Generate an impulse comb with spacing between impulses of 1.25 seconds and containing 5 impulses starting at $t = 0$.

1-6. (a) Write a MATLAB function to generate a unit parabola singularity function.
(b) Write a MATLAB function to generate a unit cubic singularity function.
(c) Use them to generate a plots of the signals shown in Problem 1-21.

CHAPTER 2

System Modeling and Analysis in the Time Domain

2-1 Introduction

Having discussed some continuous-time signal models, we now wish to consider ways of modeling systems and to analyze the effects of systems on signals. The systems analysis problem is: *given a system and an input, what is the output?* System design or synthesis is important also, but generally is more difficult than analysis. In addition, analysis procedures also suggest synthesis techniques.

We will discuss several mathematical characterizations for systems in this chapter, but our main focus throughout the book will be on *linear, time-invariant systems.* The reasons for this are threefold: First, powerful analysis techniques exist for such systems. Second, many real-world systems can be closely approximated as linear, time-invariant systems.[†] Third, analysis techniques for linear, time-invariant systems suggest approaches for the analysis of nonlinear systems.

There are three basic approaches for finding the response of a fixed, linear system to a given input: (1) Obtain a solution to the modeling equations through standard methods of solving differential equations. (2) Carry out a solution in the time domain, using the superposition integral. (3) Use frequency-domain analysis by means of the Fourier or Laplace transforms. The first method should be familiar to the student from circuits courses. The objective of this chapter is to examine the second method. The third method is dealt with in later chapters.

In passing, we note that the condition of time invariance is not necessary in many of the analysis procedures presented in this chapter, but computational problems are simplified under the assumption of time invariance.

We begin our consideration of systems in the time domain with appropriate classification procedures and modeling techniques. The primary tool for analysis of linear, time-invariant systems will be the *convolution* integral, discussed in Section 2.4.

2-2 System Modeling Concepts

Some Terminology

We now wish to look at ways of representing the effects of systems on signals; that is, we need to be able to construct appropriate *system models* that adequately represent the interaction of signals and systems and the relationship of *causes* and *effects* for that system. Usually, we refer to certain causes of in-

---

[†]The terms "linear" and "time invariant" will be given precise mathematical definitions shortly. Although it is possible to analyze systems that are time varying, it is more difficult than for time-invariant systems. Because of this, and because many systems are time invariant, we restrict our attention to this category.

terest as *inputs* and certain effects of interest as *outputs*. For example, in the accelerometer example of Chapter 1, the input of interest could be the position of the weight and the output could be a voltage proportional to velocity.

An obvious choice for input and output quantities may not always be readily apparent, easily isolated, or physically distinct. For instance, in the communications link example, the input, $x(t)$, and output of interest, $s(t)$, appear at the same physical location: the transmitter. The relationship governing their interaction was very simple: that is,

$$s(t) \simeq x(t) + \alpha\beta x(t - 2\tau), \qquad 0 \le t \le T + 2\tau \tag{1-15}$$

Usually, we will not have such simple input-output relationships.

It is convenient to visualize a system schematically by means of a box, as shown in Figure 2-1. (Sometimes symbols other than x and y may be used.) On the left-hand side of the box we represent the inputs (excitations, causes, stimuli) as a series of arrows labeled $x_1(t), x_2(t), \ldots, x_m(t)$. The outputs (responses, effects), $y_1(t), y_2(t), \ldots, y_p(t)$, are represented as arrows emanating from the right-hand side of the box. In general, the inputs and outputs vary with time, which is represented by the independent variable t; sometimes it will not be shown explicitly in order to simplify notation. The number of inputs and outputs need not be equal. Quite often, however, we will be concerned with situations where there is a single input and a single output. Such systems are referred to as *two-port* or *single-input, single-output systems* since they have one input port and one output port.

Representations for Systems

Usually, we are interested in obtaining explicit relationships between the input and output variables of a system, and we discuss systematic procedures for doing so shortly. For now, however, we will express the dependence of the system output on the input symbolically as

$$y(t) = \mathcal{H}[x(t)] \tag{2-1a}$$

Eliminating the explicit use of t, we can write this more compactly as

$$y = \mathcal{H}x \tag{2-1b}$$

which is read

$$y(t) \text{ is the response of } \mathcal{H} \text{ to } x(t)$$

where initial conditions are included if pertinent. The symbol $\mathcal{H}$, which is known as an *operator,* serves the dual role of identifying the system and specifying the operation to be performed on $x(t)$ to produce $y(t)$. (For a multiple-input, multiple-output system, $x(t)$ and $y(t)$ are vectors.)

Several examples of system input-output relationships are given below to illustrate more clearly some of the basic concepts.

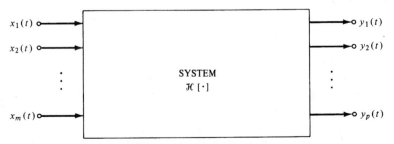

FIGURE 2-1.   Block diagram representation of a system.

1. Instantaneous (Nondynamic) Relationships. Many systems are adequately modeled for many situations by relationships such as

$$y(t) = Ax(t) + B \tag{2-2a}$$

or

$$y(t) = Ax(t) + Bx^3(t) \tag{2-2b}$$

where A and B are constants. The former equation could represent an amplifier of gain A with a dc bias of B volts at its output, and the latter equation would be a possible model for an amplifier that introduces *nonlinear distortion, $Bx^3(t)$*, into the amplified output, $Ax(t)$. The concept of nonlinearity for a system will be explicitly defined later.

2. Linear, Constant-Coefficient, Ordinary Differential Equations. The student is assumed to be familiar with obtaining the relationships between currents and voltages in an electrical circuit composed of passive elements (resistors, capacitors, and inductors) in terms of ordinary integrodifferential equations through the application of Kirchhoff's voltage and current laws. If all integrals are removed through repeated differentiation, the general form for such equations may be written as

$$a_n \frac{d^n y(t)}{dt^n} + a_{n-1} \frac{d^{n-1} y(t)}{dt^{n-1}} + \cdots + a_0 y(t) = b_m \frac{d^m x(t)}{dt^m} + b_{m-1} \frac{d^{m-1} x(t)}{dt^{m-1}} + \cdots + b_0 x(t) \tag{2-3}$$

where $a_n, a_{n-1}, \ldots, a_0, b_m, b_{m-1}, \ldots, b_0$ are constants that depend on the structure of the system and where nonzero initial conditions may be specified. For systems whose input-output relationships can be represented in this manner, the *order* of the system is defined as the order of the highest derivative of the *dependent variable,* or system output, present (integrals on the input might still be present in some cases).[†]

Three variations of (2-3) sometimes result from the modeling and analysis of systems. First, time-varying components may be present in the system (e.g., a capacitor whose capacitance varies with time). In such cases one or more of the *coefficients* of (2-3) will be *functions of time.* These are examples of *time-varying systems,* although the class of time-varying systems is broader than that described by ordinary differential equations with nonconstant coefficients.

A second situation that results in a governing differential equation of different form than (2-3) is where one or more nonlinear components are present in the system. Consider a voltage source $v_s(t)$ in series with an inductor and a resistor whose resistance depends on the current through it: for example, $R = R_0 + \alpha i(t)$, where R_0 and α are constants. Then application of Kirchhoff's voltage law results in

$$[R_0 + \alpha i(t)]i(t) + L \frac{di}{dt} = v_s(t) \tag{2-4}$$

and the relationship between the input, $v_s(t)$, and the output, $i(t)$, is no longer linear. This is an example of a *nonlinear* system. Precise definitions for time-varying and nonlinear systems are given later.

The third departure from (2-3) which may result is that of a system described by a *partial differential equation.* Such systems are said to be *distributed,* in contrast to *lumped* systems, for which the cause-effect properties of the system can be ascribed to elements modeled as infinitesimally small in the spatial dimension. This is permissible if their dimensions are small compared with the wavelength of the variable quantities within the system (voltages, currents, etc.), where the wavelength is given by

$$\lambda = \frac{c}{f} \quad \text{meters}$$

[†]The order of a multiple-input, multiple-output system can be determined through state-variable techniques, which are discussed in Chapter 7.

where $c = 3 \times 10^8$ m/s is the speed of electromagnetic radiation and f is the frequency of the oscillating waveform in hertz. A light bulb, for example, has dimensions that are small compared with $\lambda = 3 \times 10^8/60 = 5 \times 10^6$ m, which is the wavelength of the 60-Hz power-line voltage.

3. Integral Relationships. We discuss an integral representation of a system input-output relation of the form

$$y(t) = \int_{-\infty}^{\infty} h(\lambda)x(t - \lambda) \, d\lambda \tag{2-5a}$$

$$= \int_{-\infty}^{\infty} x(\lambda)h(t - \lambda) \, d\lambda \tag{2-5b}$$

which is known as a *superposition integral.* The function $h(\lambda)$ is called the *impulse response* of the system and is its response to a unit impulse applied at $\lambda = 0.$[†] All systems that can be represented by ordinary, constant-coefficient, linear differential equations can also be represented by an equation of the form (2-5). However, not all systems that can be represented by a relationship of the form (2-5) can be represented by ordinary, constant-coefficient linear differential equations.[‡]

For a time-varying system, the generalization of (2-5) is

$$y(t) = \int_{-\infty}^{\infty} h(t, \lambda)x(\lambda) \, d\lambda \tag{2-6}$$

where $h(t, \lambda)$ is the response of the system at time t to a unit impulse applied at time λ.

EXAMPLE 2-1

Consider the *RC* circuit depicted in Figure 2-2 with input $x(t)$ and output $y(t)$ as shown. We wish to obtain (a) the differential equation relating $y(t)$ to $x(t)$, and (b) convert this differential equation to the integral form (2-5). Assume that $x(t)$ is applied at $t = t_0$ and $y(t_0) = y_0$.

Solution:

(a) The differential equation relating $y(t)$ and $x(t)$ is found by writing Kirchhoff's voltage law (KVL) around the loop indicated by $i(t)$. This results in the equation

$$x(t) = Ri(t) + y(t) \tag{2-7}$$

where $y(t)$, the voltage across the capacitor, is related to $i(t)$, the current through it, by

$$i(t) = C \frac{dy(t)}{dt} \tag{2-8}$$

Eliminating $i(t)$ in the KVL equation, we obtain

$$RC \frac{dy(t)}{dt} + y(t) = x(t) \tag{2-9}$$

[†]To show this, let $x(t) = \delta(t)$, so that $x(t - \lambda) = \delta(t - \lambda)$. Then $y(t) = h(t)$ by the sifting property of the unit impulse function, and $h(t)$ is seen to be the system's response to $\delta(t)$.

[‡]An example is an ideal low-pass filter whose impulse response is

$$h(t) = B \frac{\sin(2\pi Bt)}{2\pi Bt}$$

where B is the bandwidth. This type of filter will be discussed in Chapter 4. Ideal filters are convenient mathematical models, but cannot be realized as physical circuits.

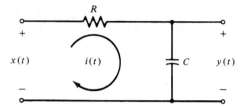

FIGURE 2-2. *RC* circuit to illustrate the derivation of input-output relationships for systems.

(b) To write the input-output relationship as an explicit equation for $y(t)$ in terms of $x(t)$, we first obtain the solution to the homogeneous differential equation

$$RC\frac{dy_h(t)}{dt} + y_h(t) = 0 \tag{2-10}$$

Assuming a solution of the form

$$y_h(t) = Ae^{pt} \tag{2-11}$$

and substituting into the homogeneous equation, we find that $p = -1/RC$, so that

$$y_h(t) = A\exp\left(\frac{-t}{RC}\right) \tag{2-12}$$

The total solution consists of $y_h(t)$ plus a particular solution for a specific $x(t)$. Assume that the input $x(t)$ is applied beginning at $t = t_0$ but is otherwise arbitrary. Also assume that the value of $y(t)$ at $t = t_0$ is y_0. To find the total solution we use the technique of *variation of parameters,* which consists of assuming a solution of the form of $y_h(t)$ but with the undetermined coefficient A replaced by a function of time to be found. Thus we assume that

$$y(t) = A(t)\exp\left(\frac{-t}{RC}\right) \tag{2-13}$$

Differentiating by means of the chain rule, we obtain

$$\frac{dy(t)}{dt} = \left[\frac{dA(t)}{dt} - \frac{A(t)}{RC}\right]\exp\left(\frac{-t}{RC}\right) \tag{2-14}$$

Substituting the assumed solution and its derivative into the nonhomogeneous differential equation, we have

$$RC\exp\left(\frac{-t}{RC}\right)\frac{dA(t)}{dt} = x(t) \tag{2-15}$$

where the term $-A(t)\exp(-t/RC)$ has been canceled by $y(t)$ in the differential equation. Solving for $dA(t)/dt$ and integrating, we get an explicit result for the unknown "varying parameter." It is given by

$$A(t) = \frac{1}{RC}\int_{t_0}^{t} x(\lambda)\exp\left(\frac{\lambda}{RC}\right)d\lambda + A(t_0) \tag{2-16}$$

where λ is a dummy variable of integration and $A(t_0)$ is the initial value of $A(t)$ at the time t_0 that the input $x(t)$ is applied. Since $y(t) = A(t)\exp(-t/RC)$, it follows that $A(t_0) = y_0\exp(t_0/RC)$. Using this result together with the expression for $A(t)$, we obtain

$$y(t) = \left[\int_{t_0}^{t} x(\lambda) \frac{\exp(\lambda/RC)}{RC} d\lambda + y_0 \exp\left(\frac{t_0}{RC}\right) \right] \exp\left(\frac{-t}{RC}\right)$$

$$= y_0 \exp\left(-\frac{t - t_0}{RC}\right) + \int_{t_0}^{t} x(\lambda) \frac{\exp[-(t - \lambda)/RC]}{RC} d\lambda \tag{2-17}$$

If we assume that the input $x(t)$ is applied at $t = -\infty$ and that $y_0 \triangleq y(-\infty) = 0$, then

$$y(t) = \int_{-\infty}^{t} x(\lambda) \frac{\exp[-(t - \lambda)/RC]}{RC} d\lambda \tag{2-18}$$

If we define

$$h(t) = \frac{\exp(-t/RC)}{RC} u(t) \tag{2-19}$$

so that

$$h(t - \lambda) = \frac{\exp[-(t - \lambda)/RC]}{RC} u(t - \lambda) \tag{2-20}$$

then the solution for $y(t)$, as determined by the variation-of-parameters approach, is exactly of the form given by (2-5b).

Properties of Systems

We now give precise definitions of several system properties, some of which were mentioned in the previous discussion.

Continuous-Time and Discrete-Time Systems. If the signals processed by a system are continuous-time signals, the system itself is referred to as a *continuous-time* system. If, on the other hand, the system processes signals that exist only at discrete times, it is called a *discrete-time* system. The signals in a system may or may not be quantized to a finite number of levels. If they are, the system is referred to as *quantized.* A quantized system may be continuous time or discrete time, although usually a quantized system is also a discrete-time system and is then referred to as a *digital system.* Discrete-time systems are discussed in Chapter 8.

Examples of continuous-time systems are electric networks composed of resistors, capacitors, and inductors that are driven by continuous-time sources. An example of a quantized, discrete-time system is the optical-wand sensing system of a department store cash register.

Fixed and Time-Varying Systems. A system is *time invariant,* or *fixed,* if its input-output relationship does not change with time. Otherwise, it is said to be *time varying.* In terms of the symbolic notation of (2-1a), a system is fixed if and only if

$$\mathcal{H}[x(t - \tau)] = y(t - \tau) \tag{2-21}$$

for *any* $x(t)$ and *any* τ. The system must be at rest[†] prior to the application of $x(t)$. In words, (2-21) states that if $y(t)$ is the response of the system to an input $x(t)$, then its response to a time-shifted version of

---

[†]A system is at rest, or relaxed, prior to some initial instant, say $t = t_0$, if all energy storage elements have zero initial energy stored. For an nth-order system described by an nth-order ordinary linear differential equation, this means that the dependent variable and all its derivatives up through the $(n - 1)$st are zero for $t < t_0$.

this input, $x(t - \tau)$, is the response of the system to $x(t)$ *time-shifted* by the same amount, or $y(t - \tau)$. For example, the system with input-output relationship

$$y(t) = x(t) + Ax(t - T) \tag{2-22}$$

is fixed if A and T are constants, and time varying if either or both are functions of time. Thus the satellite communication system is fixed if $\alpha\beta$ and τ are constants. Similarly, a system described by a differential equation of the form (2-3) is fixed if the coefficients are constant, and time varying if one or more coefficients are functions of time.

Causal and Noncausal Systems. A system is *causal* or *nonanticipatory* if its response to an input does not depend on future values of that input.‡ A precise definition is that a continuous-time system is causal if and only if the condition

$$x_1(t) = x_2(t) \qquad \text{for } t \le t_0 \tag{2-23a}$$

implies the condition

$$\mathcal{H}[x_1(t)] = \mathcal{H}[x_2(t)] \qquad \text{for } t \le t_0 \tag{2-23b}$$

for any t_0, $x_1(t)$, and $x_2(t)$. Stated another way, if the *difference* between two system inputs is zero for $t \le t_0$, the *difference* between the respective outputs must be zero for $t \le t_0$ if the system is causal. Thus the definition applies to systems for which the response to zero input is not zero, such as a system containing independent sources, as well as systems with nonzero inputs.

EXAMPLE 2-2 ⎯⎯⎯⎯⎯⎯⎯⎯⎯⎯⎯⎯⎯⎯⎯⎯⎯⎯⎯⎯⎯⎯⎯⎯⎯⎯⎯⎯⎯⎯⎯⎯⎯⎯⎯

Consider a system described by (2-2a). Let $x_1(t) = u(t)$ and $x_2(t) = r(t)$. For $t \le t_0 < 0$, both inputs are equal. The outputs for $t \le t_0$ for both inputs are $y_1(t) = y_2(t) = B$. In general, it is seen from (2-2a) that if $x_1(t) = x_2(t), t \le t_0$, then $y_1(t) = y_2(t) = B$ for $t \le t_0$ because $Ax_1(t) = Ax_2(t)$. It is easy to see that this holds for any pair of inputs for which it is true that $x_1(t) = x_2(t), t \le t_0$. Thus this system is *causal*. An easy way to see that the system is causal is to note that $y(t)$ cannot do anything in anticipation of $x(t)$.

Dynamic and Instantaneous Systems. A system for which the output is a function of the input *at the present time only* is said to be *instantaneous* (or *memoryless*, or *zero memory*). A *dynamic* system, or one which is not instantaneous, is one whose output depends on past or future values of the input in addition to the present time. If the system is also causal, the output of a dynamic system depends only on present and past values of the input. Mathematically, the input-output relationship for an instantaneous system must be of the form

$$y(t) = f[x(t), t] \tag{2-24}$$

where $f(\cdot)$ is a function, possibly time dependent, which depends on $x(t)$ *only at the present time, t*.

The systems described by (2-2a) and (2-2b) are instantaneous. Systems described by differential equations are dynamic (nonzero memory). For example, an inductor is a system with memory. This is easy to see if the input is taken as the voltage across its terminals and the output as the current through it so that

$$i(t) = \frac{1}{L} \int_{-\infty}^{t} v(\lambda)\, d\lambda \tag{2-25}$$

It is clear that $i(t)$ depends on past values of $v(t)$ through the integral.

⎯⎯⎯⎯⎯⎯⎯⎯⎯⎯⎯⎯⎯⎯⎯⎯⎯⎯⎯⎯⎯⎯

‡Another way of informally defining a causal system is that such a system does not anticipate its input.

Linear and Nonlinear Systems. From your circuits courses, you should already be familiar with the concept of linearity. It allows the analysis of circuits with multiple sources by means of *superposition*, or addition of currents or voltages calculated when each source is applied separately. A precise mathematical definition of a linear system is that *superposition holds*, where superposition for a system with any two inputs $x_1(t)$ and $x_2(t)$ is defined as

$$\mathcal{H}[\alpha_1 x_1(t) + \alpha_2 x_2(t)] = \alpha_1 \mathcal{H}[x_1(t)] + \alpha_2 \mathcal{H}[x_2(t)]$$

$$= \alpha_1 y_1(t) + \alpha_2 y_2(t) \tag{2-26}$$

In (2-26), $y_1(t)$ is the response of the system when input $x_1(t)$ is applied alone, and $y_2(t)$ is the response of the system when input $x_2(t)$ is applied alone; α_1 and α_2 are arbitrary constants.

EXAMPLE 2-3

Show that the system described by the differential equation

$$\frac{dy(t)}{dt} + ty(t) = x(t)$$

is linear.

Solution: The response to $x_1(t)$, $y_1(t)$, satisfies

$$\frac{dy_1}{dt} + ty_1 = x_1$$

and the response to $x_2(t)$, $y_2(t)$, satisfies

$$\frac{dy_2}{dt} + ty_2 = x_2$$

where the time dependence of x_1, x_2, y_1, and y_2 is omitted for simplicity. Addition of these equations multiplied, respectively, by α_1 and α_2 results in

$$\alpha_1 \frac{dy_1}{dt} + \alpha_2 \frac{dy_2}{dt} + \alpha_1 ty_1 + \alpha_2 ty_2 = \alpha_1 x_1 + \alpha_2 x_2$$

or

$$\frac{d}{dt}(\alpha_1 y_1 + \alpha_2 y_2) + t(\alpha_1 y_1 + \alpha_2 y_2) = \alpha_1 x_1 + \alpha_2 x_2$$

That is, the response to the input $\alpha_1 x_1 + \alpha_2 x_2$ is $\alpha_1 y_1 + \alpha_2 y_2$. Thus superposition holds and the system is linear.

EXAMPLE 2-4

The system described by the differential equation

$$\frac{dy(t)}{dt} + 10y(t) + 5 = x(t)$$

is *nonlinear*. This is demonstrated by attempting to apply the superposition principle. Letting $y_1(t)$ be the response to the input $x_1(t)$ and $y_2(t)$ be the response to the input $x_2(t)$, we write

$$\frac{dy_1}{dt} + 10y_1 + 5 = x_1$$

and

$$\frac{dy_2}{dt} + 10y_2 + 5 = x_2$$

where the time dependence is again dropped for simplicity. Multiplying the first equation by an arbitrary constant α_1, the second equation by the arbitrary constant α_2, adding, and regrouping terms, we obtain.

$$\frac{d}{dt}(\alpha_1 y_1 + \alpha_2 y_2) + 10(\alpha_1 y_1 + \alpha_2 y_2) + 5(\alpha_1 + \alpha_2) = \alpha_1 x_1 + \alpha_2 x_2 \qquad (2\text{-}27)$$

We *cannot* put this equation into the same form as the original differential equation for an *arbitrary choice* of α_1 and α_2. Thus the system is *nonlinear*.

2-3 The Superposition Integral for Fixed, Linear Systems

In this section we show that (2-5) can be used to obtain the output of a fixed, linear system in response to an input, $x(t)$, where the response of the system to a unit impulse applied at $t = 0$ is $h(t)$ [$h(t)$ is referred to as the *impulse response* of the system]. In this context, (2-5) is called the *superposition integral*. We must emphasize that this integral is more general than for just the computation of the response of a linear, time-invariant system. Some of the other applications of the superposition integral are discussed in Chapter 4.

A short, mathematical derivation of (2-5) is as follows. Represent the input signal to the linear, time-invariant system in terms of the sifting property of the unit impulse as

$$x(t) = \int_{-\infty}^{\infty} x(\tau)\delta(t - \tau)\, d\tau \qquad (2\text{-}28)$$

In terms of the system operator, $\mathcal{H}$, the output is

$$y(t) = \mathcal{H}[x(t)]$$

$$= \mathcal{H}\left[\int_{-\infty}^{\infty} x(\tau)\delta(t - \tau)\, d\tau\right] \qquad (2\text{-}29)$$

Assuming that the order of integration and operation by the system can be interchanged, (2-29) becomes

$$y(t) = \int_{-\infty}^{\infty} x(\tau)\mathcal{H}[\delta(t - \tau)]\, d\tau \qquad (2\text{-}30)$$

By definition, $\mathcal{H}[\delta(t - \tau)]$ is the response of the system to a unit impulse applied at $t = \tau$. For a *fixed* system, we can write this as $h(t - \tau)$, where $h(t)$ is the system's response to a unit impulse applied at $t = 0$. Substitution of $h(t - \tau)$ for $\mathcal{H}[\delta(t - \tau)]$ in (2-30) results in (2-5).

In addition to being strictly mathematical, an additional difficulty with this derivation is the blind faith required in the interchange of the improper integral and the $\mathcal{H}$-operator. Thus we consider another derivation that is more graphic in nature.

Assume for the time being that the input to the fixed, linear system is a continuous function of time in the time interval $T_1 \leq t \leq T_2$ and zero elsewhere. Figure 2-3a illustrates an arbitrary input, $x(t)$, in the time interval $T_1 \leq t \leq T_2$ which has been approximated by a stairstep function $\hat{x}(t)$. That is, $\hat{x}(t)$ can be expressed in terms of the unit pulse function $\Pi(t)$ as the summation

$$\hat{x}(t) = \sum_{n=N_1}^{N_2} x(n \, \Delta\lambda)\Pi\left(\frac{t - n \, \Delta\lambda}{\Delta\lambda}\right)$$

$$= \sum_{n=N_1}^{N_2} x(n \, \Delta\lambda)\frac{1}{\Delta\lambda}\Pi\left(\frac{t - n \, \Delta\lambda}{\Delta\lambda}\right)\Delta\lambda, \qquad T_1 \leq t \leq T_2 \qquad (2\text{-}31)$$

where $T_1 = (N_1 - 1/2) \, \Delta\lambda$ and $T_2 = (N_2 + 1/2) \, \Delta\lambda$. The reason for multiplying and dividing by $\Delta\lambda$ to obtain the second equation will be apparent shortly. Clearly, the approximation of $\hat{x}(t)$ to $x(t)$ improves as $\Delta\lambda$ gets smaller. In fact, in the limit as $\Delta\lambda \to 0$, (2-31) becomes the sifting integral for the unit impulse function since

$$\lim_{\Delta\lambda \to 0} \frac{\Pi[(t - n \, \Delta\lambda)/\Delta\lambda]}{\Delta\lambda} = \delta(t - \lambda) \qquad (2\text{-}32)$$

where $n \, \Delta\lambda$ has been replaced by the continuous variable λ.

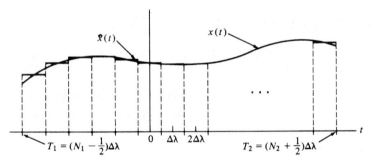

(a) Input to a fixed, linear system and a stairstep approximation

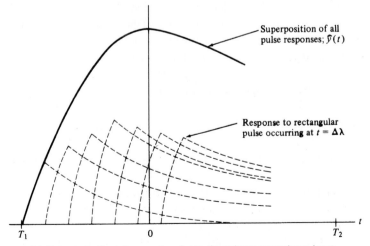

(b) Output of a fixed, linear system due to the stairstep approximate input.

FIGURE 2-3.   Input and output signals for a fixed, linear system used to derive the superposition integral.

Now consider the response of a fixed linear system to the summation of pulses (2-31). Let the response of the system to $\Pi(t/\Delta\lambda)\,\Delta\lambda$ be $\hat{h}(t)$. That is, in terms of the operator notation

$$\hat{h}(t) = \mathcal{H}\left[\frac{\Pi(t/\Delta\lambda)}{\Delta\lambda}\right] \tag{2-33}$$

where $\mathcal{H}[\cdot]$ signifies the operation that produces the system output in response to a particular input. We want the output of a fixed, linear system in response to the input (2-31). In terms of the operator $\mathcal{H}[\cdot]$, it is

$$\hat{y}(t) = \mathcal{H}[\hat{x}(t)] = \mathcal{H}\left[\sum_{n=N_1}^{N_2} x(n\,\Delta\lambda)\frac{1}{\Delta\lambda}\Pi\left(\frac{t-n\,\Delta\lambda}{\Delta\lambda}\right)\Delta\lambda\right] \tag{2-34}$$

Because $\mathcal{H}[\cdot]$ represents a linear system, superposition holds and (2-34) can be written as

$$\hat{y}(t) = \sum_{n=N_1}^{N_2} x(n\,\Delta\lambda)\mathcal{H}\left[\frac{1}{\Delta\lambda}\Pi\left(\frac{t-n\,\Delta\lambda}{\Delta\lambda}\right)\right]\Delta\lambda \tag{2-35}$$

where we assume that all initial conditions are zero. More will be said about nonzero initial conditions shortly. Although the superposition property is stated by (2-26) for inputs of the form $x(t) = \alpha_1 x_1(t) + \alpha_2 x_2(t)$, we may generalize it easily to the linear combination of any finite number of terms (Problem 2-15).

Because the system is assumed to be fixed, it follows that

$$\mathcal{H}\left[\frac{1}{\Delta\lambda}\Pi\left(\frac{t-n\,\Delta\lambda}{\Delta\lambda}\right)\right] = \hat{h}(t-n\,\Delta\lambda) \tag{2-36}$$

where $\hat{h}_t$ was defined as the response of the system to $\Pi(t/\Delta\lambda)/\Delta\lambda$ in (2-33). That is, the system's response to a pulse centered at $t = n\Delta\lambda$ is simply $\hat{h}(t)$ shifted by $t = n\Delta\lambda$. Using (2-36) in (2-35), we can express the system output, $\hat{y}(t)$, in response to the stairstep approximation to $x(t)$, $\hat{x}(t)$, as

$$\hat{y}(t) = \sum_{n=N_1}^{N_2} x(n\,\Delta\lambda)\hat{h}(t-n\,\Delta\lambda)\,\Delta\lambda \tag{2-37}$$

which shows that y is the *superposition* of $N = N_2 - N_1$ elementary repsponses, $\hat{h}(t-n\,\Delta\lambda)$, of the system to $\Pi[(t-n\,\Delta\lambda)/\Delta\lambda]/\Delta\lambda$, each of which is weighted by the value of the input signal at the time $t = n\,\Delta\lambda$. A typical superposition of elementary responses is shown in Figure 2-3b corresponding to the stairstep input of Figure 2-3a. Note that $\hat{h}(t)$ is a characteristic of the system in the sense that it tells us how the system responds to the test pulse $\Pi(t/\Delta/\lambda)/\Delta\lambda$ of width $\Delta\lambda$ and height $1/\Delta\lambda$. To remove the dependence of $\hat{h}(t)$ on $\Delta\lambda$, we *define* the impulse response of the system as

$$h(t) = \lim_{\Delta\lambda\to 0}\hat{h}(t) = \lim_{\Delta\lambda\to 0}\mathcal{H}\left[\frac{\Pi(t/\Delta\lambda)}{\Delta\lambda}\right] \tag{2-38}$$

where we recall that zero initial conditions were assumed. By interchanging the order of the limit and the operator $\mathcal{H}[\cdot]$, which is a linear operator by assumption, and recalling that $\lim_{\Delta\lambda\to 0}\Pi(t/\Delta\lambda)/\Delta\lambda$ has the properties of a unit impulse function, it is seen that *the impulse response of a system is its response to a unit impulse applied at time $t = 0$ with all initial conditions of the system zero.* That is,

$$h(t) \triangleq \mathcal{H}[\delta(t)] \qquad \text{(zero initial conditions)} \tag{2-39}$$

Returning to (2-37) and taking the limit as $\Delta\lambda \to 0$ and $n\,\Delta\lambda \to \lambda$, a continuous variable, we recognize (2-37) as an approximation to the integral

$$y(t) = \int_{T_1}^{T_2} x(\lambda)h(t - \lambda)\, d\lambda, \qquad T_1 \leq t \leq T_2 \tag{2-40}$$

Assuming that the input may have been present since the infinite past and may last indefinitely into the future, we have, in the limit as $T_1 \to -\infty$ and $T_2 \to \infty$,

$$y(t) = \int_{-\infty}^{\infty} x(\lambda)h(t - \lambda)\, d\lambda \tag{2-41}$$

for the response of a system characterized by an impulse response $h(t)$ and having an input $x(t)$.

With the lower limit of $t = -\infty$ on (2-41), we can include the response due to initial conditions by identifying it as the response due to $x(t)$ from $t = -\infty$ to an appropriately chosen starting instant, t_0, usually chosen as $t_0 = 0$. In Chapter 5 we consider another way to handle initial conditions. Making the substitution $\sigma = t - \lambda$, we obtain the equivalent result

$$y(t) = \int_{-\infty}^{\infty} x(t - \sigma)h(\sigma)\, d\sigma \tag{2-42}$$

Because these equations were obtained by superposition of a number of elementary responses due to each individual impulse, they are special cases of what are referred to as *superposition integrals.*[†] A simplification results if the system under consideration is causal. We defined such a system as being one that did not anticipate its input. For a causal system, $h(t - \tau) = 0$ for $t < \tau$, and the upper limit of (2-41) can be set equal to t. Furthermore, if $x(t) = 0$ for $t < 0$, the lower limit becomes zero, and the resulting integral is given by

$$y(t) = \int_{0}^{t} x(\lambda)h(t - \lambda)\, d\lambda, \qquad t \geq 0 \tag{2-43}$$

for a causal system with input zero for $t < 0$ and zero initial conditions.

EXAMPLE 2-5

A concrete example of the previous discussion will help to make the derivation clearer and lend credence to the stairstep approximation to $x(t)$ being convolved with $h(t)$ actually providing a good approximation to $y(t)$. The signals $x(t)$ and $h(t)$ in this example are given by

$$x(t) = (t + 10)e^{-0.4(t+10)}u(t + 10)$$

and

$$h(t) = e^{-t}u(t)$$

A MATLAB program (not given here, since the purpose of this example is to illustrate the preceding discussion) provides the plots shown in Figure 2-4. The top figure gives $x(t)$ and the stairstep approximation to it, $\tilde{x}(t)$, where the step length is 0.5. The signal $h(t)$ is shown in the second figure, and the result of convolving $\tilde{x}(t)$ and $h(t)$ is shown in the third figure. Finally, the actual convolution of $x(t)$ and $h(t)$, computed analytically, is shown in the fourth figure. The approximate and actual results are amazingly close for the step length of 0.5. As this step length is decreased, the correspondence will be closer and closer, and it is not hard to believe that (2-37) actually converges to (2-40).

[†]In this special case (i.e., a linear, time-invariant system) they are of the same form as the convolution integral to be considered in more detail in Section 2-4. A general form for the superposition integral for a linear time-varying system was given by (2-6).

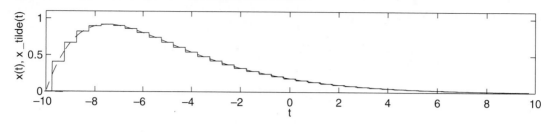

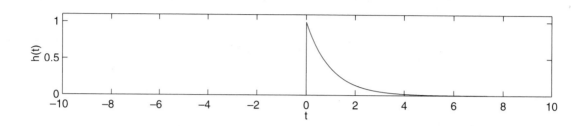

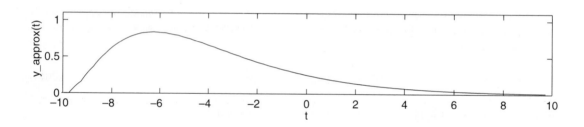

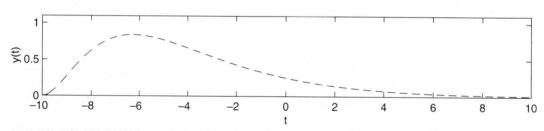

FIGURE 2-4.   Plots for Example 2-5 showing a specific example for the stairstep approximation to the derivation of the superposition integral.

EXAMPLE 2-6 _____

Consider the system with impulse response

$$h(t) = \frac{1}{RC} e^{-t/RC} u(t) \qquad (2\text{-}44)$$

Let the input be a unit step:

$$x(t) = u(t) \qquad (2\text{-}45)$$

Applying (2-42), we obtain the output

$$a(t) = \int_{-\infty}^{\infty} u(t - \sigma) \frac{1}{RC} e^{-\sigma/RC} u(\sigma) \, d\sigma$$

$$= \int_{0}^{t} \frac{1}{RC} e^{-\sigma/RC} d\sigma, \qquad t \geq 0 \tag{2-46}$$

$$= 1 - e^{-t/RC}, \qquad t \geq 0$$

Since $a(t) = 0$ for $t < 0$, this output, referred to as the *step response* of the system, can be written compactly as

$$a(t) = \left[1 - \exp\left(\frac{-t}{RC}\right) \right] u(t) \tag{2-47}$$

Note that $a(t) = \int_{-\infty}^{t} h(\lambda) \, d\lambda$.

2-4 Examples Illustrating Evaluation of the Convolution Integral

The convolution of two signals, $x(t)$ and $h(t)$, is a new function of time, $y(t)$, which is given by (2-5). A useful symbolic notation often employed to denote (2-5a) and (2-5b), respectively, is

$$y(t) = h(t) * x(t) \tag{2-48}$$

$$y(t) = x(t) * h(t) \tag{2-49}$$

It is possible to prove several properties of the convolution integral. These are listed below, but their proofs are left to the problems.

1. $h(t) * x(t) = x(t) * h(t)$.
2. $h(t) * [\alpha x(t)] = \alpha[h(t) * x(t)]$, where α is a constant.
3. $h(t) * [x_1(t) + x_2(t)] = h(t) * x_1(t) + h(t) * x_2(t)$.
4. $h(t) * [x_1(t) * x_2(t)] = [h(t) * x_1(t)] * x_2(t)$.
5. If $h(t)$ is time-limited to (a,b) and $x(t)$ is time-limited to (c,d), then $h(t) * x(t)$ is time-limited to $(a + c, b + d)$.
6. If A_1 is the area under $h(t)$ and A_2 is the area under $x(t)$, then the area under $h(t) * x(t)$ is $A_1 A_2$.

In these relations, $h(t)$, $x_1(t)$, and $x_2(t)$ are signals. The parentheses and brackets show the order of carrying out the operations.

The integrand of (2-5a) is found by three operations: (1) reversal in time, or folding, to obtain $x(-\lambda)$; (2) shifting, to obtain $x(t - \lambda)$; and (3) multiplication of $h(\lambda)$ and $x(t - \lambda)$, to obtain the integrand. A similar series of operations is required for (2-5b).

Four examples will be given to illustrate these three operations and the subsequent evaluation of the integral to form $y(t)$.

EXAMPLE 2-7 _____

Consider the convolution of the two rectangular-pulse signals

$$x(t) = 2\Pi\left(\frac{t - 5}{2}\right) \tag{2-50}$$

and

$$h(t) = \Pi\left(\frac{t-2}{4}\right) \tag{2-51}$$

These signals are sketched in Figures 2-5a and b.

We base our evaluation of the convolution of $x(t)$ and $h(t)$ on (2-5a). Thus $x(t)$ is reversed and the variable changed to λ, which results in $x(-\lambda)$, also shown in Figure 2-5. Finally, $x(-\lambda)$ is shifted so that what was the origin now appears at $\lambda = t$. It is sometimes helpful to think of $x(t - \lambda)$ as $x[-(\lambda - t)]$; that is, t in $x(t)$ is replaced by $\lambda - t$, with λ thought of as the independent variable,

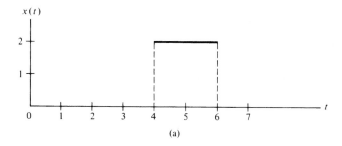

(a)

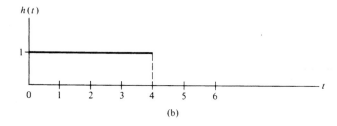

(b)

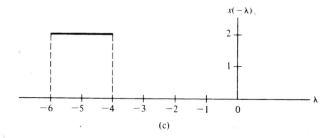

(c)

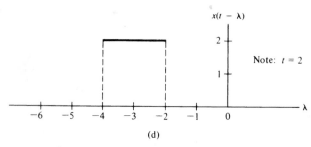

(d)

FIGURE 2-5. Signals to be convolved in Example 2-7.

whereupon the function $x(\lambda - t)$ is reversed or folded (its mirror image about the ordinate is taken). The result of these operations is $x(t - \lambda)$, which is shown in Figure 2-5d for $t = 2$.

The product of $x(t - \lambda)$ and $h(\lambda)$ is formed, point by point, and this product integrated to form $y(t)$. Clearly, the product will be zero if $t < 4$ in this example. If $t \geq 4$, the rectangles forming $h(\lambda)$ and $x(t - \lambda)$ overlap until $t > 10$, whereupon the product is again zero.

After a little thought, we see that three different cases of overlap result. These are illustrated in Figure 2-6. In each case, the area of integration is shaded; because the height of $h(t)$ is unity, the area is simply the width of the overlap times the height of $x(t - \lambda)$. Thus the area for Case I is $2(t - 4)$;

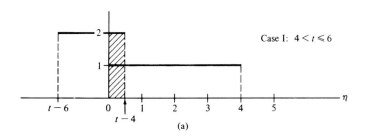

(a)

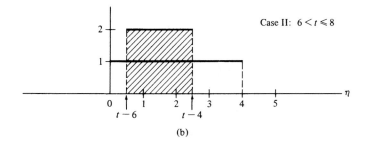

(b)

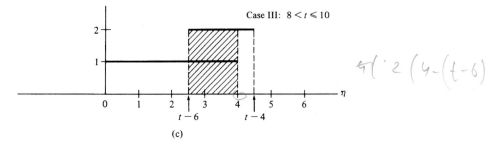

(c)

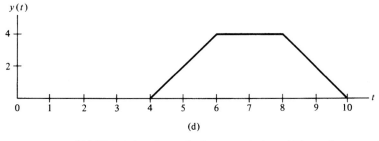

(d)

FIGURE 2-6.   Steps in the convolution of Example 2-7 and the final result.

that is, it is linearly increasing with t. For Case II, it is $2[(t - 4) - (t - 6)] = 4$. Finally, for Case III, it is $2[4 - (t - 6)] = 2(10 - t)$; that is, the area linearly decreases with t. The result for $y(t)$ is sketched in Figure 2-6d.

EXAMPLE 2-8

As a second example of the convolution operation, we consider the signals shown in Figure 2-7a. We use the convolution integral (2-5b) in obtaining the result. This example differs from Example

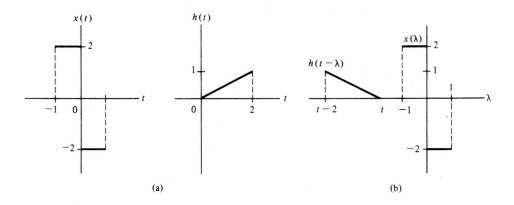

(a)   (b)

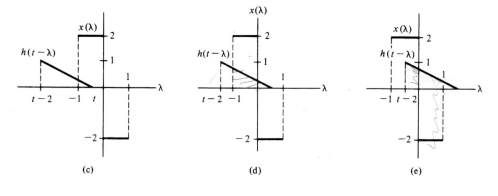

(c)   (d)   (e)

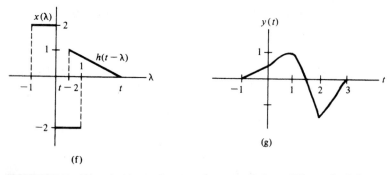

(f)   (g)

FIGURE 2-7 Waveforms pertinent to the convolution of Example 2-8.

2-7 in that part of the area under the product $x(\lambda)h(t - \lambda)$ is negative. Our first step is to express $h(t - \lambda)$ mathematically. Since

$$h(t) = \begin{cases} \dfrac{t}{2}, & 0 \le t \le 2 \\ \\ 0, & \text{otherwise} \end{cases} \tag{2-52}$$

we obtain

$$h(t - \lambda) = \begin{cases} \dfrac{t - \lambda}{2}, & 0 \le t - \lambda \le 2 \quad \text{or} \quad t - 2 \le \lambda \le t \\ \\ 0, & \text{otherwise} \end{cases} \tag{2-53}$$

The signals $x(\lambda)$ and $h(t - \lambda)$ are sketched in Figure 2-7b. Some thought on the part of the student should result in the four cases of nonzero overlap illustrated in Figures 2-7c through f. The resulting convolutions for these cases are expressed mathematically by the following equations:

$$y(t) = \begin{cases} 0, & t < -1 \quad \text{or} \quad t > 3 \\ \displaystyle\int_{-1}^{t} (t - \lambda)\, d\lambda, & -1 \le t < 0 \\ \displaystyle\int_{-1}^{0} (t - \lambda)\, d\lambda - \int_{0}^{t} (t - \lambda)\, d\lambda, & 0 \le t < 1 \\ \displaystyle\int_{t-2}^{0} (t - \lambda)\, d\lambda - \int_{0}^{1} (t - \lambda)\, d\lambda, & 1 \le t < 2 \\ -\displaystyle\int_{t-2}^{1} (t - \lambda)\, d\lambda, & 2 \le t < 3 \end{cases} \tag{2-54}$$

Integration of these expressions results in the following for $y(t)$:

$$y(t) = \begin{cases} 0, & t < -1 \quad \text{or} \quad t > 3 \\ \frac{1}{2}(t + 1)^2, & -1 \le t < 0 \\ \frac{1}{2}(-t^2 + 2t + 1), & 0 \le t < 1 \\ \frac{1}{2}(-t^2 - 2t + 1) + 2, & 1 \le t < 2 \\ \frac{1}{2}(t^2 - 2t + 1) - 2, & 2 \le t < 3 \end{cases} \tag{2-55}$$

The resulting signal, which is the convolution of $x(t)$ with $h(t)$, is shown in Figure 2-7g.

MATLAB Application

MATLAB can be used to do convolutions with the `conv(x,h)` function. The MATLAB program given below implements the convolution of the signals of Example 2-8 (echo on):

```
EDU» c2ex8
%       MATLAB plot of Example 2-8
%
del_t=.005;
t=-2:del_t:4;
```

```
L=length(t);
tp=[2*t(1):del_t:2*t(L)];
x=2*(pls_fn(t+.5)-pls_fn(t-.5));  % Defined in Chapter 1
h=.5*rmp_fn(t).*stp_fn(2-t);      % Defined in Chapter 1
y=del_t*conv(x,h);                % Multiply by step size to approximate
                                  % rectangular rule integration.
subplot(3,1,1), plot(t,x), xlabel('t'), ylabel ('x(t)'),
axis([t(1) t(L) -3 3])
subplot(3,1,2), plot(t,h), xlabel('t'), ylabel('h(t)'),
axis[(t(1) t(L) 0 2])
subplot(3,1,3), plot(tp,y), xlabel('t'), ylabel('y(t)'),
axis([t(1) t(L) -2 2])
```

The plot output of the program, given in Figure 2-8, is seen to be identical to Figure 2-7g.

EXAMPLE 2-9

In this example, we consider the convolution of a ramp and the decaying exponential signal of the form $x(t) = \exp(-\alpha t)u(t)$. The convolution integral in this case is

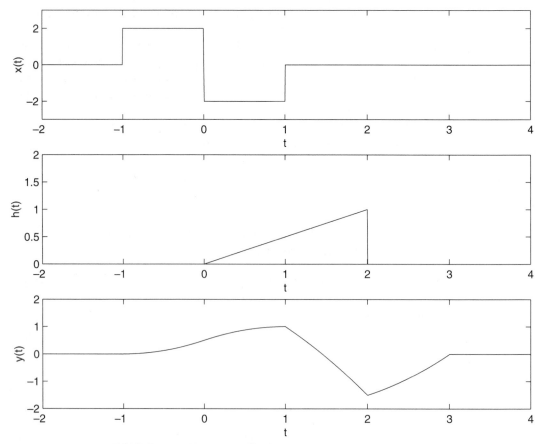

FIGURE 2-8. MATLAB verification of the result of Example 2-8.

$$y(t) = \int_{-\infty}^{\infty} r(\tau)e^{-\alpha(t-\tau)}u(t-\tau)\,d\tau$$

$$= \int_{0}^{t} \tau e^{-\alpha(t-\tau)}\,d\tau, \qquad t \ge 0$$

$$= e^{-\alpha t}\int_{0}^{t} \tau e^{\alpha\tau}\,d\tau, \qquad t \ge 0$$

$$= \frac{t}{\alpha} - \frac{1}{\alpha^2}(1 - e^{-\alpha t}), \qquad t \ge 0 \qquad (2\text{-}56)$$

Note that the first term increases linearly like the ramp, but with a slope of $1/\alpha$. The second term approaches $-1/\alpha^2$ as $t \to \infty$. The student should sketch the result of the convolution along with the components of the integrand for different values of t to provide justification for the result.

2-5 Impulse Response of a Fixed, Linear System

We have shown in Section 2-3 that the response of a fixed, linear system with impulse response $h(t)$ to an input $x(t)$ is the convolution of $h(t)$ and $x(t)$. In this section we therefore consider the impulse response of a linear system. Although the impulse response of a time-varying system can be defined also, we limit our consideration to *fixed* linear systems.

We have defined the *impulse response,* denoted $h(t)$, of a fixed, linear system, assumed initially unexited, to be *the response of the system to a unit impulse applied at time $t = 0$.* Since the system is assumed to be fixed, the response to an impulse applied at some time other than zero, say $t = \tau$, is simply $h(t - \tau)$. We now illustrate by example two techniques for obtaining the impulse response of a system.

EXAMPLE 2-10 _____

Find the impulse response of a system modeled by the differential equation

$$\tau_0 \frac{dy(t)}{dt} + y(t) = x(t), \qquad -\infty < t < \infty \qquad (2\text{-}57)$$

where $x(t)$ is the input and $y(t)$ the output.

Solution: Setting $x(t) = \delta(t)$ results in the response $y(t) = h(t)$. For $t > 0$, $\delta(t) = 0$, so that the governing differential equation for the impulse response is

$$\tau_0 \frac{dh(t)}{dt} + h(t) = 0, \qquad t > 0 \qquad (2\text{-}58)$$

Assuming a solution of the form

$$h(t) = Ae^{pt} \qquad (2\text{-}59)$$

and substituting into the equation for $h(t)$, we obtain

$$(\tau_0 p + 1)Ae^{pt} = 0 \qquad (2\text{-}60)$$

which is satisfied if

$$p = \frac{-1}{\tau_0} \qquad (2\text{-}61)$$

Thus

$$h(t) = Ae^{-t/\tau_0}, \qquad t > 0 \tag{2-62}$$

To fix A, we require an initial condition for $h(t)$. The system is unexcited for $t < 0$. Therefore, from the definition of the impulse response,

$$h(t) = 0, \qquad t < 0 \tag{2-63}$$

From (2-57) it follows that the impulse response for $t \geq 0$ obeys the differential equation

$$\tau_0 \frac{dh(t)}{dt} + h(t) = \delta(t), \qquad t \geq 0 \tag{2-64}$$

For the left-hand side to be identically equal to the right-hand side, one of the terms on the left-hand side must contain an impulse at $t = 0$. It cannot be $h(t)$, for then $\tau_0[dh(t)/dt]$ would contain a doublet,[†] and there is no doublet on the right-hand side [recall (1-68)]. Thus $h(t)$ is discontinuous at $t = 0$, but no impulse is present in $h(t)$. Therefore, its integral through $t = 0$ must be zero,

$$\int_{t=0^-}^{0^+} h(t) \, dt = 0 \tag{2-65}$$

and the integral of (2-64) from $t = 0^-$ to $t = 0^+$ gives

$$\int_{t=0^-}^{0^+} \tau_0 \frac{dh(t)}{dt} \, dt + \int_{t=0^-}^{0^+} h(t) \, dt = \int_{t=0^-}^{0^+} \delta(t) \, dt$$

$$\tau_0[h(0^+) - h(0^-)] + \qquad 0 \qquad = \qquad 1 \tag{2-66}$$

Since $h(0^-) = 0$, it follows that

$$h(0^+) = \frac{1}{\tau_0} \tag{2-67}$$

Setting this result equal to (2-62) with $t = 0^+$ shows that

$$A = \frac{1}{\tau_0} \tag{2-68}$$

Thus

$$h(t) = \begin{cases} 0, & t < 0 \\ \dfrac{1}{\tau_0} e^{-t/\tau_0}, & t \geq 0 \end{cases} \tag{2-69}$$

is the impulse response of the system.

EXAMPLE 2-11

Considering next the *RC* circuit shown in Figure 2-9, we see that the governing differential equation is (2-57) with $\tau_0 = RC$. We will find the impulse response by considering the physical properties of the resistor and capacitor.

[†]A doublet was defined as $d\delta(t)/dt$ in Chapter 1.

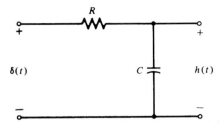

FIGURE 2-9.   *RC* circuit to illustrate the calculation of impulse response.

At $t = 0^-$, the capacitor is uncharged and acts as a short circuit (infinite charge sink) to any current flowing through R. With $x(t) = \delta(t)$, the current through R at $t = 0$ is given by

$$i(t) = \frac{\delta(t)}{R}, \qquad 0^- \le t \le 0^+ \tag{2-70}$$

The charge stored in C due to this initial current flow is

$$q(0^+) = \int_{0^-}^{0^+} \frac{\delta(t)}{R} \, dt = \frac{1}{R} \tag{2-71}$$

Thus the capacitor voltage is $v_c(0^+) = 1/RC$. For $t > 0$, $x(t) = 0$; that is, the input is a short circuit and C discharges. The voltage across C exponentially decays according to

$$v_c(t) = Ae^{-t/RC}, \qquad t > 0 \tag{2-72}$$

Setting $v_c(0^+) = A = 1/RC$ and noting that $h(t) = v_c(t)$, we obtain the same result for $h(t)$ as in Example 2-10 with $\tau_0 = RC$.

EXAMPLE 2-12 _____

Find the impulse response of the *LC* circuit shown in Figure 2-10.

Solution: For the unit impulse input, the differential equation governing the response of the circuit is

$$LC \frac{d^2h(t)}{dt^2} + h(t) = \delta(t) \tag{2-73}$$

For $t > 0$ the right-hand side is zero, and the solution to the homogeneous differential equation is

$$h(t) = (A \cos \omega_0 t + B \sin \omega_0 t)u(t) \tag{2-74}$$

where $\omega_0 = 1/LC$ and the fact that $h(t) = 0$ for $t < 0$ has been incorporated into the expression for $h(t)$ by multiplying by a unit step function. To obtain the unknown constants A and B, we

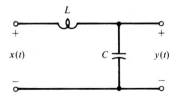

FIGURE 2-10.   Circuit for Example 2-12.

differentiate $h(t)$ twice with respect to t and substitute the result back into (2-73). The first differentiation gives

$$\frac{dh(t)}{dt} = (-A\omega_0 \sin \omega_0 t + B\omega_0 \cos \omega_0 t)u(t) + (A \cos \omega_0 t + B \sin \omega_0 t)\delta(t)$$

$$= -\omega_0(A \sin \omega_0 t - B \cos \omega_0 t)u(t) + A\delta(t) \tag{2-75}$$

where we used the facts that $0 \cdot \delta(t) = 0$ (see Problem 1-25 and the footnote on page 27). A second differentiation gives

$$\frac{d^2h(t)}{dt^2} = -\omega_0^2(A \cos \omega_0 t + B \sin \omega_0 t)u(t) - \omega_0(A \sin \omega_0 t - B \cos \omega_0 t)\delta(t) + A\frac{d\delta(t)}{dt}$$

$$= -\omega_0^2(A \cos \omega_0 t + B \sin \omega_0 t)u(t) + \omega_0 B\delta(t) + A\frac{d\delta(t)}{dt} \tag{2-76}$$

Substitution of this result into (2-73) along with the assumed $h(t)$, (2-74), gives

$$LC\left[-\omega_0^2(A \cos \omega_0 t + B \sin \omega_0 t)u(t) - \omega_0 B\delta(t) + A\frac{d\delta(t)}{dt}\right]$$

$$+ (A \cos \omega_0 t + B \sin \omega_0 t)u(t) = \delta(t) \tag{2-77}$$

But $\omega_0^2 = (LC)^{-1}$. Canceling like terms, we have

$$LC\left[\omega_0 B\delta(t) + A\frac{d\delta(t)}{dt}\right] = \delta(t) \tag{2-78}$$

By equating coefficients of like derivatives of $\delta(t)$, we obtain $A = 0$ and $B = \omega_0$. Thus the result for the impulse response of this circuit is

$$h(t) = \omega_0 \sin(\omega_0 t) u(t) \tag{2-79}$$

2-6 Superposition Integrals in Terms of Step Response

We saw in Example 2-6 that the step response of the RC circuit shown in Figure 2-11 is simply the integral of the impulse response of the circuit. We now ask: Can the response of a system to an arbitrary input be expressed in terms of its step response? The answer is yes.

To find the appropriate relationship, consider the superposition integral in terms of impulse response. Repeating (2-5a) here for convenience, we have

$$y(t) = \int_{-\infty}^{\infty} x(t - \lambda)h(\lambda)\, d\lambda \tag{2-5}$$

We employ the formula for integration by parts:

$$\int_a^b u\, dv = uv \Big|_a^b - \int_a^b v\, du \tag{2-80}$$

with $u = x(t - \lambda)$ and $dv = h(\lambda)\, d\lambda$. Thus

$$v(\lambda) = \int_{-\infty}^{\lambda} h(\zeta)\, d\zeta \triangleq a(\lambda) \tag{2-81}$$

is simply the step response, and

$$du(\lambda) = -\dot{x}(t - \lambda) \, d\lambda \tag{2-82}$$

where the overdot denotes differentiation with respect to the argument. Substituting these expressions into (2-80), we obtain

$$y(t) = a(\lambda)x(t - \lambda) \big|_{\lambda=-\infty}^{\infty} + \int_{-\infty}^{\infty} \dot{x}(t - \lambda)a(\lambda) \, d\lambda \tag{2-83}$$

The system is initially unexcited, so that $a(-\infty) = 0$ and $x(t - \lambda)|_{\lambda = \infty} = 0$. Thus the first term is zero when the limits are substituted, and the final result is

$$y(t) = \int_{-\infty}^{\infty} \dot{x}(t - \lambda)a(\lambda) \, d\lambda \tag{2-84}$$

The change of variables $\eta = t - \lambda$ results in the alternative form

$$y(t) = \int_{-\infty}^{\infty} \dot{x}(\eta)a(t - \eta) \, d\eta \tag{2-85}$$

Thus in terms of the step response $a(t)$, the response of a system to an input $x(t)$ is the *convolution of the derivative of the input with the step response*. Equations (2-84) and (2-85) are known as *Duhamel's integrals*. Note the similarity to (2-5a) and (2-5b), respectively.

EXAMPLE 2-13 _____

Consider a system with a ramp input for which

$$x(t) = tu(t) \tag{2-86a}$$
$$\dot{x}(t) = u(t) \tag{2-86b}$$

Applying (2-85), we obtain the response to a ramp as

$$y_R(t) = \int_{-\infty}^{\infty} u(t - \lambda)a(\lambda) \, d\lambda$$

$$= \int_{-\infty}^{t} a(\lambda) \, d\lambda \tag{2-87}$$

Thus the response of a system to a unit ramp, which is the integral of the unit step, is the integral of the step response.

Generalizing, we conclude that *for a fixed, linear system, any linear operation on the input produces the same linear operation on the output.*

EXAMPLE 2-14 _____

Find the response of the *RC* circuit of Figure 2-9 to the triangular signal

$$x_\Delta(t) = r(t) - 2r(t - 1) + r(t - 2) \tag{2-88}$$

Solution: We first substitute (2-47) into (2-87) to obtain the ramp response. Thus

$$y_R(t) = \int_{-\infty}^{t} \left[1 - \exp\left(\frac{-\lambda}{RC}\right) \right] u(\lambda) \, d\lambda$$

$$= r(t) + RC \left[\exp\left(\frac{-\lambda}{RC}\right) \right]_0^t u(t)$$

$$= r(t) - RC \left[1 - \exp\left(\frac{-t}{RC}\right) \right] u(t) \tag{2-89}$$

By superposition, the response to the triangle $x_\Delta(t)$ is $y_\Delta(t) = y_R(t) - 2y_R(t-1) + y_R(t-2)$. This input and output are compared in Figure 2-11. Note that for $RC << 1$, the output closely approximates the input, whereas if $RC = 1$ the output does not resemble the input.

MATLAB *Application*

In this example, we verify the result of Example 2-14, shown in Figure 2-11, by writing a MATLAB program to carry out a Duhamel's integral evaluation. The program listing is given below:

```
EDU>>c2ex14
%    DuHammel's computation of the output for Example 2-14
%
RC=0.1;
t_min=-0.5;                                    % Minimum t-value for
                                               % computation window
t_max=2.5;                                     % Maximum t-value for
                                               % computation window
del_t=0.001;                                   % Step size for 5
t=t_min:del_t:t_max;                           % Vector of t-values
L=length(5);
tp=[2*t(1):del_t:2*t(L)];                      % t-variable defined for
                                               % plotting output
x_del_dot=stp_fn(5)-2*stp_fn(t-1)-stp_fn(t-2); % Derivative of the input
a=(1 - exp(-t/RC)).*stp_fn(t);                 % Step response of the
                                               % system
y=del_t*conv(x_del_dot,a);                     % The convolution;
                                               % multiply by del_t to
                                               % scale
```

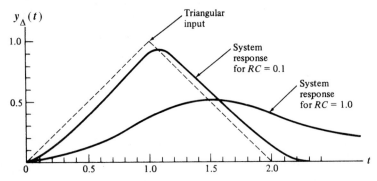

FIGURE 2-11. Response of an *RC* circuit to a triangular input signal.

```
subplot(3,1,1),plot(t,x_del_dot),xlabel('t'),ylabel('der.of input'),
axis([t_min t_max -1.5 1.5])
subplot(3,1,2),plot(t,a'),xlabel('t'),ylabel('step resp.'),axis([t_min
t_max 0 1.5])
subplot(3,1,3),plot(tp,y),xlabel('t'),ylabel('output'),axis([t_min
t_max 0 1])
```

A plot of the output of the system in response to the triangle input is shown in Figure 2-12 for $RC = 0.1$. Note that it is the same as the corresponding response plotted in Figure 2-11.

EXAMPLE 2-15

We again consider the RC circuit of Figure 2-9 and compute its impulse response by first finding its step response and then differentiating it. The differential equation relating input and output is

$$RC \frac{dy(t)}{dt} + y(t) = x(t) \tag{2-90}$$

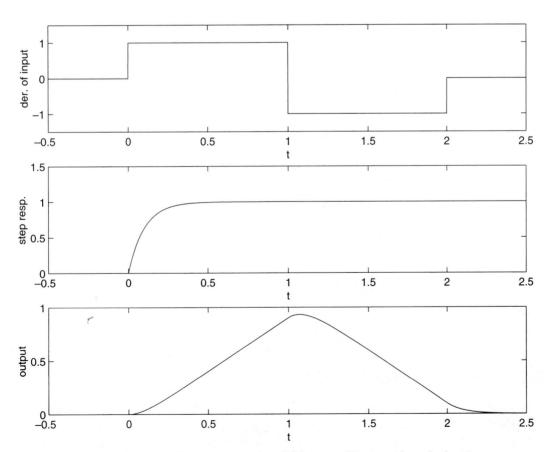

FIGURE 2-12.   The response of an RC lowpass filter to a triangular input.

which is (2-57) with $\tau_0 = RC$. With a unit step input, this equation can be written as

$$RC\frac{da(t)}{dt} + a(t) = 1, \qquad t \geq 0 \tag{2-91}$$

where $a(t)$ is the step response. We have already considered the solution of this differential equation in Example 2-1 for an arbitrary input. As mentioned in that example, the solution consists of the sum of a homogeneous solution, which is

$$a_h(t) = A \exp\left(\frac{-t}{RC}\right) \tag{2-92}$$

plus a particular solution for the forcing function $u(t)$. This particular solution is simply $a_p(t) = 1$, which can be seen by direct substitution into the nonhomogeneous differential equation for $a(t)$. Thus

$$a(t) = a_h(t) + a_p(t) = A \exp\left(\frac{-t}{RC}\right) + 1, \qquad t \geq 0 \tag{2-93}$$

The initial charge on the capacitor is zero for $t < 0$. Since the forcing function does not contain impulses (it is a step and therefore cannot change the capacitor charge instantaneously), it follows that the charge on the capacitor at $t = 0^+$ is zero. Therefore, $a(0^+) = 0$ or

$$a(0^+) = A \exp\left(\frac{-t}{RC}\right)_{t=0} + 1 = A + 1 = 0 \tag{2-94}$$

and

$$a(t) = 1 - \exp\left(\frac{-t}{RC}\right), \qquad t \geq 0 \tag{2-95a}$$

Using the fact that $a(t) = 0$ for $t < 0$, we may write this as

$$a(t) = \left[1 - \exp\left(\frac{-t}{RC}\right)\right]u(t) \tag{2-95b}$$

Since $\delta(t) = du(t)/dt$ and using the principle stated after Example 2-13 that for a fixed, linear system, any linear operation on the input produces the same linear operation on the output, we can obtain the impulse response by differentiating $a(t)$ with respect to t. The result is

$$h(t) = \frac{da(t)}{dt} = \left[1 - \exp\left(\frac{-t}{RC}\right)\right]\delta(t) + \frac{1}{RC}\exp\left(\frac{-1}{RC}\right)u(t)$$

$$= \frac{1}{RC}\exp\left(\frac{-t}{RC}\right)u(t) \tag{2-96}$$

which follows by using the chain rule for differentiation and noting that $1 - \exp(-t/RC) = 0$ for $t = 0$.

2-7 Frequency Response Function of a Fixed, Linear System

If the input to a fixed, linear system is a sinusoid of frequency ω rad/s, the steady-state response (i.e., the response after all transients have approached negligible values) is a sinusoid of the same frequency, but with amplitude multiplied by a factor $A(\omega)$ and phase-shifted by $\theta(\omega)$ radians. We prove this assertion in this section and illustrate it with an example.

The complex function of frequency

$$H(\omega) = A(\omega)e^{j\theta(\omega)} \tag{2-97}$$

is called the *frequency-response function* of the system; $A(\omega)$ and $\theta(\omega)$ are referred to as the *amplitude-response* and *phase-response functions* of the system, respectively.

The frequency-response function, (2-97), completely characterizes the steady-state response of a fixed, linear system to a sinusoid or, equivalently, a rotating phasor, $e^{j\omega t}$. That is, the steady-state output of a fixed, linear system is of the same form as its input when its input is $e^{j\omega t}$. In mathematical terms, $e^{j\omega t}$ is said to be an *eigenfunction* of the system. That this is the case can be shown by considering the superposition integral with the input $x(t) = e^{j\omega t}$. Then the output is

$$y(t) = \int_{-\infty}^{\infty} e^{j\omega(t-\lambda)}h(\lambda)\, d\lambda \tag{2-98}$$

where $h(\lambda)$ is the impulse response. Since $e^{j\omega t}$ can be factored out of the integral, the output can be written as

$$y(t) = e^{j\omega t}\int_{-\infty}^{\infty} h(\lambda)e^{-j\omega\lambda}\, d\lambda \triangleq e^{j\omega t}H(\omega) \tag{2-99}$$

We identify the integral multiplying $e^{j\omega t}$ as the frequency response $H(\omega)$. That is,

$$H(\omega) = \int_{-\infty}^{\infty} h(\lambda)e^{-j\omega\lambda}\, d\lambda \tag{2-100}$$

Later we shall see that $H(\omega)$ corresponds to the *Fourier transform* of the impulse response $h(t)$.

EXAMPLE 2-16

To illustrate these remarks, consider the *RC* filter of Example 2-1 as illustrated in Figure 2-2. Its impulse response was found in Examples 2-10 and 2-11 to be

$$h(t) = \frac{1}{RC}e^{-t/RC}u(t) \tag{2-101}$$

From (2-100), the frequency response function is found to be

$$H(\omega) = \int_{-\infty}^{\infty} \frac{1}{RC}e^{-t/RC}e^{-j\omega t}u(t)\, dt$$

$$= \int_{0}^{\infty} \frac{1}{RC}e^{-(j\omega+1/RC)t}\, dt$$

$$= \left[-\frac{1/RC}{j\omega + 1/RC}e^{-(j\omega+1/RC)t} \right]_{0}^{\infty}$$

$$= \frac{1}{1 + j\omega RC} \tag{2-102}$$

We write this in polar form (2-97) as

$$H(\omega) = \frac{1}{\sqrt{1 + (\omega RC)^2}}e^{-j\tan^{-1}(\omega RC)} \tag{2-103}$$

Thus, the amplitude and phase responses are, respectively, given by

$$A(\omega) = \frac{1}{\sqrt{1 + (\omega RC)^2}} \tag{2-104a}$$

and

$$\theta(\omega) = -\tan^{-1}(\omega RC) \tag{2-104b}$$

The output, for an input of the form $\alpha e^{j\omega t}$, is

$$y(t) = \frac{\alpha}{\sqrt{1 + (\omega RC)^2}} e^{j[\omega t - \tan^{-1}(\omega RC)]} \tag{2-105}$$

For a cosinusoidal input, we may find the output by writing the cosine in exponential form and using superposition to obtain the output

$$y(t) = \frac{\alpha}{\sqrt{1 + (\omega RC)^2}} \cos[\omega t - \tan^{-1}(\omega RC)] \tag{2-106}$$

Note that the same result could have been obtained by taking the real part of (2-105). This is not surprising, since the cosinuosidal input is obtained by taking the real part of the rotating phasor input, and the operations of taking the real part and convolution are interchangeable since both are linear operations.

Some numerical values will illustrate the behavior of $A(\omega)$ and $\theta(\omega)$ versus ω. For $(2\pi RC)^{-1} = 1$ kHz and $\alpha = 1$, the values in Table 2-1 result. For example, for $x(t) = \cos(2,000\pi t)$, the output is

$$y(t) = 0.707 \cos(2,000\pi t - 45°) \tag{2-107}$$

Using the superposition property of the linear RC filter, we obtain the output for an input of the form

$$x(t) = \cos(2,000\pi t) + \cos(20,000\pi t) \tag{2-108}$$

as

$$y(t) = 0.707 \cos(2,000\pi t - 45°) + 0.0995 \cos(20,000\pi t - 84.29°) \tag{2-109}$$

Applying the superposition property an indefinite number of times, one could find a series expression for the output due to any periodic input simply by representing the input in terms of its Fourier series, as we will show in the next chapter.

TABLE 2-1
Amplitude and Phase Response Values
for an RC Filter

ω	$A(\omega)$	$\theta(\omega)$
200π	0.995	$-5.71°$
$1,000\pi$	0.894	$-26.57°$
$2,000\pi$	0.707	$-45.00°$
$10,000\pi$	0.196	$-78.69°$
$20,000\pi$	0.0995	$-84.29°$

2-8 Stability of Linear Systems

One of the considerations in any system design is the question of stability. Although there are various definitions of stability in common usage, the one we use is referred to as *bounded-input, bounded-output* (BIBO) *stability.* By definition, a *system is BIBO stable if and only if every bounded input results in a bounded output.* Clearly, stability is a desirable property of most systems.

For a fixed, linear system we may obtain a condition on the impulse response which guarantees BIBO stability. To derive this condition, consider (2-41), which relates the input and output of a fixed, linear system:

$$y(t) = \int_{-\infty}^{\infty} x(\lambda) h(t - \lambda) \, d\lambda$$

It follows that

$$|y(t)| = \left| \int_{-\infty}^{\infty} x(\lambda) h(t - \lambda) \, d\lambda \right| \leq \int_{-\infty}^{\infty} |x(\lambda)| \, |h(t - \lambda)| \, d\lambda \qquad (2\text{-}110)$$

If the input is bounded, then

$$|x(\lambda)| \leq M < \infty \qquad (2\text{-}111)$$

where M is a finite constant. Replacing $|x(\lambda)|$ by M in (2-110), we have the inequality

$$|y(t)| \leq M \int_{-\infty}^{\infty} |h(t - \lambda)| \, d\lambda \qquad (2\text{-}112\text{a})$$

or

$$|y(t)| \leq M \int_{-\infty}^{\infty} |h(\eta)| \, d\eta \qquad (2\text{-}112\text{b})$$

which follows by the change of variables $\eta = t - \lambda$. Thus the output is bounded provided that

$$\int_{-\infty}^{\infty} |h(t)| \, dt < \infty \qquad (2\text{-}113)$$

That is, (2-113) is a sufficient condition for stability.

To show that it is also a necessary[†] condition, consider the input

$$x(\lambda) = \begin{cases} +1 & \text{if} \quad h(t - \lambda) > 0 \\ 0 & \text{if} \quad h(t - \lambda) = 0 \\ -1 & \text{if} \quad h(t - \lambda) < 0 \end{cases} \qquad (2\text{-}114)$$

For this input it follows that

$$|y(t)| = \left| \int_{-\infty}^{\infty} |h(t - \lambda)| \, d\lambda \right| = \int_{-\infty}^{\infty} |h(\eta)| \, d\eta \qquad (2\text{-}115)$$

for any fixed value of t since the integrand is always nonnegative. Thus the output will be unbounded if (2-113) is not satisfied. That is, it is also a necessary condition for BIBO stability.

[†]A condition from which a given statement logically follows is said to be a *sufficient condition;* a condition which is a logical consequence of a given statement is said to be a *necessary condition.* A condition may be necessary but not sufficient, and vice versa.

EXAMPLE 2-17 _____

The system of Example 2-11 is BIBO stable since

$$\int_{-\infty}^{\infty} \frac{1}{RC} e^{-t/RC} u(t)\, dt = \int_{0}^{\infty} \exp(-v)\, dv$$

$$= -\exp(-v)\Big|_{0}^{\infty}$$

$$= 1 < \infty \qquad (2\text{-}116)$$

EXAMPLE 2-18 _____

Is the system whose impulse response was derived in Example 2-12 BIBO stable?

Solution: Substituting the impulse response for this system into (2-113), we obtain

$$\int_{-\infty}^{\infty} |h(t)|\, dt = \omega_0 \int_{-\infty}^{\infty} |\sin(\omega_0 t)\, u(t)|\, dt = \omega_0 \int_{0}^{\infty} |\sin \omega_0 t|\, dt \qquad (2\text{-}117)$$

The integral does not converge, so the condition for BIBO stability does not give a bounded result. This system is BIBO unstable.

We will return to the idea of stability of a system in Chapter 6, where additional methods for determining the stability of systems describable by constant coefficient, linear differential equations, are discussed.

2-9 System Modeling and Simulation†

We briefly consider in this section how systems can be modeled and simulated in terms of several basic building blocks. These basic building blocks are given in Table 2-2, and consist of ideal adders or subtractors, constant multipliers, and integrators. Any nth-order constant-coefficient differential equation can be simulated by means of these components. It may seem that we are simply replacing one unsolved problem with another in so doing, or that we really don't need such simulations at all because linear, constant-coefficient differential equations are amenable to solution. In answer to the first thought, when a simulation block diagram is available for a system, the realization of such a simulation can take one of two forms: (1) The realization can be accomplished in analog fashion by means of operational amplifier circuits that approximate the ideal summers and integrators needed (such circuits are also shown in Table 2-2); (2) the simulation can be realized by means of a digital computer program where the integration is done numerically, as in Example 1-3. In regard to simulations being unnecessary because the differential equation representing a system is analytically solvable, we simply point out that the system realizations and simulation procedures touched on in this section also apply to cases where the system is nonlinear or time varying as well, and these problems are often not solvable by analytical means.

To see how one might simulate a system, either by means of operational amplifier circuits or by digital computer algorithms, consider the first-order differential equation

$$\frac{dy}{dt} + ay = b_1 \frac{dx}{dt} + b_0 x \qquad (2\text{-}118)$$

_____

†Systematic modeling procedures for electrical and mechanical systems are discussed in Appendix B.

TABLE 2-2
Building Blocks for System Modeling

Operation	Symbol	Op Amp Realization
Multiplication by a constant	$x(t) \rightarrow \boxed{a} \rightarrow y(t)$ $y(t) = ax(t)$	$-a = R_2/R_1$
Integration	$x(t) \rightarrow \boxed{a\int^t} \rightarrow y(t)$ $y(t) = a\int^t x(\lambda)\,d\lambda$	$-a = 1/RC$
Summation	$x_1(t) \rightarrow \Sigma \rightarrow y(t)$, $x_2(t)$ $-y(t) = x_1(t) + x_2(t)$	$R_3/R_2 = 1$ $R_3/R_1 = 1$

It is realized by the block diagram of Figure 2-13. To show that this is the case, let the output of the integrator be $q(t)$. Its input is then dq/dt. Equating dq/dt to the sum of the inputs to the left-hand summer, we obtain

$$\frac{dq}{dt} = -ay + b_0 x \qquad (2\text{-}119)$$

Considering the output of the right-hand summer, we find that

$$y = q + b_1 x \qquad (2\text{-}120)$$

Differentiating both sides of (2-120) and substituting (2-119), we obtain (2-118).

In a similar manner to the first-order case, we can realize an nth-order system by a block diagram similar to Figure 2-12. For the differential equation

$$\frac{d^n y(t)}{dt^n} + \sum_{i=0}^{n-1} a_i \frac{d^i y(t)}{dt^i} = \sum_{i=0}^{n} b_i \frac{d^i x(t)}{dt^i} \qquad (2\text{-}121)$$

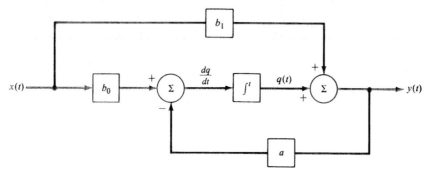

FIGURE 2-13. Integrator realization of a general first-order system.

we have the realization diagram shown in Figure 2-14. That this is the realization of (2-121) can be demonstrated in a similar, although longer, procedure to that used for the first-order case. [Note that this is only one of many ways to realize this system. Also note that the number of derivatives on either side of (2-121) can differ by having some a_i's or b_i's zero.]

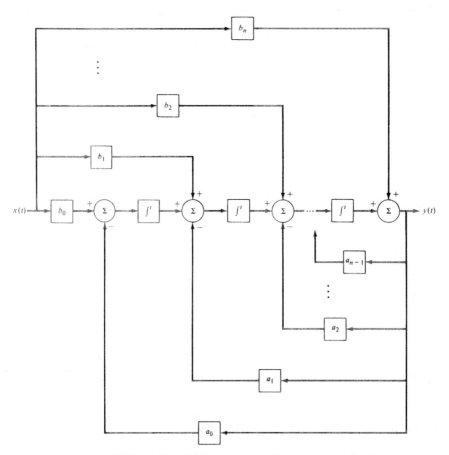

FIGURE 2-14. Integrator realization of a general nth-order system.

EXAMPLE 2-19

Consider the *RLC* circuit shown in Figure 2-15. Obtain an integrator realization of this circuit.

Solution: The differential equation describing the voltage response of the circuit is

$$\frac{v(t)}{R} + \frac{1}{L} \int^t v(\lambda) \, d\lambda + C \frac{dv(t)}{dt} = x(t) \tag{2-122}$$

Differentiating once and rearranging, we obtain

$$\frac{d^2v}{dt^2} + \frac{1}{RC} \frac{dv}{dt} + \frac{1}{LC} v = \frac{1}{C} \frac{dx}{dt} \tag{2-123}$$

This is of the form (2-121) with $y = v, n = 2, a_0 = 1/LC, a_1 = 1/RC, b_0 = 0, b_1 = 1/C$, and $b_2 = 0$. The integrator realization is shown in Figure 2-16.

MATLAB APPLICATION

An integrator (or state variable) simulation of a system can be carried out using SIMULINK, which is a block diagram oriented toolbox. There is a limited version available in *The Student Version of* MATLAB. One invokes SIMULINK by typing "simulink" after the command window prompt. One can then construct a block diagram realization of the desired system using the menus available. Such a SIMULINK block diagram realization is shown in Fig. 2-17 for the RLC circuit of Example 2-19.

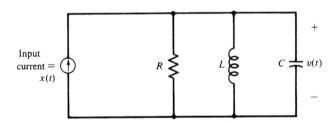

FIGURE 2-15. Circuit for Example 2-19.

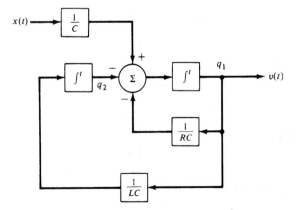

FIGURE 2-16. Integrator realization of a parallel *RLC* circuit.

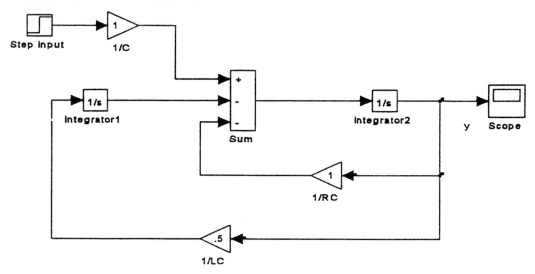

FIGURE 2-17. A SIMULINK simulation block diagram for the RLC circuit of Example 2-19.

The simulation can be carried out directly in the model window or it can be carried out in the MATLAB command window using the function lsim as suggested by the script below:

```
EDU>>c2ex19p1
%       Plot of simulation output for Example 2_19
%
[t,q]=lsim('c2ex19',20);
plot(t, q(:,1)), xlabel('t'), ylabel('output')
```

The output variables are defined by dropping down the menu under "Simulation" in the simulation window, and then the menu under "parameters." This brings up the "SIMULINK Control Panel" in which various simulation variables are specified such as start and end time of the simulation. The last blank to be filled out is labeled "Return Variables" which can be left blank if desired and the return variables default to time (t) and the state variables or integrator outputs (x). In our example here we gave the state variables the label q. This is a matrix $q(m,n)$ with m denoting the time index and n indicating the number of the state variable. Thus by specifying $q(:, 1)$ we have specified the output in this example, which is shown in Figure 2-18. The step size was set to 0.1.

A useful way to write the equations for a system are in terms of *state variables,* to be discussed further in Chapter 7. For now, we simply point out that the state variables for a system are the outputs of the integrators of the integrator realization of a system. The state variable equations are then written down by taking the input to each integrator (the derivative of that state variable) and setting it equal to the summer output appearing there.

The advantage of writing the system equations in terms of state variables is that the same form is obtained no matter what the order of the system. For orders greater than one, the equations become matrix equations, as will be pointed out in Chapter 7. For example, the state equations for the system of Example 2-21 are

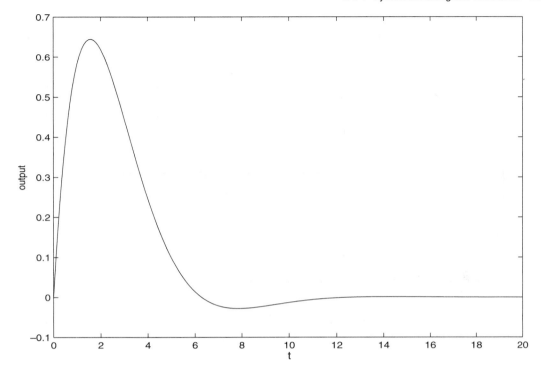

FIGURE 2-18. Output of the parallel RLC circuit of Figure 2-15 obtained by SIMULINK for parameter values shown in Figure 2-17.

$$\frac{dq_1(t)}{dt} = -\frac{1}{RC}q_1(t) - q_2(t) + \frac{1}{C}x(t)$$

$$\frac{dq_2(t)}{dt} = \frac{1}{LC}q_1(t) \qquad\qquad (2\text{-}124)$$

and the output equation is $v(t) = q_1(t)$. This can be put into the form of the matrix equation

$$\dot{\mathbf{Q}}(t) = \mathbf{A}\mathbf{Q}(t) + \mathbf{B}x(t) \qquad\qquad (2\text{-}125)$$

where **A** and **B** are matrices and $\mathbf{Q}(t)$ is a column matrix or vector with elements $q_1(t)$ and $q_2(t)$.

EXAMPLE 2-20

Reconsider the accelerometer of Example 1-1 and the integrator circuit for determining velocity of Example 1-2. We make one addition to the accelerometer model to make it more realistic—the addition of viscous friction between the weight and its housing. Such friction is often modeled as being proportional to velocity so that (1-1) now is modified to

$$Ma = Kx + B\frac{dx}{dt} \qquad\qquad (2\text{-}126)$$

where B is the constant of proportionality. A small amount of friction will hardly affect the accelerometer action at all, whereas a large amount will make its response sluggish and the indicated velocity at the integrator circuit output will lag that of the actual velocity. To investigate this effect, we model the rest of the system and simulate it with SIMULINK. The remaining equations necessary for constructing the SIMULINK model are (1-8), which is repeated here

$$v_o(t) = -\frac{1}{RC} \int_0^t v_s(\lambda) \, d\lambda \quad \text{or} \quad \frac{dv_o(t)}{dt} = -\frac{1}{RC} x(t) \tag{2-127}$$

and the equation relating v_s to the mass excursion from equilibrium, x. To make the maximum of $v_s = 0.1$ V when the mass is at its maximum excursion of 1 cm = 0.01 m, this equation is simply

$$v_s(t) = 10x(t) \tag{2-128}$$

We define *auxiliary variables* (called state variables) as the outputs of the integrators in the SIMULINK model. Thus $q_1 = x$ and $q_2 = v_o$, and the system governing equations may be written as

$$\frac{dq_1(t)}{dt} = -\frac{K}{B} q_1(t) + \frac{M}{B} a(t)$$

and

$$\frac{dq_2(t)}{dt} = -\frac{1}{RC} v_s(t) = -\frac{10}{RC} x(t) \tag{2-129}$$

The desired output is $q_2(t) = v_o(t)$ which is proportional to the velocity of the rocket (recall that the integrator circuit was designed in Example 1-2 such that when the operational amplifier output voltage is -10 volts, the rocket's velocity is the burnout value of 1500 m/s).

A SIMULINK block diagram modeling the accelerometer and integrator circuit is shown in Figure 2-19. Recall that the following parameter values were used in Examples 1-1 and 1-2:

$$M = 2 \text{ grams} = 0.002 \text{ kg}; \quad K = 8 \text{ kg/s}^2; \quad RC = 0.36 \text{ s}$$

We will take three values for B, in particular 1, 10, and 100 kg/s which corresponds to low, moderate, and high damping. Note that the ramp acceleration input has been modeled as a step integrated.

The SIMULINK simulation can take place either in the SIMULINK model window itself, in which variables of interest can be monitored on scopes strategically placed, or it can be run in the MATLAB command window with the statement below:

```
EDU>>[t, q]=sim('c2ex20',72);
```

Various plots can then be made with the plot command. In this case, three simulations were run for the three different values of B (1, 10, and 100) and the plot held in each case so that the integrator output for each succeeding case could be plotted on top of the previous cases. The resulting graph is shown in Figure 2-20. Note that the largest value of $B = 100$ gives considerable lag in the output.

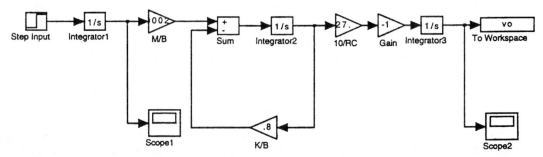

FIGURE 2-19. SIMULINK model for the accelerometer/integrator circuit of Examples 1-1 & 1-2.

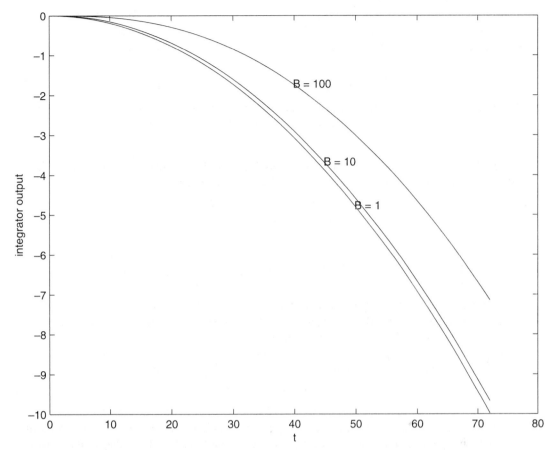

FIGURE 2-20 Output voltages for the accelerometer/integrator system of Examples 1-1 and 1-2 for the three damping parameter values of 1, 10, and 100 kg/s.

EXAMPLE 2-21

As a final example using SIMULINK, we do another system that involves both mechanical and electrical subsystems. Figure 2-21 represents a direct current (dc) motor with separately excited field coils (a separate source supplies the dc current to provide the magnetic field in which the armature rotates). The applied voltage to the armature, $v_s(t)$, is assumed to be a step applied at $t = 0$. The angular velocity (rotational speed), radians/second, of the armature as a function of time is desired.

The left-hand side of the block diagram of Figure 2-21, representing the armature circuit, includes the applied source, a resistor representing the resistance of the armature windings, an inductor representing the inductance of the armature windings, and a controlled source $v_b = K \, d\theta/dt = K\Omega$ representing the back-electromotive force (emf) generated because the armature turning in the magnetic field also acts as a generator. The latter is proportional to the angular velocity of the armature.

The right-hand side of the block diagram of Figure 2-21 represents the rotating armature of the motor. It consists of an ideal torque source $T_s = Ki$ (note that the same constant relates torque to current as relates back emf to angular velocity) which is proportional to the armature current, an

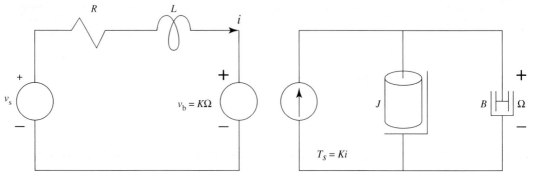

FIGURE 2-21. Schematic diagram of a separately field-excited dc motor with viscous friction loading.

angular momentum J which is due to the mass of the armature, and a rotational damper B which represents damping due to air and mechanical friction (part of which may be the load).

Since both the electrical and mechanical parts of the system are represented by circuit diagrams (the circuit diagram for the mechanical side is perhaps easier for electrical engineers to understand than mechanical diagrams), we can apply Kirchhoff's circuit laws to write down the governing differential equations which are

$$v_s(t) = Ri(t) + L\frac{di(t)}{dt} + K\Omega(t)$$

$$T_s(t) = Ki(t) = J\frac{d\Omega(t)}{dt} + B\Omega(t) \tag{2-130}$$

If we are interested in the angular velocity $\Omega(t)$, we could eliminated $i(t)$ between these two equations and solve the resulting second-order differential equation for $\Omega(t)$ using standard techniques. The purpose of this example is to illustrate the application of SIMULINK, however. Easier techniques for solving system equations based on Laplace transforms will be presented in Chapters 5 and 6. In order to easily construct the SIMULINK block diagram, we solve (2-130) for di/dt and $d\Omega/dt$. These will then be the inputs to the two required integrators.

$$\frac{di(t)}{dt} = -\frac{R}{L}i(t) - \frac{K}{L}\Omega(t) + \frac{1}{L}v_s(t)$$

$$\frac{d\Omega(t)}{dt} = -\frac{B}{J}\Omega(t) + \frac{K}{J}i(t) \tag{2-131}$$

Using these equations, the SIMULINK diagram shown in Figure 2-22 can be constructed. We have two options for running the simulation—in particular, we can simulate directly in the SIMULINK window, or we can use lsim in the MATLAB command window. The latter is convenient for generating plots that can be stored in files or printed. Note that the coupling between the electrical and mechanical parts of the system is controlled by the parameter K. We are interested in studying the effects of the dissipative elements on the system operation—in particular R and B. We intuitively expect the final angular velocity of the motor to be less the higher B, and we expect it to take longer in getting to its final angular velocity the higher B. These outputs for three values of B are illustrated in Figure 2-23 where it is seen that this is indeed the case.

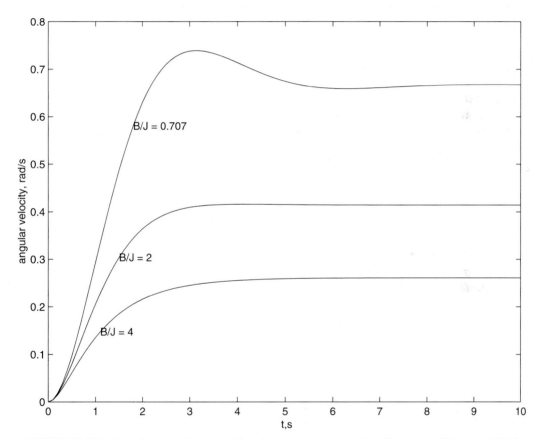

FIGURE 2-22. SIMULINK diagram for the separately excited dc motor system of Figure 2-21 (L = 1).

FIGURE 2-23. Angular velocity versus time for motor armature for different coefficients of friction.

Summary

In this chapter, the response of a fixed, linear system to an arbitrary input has been considered. The basic tool for obtaining the response is the superposition integral, which provides the output of the system as a superposition of elemental impulse responses weighted by the input signal values at various delays. The linearity of the system allowed the superposition of the elemental responses, while the fixed property means that the response to an impulse applied at $t = 0$ need only be considered with all other possible delays obtained by delaying the impulse response due to the fixed nature of the system. The following are the main points made in this chapter.

1. A system is said to be *continuous time* if the signals processed by the system are continuous-time signals. It is said to be a *discrete-time system* if it processes discrete-time signals. It is said to be *quantized* if the signals processed by it are quantized signals. If a system is built to process signals that are both discrete time and quantized, it is said to be a *digital system*. Continuous-time systems are dealt with almost exclusively in this chapter. Discrete-time systems will be the subject of Chapters 8-10. Furthermore, the systems considered in this chapter are all single-input, single-output; that is, they have a single input and a single output. Systems with multiple inputs and outputs can be handled by the techniques to be explored in Chapter 7.

2. A system is *time invariant*, or *fixed*, if its input-output relationship does not change with time. Thus, for a time-invariant system, if the input $x(t)$ produces the output $y(t)$, then the delayed input $x(t - \tau)$, where τ is a constant delay, produces the output $y(t - \tau)$.

3. A system is *causal*, or *nonanticipatory*, if its response to an input does not depend on future values of that input.

4. A system is *instantaneous*, or *zero memory*, if its output depends only on present values of the input, and not on past or future values.

5. If a single-input, single-output system is represented symbolically as $y(t) = \mathcal{H}[x(t)]$, where $\mathcal{H}[\cdot]$ is an operator, the system is linear if superposition holds. That is, a system is linear if, for the arbitrary linear combination of two inputs $\alpha_1 x_1(t) + \alpha_2 x_2(t)$, the output can be expressed as

$$y(t) \triangleq \mathcal{H}[\alpha_1 x_1(t) + \alpha_2 x_2(t)]$$
$$= \alpha_1 \mathcal{H}[x_1(t)] + \alpha_2 \mathcal{H}[x_2(t)]$$
$$= \alpha_1 y_1(t) + \alpha_2 y_2(t)$$

where $y_1(t)$ is the response to $x_1(t)$ and $y_2(t)$ is the response to $x_2(t)$ and α_1 and α_2 are arbitrary constants.

6. For a linear system that is not fixed, the superposition integral relating the input $x(t)$ to the output $y(t)$ is

$$y(t) = \int_{-\infty}^{\infty} h(t, \lambda)x(\lambda) \, d\lambda$$

where $h(t, \lambda)$ is the response of the system at time t to a unit impulse applied at time λ. If, in addition, the system is fixed, then $h(t, \lambda) = h(t - \lambda)$, and the superposition integral becomes

$$y(t) = \int_{-\infty}^{\infty} h(t - \lambda)x(\lambda) \, d\lambda$$

$$= \int_{-\infty}^{\infty} h(\tau)x(t - \tau) \, d\tau$$

where the second form follows from the first by substituting $\tau = t - \lambda$. These integrals are known as convolution integrals.

7. In addition to its use for finding the output of a fixed, linear system, the convolution integral has many other uses. We denote the convolution operation by $y(t) = h(t) * x(t) = x(t) * h(t)$. The following properties of the convolution integral hold:

1. $h(t) * x(t) = x(t) * h(h)$.
2. $h(t) * [\alpha x(t)] = \alpha[h(t) * x(t)]$, where α is a constant.
3. $h(t) * [x_1(t) + x_2(t)] = h(t) * x_1(t) + h(t) * x_2(t)$.
4. $h(t) * [x_1(t) * x_2(t)] = [h(t) * x_1(t)] * x_2(t)$.
5. If $h(t)$ is time-limited to (a, b), and $x(t)$ is time-limited to (c, d), then $h(t) * x(t)$ is time-limited to $a + c, b + d)$.
6. If A_1 is the area under $h(t)$ and A_2 is the area under $x(t)$, then the area under $h(t) * x(t)$ is $A_1 A_2$.

8. The integrand of the convolution operation is found by three operations: (1) reversal in time, or folding, to obtain $x(-\lambda)$; (2) shifting, to obtain $x(t - \lambda)$; (3) multiplication of $h(\lambda)$ and $x(t - \lambda)$, to obtain the integrand. A similar series of operations is required to obtain the second form of the convolution integrand.
9. The impulse response of a fixed, linear system is the response of the system to a unit impulse applied at $t = 0$ with zero initial conditions.
10. For systems described by constant-coefficient, linear differential equations, the time-domain impulse response can be found by solving the differential equation in response to a unit impulse with appropriate initial conditions. Since the unit impulse is zero for $t > 0$, this reduces to solving the homogeneous equation and finding the initial conditions by integrating the differential equation with unit impulse forcing function through $t = 0$.
11. A second method of finding the impulse response of a lumped-element circuit by using time-domain techniques is to look at the properties of the lumped circuit elements in response to impulse forcing functions. To this end, capacitors are infinite charge sinks (short circuits) and inductors are open circuits at the instant of application of the impulse forcing function.
12. The output of a fixed, linear system can also be found in terms of a superposition of its step response. The resulting integrals, called Duhamel's integrals, are given by

$$y(t) = \int_{-\infty}^{\infty} \dot{x}(t - \lambda)a(\lambda)\, d\lambda$$

$$= \int_{-\infty}^{\infty} \dot{x}(\eta)a(t - \eta)\, d\eta$$

where $a(t)$ is the response of the system to a unit step and the overdot denotes differentiation with respect to time.
13. For a fixed, linear system, *any linear operation on the input produces the same linear operation on the output*. Thus, for example, the step response is the integral of the impulse response, since a unit step is the integral of a unit impulse. The ramp response is the integral of the step response, since a unit ramp is the integral of the unit step.
14. For a fixed, linear system, the steady-state response (i.e., the response after all transients have approached negligible values) to a rotating phasor input signal is a rotating phasor, but usually with different amplitude and different phase. If the rotating phasor input has unit amplitude and zero phase, then the complex proportionality constant, denoted $H(\omega)$, relating output phasor to input phasor is called the *frequency response function* of the system. The magnitude of $H(\omega)$ is called the *amplitude response* of the system, and the argument, or angle, of $H(\omega)$ is called the *phase response* of the system. For example, through superposition, the steady-state output of a fixed, linear system in response to the cosinusoidal input $A \cos(\omega_0 t + \theta)$ is

$$y(t) = A|H(\omega_0)|\cos[\omega_0 t + \theta + \underline{/H(\omega_0)}]$$

Likewise, the steady-state response of the system to any sum of sinusoids could be found by using superposition.

15. Bounded-input, bounded-output (BIBO) stability of a system means that every bounded input produces a bounded output. For a fixed, linear, system a necessary and sufficient condition for BIBO stability is

$$\int_{-\infty}^{\infty} \left| h(t) \right| dt < \infty$$

where $h(t)$ is the impulse response of the system.

16. Fixed, linear systems can be modeled by interconnections of operational amplifier circuits configured as integrators, summers, inverters, and scale changers (amplifiers). A standard configuration is given in Figure 2-14 for a system described by the differential equation

$$\frac{d^n y(t)}{dt^n} + \sum_{i=0}^{n-1} a_i \frac{d^i y(t)}{dt^i} = \sum_{i=0}^{n} b_i \frac{d^i x(t)}{dt^i}$$

In Chapter 7, such configurations will be found systematically by state-variable techniques. State-variable techniques are very powerful in that they provide a means to represent multiple-input, multiple-output systems, time-varying systems, and nonlinear systems. A block diagram, such as Figure 2-14, is called an analog computer simulation of the system.

17. Analog computer simulations are very seldom used nowadays. The reason is that digital computers have become very powerful and reasonably priced. Even a desktop computer or workstation is capable of simulating very complex systems. The last section of this chapter introduced the topic of numerical simulation of systems using MATLAB SIMULINK. Several linear systems are employed as examples. The SIMULINK feature of MATLAB makes numerical solution straight forward.

Further Reading

In addition to the references listed in Chapter 1, the following books provide alternative reading to systems modeling in the time domain.

T. KAILATH, *Linear Systems.* Englewood Cliffs, NJ: Prentice-Hall, 1980.

E. W. KAMEN, *Introduction to Signals and Systems.* 2nd ed. New York: Macmillan, 1990.

N. K. SINHA, *Linear Systems.* New York: Wiley, 1991.

H. KWAKERNAAK and R. SIVAN, *Modern Signals and Systems.* Englewood Cliffs, NJ: Prentice-Hall, 1991.

The following is an old book and out of print. Yet, it is referenced because many of the definitions and examples in this chapter are patterned after material found in it.

R. J. SCHWARZ and B. FRIEDLAND, *Linear Systems.* New York: McGraw-Hill, 1965.

The following references provide extensive treatments on modeling physical systems.

W. A. BLACKWELL, *Mathematical Modeling of Physical Networks.* New York: Macmillan, 1968.

C. M. CLOSE and D. K. FREDERICK, *Modeling and Analysis of Dynamic Systems.* Boston: Houghton-Mifflin, 1978.

Material on numerical solution of system equations can be found in the following book:

J. M. SMITH, *Mathematical Modeling and Digital Simulation for Engineers and Scientists,* 2nd ed. New York: Wiley, 1987.

For guidance on the use of SIMULINK, see the following reference:

The Student Edition of SIMULINK, Version 2: User's Guide. Upper Saddle River, NJ: Prentice Hall, 1998.

Problems

Section 2-2

2-1. A system is defined by the set of algebraic equations

$$y_1(t) = 2x_1(t) - x_2(t)$$
$$y_2(t) = 5x_2(t) + 3x_3(t)$$

Represent this system in terms of a matrix equation of the form

$$y(t) = Hx(t)$$

That is, give $y(t)$, H, and $x(t)$ explicitly.

2-2. What is the order of each system defined by the following equations?

(a) $2\dfrac{dy(t)}{dt} + 3y(t) = \dfrac{d^2x(t)}{dt^2} + x(t)$

(b) $3y(t) + \displaystyle\int_{-\infty}^{t} y(\lambda)\, d\lambda = x(t)$

(c) $4y(t) + 10 = \dfrac{dx(t)}{dt} + 5x(t)$

(d) $\dfrac{dy(t)}{dt} + t^2y(t) = \displaystyle\int_{-\infty}^{t} x(\lambda)\, d\lambda$

(e) $\dfrac{d^2y(t)}{dt^2} + y(t)\dfrac{dy(t)}{dt} + y(t) = 5x(t)$

2-3. Which of the systems defined by the equations of Problem 2-2 are fixed? Justify your answers.

2-4. Which of the systems defined by the equations of Problem 2-2 are nonlinear? Justify your answers.

2-5. A system is defined by the input-output relationship

$$y(t) = x(t^{1/2})$$

Is this system causal or noncausal? Justify your answer by choosing a specific pair of inputs that will or will not satisfy (2-23).

2-6. A system is defined by the input-output relationship

$$y(t) = 10x(t + 2) + 5$$

(a) Is this system linear? Prove your answer.
(b) Is it causal or noncausal? Why?

2-7. A system is defined by the input-output relationship

$$y(t) = x(t^2)$$

Is this system:
(a) Linear?
(b) Causal?
(c) Fixed?
Prove your answers.

2-8. An echo system is defined by the input-output relationship $y(t) = x(t) + \alpha x(t - \tau_0)$.
   (a) Show that it is a linear system.
   (b) Is it zero memory?
   (c) Show that it is causal if $\tau_0 \geq 0$.
   (d) Sketch its output for $\alpha = 0.5$ and 1.5 if $x(t) = u(t) - u(t - 1)$ and $\tau_0 = 1$.

2-9. An averager is defined by the input-output relationship

$$y(t) = \frac{1}{T_1 + T_2} \int_{t-T_1}^{t+T_2} x(\lambda)\, d\lambda$$

where T_1 and T_2 are positive constants.
   (a) Show that this system is linear.
   (b) What conditions on T_1 and T_2 make this a causal system?

2-10. **(a)** Write a differential equation relating $y(t)$ to $x(t)$ for the circuit shown in Figure P2-10.
   (b) Show that this system is linear.
   (c) Show that this system is fixed.
   (d) Convert the differential equation found in part (a) to an integral equation of the form (2-5). Assume that the input $x(t)$ is applied at $t = 0$, and that the current through the inductor is initially zero.

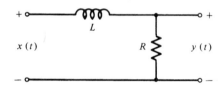

FIGURE P2-10

2-11. Fill in the table at the end of this problem to state whether the systems specified are linear or nonlinear, causal or noncausal, fixed or time-varying, and dynamic or instantaneous. Also give their order.

   (a) $\dfrac{dy}{dt} + 3y + 2 \int_{-\infty}^{t} y(\lambda)\, d\lambda = x(t)$

   (b) $\dfrac{d^3y}{dt^3} + 4\dfrac{d^2y}{dt^2} + 5\dfrac{dy}{dt} + 2y^2(t) = x(t)$

   (c) $\dfrac{d^2y}{dt^2} + 3t\dfrac{dy}{dt} + y(t) = x(t)$

   (d) $y(t)\dfrac{d^2y}{dt^2} + 3t\dfrac{dy}{dt} + y(t) = x(t)$

   (e) $y(t) = x(t^2) + x(t)$

   (f) $y(t)\dfrac{d^2y}{dt^2} + 3t\dfrac{dy}{dt} + y(t) = x(t + 5)$

If true, check the appropriate box. Fill in the bottom row with system order.

		System				
Property	**a**	**b**	**c**	**d**	**e**	**f**
Linear						
Causal						
Fixed						
Dynamic						
Order						

2-12. Classify the following input-output relations for systems as to linearity, order, causality, and time invariance. State the order of each.

(a) $\dfrac{dy}{dt} + \dfrac{1}{RC} y(t) = x(t); \quad RC = \text{constant}$

(b) $\dfrac{dy}{dt} + t^2 y(t) = x(t)$

(c) $\dfrac{d^2y}{dt^2} + y(t) \dfrac{dy}{dt} + y(t) = x(t)$

(d) $y(t) = x^2(t) + 1$

2-13. Given the system

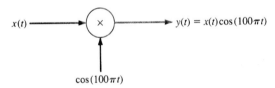

$$y(t) = x(t)\cos(100\pi t)$$

FIGURE P2-13

(a) Show that it is linear.
(b) Is it time varying? Why?
(c) Is it causal? Why?
(d) Is it instantaneous? Why?

2-14. A fixed, linear system responds to the input $x(t) = \Pi[(t - 1)/2]$ with the output $y(t) = (t + 1)\exp[2\,(t + 1)]u(t + 1)$.
(a) Is it causal? Why or why not?
(b) Write down the response to the input $2\,\Pi[(t - 1)/2] + 3\,\Pi[(t - 3)/2]$. Sketch input and output signals.

Section 2-3

2-15. Assuming the superposition property for the linear combination of two signals, extend this expression to the linear combination of N signals.

2-16. If $x_1(t)$, $x_2(t)$, and $h(t)$ are arbitrary signals, and α is a constant, show the following:

  (a) $h(t) * [x_1(t) + x_2(t)] = h(t) * x_1(t) + h(t) * x_2(t)$

  (b) $h(t) * [x_1(t) * x_2(t)] = [h(t) * x_1(t)] * x_2(t)$

  (c) $h(t) * [\alpha x_1(t)] = \alpha h(t) * x_1(t)$

  (d) If $h(t)$ is time-limited to (a, b) and $x(t)$ is time-limited to (c, d), then $h(t) * x(t)$ is time-limited to $(a + c, b + d)$.

  (e) If A_1 is the area under $h(t)$ and A_2 is the area under $x(t)$, then the area under $h(t) * x(t)$ is A_1A_2.

Section 2-4

2-17. Find and sketch the signal $y(t)$, which is the convolution of the following pairs of signals.

  (a) $x(t) = 2 \exp(-10t)u(t)$ and $h(t) = \Pi(t/2)$

  (b) $x(t) = \Pi[(t - 1)/2]$ and $h(t) = u(t - 10)$

  (c) $x(t) = 2 \exp(-10t)u(t)$ and $h(t) = u(t - 2)$

  (d) $x(t) = [\exp(-2t) - \exp(-10t)]u(t)$ and $h(t) = \Pi\left(\dfrac{t - 1}{2}\right)$

  (e) $x(t) = r(t)$ and $h(t) = 2 \exp(-2t)u(t)$

2-18. Given the signals

$$x(t) = \Pi(t - 0.5)$$

and

$$h(t) = \sum_{n=-\infty}^{\infty} \delta(t - 2n)$$

Write down an expression for their convolution. Sketch the result.

Section 2-5

2-19. Obtain the impulse response of the system shown by:

  (a) Using the method of Example 2-10.

  (b) Using the method of Example 2-11.

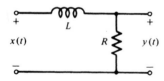

FIGURE P2-19

2-20. Find the impulse response for the circuit

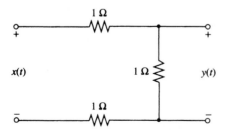

FIGURE P2-20

2-21. Given the circuit

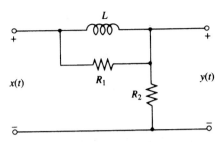

FIGURE P2-21

(a) Show that the differential equation relating the output to the input is

$$\frac{dy(t)}{dt} + \frac{R_2}{R_1 + R_2}\frac{R_1}{L}y(t) = \frac{R_2}{R_1 + R_2}\left[\frac{dx(t)}{dt} + \frac{R_1}{L}x(t)\right]$$

(b) Obtain the impulse response of this circuit.

2-22. Obtain the impulse response of the system shown.

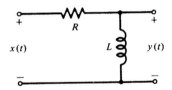

FIGURE P2-22

2-23. Obtain the impulse response of the system shown.

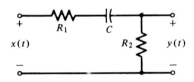

FIGURE P2-23

2-24. (a) Show that the impulse response $h(t)$ of the operational amplifier circuit shown is $h(t)$ proportional to $u(t)$.
(b) Express the output of the circuit in terms of the superposition integral for an arbitrary input. Is it the expected result given the function of this circuit?

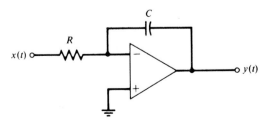

FIGURE P2-24

Section 2-6

2-25. (a) What is the step response of the circuit in Problem 2-24?
   (b) Given the input $x(t) = u(t) + u(t - 1) - 2u(t - 2)$. Find the output for this input. Sketch both the input and the output. Label carefully.

2-26. Obtain the output of the system of Problem 2-19 in response to the input

$$x(t) = \Pi(t - 0.5)$$

Sketch for $R/L = 0.1$ and 1.0. Use (2-41).

2-27. Obtain the output of the system of Problem 2-22 in response to the input

$$x(t) = \Pi(t - 0.5)$$

Sketch for $R/L = 0.1$ and 1.0.

2-28. Using (2-85), obtain the output of the system of Problem 2-19 in response to the input

$$x(t) = \Pi(t - 0.5)$$

Sketch for $R/L = 0.1$ and 1.0.

2-29. Given that the RL filter shown has impulse response

$$h(t) = \delta(t) - (R/L)\, e^{-(R/L)t}u(t)$$

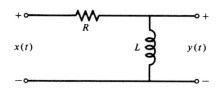

$x(t)$ R L $y(t)$

FIGURE P2-29

   (a) Find the step response.
   (b) Find the ramp response.

2-30. Obtain and sketch the response of the filter of Problem 2-29 to the input shown for:
   (a) $R/L = 0.1$
   (b) $R/L = 1.0$
   (*Hint:* Use the answers obtained for Problem 2-29 together with superposition.)

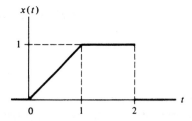

$x(t)$

FIGURE P2-30

2-31. (a) Using the superposition property, find the response of the *RC* circuit of Example 2-11 to the input

$$x(t) = r(t) - 2r(t - 1) + r(t - 2)$$

Sketch the input and the output for $RC = 0.1$.

(b) Find and sketch the response to $\dfrac{dx}{dt}$.

2-32. Given the circuit shown

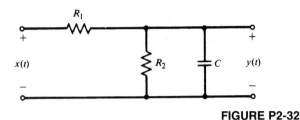

FIGURE P2-32

(a) Show that the input-output relationship is

$$R_1 C \frac{dy}{dt} + \left[1 + \frac{R_1}{R_2} \right] y = x$$

(b) Show that the impulse response is

$$h(t) = (R_1 C)^{-1} \exp\left(-\frac{t}{\tau} \right) u(t)$$

where $\tau = R_1 R_2 C / (R_1 + R_2)$.

(c) Show that the step response is

$$a(t) = \frac{R_2}{R_1 + R_2} \left[1 - \exp\left(-\frac{t}{\tau} \right) \right] u(t)$$

(d) If the input is $x(t) = \Pi[(t - 1)/2]$, obtain the output.

Answer: $y(t) = \dfrac{R_2}{R_1 + R_2} \left\{ \left[1 - \exp\left(-\dfrac{t}{\tau} \right) \right] u(t) - \left[1 - \exp\left(-\dfrac{t - 2}{\tau} \right) \right] u(t - 2) \right\}$

(e) If the input is $x(t) = r(t)$, find the output. (*Hint:* Use Duhamel's integral.)

Answer: $y_r(t) = \dfrac{R_2}{R_1 + R_2} \left\{ t - \tau \left[1 - \exp\left(-\dfrac{t}{\tau} \right) \right] \right\} u(t)$

(f) If the input is $x(t) = r(t) - 2r(t - 1) + r(t - 2)$, find the output in terms of the answer of part (e).

Section 2-7

2-33. (a) Obtain the frequency-response function for the system of Problem 2-29.

(b) Obtain the amplitude-response function for this system. Put it in terms of the parameter $f_3 = R/2\pi L$ and plot, dimensioning carefully.

(c) Obtain the phase-response function for this system. Put it in terms of the parameter $f_3 = R/2\pi L$ and plot, dimensioning carefully.

2-34. (a) Obtain the frequency-response function for the system of Problem 2-32.
   (b) Obtain the amplitude- and phase-response functions for this system.
   (c) If $R_1 = R_2 = 1\Omega$ and $C = 1\ F$, obtain the steady-state response of this system to the input $x(t) = \cos 2\pi t + \sin 5\pi t$.

Section 2-8

2-35. Show that the system of Problem 2-19 is BIBO stable.

2-36. Show that the system of Problem 2-22 is BIBO stable.

2-37. Is the system of Problem 2-24 BIBO stable? Prove your answer.

2-38. Is the system of Problem 2-20 BIBO stable? Prove your answer.

Section 2-9

2-39. Given the system described by the differential equation

$$\frac{d^2y}{dt^2} + \omega_0^2 y = x(t), \qquad y(0^-) = 0 \quad \text{and} \quad \frac{dy}{dt}\bigg|_{t=0^-} = 0$$

(a) Show that the impulse response of the corresponding system is

$$h(t) = \omega_0^{-1} \sin(\omega_0 t)\, u(t)$$

(b) Show that this second-order differential equation is equivalent to the system of first-order differential equations

$$\frac{dq_1}{dt} = q_2$$

$$\frac{dq_2}{dt} = -\omega_0^2 q_1 + x$$

$$y = q_1$$

(c) Using only integrators, summers, and inverting amplifiers, draw a block diagram that realizes the equations given in part (b).

(d) Using the block diagram of part (c) and noting the impulse response derived in part (a), design an oscillator for the frequency range $100\ \text{Hz} \le \omega_0/2\pi \le 10{,}000\ \text{Hz}$.

2-40. Show that the state equations for the circuit of Figure 2-12 are

$$\frac{dq_1(t)}{dt} = -\frac{1}{RC} q_1(t) - q_2(t) + \frac{1}{C} x(t)$$

$$\frac{dq_2(t)}{dt} = \frac{1}{LC} q_1(t)$$

and that the output equation is $y(t) = q_1(t)$.

Problem Extending Text Material

2-41. To illustrate the impulse response for a time-varying system, consider a boat that is spreading chemicals into a lake. Thus its mass changes according to $M(t) = M_0 - kt$, where M_0 is the initial mass and k is a constant. The water exerts a drag force of $\alpha v(t)$ on the boat. Therefore, the equation of motion is

$$\frac{d}{dt}[Mv(t)] + \alpha v(t) = F(t)$$

or

$$(M_0 - kt)\frac{dv(t)}{dt} + (\alpha - k)v(t) = F(t)$$

where $F(t)$ is the applied force. If this force is a unit impulse applied at time $t = \tau$, show that the impulse response is

$$h(t, \tau) = \frac{(M_0 - kt)^{(\alpha-k)/k}}{(M_0 - k\tau)^{\alpha/k}}, \qquad t > \tau > 0$$

(b) If $M = 10$ kg, $\alpha = 2$, and $k = 1$, find the response of the boat to a step applied at time $t = 0$. (See Schwartz and Friedland, pp. 84–85).

Computer Exercises

2-1. (a) Following the MATLAB program of Example 2-8, write a program to convolve the signals

$$x(t) = \Pi\left[\frac{t - 1}{2}\right]$$

and $\quad h(t) = e^{-0.5t}u(t)$

Plot the two signals $x(t)$ and $h(t)$ along with the result of the convolution, $y(t)$. For this purpose, use the subplot capability of MATLAB. Note that the result of the convolution will have twice as many t-values minus 1 as defined for either input signal, so you will have to use the axis function on the plot for $y(t)$ to make its plot have the same t-range as for $x(t)$ and $h(t)$.

(b) Analytically compute the convolution and compare the exact result with the numerically computed result. Vary the step size for t in the numerical computation and try to make a statement as to what size step size leads to unsatisfactory numerical results.

2-2. Assuming $x(t)$ and $h(t)$ to be the same as Computer Exercise 2-1, where the input and output of a fixed, linear system use Duhamel's integral, (2-85), to find the output of the system after finding the step response from $h(t)$ (recall that you integrate the impulse response to get step response). Compare your result with the result of Computer Exercise 2-1. Experiment with the width of the impulse approximation used for dx/dt to make a judgment on when it is too wide to give satisfactory results.

2-3. Using the MATLAB program developed in Computer Exercise 1-5, find the result of convolving a train of seven unit impulses symmetrically spaced about $t = 0$ at intervals of 2 seconds with a triangle function $\Lambda(t)$. Plot the result and compare with the analytically computed result.

2-4. Use your MATLAB program of Computer Exercise 2-1 to demonstrate the fixed property of a fixed, linear system. You have the result for an input $x(t) = \Pi[(t - 1)/2]$. Now delay $x(t)$ some arbitrary amount, τ, and obtain the convolution of this delayed $x(t)$ with $h(t)$. Show that it is the same result as obtained by delaying the original output (that obtained in Computer Exercise 2-1) the same amount τ.

2-5. Use your MATLAB program of Computer Exercise 2-1 to demonstrate the superposition of property of a fixed, linear system. You have the result for an input $x(t) = \Pi[(t - 1)/2]$. Now rewrite $x(t)$ as the difference of steps and obtain the convolutions of the two steps with $h(t)$. Show that the difference of the two outputs due to the steps is the same result as obtained in Computer Exercise 2-1.

2-6. Write a SIMULINK program to verify the results of Problem 2-39. Note that you can obtain an impulse source in SIMULINK by using a step function source followed by a differentiator. In part (d) you will want to use lower frequencies than specified there for plotting purposes.

2-7. Write a SIMULINK program to verify the result stated in Problem 2-41.

CHAPTER 3

The Fourier Series

3-1 Introduction

After considering some simple signal models in Chapter 1, we turned to systems characterization and analysis in the time domain in Chapter 2. The principal tool developed there was the superposition integral that resulted from resolving the input signal into a continuum of unit impulses and superimposing the elemental system outputs due to each impulsive input—thus the name "superposition integral." The application of the superposition integral involves first finding the impulse response of the system and then carrying out the actual integration—tasks that often are not simple.

In this chapter and in Chapter 4 we consider procedures for resolving certain classes of signals into superpositions of sines and cosines or, equivalently, complex exponential signals of the form $\exp(j\omega t)$. For periodic power signals this resolution results in the Fourier series coefficients of the signal and the resulting representation of such signals, known as the Fourier series of the signal, is considered in this chapter. For energy signals of finite or infinite extent, the Fourier integral provides the desired resolution with the signal representation for such signals given by the inverse Fourier transform integral. This resolution technique is examined in Chapter 4. Signals of the *exponential class*, of which energy and power signals are subclasses, may be resolved with complex exponential signals of the form $\exp[(\sigma + j\omega t)]$ by means of the Laplace transform integral. We deal with the Laplace transform in Chapter 5.

Advantages of Fourier series and Fourier transform representations for signals are twofold. First, in the analysis and design of systems it is often useful to characterize signals in terms of *frequency-domain* parameters such as bandwidth or spectral content. Second, the superposition property of linear systems, and the fact that the steady-state response of a fixed, linear system to a sinusoid of a given frequency is itself a sinusoid of the same frequency, provide a means of solving for the response of such systems. Indeed, the Fourier or Laplace transform, as appropriate, allows one to convert the constant-coefficient linear differential equation representation for a lumped system to an algebraic expression that considerably simplifies the solution for the system output. We consider this method of solution in Chapters 4 through 7.

We begin this chapter with some simple examples to convince ourselves that a sine-cosine series is a useful representation for a periodic signal.

3-2 Trigonometric Series[†]

Recalling Example 1-6, parts (d) and (e), we note that the sum of two sinusoids is periodic provided that their frequencies are commensurable. Stated another way, the sum of two sinusoids is periodic provided that their frequencies are integer multiples of a fundamental frequency.

[†]An alternative approach to Fourier series is provided in Section 3-10 in terms of generalized vector-space concepts.

To examine the fruitfulness of representing periodic signals by sums of sinusoids whose frequencies are harmonics, or integer multiples, of a fundamental frequency, we consider two such series and plot their partial sums.

EXAMPLE 3-1

As a first example, consider the trigonometric series, assumed to hold for all t, given by

$$x(t) = \sin \omega_0 t + \frac{1}{3} \sin 3\omega_0 t + \frac{1}{5} \sin 5\omega_0 t + \cdots \qquad (3\text{-}1)$$

where $2\pi/\omega_0$ is the period. Its partial sums are

$$s_1 = \sin \omega_0 t$$
$$s_2 = \sin \omega_0 t + \frac{1}{3} \sin 3\omega_0 t$$

and so on. Several of them are plotted in Figure 3-1 with $\omega_0 t$ as the independent variable. Stretching our imagination, we see that a square wave appears to be taking shape as more terms are included in the partial sum. As each harmonic is added, the ripple in the flat-top approximation of the square wave becomes smaller and possesses a frequency equal to that of the highest harmonic included in the series.

Somewhat disconcerting, however, is the appearance of "ears" (or overshoot) in the plot of the partial sums where the square wave is discontinuous. This phenomenon, referred to as the Gibbs phenomenon, is examined in more detail in Section 4-10.

The convergence of the series at any particular point may be examined by substituting the appropriate value of $\omega_0 t$. For example, setting $\omega_0 t = \pi/2$ in (3-1), we obtain the alternating series

$$1 - \frac{1}{3} + \frac{1}{5} - \frac{1}{7} + \cdots = \frac{\pi}{4} \qquad (3\text{-}2)$$

which may be checked by consulting a table of sums of series.[‡] We can normalize the series (3-1) to have a value of unity at $\omega_0 t = \pi/2$ by multiplying it by $4/\pi$. Later we will show that a unit-amplitude square wave can be represented by the series

$$x_{sq}(t) = \frac{4}{\pi} \left(\sin \omega_0 t + \frac{1}{3} \sin 3\omega_0 t + \frac{1}{5} \sin 5\omega_0 t + \cdots \right), \qquad -\infty < t < \infty \qquad (3\text{-}3)$$

EXAMPLE 3-2

As a second example, consider the partial sums of the trigonometric series

$$y(t) = \sin \omega_0 t - \frac{1}{2} \sin 2\omega_0 t + \frac{1}{3} \sin 3\omega_0 t \cdots, \qquad -\infty < t < \infty \qquad (3\text{-}4)$$

Plots showing its partial sums are given in Figure 3-2. Again, with a little imagination, we can see the beginnings of a sawtooth waveform. Each term added decreases the ripple in magnitude. The frequency of the ripple is equal to that of the highest-frequency term included in the series. The overshoot phenomenon, present at the discontinuities of the square wave considered previously, is also present with the sawtooth waveform.

Having convinced ourselves of the possibility of building up arbitrary periodic waveforms from sums of harmonically related sinusoidal terms, we turn to the crux of the matter: *Given a specific periodic waveform, how do we find its trigonometric series representation?* The resulting series, which is unique for each periodic signal, is called the *trigonometric Fourier series* of the signal.

[‡]See, for example, *Mathematical Tables from Handbook of Chemistry and Physics* (Boca Raton, Fla.: CRC Press, 1962), p. 325.

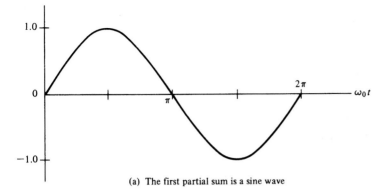

(a) The first partial sum is a sine wave

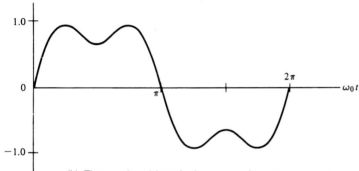

(b) The second partial sum begins to approximate a square wave

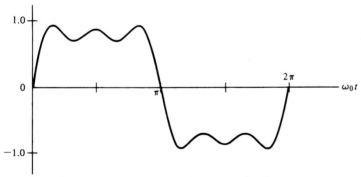

(c) The third partial sum provides a better approximation to a square wave

FIGURE 3-1. The series of Example 3-1 shows how a trigonometric series approximates a square wave (only one period shown).

3-3 Obtaining Trigonometric Fourier Series Representations for Periodic Signals

We begin by writing down the general form that the trigonometric Fourier series representation of a periodic signal may have:

$$
\begin{aligned}
x(t) = a_0 &+ a_1 \cos \omega_0 t + a_2 \cos 2\omega_0 t + \cdots \\
&+ b_1 \sin \omega_0 t + b_2 \sin 2\omega_0 t + \cdots, \qquad -\infty < t < \infty
\end{aligned}
\tag{3-5}
$$

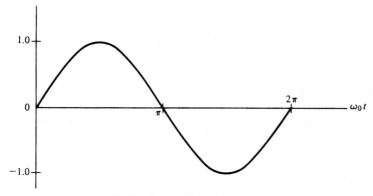

(a) The first partial sum is a sine wave

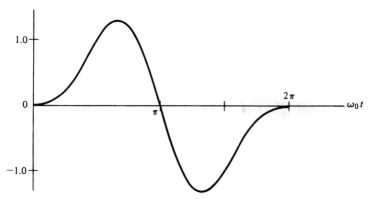

(b) The second partial sum begins to approximate a sawtooth waveform

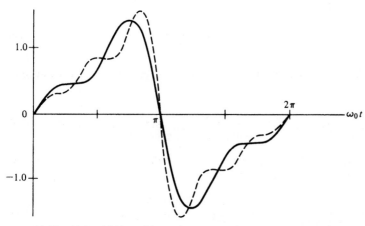

(c) The third and fifth partial sums better approximate a sawtooth waveform

FIGURE 3-2. The trigonometric series of Example 3-2 illustrates approximation of a sawtooth waveform (only one period shown).

This may be written compactly in terms of summations as

$$x(t) = a_0 + \sum_{n=1}^{\infty} a_n \cos n\omega_0 t + \sum_{n=1}^{\infty} b_n \sin n\omega_0 t \tag{3-6}$$

where $-\infty < t < \infty$.[†] Since the right-hand side is a sum of harmonically related sinusoids, which themselves are periodic functions, the left-hand side is periodic. The problem that faces us is to find $a_0, a_1, a_2, \ldots, b_1, b_2, \ldots$ for a given $x(t)$.

To obtain relationships for finding the a_n's and b_n's for an arbitrary (periodic) $x(t)$ we begin with a_0. Integrating the series (3-5) term by term over one period of $x(t)$, we obtain

$$\int_{T_0} x(t) \, dt = a_0 \int_{T_0} dt + a_1 \int_{T_0} \cos \omega_0 t \, dt + a_2 \int_{T_0} \cos 2\omega_0 t \, dt + \cdots$$

$$+ b_1 \int_{T_0} \sin \omega_0 t \, dt + b_2 \int_{T_0} \sin 2\omega_0 t \, dt + \cdots \tag{3-7}$$

where $\int_{T_0} (\cdot) \, dt$ denotes integration over any period. All terms except the first involve the integration of a sine or cosine over an integral number of periods and are therefore zero. (For $\sin n\omega_0 t$ or $\cos n\omega_0 t$, as much area appears above the t-axis as below it in one period.) The first term yields $a_0 T_0$. Thus the coefficient a_0 is given by

$$a_0 = \frac{1}{T_0} \int_{T_0} x(t) \, dt \tag{3-8}$$

which is the average value of the waveform.

Using the series form (3-6) we derive a general expression valid for any of the a_n's. The derivation proceeds as follows for any a_n except a_0. Multiplying both sides of (3-6) by $\cos m\omega_0 t$ (the use of m allows us to choose any a_n coefficient), and integrating over a period of $x(t)$, we obtain

$$\int_{T_0} x(t) \cos m\omega_0 t \, dt = a_0 \int_{T_0} \cos m\omega_0 t \, dt + \int_{T_0} \left(\sum_{n=1}^{\infty} a_n \cos n\omega_0 t \right) \cos m\omega_0 t \, dt$$

$$+ \int_{T_0} \left(\sum_{n=1}^{\infty} b_n \sin n\omega_0 t \right) \cos m\omega_0 t \, dt \tag{3-9}$$

The first term integrates to zero. We multiply each term of the series in parentheses by $\cos m\omega_0 t$ and integrate term-by-term to obtain

$$\int_{T_0} x(t) \cos m\omega_0 t \, dt = \sum_{n=1}^{\infty} a_n \int_{T_0} \cos n\omega_0 t \cos m\omega_0 t \, dt + \sum_{n=1}^{\infty} b_n \int_{T_0} \sin n\omega_0 t \cos m\omega_0 t \, dt \tag{3-10}$$

To continue, we note some rather interesting properties of integrals involving products of sines and cosines. Three possible cases occur. The resulting integrals are

$$I_1 = \int_{T_0} \sin m\omega_0 t \sin n\omega_0 t \, dt = \begin{cases} 0, & m \neq n \\ T_0/2, & m = n \neq 0 \end{cases} \tag{3-11}$$

$$I_2 = \int_{T_0} \cos m\omega_0 t \cos n\omega_0 t \, dt = \begin{cases} 0, & m \neq n \\ T_0/2, & m = n \neq 0 \end{cases} \tag{3-12}$$

[†]Since most signals in this chapter are defined over the entire t-axis, we dispense with giving the range of definition unless it is other than $-\infty < t < \infty$.

and

$$I_3 = \int_{T_0} \sin m\omega_0 t \cos n\omega_0 t \, dt = 0, \qquad \text{all } m, n \tag{3-13}$$

where $T_0 = 2\pi/\omega_0$ is a period of the fundamental and n and m are integers.[†] We note that with $n = 0$ in (3-12) and (3-13), we have the integral of $\cos m\omega_0 t$ and $\sin m\omega_0 t$, respectively. We also note that the product of two sinusoids with harmonically related frequencies is periodic, so that the integrals could be taken over *any* period.

Applying (3-13), we see that each term of the second series on the right-hand side of (3-10) is zero. Applying (3-12), we see that all terms of the first series on the right-hand side of (3-10) are also zero except for the term for which $n = m$. For $n = m$, the integral gives $T_0/2$, which yields

$$\int_{T_0} x(t) \cos m\omega_0 t \, dt = a_m\left(\frac{T_0}{2}\right) \tag{3-14}$$

Multiplying both sides of (3-14) by $2/T_0$ gives

$$a_m = \frac{2}{T_0} \int_{T_0} x(t) \cos m\omega_0 t \, dt, \qquad m \neq 0 \tag{3-15}$$

In a similar manner, by employing (3-6), (3-11), and (3-13), we may show that

$$b_m = \frac{2}{T_0} \int_{T_0} x(t) \sin m\omega_0 t \, dt \tag{3-16}$$

Without concerning ourselves about the validity of the steps in obtaining (3-8), (3-15), and (3-16), we *define* a Fourier series to be any trigonometric series of the form (3-6), where the coefficients are found according to the formulas just derived. To find the coefficients, we require only that the integrals involved exist.[‡]

[†]To show (3-11) through (3-13), trigonometric identities could be used. However, Euler's theorem gives a more easily remembered derivation. Since, by Euler's theorem, we have

$$e^{\pm jn\omega_0 t} = \cos n\omega_0 t \pm j \sin n\omega_0 t$$

we can add and subtract the expressions for $e^{jn\omega_0 t}$ and $e^{-jn\omega_0 t}$ to obtain

$$\sin n\omega_0 t = \frac{1}{2j} (e^{jn\omega_0 t} - e^{-jn\omega_0 t}) \qquad \text{and} \qquad \cos n\omega_0 t = \frac{1}{2} (e^{jn\omega_0 t} + e^{-jn\omega_0 t})$$

(The student should memorize these expressions as well as Euler's theorem.) Substituting for $\sin n\omega_0 t$ in (3-11), and using the exponential form for sine above, we obtain

$$I_1 = \int_0^{T_0} \frac{1}{2j} (e^{jm\omega_0 t} - e^{-jm\omega_0 t}) \frac{1}{2j} (e^{jn\omega_0 t} - e^{-jn\omega_0 t}) \, dt$$

$$= -\frac{1}{4} \int_0^{T_0} [e^{j(m+n)\omega_0 t} - e^{j(n-m)\omega_0 t} - e^{j(m-n)\omega_0 t} + e^{-j(n+m)\omega_0 t}] \, dt$$

Now $\int_0^{T_0} e^{jk\omega_0 t} \, dt = 0$ for k an integer not equal to zero by Euler's theorem (the integral of a sine or cosine over an integer number of periods is zero). If $k = 0$, we have $\int_0^{T_0} dt = T_0$. Using this in the expression for I_1, we have

$$I_1 = \begin{cases} 0, & n \neq m \\ T_0/2 & n = m \end{cases}$$

where the nonzero result for $n = m$ results from the second and third terms in the integrand for I_1. A similar series of steps can be used to prove (3-12) and (3-13).

[‡]We give only an informal discussion of convergence here. Conditions concerning the possibility of expanding a function in a Fourier series are given in a theorem proved by Dirichlet in 1829. These sufficient conditions are that $x(t)$ be defined and bounded on the range $(t_0, t_0 + T_0)$ and have only a finite number of maxima and minima and a finite number of discontinuities on this range. These conditions, being *sufficient conditions,* are more restrictive than necessary. Fejér, in 1904, published two theorems giving conditions for the convergence of Fourier series that are less restrictive than the Dirichlet conditions.

In going from (3-9) to (3-10) it was necessary to interchange the order of summation and integration. We did not worry about the validity of this step. Indeed, the whole development so far has been based on the assumption that (3-6) is truly an equality; that is, an arbitrary periodic function can be represented in terms of a sum of sines and cosines. There is no reason to believe that this is always the case. However, we noted in the previous paragraph that the coefficients a_n and b_n will exist if $x(t)$ is integrable. We can therefore find the coefficients for an appropriate $x(t)$ and then examine the convergence properties of the series formed with these coefficients. It is too much to expect that the Fourier series will converge to a periodic, but otherwise arbitrary, $x(t)$ for every t. For example, in Example 3-1 it is apparent that each partial sum is zero at $\omega_0 t = n\pi$, where n is an integer. If $x(t)$ has been defined to be $\pi/4$ at these points (or any other desired value), the series cannot represent $x(t)$ at $\omega_0 t = n\pi$. As another example, consider two periodic signals $x(t)$ and $y(t)$ which are identical except at a single point, where they differ by a finite amount. This finite difference will not affect the value of the integrals for the coefficients given by (3-15) and (3-16). Therefore, both $x(t)$ and $y(t)$ have identical Fourier series, and we conclude that the Fourier series cannot represent both functions at the point in question. In general, functions that are equal "almost everywhere"—that is, everywhere except at a number of isolated points—will have identical Fourier series representations. Fourier series, therefore, do not converge in a *pointwise* sense.

The question remains as to what conditions are required to ensure that a Fourier series *will* converge to $x(t)$. Indeed, very little is required to guarantee convergence. It turns out that if $x(t)$ has period equal to T_0, and has continuous first and second derivatives for all t, *except possibly at a finite number of points where it may have finite jump discontinuities,* the Fourier series for $x(t)$ converges uniformly to $x(t)$ for every t except at the points of discontinuity.[‡] Even at points of discontinuity, the behavior of a Fourier series is very reasonable. The limit of the Fourier series sum at a point of discontinuity is simply the average of the left- and right-hand limits. This is illustrated in Figure 3-3.

Although we have discussed only periodic functions so far, it should be noted that the basic coefficient formulas *use* only the values of $x(t)$ in a T_0-second interval. Thus, if $x(t)$ is given only in some interval of length T_0, the corresponding Fourier series can be formed and converges uniformly to $x(t)$ within this interval at all points of continuity. Outside this interval, the Fourier series converges to a signal which is the *periodic extension of $x(t)$*. This is illustrated in Figure 3-4.

We close this section with an example that illustrates the representation of a function over a finite interval by a Fourier series and an example that involves computation of the Fourier series coefficients for a square wave.

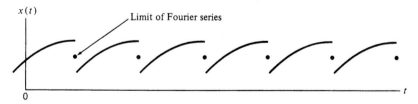

FIGURE 3-3.   The Fourier series of a waveform converges to the mean of the left- and right-hand limits at a point of discontinuity.

[‡]By uniform convergence of an infinite series whose terms are functions of a variable, it is meant that the numerical value of the remainder after the first n terms is as small as desired *throughout the given interval* for n greater than a sufficiently large chosen number.

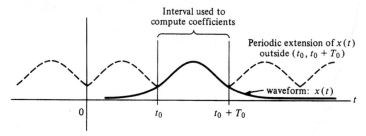

FIGURE 3-4.   Expansion of a nonperiodic signal in terms of a Fourier series.

EXAMPLE 3-3

A violin string, to be plucked by a musician, has the initial shape shown in Figure 3-5. (a) Given that the trigonometric Fourier series coefficients of such a triangular wave of amplitude A are given by

$$a_0 = A/2,$$

$$a_n = 4A/\pi^2 n^2, \qquad n \neq 0 \text{ and odd}$$

$$a_n = 0, \qquad n \text{ even} \tag{3-17}$$

$$b_n = 0, \qquad \text{all } n$$

write out the first four nonzero terms of the Fourier series describing the initial shape of the violin string in the interval $|x| \leq 9$ inches. (b) Sketch the function to which the Fourier series converges for $-\infty < x < \infty$.

Solution:

(a) Note that the independent variable is x, not t. Assume that the $y(x)$ shown is to be represented by one full period of the Fourier series for a triangular waveform. Since the b_n's are zero (even function), the Fourier series is of the form

$$y(x) = a_0 + \sum_{n=1}^{\infty} a_n \cos n\omega_0 x \tag{3-18}$$

The fundamental frequency in rad/inch of this series is $\omega_0 = 2\pi/T_0 = 2\pi/18 = \pi/9$. The Fourier series coefficients are

$$a_0 = 0.5/2 = 1/4$$

$$a_1 = 4(0.5)/\pi^2 = 2/\pi^2$$

$$a_2 = 0$$

$$a_3 = 4(0.5)/9\pi^2 = 2/9\pi^2 \tag{3-19}$$

$$a_4 = 0$$

$$a_5 = 4(0.5)/25\pi^2 = 2/25\pi^2$$

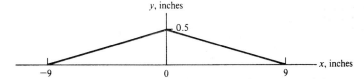

FIGURE 3-5. Initial position of a violin string to be represented by a Fourier series.

Thus, in the interval $|x| \leq 9$ inches, the approximating series is

$$\hat{y}(x) = \frac{1}{4} + \frac{2}{\pi^2} \cos \frac{\pi x}{9} + \frac{2}{9\pi^2} \cos \frac{\pi x}{3} + \frac{2}{25\pi^2} \cos \frac{5\pi x}{9} + \cdots \tag{3-20}$$

where the circumflex indicates that this is an approximation to the true initial shape of the string. Figure 3-6 shows the series approximation to the initial shape of the string for the first two terms only, and all four terms of the approximating series. Only the right half of the string and approximating series are shown, since the left half is the mirror image of the right half. Also shown is the error in the approximation, which is the difference between the actual string position and the approximating series. Note that the error is largest for $x = 9$ inches (i.e., at the end of the string). For a two-term approximation, the error is about 10% of the maximum deflection, while for a four-term approximation, the error is about 3.5% of the maximum deflection.

(b) The actual Fourier series converges to a triangular waveform infinite in extent.

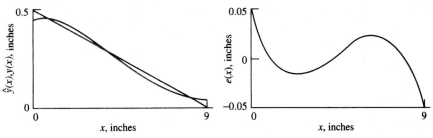

(a) A two-term approximation to the string initial position (left) and the error (right) as a function of distance

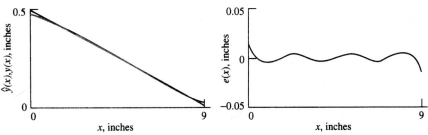

(b) A four-term approximation to the string initial position (left) and the error (right) as a function of distance

FIGURE 3-6. Fourier series representations and the error for the initial position of the violin string. (Only the right half is shown due to symmetry.)

EXAMPLE 3-4 _____

Consider the square wave defined by

$$x(t) = \begin{cases} A, & 0 < t < \dfrac{T_0}{2} \\ -A, & \dfrac{T_0}{2} < t < T_0 \end{cases} \tag{3-21}$$

and periodically extended outside this interval. The average value is zero, so $a_0 = 0$. From (3-15), the a_m's are

$$
\begin{aligned}
a_m &= \frac{2}{T_0} \int_0^{T_0/2} A \cos m\omega_0 t \, dt + \frac{2}{T_0} \int_{T_0/2}^{T_0} (-A) \cos m\omega_0 t \, dt \\
&= \frac{2A}{T_0} \left[\frac{\sin m\omega_0 t}{m\omega_0} \bigg|_0^{T_0/2} - \frac{\sin m\omega_0 t}{m\omega_0} \bigg|_{T_0/2}^{T_0} \right]
\end{aligned} \tag{3-22}
$$

Recalling that $\omega_0 = 2\pi/T_0$, we see that

$$\sin \frac{m\omega_0 T_0}{2} = \sin m\pi = 0$$

and

$$\sin m\omega_0 T_0 = \sin 2m\pi = 0$$

Thus all the a_m coefficients are zero. The b_m coefficients result by applying (3-16). This gives

$$
\begin{aligned}
b_m &= \frac{2}{T_0} \int_0^{T_0/2} A \sin m\omega_0 t \, dt + \frac{2}{T_0} \int_{T_0/2}^{T_0} (-A) \sin m\omega_0 t \, dt \\
&= \frac{2A}{T_0} \left[-\frac{\cos m\omega_0 t}{m\omega_0} \bigg|_0^{T_0/2} + \frac{\cos m\omega_0 t}{m\omega_0} \bigg|_{T_0/2}^{T_0} \right]
\end{aligned} \tag{3-23}
$$

Again, recalling that $\omega_0 = 2\pi/T_0$, we find upon substitution of the limits that

$$b_m = \frac{2A}{m\pi} (1 - \cos m\pi)$$

which results because $\cos 2m\pi = 1$ for all m. Finally, noting that $\cos m\pi = -1$ for m odd and $\cos m\pi = 1$ for m even, we obtain

$$b_m = \begin{cases} \dfrac{4A}{m\pi}, & m \text{ odd} \\ 0, & m \text{ even} \end{cases} \tag{3-24}$$

The Fourier series of a square wave of amplitude A with odd symmetry is therefore

$$x(t) = \frac{4A}{\pi} \left(\sin \omega_0 t + \tfrac{1}{3} \sin 3\omega_0 t + \tfrac{1}{5} \sin 5\omega_0 t + \cdots \right) \tag{3-25}$$

We note that only odd harmonic terms are present. This is a consequence of $x(t)$ having a special type of symmetry, referred to as half-wave odd symmetry. More will be said about special symmetry cases later. If $A = 1$, (3-25) becomes the series first shown in (3-3) and verifies the previous assertion that (3-3) represents a unity amplitude square wave.

MATLAB *Application*

It is interesting to examine the convergence of a Fourier series using MATLAB. The program given below accomplishes this for the square wave considered in this example.

```
%      Program to give partial Fourier sums of an
%      odd square wave of unit amplitude
%
n_max=input('Enter vector of highest harmonic values desired (odd)');
N=length(n_max);
t=0:.002:1;
omega_0=2*pi;
for k=1:N
       n=[];
       n=[1:2:n_max(k)];
       b_n=4./(pi*n);           % Form vector of Fourier sine-coefficients
       x=b_n*sin(omega_0*n'*t); % Rows of sine matrix are versus time;
                                % columns are versus n,
                                % so matrix multiply sums over n and a
                                % vector for x(t) results
       subplot(N,1,k),plot(t,x), xlabel('t'), ylabel('partial sum'),...
       axis([0 1 -1.5 1.5)], text(.05,-5, ['max. har.=',
       num2str(n_max(k))])
end
```

Note that matrix multiplication is used to form the partial sums of the Fourier series. This saves considerable computation time over using a for loop because MATLAB is an interpretive language which compiles each statement separately. The for loop simply runs through different cases for the maximum number of harmonics summed. Note that any number of cases for the maximum number of harmonics can be run, but the plots would get rather small for more than three or four.

In this case, the program is not run with echo on because one would see it go through the for loop three times and this would be rather boring and long. The entry in the command window is simply

```
EDU»c3ex4
Enter vector of highest harmonic values desired (odd) [11 21 31]
```

where the second line comes up after the program begins running and the [11 21 31] is entered by the user to specify the three cases to be run for the maximum harmonic. If four cases are desired, this would be a 1 by 4 vector. The resulting plots are shown in Figure 3-7. Note that the number of humps on the flat part of the waveform is determined by the number of harmonics included in the partial sum, and also note the presence of ears at the discontinuity points of the square wave. The latter are known as the Gibbs phenomenon, to be discussed in the Chapter 4, Section 4-10.

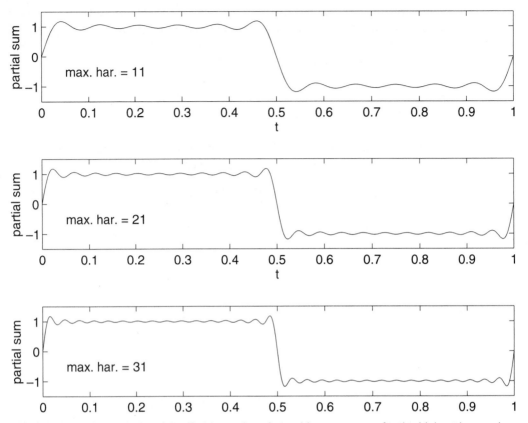

FIGURE 3-7.   Partial sums of the Fourier series of an odd square wave for the highest harmonic being 11, 21, and 31, top to bottom.

3-4 The Complex Exponential Fourier Series

Another form of the Fourier series that involves complex exponential functions can be obtained by substituting the complex exponential forms of $\sin \omega_0 t$ and $\cos \omega_0 t$ into (3-6). Thus, letting

$$\sin n\omega_0 t = \frac{e^{jn\omega_0 t} - e^{-jn\omega_0 t}}{2j} \tag{3-26a}$$

and

$$\cos n\omega_0 t = \frac{e^{jn\omega_0 t} + e^{-jn\omega_0 t}}{2} \tag{3-26b}$$

we obtain from (3-6) the series

$$
\begin{aligned}
x(t) &= a_0 + \sum_{n=1}^{\infty} a_n \frac{e^{jn\omega_0 t} + e^{-jn\omega_0 t}}{2} + \sum_{n=1}^{\infty} b_n \frac{e^{jn\omega_0 t} - e^{-jn\omega_0 t}}{2j} \\
&= a_0 + \sum_{n=1}^{\infty} \frac{1}{2}(a_n - jb_n)e^{jn\omega_0 t} + \sum_{n=1}^{\infty} \frac{1}{2}(a_n + jb_n)e^{-jn\omega_0 t}
\end{aligned}
\tag{3-27}
$$

This is known as the complex exponential form of the Fourier series.

We see that terms involving both $\exp(jn\omega_0 t)$ and $\exp(-jn\omega_0 t)$ are present in the series. This is in keeping with our observation in Chapter 1 that the sum of two complex conjugate rotating phasors is required to produce a real, sinusoidal signal. To generalize the derivation of the coefficients in the complex exponential Fourier series, we write the series in the form

$$x(t) = \cdots X_{-2}e^{-j2\omega_0 t} + X_{-1}e^{-j\omega_0 t} + X_0 + X_1 e^{j\omega_0 t} + X_2 e^{j2\omega_0 t} + \cdots$$

$$= \sum_{n=-\infty}^{\infty} X_n e^{jn\omega_0 t} \tag{3-28}$$

where the X_n's are, in general, complex constants. To find an expression for computing the X_n's, we multiply both sides of (3-28) by $e^{-jm\omega_0 t}$ and integrate over any period of $x(t)$. This results in

$$\int_{T_0} x(t)e^{-jm\omega_0 t}\, dt = \int_{T_0} \left(\sum_{n=-\infty}^{\infty} X_n e^{jn\omega_0 t} \right) e^{-jm\omega_0 t}\, dt \tag{3-29}$$

We next multiply each term in the sum by $e^{-jm\omega_0 t}\, dt$ and integrate the result term by term without worrying about the validity of these steps at this point.[†] This results in

$$\int_{T_0} x(t)e^{-jm\omega_0 t}\, dt = \sum_{n=-\infty}^{\infty} X_n \int_{T_0} e^{j(n-m)\omega_0 t}\, dt \tag{3-30}$$

If $m \neq n$, the integral

$$\int_{T_0} e^{j(n-m)\omega_0 t}\, dt$$

is zero because

$$e^{j(n-m)\omega_0 t} = \cos(n-m)\omega_0 t + j \sin(n-m)\omega_0 t$$

is a periodic function with period T_0 and is symmetric about the t-axis. For $n = m$, $e^{j(n-m)\omega_0 t} = 1$, and the integral evaluates to T_0. Thus every term in the sum on the right-hand side of (3-30) is zero except the one for $n = m$; (3-30) therefore reduces to

$$\int_{T_0} x(t)e^{-jm\omega_0 t}\, dt = T_0 X_m$$

or, solving for X_m, we obtain

$$X_m = \frac{1}{T_0} \int_{T_0} x(t)e^{-jm\omega_0 t}\, dt \tag{3-31}$$

As with the trigonometric Fourier series, we *define* an exponential Fourier series to be a series of the form (3-28) with the coefficients calculated according to (3-31). The computation of the coefficients is usually considerably simpler than the computation of the trigonometric Fourier series coefficients. In addition, the exponential Fourier series provides us with a direct means of plotting the two-sided amplitude and phase spectra of a signal.

We note, by comparing (3-27) and (3-28), that the coefficients of the trigonometric Fourier series and the complex coefficients (3-31) are related by

$$X_n = \begin{cases} \frac{1}{2}(a_n - jb_n), & n > 0 \\ \frac{1}{2}(a_{-n} + jb_{-n}), & n < 0 \end{cases} \tag{3-32a}$$

[†]A discussion of convergence for this form of the series would be identical to the one given for the trigonometric form.

and

$$X_0 = a_0 \tag{3-32b}$$

This section will be concluded with two examples to illustrate the calculation of complex Fourier series coefficients.

EXAMPLE 3-5

Consider the periodic sequence of asymmetrical pulses (referred to as a pulse train)

$$x(t) = \sum_{m=-\infty}^{\infty} A\Pi\left(\frac{t - t_0 - mT_0}{\tau}\right), \qquad \tau < T_0 \tag{3-33}$$

whose period is T_0. The exponential Fourier series coefficients are calculated by evaluating the integral

$$X_n = \frac{1}{T_0} \int_{t_0-\tau/2}^{t_0+\tau/2} A e^{-jn\omega_0 t} \, dt \tag{3-34}$$

where $\omega_0 = 2\pi/T_0$. The evaluation is straightforward. Integration gives

$$X_n = \frac{-A}{jn\omega_0 T_0} e^{-jn\omega_0 t} \Big|_{t_0-\tau/2}^{t_0+\tau/2}$$

$$= \frac{2A}{n\omega_0 T_0} e^{-jn\omega_0 t_0} \left(\frac{e^{jn\omega_0\tau/2} - e^{-jn\omega_0\tau/2}}{2j}\right), \qquad n \neq 0$$

$$= \frac{2A}{n\omega_0 T_0} e^{-jn\omega_0 t_0} \sin\frac{n\omega_0\tau}{2}, \qquad n \neq 0 \tag{3-35a}$$

Letting $n = 0$ in (3-34), we find X_0 to be

$$X_0 = \frac{A\tau}{T_0} \tag{3-35b}$$

Substituting $\omega_0 = 2\pi f_0$ into (3-35a), where f_0 is the fundamental frequency in hertz, we obtain the form

$$X_n = \frac{A\tau}{T_0} e^{-j2\pi n f_0 t_0} \frac{\sin \pi n f_0 \tau}{\pi n f_0 \tau} \tag{3-35c}$$

where the numerator and denominator have been multiplied by τ.

It is convenient to define

$$\text{sinc } z = \frac{\sin \pi z}{\pi z} \tag{3-36}$$

which is referred to as the *sinc function*. Values for sinc z and sinc^2 z are tabulated in Appendix F. Examining (3-36), we see that the sinc function is zero whenever

$$\sin \pi z = 0$$

This happens for $z = \pm1, \pm2, \pm3, \ldots$. Furthermore, the sinc function is even (substitute $-z$ for z and obtain the same function) and has a maximum of unity at $z = 0$. The latter may be shown by using L'Hospital's rule. Also, sinc z is oscillatory with the amplitude of the oscillations decreasing as $|z| \to \infty$. Thus sketches of sinc z and sinc^2 z appear as shown in Figure 3-8.

It is straightforward to express (3-35c) in terms of the sinc function. Table 3-1 summarizes the Fourier coefficients for the pulse train as well as those for several other waveforms of interest.

The next example illustrates the derivation of the coefficients of the complex exponential Fourier series for a half-rectified sine wave.

EXAMPLE 3-6

Find the coefficients of the complex exponential Fourier series for a half-rectified sine wave, defined by

$$x(t) = \begin{cases} A \sin \omega_0 t, & 0 \le t \le T_0/2 \\ 0, & T_0/2 \le t \le T_0 \end{cases} \tag{3-37}$$

with $x(t) = x(t + T_0)$.

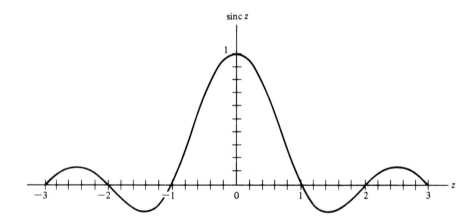

sinc z

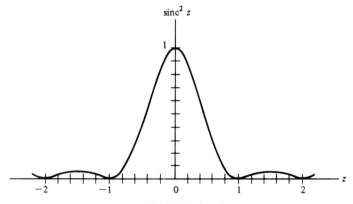

sinc^2 z

FIGURE 3-8. Sinc z and sinc^2 z.

TABLE 3-1
Coefficients for the Complex Exponential Fourier Series of Several Signals

1. Half-rectified sine
wave

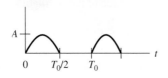

$$X_n = \begin{cases} \dfrac{A}{\pi(1-n^2)}, & n = 0, \pm 2, \pm 4, \ldots \\ 0, & n \text{ odd and} \neq \pm 1 \\ -\dfrac{1}{4}jnA, & n = \pm 1 \end{cases}$$

2. Full-rectified sine
wave*

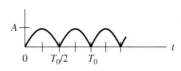

$$X_n = \begin{cases} \dfrac{2A}{\pi(1-n^2)}, & n \text{ even} \\ 0 & n \text{ odd} \end{cases}$$

3. Pulse-train signal

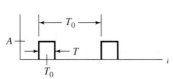

$$X_n = \frac{A\tau}{T_0}\, \text{sinc}\, nf_0\tau\, e^{-j2\pi nf_0t_0}, \quad f_0 = T_0^{-1}$$

4. Square wave

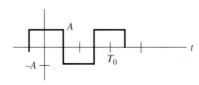

$$X_n = \begin{cases} \dfrac{2A}{|n|\pi}, & n = \pm 1, \pm 5, \ldots \\ \dfrac{-2A}{|n|\pi}, & n = \pm 3, \pm 7, \ldots \\ 0, & n \text{ even} \end{cases}$$

5. Triangular wave

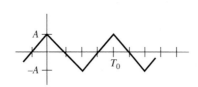

$$X_n = \begin{cases} \dfrac{4A}{\pi^2 n^2}, & n \text{ odd} \\ 0, & n \text{ even} \end{cases}$$

*Note that T_0 is *not* the period of the full-wave rectified sinewave. Rather, its period is $T_0/2$.

Solution: From (3-31), the Fourier coefficients are given by the integral

$$X_n = \frac{1}{T_0}\int_0^{T_0/2} A \sin \omega_0 t\, e^{-jn\omega_0 t}\, dt$$

$$= \frac{A}{2jT_0}\left[\int_0^{T_0/2}(e^{j\omega_0 t} - e^{-j\omega_0 t})e^{-jn\omega_0 t}\, dt\right]$$

$$= \frac{A}{2jT_0}\left[\int_0^{T_0/2}e^{j\omega_0(1-n)t}\, dt - \int_0^{T_0/2}e^{-j\omega_0(1+n)t}\, dt\right]$$

$$= -\frac{A}{4\pi}\left[\frac{e^{j(1-n)\pi}-1}{1-n} + \frac{e^{j(1+n)\pi}-1}{1+n}\right], \quad n \neq 1 \text{ or } -1 \tag{3-38}$$

where use has been made of the complex exponential form for $\sin \omega_0 t$ and $\omega_0 = 2\pi/T_0$. We note that $e^{j(1\pm n)\pi} = \cos(1\pm n)\pi + j\sin(1\pm n)\pi = -(-1)^n$, where n is an integer. Thus, (3-38) simplifies to

$$X_n = 0, \qquad n \text{ odd}, \quad n \neq \pm 1$$

and

$$X_n = \frac{A}{\pi} \frac{1}{1 - n^2}, \qquad n \text{ even} \tag{3-39}$$

The special cases for $n = 1$ and $n = -1$ must be handled separately. For $n = 1$, the calculation is

$$X_1 = \frac{A}{2jT_0} \int_0^{T_0/2} (e^{j\omega_0 t} - e^{-j\omega_0 t})e^{-j\omega_0 t} \, dt$$

$$= \frac{A}{2jT_0} \int_0^{T_0/2} (1 - e^{-j2\omega_0 t}) \, dt$$

$$= \frac{A}{4j} \tag{3-40}$$

For $n = -1$, a similar integration gives $X_1 = -A/4j$. Putting all these results together, we see that the first entry in Table 3-1 is correct. A similar series of integrations would be carried out to prove the second entry for the full-rectified sine wave.

MATLAB *Application*

To show how fast the partial sums of the Fourier series approaches a half-rectified sinewave, we provide a MATLAB program below for computing the complex exponential Fourier series.

```
%       Program to give partial complex exponential Fourier sums of a
%       half rectified sine wave of unit amplitude
%
n_max=input('Enter vector of highest harmonic values desired (even)');
N=length(n_max);
t=0:.005:1;
omega_0=2*pi;
for k=1:N
        n=[];
        n=[-n_max(k):n_max(k)];
        L_n=length(n);
        X_n=zeros(1, L_n);          % Form vector of Fourier sine-
                                    % coefficients; all
        X_n((L_n+1)/2+1)=-0.25*j;   % odd-order terms are zero except X(1)
                                    % and X(-1)
        X_n((L_n+1)/2-1)=0.25*j;    % so define coefficient array as a zero
                                    % array and
        for i=1:2:L_n               % then fill in nonzero values
            X_n(i)=1/(pi*(1-n(i)^2));
        end
        x=X_n*exp(j*omega_0*n'*t); % Rows of exponential matrix are versus
                                    % time;
                                    % columns are versus n; matrix multiply
                                    % sums over n
%       Plot real part of x to get rid of small imaginary part due to
%       computational error
```

```
        subplot(N,1,k),plot(t,real(x)), xlabel('t'), ylabel('partial
        sum'),...
        axis([0 1 -0.5 1.5]), text(.05,-.25, ['max. har.=',
        num2str(n_max(k))])
end
```

As with the MATLAB program for Example 3-4, a vector of values can be entered for the highest harmonic present in the sum. The command window appears as follows:

```
EDU»c3ex6
Enter vector of highest harmonic values desired (even) [4 8 12]
```

The resulting plot is shown in Figure 3-9. Note that very few terms are required to make the resulting plot look very much like a half-rectified sine wave. This is because there are no discontinuities present in this waveform.

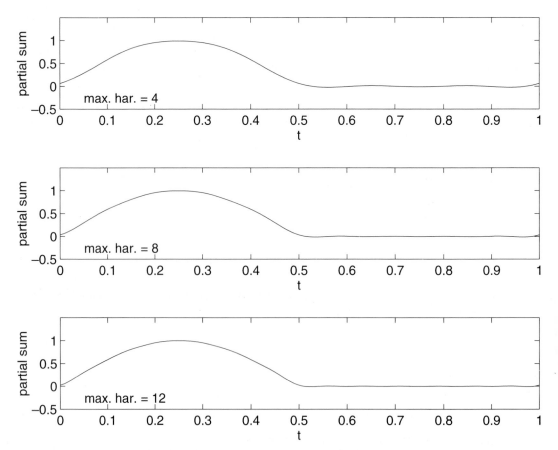

FIGURE 3-9. Partial sums of the complex exponential Fourier series of a half-rectified sine wave. Maximum harmonic is 4, 8, and 12, top to bottom.

EXAMPLE 3-7

Suppose we desire the complex exponential Fourier series for an odd square wave, whereas Table 3-1 has the Fourier series coefficients only for an even square wave. In other words, we want to know the change in the Fourier coefficients due to a time shift in the waveform, in particular, a shift of $T_0/4$ in this case. Consider the general case first. Let the Fourier coefficients for the unshifted waveform be denoted as X_n and those for the time-shifted waveform be denoted as Y_n. Thus

$$X_n = \frac{1}{T_0} \int_{T_0} x(t) e^{-jn\omega_0 t}\, dt$$

and

$$Y_n = \frac{1}{T_0} \int_{T_0} x(t - \tau_0) e^{-jn\omega_0 t}\, dt, \quad \omega_0 = \frac{2\pi}{T_0} \tag{3-41}$$

Make the change of variables in the second integral $t' = t - \tau_0$ to obtain

$$Y_n = \left[\frac{1}{T_0} \int_{T_0} x(t') e^{-jn\omega_0 t'}\, dt \right] e^{-jn2\pi\tau_0/T_0} = X_n e^{-j2\pi n\tau_0/T_0} \tag{3-42}$$

where we remember that the integration can be over any T_0-second interval.

For the case of going from an even to an odd square wave, $\tau_0 = T_0/4$ so that (3-42) becomes

$$X_{n,\, odd} = X_{n,\, even} e^{-j\pi n/2} \ (n \text{ odd}) = \begin{cases} -jX_{n,\, odd}, & n = \pm 1, \pm 5, \cdots \\ jX_{n,\, odd}, & n = \pm 3, \pm 7, \cdots \end{cases} \tag{3-43}$$

where $X_{n,\, odd}$ are the Fourier coefficients for the square wave given in Table 3-1, item 4, and only n odd need be considered because the even-order Fourier coefficients of a square wave are zero.

One of our objectives in this chapter is to be able to plot amplitude and phase spectra for periodic signals. Before doing this, however, we discuss symmetry properties of the Fourier coefficients. This will simplify the plotting of spectra.

3-5 Symmetry Properties of the Fourier Series Coefficients

The expression for the coefficients of the complex exponential Fourier series (3-31), through use of Euler's theorem, can be written as

$$X_m = \frac{1}{T_0} \int_{T_0} x(t) \cos m\omega_0 t\, dt - \frac{j}{T_0} \int_{T_0} x(t) \sin m\omega_0 t\, dt \tag{3-44}$$

If $x(t)$ is real, the first term is the real part of X_m and the second term is the imaginary part of X_m. Comparing this with (3-15) and (3-16), we see that

$$X_m = \begin{cases} \frac{1}{2}(a_m - jb_m), & m > 0 \\ \frac{1}{2}(a_{-m} + jb_{-m}), & m < 0 \end{cases} \tag{3-45}$$

which is another way to obtain (3-32a).

Solving (3-45) for a_m and b_m, we obtain

$$a_m = 2\, \text{Re}\, X_m \quad \text{and} \quad b_m = -2\, \text{Im}\, X_m, \quad m > 0 \tag{3-46}$$

If we replace m by $-m$ in (3-44) or (3-45), it is clear that

$$X_m = X_{-m}^* \tag{3-47}$$

for $x(t)$ real. Writing X_m in polar form as

$$X_m = |X_m|e^{j\theta_m} \tag{3-48}$$

we conclude from (3-44) that

$$|X_m| = |X_{-m}| \quad \text{and} \quad \theta_m = -\theta_{-m} \tag{3-49}$$

That is, for real signals, the magnitude of the Fourier coefficients is an even function of the index, m, and the argument is an odd function of m.

Considering the Fourier coefficients for a real, even signal, that is $x(t) = x(-t)$, we see from (3-44) that the imaginary part will be zero since $x(t) \sin m\omega_0 t$ is an odd function that integrates to zero over an interval symmetrically placed about $t = 0$. That is, the complex exponential Fourier series coefficients X_m of a real, even signal are real (see Example 3-5 with $t_0 = 0$). Furthermore, since the dependence on m is through the function $\cos m\omega_0 t$, they are also even functions of m. From (3-44) and (3-46) it follows that the b_m's are zero for the trigonometric series of real, even signals.

Similar reasoning shows that the X_m's are imaginary and odd functions of m if $x(t)$ is odd, that is, if $x(t) = -x(-t)$. Also $a_m = 0$, all m, if $x(t)$ is odd, so that the trigonometric Fourier series of an odd function consists only of sine terms (see Example 3-4). The proofs of these statements are left to the problems.

Another type of symmetry is *half-wave* (odd) symmetry defined as

$$x\left(t \pm \frac{T_0}{2}\right) = -x(t) \tag{3-50}$$

where T_0 is a period of $x(t)$.[†] For signals with half-wave symmetry, it turns out that

$$X_n = 0, \qquad n = 0, \pm 2, \pm 4, \ldots \tag{3-51}$$

The proof is left to the problems. Thus the Fourier series of a half-wave (odd) symmetrical signal contains only odd harmonics (see Example 3-4 and Table 3-1, waveforms 4 and 5). We note the half-wave even symmetry, defined by

$$x\left(t \pm \frac{T_0}{2}\right) = x(t) \tag{3-52}$$

simply means that the fundamental period of the waveform is $T_0/2$ rather than T_0.

Table 3-2 summarizes the properties of the two types of Fourier series that we have considered.

3-6 Parseval's Theorem

In Chapter 1 the average normalized power of a periodic waveform $x(t)$ was written as

$$P_{av} = \frac{1}{T_0} \int_{T_0} |x(t)|^2 \, dt = \frac{1}{T_0} \int x(t)x*(t) \, dt \tag{3-53}$$

[†]To illustrate half-wave symmetry, consider waveform 4 in Table 3-1. Pick an arbitrary point on the t-axis. The value of the signal at this time is equal in magnitude, but opposite in sign, to the value of the signal one-half period earlier or later. The same argument holds for waveform 5 in Table 3-1.

TABLE 3-2
Summary of Fourier Series Properties[a]

Series	Coefficients[b]	Symmetry Properties
1. Trigonometric sine-cosine		
$x(t) = a_0 + \sum\limits_{n=1}^{\infty} (a_n \cos n\omega_0 t + b_n \sin n\omega_0 t)$	$a_0 = \dfrac{1}{T_0} \displaystyle\int_{T_0} x(t)\, dt$	$a_0 = $ Average value of $x(t)$
	$a_n = \dfrac{2}{T_0} \displaystyle\int_{T_0} x(t) \cos n\omega_0 t\, dt$	$a_n = 0 \quad$ for $x(t)$ odd,
		$b_n = 0 \quad$ for $x(t)$ even
	$b_n = \dfrac{2}{T_0} \displaystyle\int_{T_0} x(t) \sin n\omega_0 t\, dt$	$a_n, b_n = 0, \quad n$ even, for $x(t)$ odd, half-wave symmetrical
2. Complex exponential		
$x(t) = \sum\limits_{n=-\infty}^{\infty} X_n e^{jn\omega_0 t}$	$X_n = \dfrac{1}{T_0} \displaystyle\int_{T_0} x(t) e^{-jn\omega_0 t}\, dt$	$X_0 = $ Average value of $x(t)$
		X_n real for $x(t)$ even
	$X_n = \begin{cases} \frac{1}{2}(a_n - jb_n), & n > 0 \\ \frac{1}{2}(a_{-n} + jb_{-n}), & n < 0 \end{cases}$	X_n imaginary for $x(t)$ odd
		$X_n = 0, \quad n$ even, for $x(t)$ odd half-wave symmetrical
	$X_n = X_{-n}^{*}$ for $x(t)$ real	

[a] $x(t)$ even means that $x(t) = x(-t)$; $x(t)$ odd means that $x(t) = -x(-t)$; $x(t)$ odd half-wave symmetrical means that $x(t) = -x(t \pm T_0/2)$.

[b] $\int_{T_0} (\cdot)\, dt$ means integration over any period T_0 of $x(t)$.

We can express P_{av} in terms of the Fourier coefficients of $x(t)$ by replacing $x*(t)$ in (3-53) with its Fourier series representation (3-28).[‡] The substitution yields

$$P_{av} = \frac{1}{T_0} \int_{T_0} x(t) \left(\sum_{n=-\infty}^{\infty} X_n^* e^{-jn\omega_0 t} \right) dt$$

$$= \sum_{n=-\infty}^{\infty} X_n^* \left[\frac{1}{T_0} \int_{T_0} x(t) e^{-jn\omega_0 t} \, dt \right] \tag{3-54}$$

where the order of integration and summation have been interchanged. We recognize the term in brackets as X_n [see (3-31)]. Thus we may write

$$P_{av} = \frac{1}{T_0} \int_{T_0} |x(t)|^2 \, dt = \sum_{n=-\infty}^{\infty} |X_n|^2 \tag{3-55a}$$

$$= X_0^2 + 2 \sum_{n=1}^{\infty} |X_n|^2 \tag{3-55b}$$

where (3-55b) results by applying (3-47). In words, (3-55a) simply states that the average power of a periodic signal $\hat{x}(t)$ is the sum of the powers in the phasor components of its Fourier series (recall Example 1-13, in which the power of a phasor was computed). Equivalently, (3-55b) states that the average power in a periodic signal is the sum of the power in its dc component plus the powers in its harmonic components (recall that the average power of a sinusoid is one-half the square of its amplitude).

EXAMPLE 3-8

The average power of the sine wave $x(t) = 4 \sin 50\pi t \, dt$ is

$$P_{av} = 25 \int_0^{0.04} 16 \sin^2 50\pi t \, dt = \frac{16}{2} = 8 \, W \tag{3-56}$$

computed from (3-53). Its exponential Fourier series coefficients are $X_{-1} = -X_1 = 2j$ and its power, when computed from (3-55a), is $P_{av} = 4 + 4 = 8$ W.

EXAMPLE 3-9

Consider the use of a square wave to test the fidelity of an amplifier. Suppose that the amplifier ideally passes all Fourier components of the input with frequencies less than 51 kHz. By ideal, it is meant that the individual Fourier components of the input with frequencies less than 51 kHz suffer zero phase shift in passing through the amplifier and all are multiplied by the same factor, called the *voltage gain*, G_v, while all Fourier components above 51 kHz do not appear at the output.

Suppose that $G_v = 10$ and the input, $x(t)$, is a 10-kHz square wave with amplitudes ± 1 V. (a) What is the average power P_x of the input signal? (b) What is the average power P_y of the output signal, $y(t)$? (c) What would the average power of the output signal be if the amplifier passed all frequency components at its input?

Solution: (a) The average power of the input is found by using (1-84) for the average power of a periodic signal. Letting T_0 be the period of the square wave and A its amplitude, we obtain

---

[‡]Recall that the conjugate of a sum is obtained by conjugating each term in the sum. Also, the conjugate of a product of factors is the product of the conjugate of each factor.

$$P_x = \frac{1}{T_0} \int_{t_0}^{t_0+T_0} x^2(t)\, dt$$

$$= \frac{1}{T_0} \int_{t_0}^{t_0+T_0} A^2\, dt \tag{3-57}$$

Equation (3-57) follows by virtue of the square-wave amplitude being either A or $-A$. Thus the squared amplitude of $x(t)$ is A^2, and the average power becomes

$$P_x = A^2 \qquad \text{(square wave of amplitudes } \pm A\text{)}$$

$$= 1 \quad \text{W} \tag{3-58}$$

Parseval's theorem could be used to obtain P_x by summing the powers in the spectral components provided that one has the patience to sum an infinite series!

(b) The exponential Fourier series of an even-symmetry square wave, $x(t)$, with amplitude $A = 1$ V is

$$x(t) = \frac{2}{\pi} (\cdots -\frac{1}{7} e^{-j7\omega_0 t} + \frac{1}{5} e^{-j5\omega_0 t} - \frac{1}{3} e^{-j3\omega_0 t} + e^{-j\omega_0 t} +$$

$$+ e^{j\omega_0 t} - \frac{1}{3} e^{j3\omega_0 t} + \frac{1}{5} e^{j5\omega_0 t} - \frac{1}{7} e^{j7\omega_0 t} + \cdots) \tag{3-59}$$

where $\omega_0 = 2\pi/T_0 = 2\pi f_0$ with $f_0 = 10$ kHz. The amplifier multiplies all spectral components with frequencies less than 51 kHz by the factor $G_v = 10$ and blocks all other spectral components. Thus the Fourier series of the output of the amplifier is

$$y(t) = \frac{20}{\pi} (\frac{1}{5} e^{-j5\omega_0 t} - \frac{1}{3} e^{-j3\omega_0 t} + e^{-j\omega_0 t} + e^{j\omega_0 t} - \frac{1}{3} e^{j3\omega_0 t} + \frac{1}{5} e^{j5\omega_0 t})$$

$$\equiv \sum_{n=-\infty}^{\infty} Y_n e^{jn\omega_0 t} \tag{3-60}$$

The amplitudes of the Fourier components of the output are therefore

$$Y_n = 0, \qquad |n| > 5$$

$$Y_n = 0, \qquad n \text{ even}$$

$$Y_{-1} = Y_1 = \frac{20}{\pi} \tag{3-61}$$

$$Y_{-3} = Y_3 = -\frac{20}{3\pi}$$

$$Y_{-5} = Y_5 = \frac{20}{5\pi}$$

Using Parseval's theorem, the output power is

$$P_y = Y_0^2 + 2 \sum_{n=1}^{\infty} |Y_n|^2 = 2\left[\left(\frac{20}{\pi}\right)^2 + \left(\frac{20}{3\pi}\right)^2 + \left(\frac{20}{5\pi}\right)^2\right]$$

$$= \frac{8 \times 10^2}{\pi^2} \left(1 + \frac{1}{9} + \frac{1}{25}\right)$$

$$= 93.31 \quad \text{W} \tag{3-62}$$

(c) If the amplifier passes $x(t)$ perfectly, then

$$\hat{y}(t) = G_v x(t) \tag{3-63}$$

and

$$P_{\hat{y}} = G_v^2 P_x$$
$$= 10^2 \quad \text{W} \tag{3-64}$$

Note that $P_y/P_{\hat{y}} = 93.31\%$. This result tells us that only the first through fifth harmonics of the square wave need to be included to obtain 93.3% of the average signal power possible at the amplifier output. The student should compute the percent of average signal power possible at the amplifier output if its cutoff frequency is such that the 7th, 9th, 11th, and so on, harmonics are passed by the amplifier but all others rejected. (*Answers:* 94.96%; 95.96%; 96.63%.)

3-7 Line Spectra

The complex exponential Fourier series (3-28) of a signal consists of a summation of rotating phasors. In Chapter 1 we showed how sums of rotating phasors can be characterized in the frequency domain by two plots, one showing their amplitudes as a function of frequency and one showing their phases. This observation allows a periodic signal to be characterized graphically in the frequency domain by making two plots. The first, showing amplitudes of the separate phasor components versus frequency, is known as the *amplitude spectrum* of the signal. The second, showing the relative phases of each component versus frequency, is called the *phase spectrum* of the signal. Since these plots are obtained from the complex exponential Fourier series, spectral components, or *lines* as they are called, are present at both positive and negative frequencies. Such spectra are referred to as *two-sided*. From (3-49) it follows that, for a real signal, the amplitude spectrum is even and the phase spectrum is odd, which simply is a result of the necessity to add complex conjugate phasors to get a real sinusoidal signal. Figure 3-10 shows the two-sided spectra for a half-rectified sine wave. (See Table 3-1 for the Fourier coefficients.) In order that the phase be odd, X_n is represented as

$$X_n = -\left| \frac{A}{\pi(1 - n^2)} \right| = \frac{A}{\pi(n^2 - 1)} e^{-j\pi} \tag{3-65}$$

for $n = 2, 4, \ldots$, and as

$$X_n = -\left| \frac{A}{\pi(1 - n^2)} \right| = \frac{A}{\pi(n^2 - 1)} e^{j\pi} \tag{3-66}$$

for $n = -2, -4, \ldots$. For $n = \pm 1$, the amplitude spectrum is $A/4$ with a phase shift of $-\pi/2$ for $n = 1$ and a phase shift of $\pi/2$ for $n = -1$, which again ensures that the phase is odd.

The *single-sided line spectra* are obtained by writing the complex exponential form of the Fourier series (3-28) as a cosine series by representing the Fourier coefficients in polar form given by (3-48). For a real signal, the magnitudes and phases of the Fourier coefficients have the symmetry properties given by (3-49). Substituting the polar form of the Fourier coefficients into (3-28) and using the symmetry properties of the magnitudes and phases, we obtain

$$x(t) = \sum_{n=-\infty}^{\infty} |X_n| e^{j\theta_n} e^{jn\omega_0 t}$$

$$= \sum_{n=-\infty}^{-1} |X_n| e^{j(n\omega_0 t + \theta_n)} + X_0 + \sum_{n=1}^{\infty} |X_n| e^{j(n\omega_0 t + \theta_n)}$$

$$= X_0 + \sum_{n=1}^{\infty} [|X_n| e^{j(n\omega_0 t + \theta_n)} + |X_{-n}| e^{j(-n\omega_0 t + \theta_{-n})}]$$

$$= X_0 + \sum_{n=1}^{\infty} 2|X_n| \frac{e^{j(n\omega_0 t + \theta_n)} + e^{-j(n\omega_0 + \theta_n)}}{2}$$

$$= X_0 + \sum_{n=1}^{\infty} 2|X_n| \cos(n\omega_0 t + \theta_n) \qquad (3\text{-}67)$$

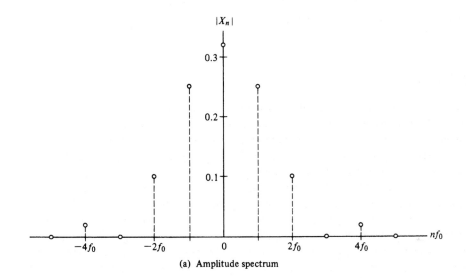

(a) Amplitude spectrum

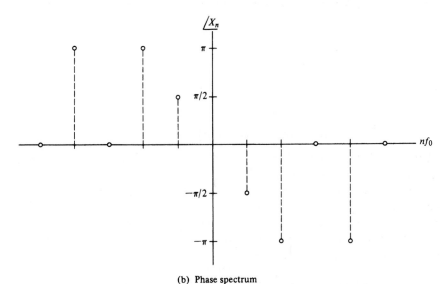

(b) Phase spectrum

FIGURE 3-10.   Two-sided line spectra for a half-rectified sine wave.

From this it is seen that the *single-sided amplitude spectrum* has lines that are twice as high as the lines in the double-sided amplitude spectrum for $f > 0$, with the line at $f = 0$ being the same in both spectra, and that the *single-sided phase spectrum* is identical to the double-sided phase spectrum for $f \geq 0$. We reached the same conclusions in Chapter 1 with regard to spectra of sums of sinusoids. Since the Fourier series of a periodic function is the sum of sinusoids as shown by (3-67), these observations should come as no surprise. The student should plot the single-sided amplitude and phase spectra for the half-rectified sine wave.

As a second example, consider the periodic pulse train shown in Table 3-1 as waveform 3. We consider the special case of $t_0 = \tau/2$. From (3-35c), the Fourier coefficients are

$$X_n = \frac{A\tau}{T_0} \operatorname{sinc} nf_0\tau \, e^{-j\pi n f_0 \tau} \tag{3-68}$$

The Fourier coefficients can be put in the form $|X_n| \exp(j\underline{/X_n})$, where

$$|X_n| = \frac{A\tau}{T_0} |\operatorname{sinc} nf_0\tau| \tag{3-69a}$$

and

$$\underline{/X_n} = \begin{cases} -\pi nf_0\tau & \text{if } \operatorname{sinc} nf_0\tau > 0 \\ -\pi nf_0\tau + \pi & \text{if } nf_0 > 0 \text{ and } \operatorname{sinc} nf_0\tau < 0 \\ -\pi nf_0\tau - \pi & \text{if } nf_0 < 0 \text{ and } \operatorname{sinc} nf_0\tau < 0 \end{cases} \tag{3-69b}$$

The $\pm\pi$ on the right-hand side of (3-69b) accounts for $|\operatorname{sinc} nf_0\tau| = -\operatorname{sinc} nf_0\tau$ whenever $\operatorname{sinc} nf_0\tau < 0$. Since the phase spectrum must have odd symmetry if $x(t)$ is real, π is subtracted if $nf_0 < 0$ and added if $nf_0 > 0$. The two-sided amplitude and phase spectra are shown in Figure 3-11 for several choices of τ and T_0. Note that whenever a phase line is 2π or greater in magnitude, 2π is subtracted. Two important observations can be made from Figure 3-11. First, comparing Figures 3-11a and b we note that the width of the envelope of the amplitude spectrum increases as the pulse width decreases. That is, *the pulse width of the signal and its corresponding spectral width are inversely proportional*. Second, comparing Figures 3-11a and c, we note that the separation between lines in the spectrum is $1/T_0$, and therefore *the density of the spectral lines with frequency increases as the period of x(t) increases*.

3-8 Steady-State Response of a Fixed, Linear System to a Periodic Input; Distortionless Systems

It was shown in Chapter 2 that if the input to a fixed, linear system is a rotating phasor signal of frequency ω rad/s, then the steady-state output is also a sinusoid of the same frequency but, in general, with different amplitude and phase from the input. The complex proportionality constant that multiplies the rotating phasor input [see (2-99)] to produce the output is called the *frequency response function* of the system, and is given by the integral

$$H(\omega) = \int_{-\infty}^{\infty} h(t) \, e^{-j\omega t} \, dt \tag{3-70}$$

where $h(t)$ is the impulse response of the system. It will be shown in Chapter 4 that this integral is the *Fourier transform* of the impulse response. For any periodic input expressed in the form of a complex

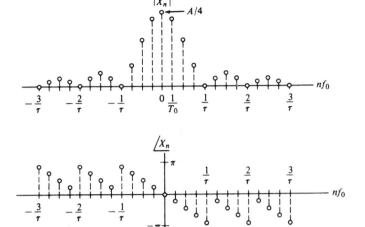

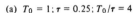

(a) $T_0 = 1; \tau = 0.25; T_0/\tau = 4$

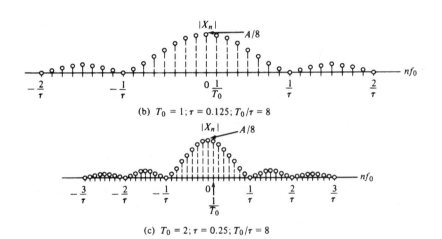

(b) $T_0 = 1; \tau = 0.125; T_0/\tau = 8$

(c) $T_0 = 2; \tau = 0.25; T_0/\tau = 8$

Figure 3-11. Spectra for a periodic pulse-train signal.

exponential Fourier series, and using superposition, we may express the steady-state output of a fixed, linear system as

$$y(t) = \sum_{n=-\infty}^{\infty} X_n H(n\omega_0) e^{jn\omega_0 t} \qquad (3\text{-}71)$$

If $H(\omega)$ is represented in polar form as

$$H(\omega) = A(\omega) e^{j\theta(\omega)} \qquad (3\text{-}72)$$

where $A(\omega)$ is called the amplitude response and $\theta(\omega)$ is called the phase response of the system, it was also shown in Chapter 2 that for a sinusoidal input of the form $\cos \omega_0 t$ the steady-state output is $A(\omega_0) \cos[\omega_0 t + \theta(\omega_0)]$ (see Example 2-16 for a specific case). For any periodic input, it follows from

(3-67) and the superposition property of a fixed, linear system that its steady-state output can be expressed as

$$y(t) = X_0 H(0) + \sum_{n=1}^{\infty} 2|X_n| A(n\omega_0) \cos[n\omega_0 t + \underline{/X_n} + \theta(n\omega_0)] \tag{3-73}$$

The same result could have been obtained by expressing the series in (3-71) as a sum of cosinusoids, using the same steps as were used in deriving (3-67). In so doing, it would have been necessary to invoke the following symmetry properties for $A(\omega)$ and $\theta(\omega)$ for $h(t)$ real:

$$A(\omega) = A(-\omega)$$

$$\theta(\omega) = -\theta(-\omega) \tag{3-74}$$

That these relationships hold can be shown from the defining expression (3-70) by substituting $e^{-j\omega t} = \cos \omega t - j \sin \omega t$, writing the integral as the sum of two integrals, and noting that the first integral is the real part of $H(\omega)$ and that the second integral is its imaginary part. By replacing ω by $-\omega$ in these integrals, it is clear that $H(\omega) = H^*(-\omega)$, which is similar to (3-47). The symmetry properties given by (3-74) then follow by expressing $H(\omega)$ in polar form.

EXAMPLE 3-10

The application of (3-71) can be illustrated by finding the output of the RC lowpass filter of Chapter 2 (see Example 2-16) to an odd square wave of unit amplitude. Recall Example 3-7 for finding the complex exponential Fourier coefficients of an odd square wave, and recall (2-102) where the transfer function of the RC lowpass filter was given as

$$H(j\omega) = \frac{1}{1 + j\omega RC} \tag{3-75}$$

This is substituted into (3-71) along with the Fourier coefficients of the odd square wave to obtain the Fourier series expression for steadystate output of the filter. A MATLAB program for computing it is given below. Note that matrix computation is used throughout (with the exception of the for loop to run the different cases for the number of terms in the Fourier series) to keep the program as efficient as possible.

```
%       Plots for Example 3-10
%
RC=input('Enter RC time constant of filter (square wave period=1)');
n_max=input('Enter vector of highest harmonic values desired (odd)');
N=length(n_max);
t=0:.005:1;
omega_0=2*pi;
for k=1:N                          % Loop for various numbers of highest
                                   % harmonics
        n=[-n_max(k):2:n_max(k)];
        L_n=length(n);
        nn=2:L_n/2+1;
        sgn=(-1).^nn;              % This will make the 1,5, . . . terms + and
                                   % the 3,7, . . . terms -
%       Fourier coefficients for odd square wave; for even square wave,
%       leave exp() off
        X_n=(2./(pi*abs(n))).*[fliplr(sgn)sgn].*exp(-j*0.5*pi*n);
```

```
%       Input to filter
        x=X_n*exp(j*omega_0*n'*t);      % Rows of exponential matrix are
                                        % versus time;
                                        % columns are versus n; matrix
                                        % multiply sums over n
%       Fourier coefficients of output
        Y_n=X_n.*(1./(1+j*n*omega_0*RC));
        y=Y_n*exp(j*omega_0*n'*t);      % The output
%       Plot real part of y to get rid of small imaginary part due to
%       computational error
        subplot(N,1,k),plot(t,real(y),xlabel('t'),ylabel('approx.
        output'), . . .
        axis([0 1 -1.5 1.5]), text(.1,-.25,['max.har.=',num2str(n_max(k))])
        if k == 1
                title(['Filter output for RC = ',num2str(RC),' seconds'])
        end
end
```

To run the program, the following is entered in the command window, where two prompts are printed in this case—one for RC time constant of the filter and one for the maximum harmonic used in the Fourier series. A plot of the output of the program is shown in Figure 3-11. The plot is most inaccurate at the discontinuity in slope where the square wave input switches sign.

```
EDU»c3ex10
Enter RC time constant of filter (square wave period=1).2
Enter vector of highest harmonic values desired (odd) [7 15 27]
```

EXAMPLE 3-11

The input and output spectra of the RC filter for the previous example will reveal information on why the output time signals look as they do. The MATLAB program below provides plots for the these spectra for the two RC values used to plot Figure 3-12. Note that since we haven't plotted the phase spectrum, the spectral plots will be the same for both the even and odd square wave cases. Figure 3-13 shows the input and output spectra. Since the output spectrum is the product of the input spectrum and the transfer function (magnitude) at each frequency, the plot clearly shows how the filter affects the output spectra.

```
%       Plots for Example 3-11
%
RC=input('Enter vector of RC time constant of filter (sq. wave
period=1)');
n_max=input('Enter highest harmonic value desired (odd)');
N_RC=length(RC);
omega_0=2*pi;
f0=1;
for k=1:N_RC
        n=[-n_max:2:n_max];
        f=[-n_max*f0:.02:n_max*f0];
        H=1./(1+j*2*pi*f*RC(k));
        L_n=length(n);
        nn=2:L_n/2+1;
```

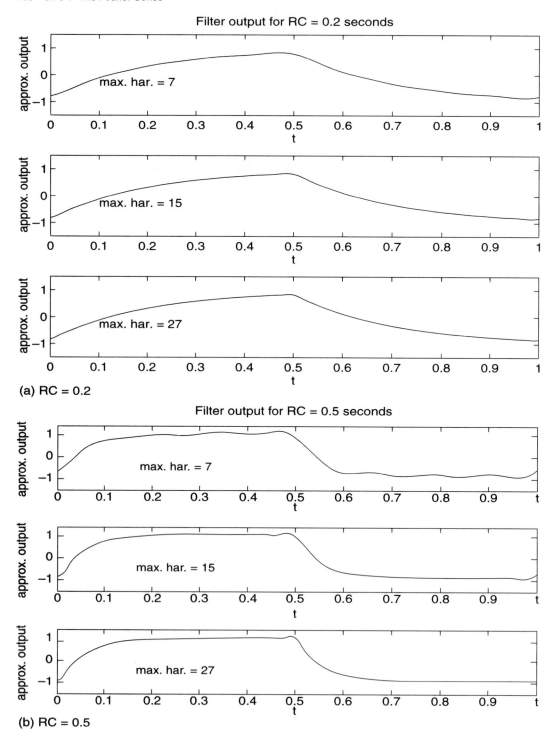

Figure 3-12. Partial sum approximations to the Fourier series for the steady state output of an RC lowpass filter to a square wave.

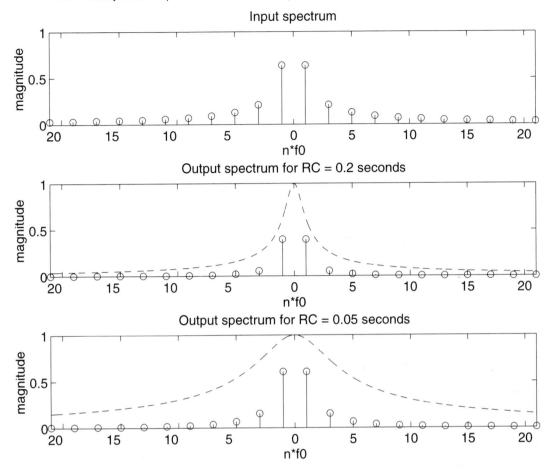

FIGURE 3-13. Input and output amplitude spectra for an RC filter for $RC = 0.2$ and 0.05. The transfer function magnitude is shown superimposed on the spectra.

```
         sgn=(-1).^nn;
%        Fourier coefficients for square wave;
         X_n=(2./(pi*abs(n))).*[fliplr(sgn) sgn];
%        Fourier coefficients of output
         Y_n=X_n.*(1./(1+j*n*omega_0*RC(k)));
%        Plot magnitude of X_n and Y_n
         if k==1
         subplot(N_RC+1,1,1),stem(n*f0,abs(X_n)), xlabel('n*f0'),
         ylabel('magnitude'),...
         title('Input spectrum'),axis(-n_max*f0 n_max*f0 01])
         end
         subplot(N_RC+1,1,k+1),stem(n*f0,abs(Y_n)),xlabel('n*f0'),
         ylabel('magnitude'),...
         title(['Output spectrum for RC= ', num2str(RC(k)),' seconds']),
         axis([-n_max*f0 n_max*f0 0 1]), hold on
         subplot(N_RC+1,1,k+1), plot(f,abs(H),'-')
end
```

The entries into the command window look as follows:

```
EDU»c3ex11
Enter vector of RC time constant of filter (sq. wave period=1)[.2 .05]
Enter highest harmonic value desired (odd) 21
```

If a fixed, linear system is to have no effect on the input except, possibly, for an amplitude scaling and delay which are the same for all input-frequency components, the system is called *distortionless*. For a distortionless system with periodic input of the form (3-67), the steady-state output is of the form

$$y(t) = K\left\{ X_0 + \sum_{n=1}^{\infty} 2|X_n| \cos[n\omega_0(t - \tau_0) + \theta_n) \right\}$$

$$= Kx(t - \tau_0) \tag{3-76}$$

where K is the scaling factor and τ_0 is the delay. From (3-73) it follows that

$$A(n\omega_0) = K \quad \text{and} \quad \theta(n\omega_0) = -\tau_0 n\omega_0$$

or

$$A(\omega) = K \quad \text{and} \quad \theta(\omega) = -\tau_0\omega \tag{3-77}$$

respectively. That is, the amplitude response of a distortionless system is the same, or *constant*, for all frequency components of the input and the phase response is a *linear* function of frequency. In other words, the frequency response function of a distortionless system is of the form

$$H_d(\omega) = K \exp(-j\tau_0\omega) \tag{3-78}$$

In general, three major types of distortion can be introduced by a system. First, if a system is linear but its amplitude response is not constant with frequency, the system introduces *amplitude distortion*. Second, if a system is linear but its phase shift is not a linear function of frequency, it introduces *phase, or delay, distortion*. Third, if the system is not linear, *nonlinear* distortion results. These three types of distortion may occur in combination with each other.

EXAMPLE 3-12

To illustrate the ideas of amplitude and phase distortion, consider a system with amplitude response and phase response as shown in Figure 3-14[†] and the following three inputs:

1. $x_1(t) = 2 \cos 10\pi t + \sin 12\pi t$
2. $x_2(t) = 2 \cos 10\pi t + \sin 26\pi t$
3. $x_3(t) = 2 \cos 26\pi t + \sin 34\pi t$

Although this system has idealized frequency response characteristics, it can be used to illustrate various combinations of amplitude and phase distortion. Using (3-73), we find the outputs for the given inputs to be:

[†]The filter amplitude- and phase-response functions used here are not realizable, but are employed only to easily illustrate amplitude and phase distortion.

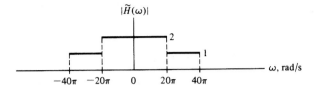

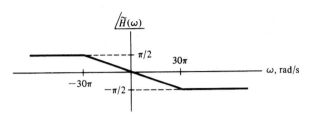

FIGURE 3-14. Amplitude response and phase response of the filter for Example 3-12.

1. $y_1(t) = 4\cos(10\pi t - \pi/6) + 2\sin(12\pi t - \pi/5)$
$= 4\cos 10\pi(t - 1/60) + 2\sin 12\pi(t - 1/60)$

2. $y_2(t) = 4\cos(10\pi t - \pi/6) + \sin(26\pi t - 13\pi/30)$
$= 4\cos 10\pi(t - 1/60) + \sin 26\pi(t - 1/60)$

3. $y_3(t) = 2\cos(26\pi t - 13\pi/30) + \sin(34\pi t - \pi/2)$
$= 2\cos 26\pi(t - 1/60) + \sin 34\pi(t - 1/68)$

Comparison with (3-76) shows that only the input $x_1(t)$ is passed without distortion by the system. The system imposes amplitude distortion on $x_2(t)$ and phase (delay) distortion on $x_3(t)$.

EXERCISE

A second-order low-pass Butterworth filter is one having frequency-response function

$$H(\omega) = \frac{\omega_c^2}{\omega_c^2 - \omega^2 + j\sqrt{2}\,\omega_c\omega} \tag{3-79}$$

where ω_c is a filter parameter.
(a) Show that its amplitude- and phase-response functions are

$$A(\omega) = \left[1 + \left(\frac{\omega}{\omega_c}\right)^4\right]^{-1/2} \tag{3-80}$$

and

$$\theta(\omega) = \begin{cases} -\tan^{-1}\dfrac{\sqrt{2}\,(\omega/\omega_c)}{1 - \omega^2/\omega_c^2}, & \omega < \omega_c \\[2ex] -\pi + \tan^{-1}\dfrac{\sqrt{2}\,(\omega/\omega_c)}{\omega^2/\omega_c^2 - 1}, & \omega > \omega_c \end{cases} \tag{3-81}$$

respectively, where the value of the arctangent function is taken as its principal value.
(b) If $\omega_c = 600\pi$ rad/s above and $\omega_0 = 200\pi$ rad/s in (3-25), obtain the attenuations and phase shifts introduced by the filter in the first three terms of the Fourier series of a square-wave input to the filter. (*Answer:* attenuations: 0.994, 0.707, 0.339; phase shifts: $-27.9°$, $-90°$, $-127°$.)

(c) Write out the first three nonzero terms of the Fourier series of the output and compare with the input by plotting.

Answer: $y(t) = 4/\pi[0.994\sin(200\pi t - 27.9°) + 0.236\sin(600\pi t - 90°) + 0.068\sin(1000\pi t - 127°)]$.

3-9 Rate of Convergence of Fourier Spectra

It is handy to be able to predict how rapidly the amplitude spectral lines of a periodic signal approach zero as frequency (or harmonic order) increases. We may determine this from the complex exponential Fourier series expression (3-28) and the formula for finding the Fourier coefficients (3-31). Consider a periodic signal $x(t)$ that has a kth derivative which is piecewise continuous (i.e., is has finite jumps present, but it otherwise continuous). Thus its $(k + 1)$st derivative contains impulses wherever the kth derivative has jumps. Now, by differentiating the exponential Fourier series (3-28) $k + 1$ times, we obtain

$$\frac{d^{k+1}x(t)}{dt^{k+1}} = \sum_{n=-\infty}^{\infty} (jn\omega_0)^{k+1}X_n e^{jn\omega_0 t} \tag{3-82}$$

Therefore the Fourier coefficients of the $(k + 1)$st derivative of $x(t)$ are the Fourier coefficients of $x(t)$ multiplied by $(jn\omega_0)^{k+1}$.

Now consider the computation of the Fourier coefficients of the $(k + 1)$st derivative of $x(t)$. When the expression for the $(k + 1)$st derivative of $x(t)$ is substituted into (3-31), we may evaluate the impulse terms immediately by using the sifting property of the unit impulse function. These terms will give constants in the expression for the Fourier coefficients for the $(k + 1)$st derivative of $x(t)$. This means that, given our observation in relation to (3-82), the Fourier coefficients for $x(t)$ will have terms proportional to $(jn\omega_0)^{-(k+1)}$, since we divide the expression for the Fourier coefficients for the $(k + 1)$st derivative of $x(t)$, which have constant components, by $(jn\omega_0)^{k+1}$ to get X_n.

We conclude, therefore, that a periodic signal $x(t)$ that has a kth derivative which is piecewise continuous has Fourier coefficients that decrease with frequency as $(n\omega_0)^{-(k+1)}$. As an example, a square wave is piecewise continuous (zeroth derivative), so its first derivative contains impulses. Therefore, the Fourier coefficients of a square wave decrease inversely with increasing frequency, or order of harmonic.

This is a very useful result for design of waveforms. If we want a signal that has its spectral content concentrated about zero frequency, we think of those waveforms that have many continuous derivatives. An example of a fairly simple waveform of this nature is given in the next example.

EXAMPLE 3-13

Consider a periodic waveform consisting of raised cosine pulses, defined by the equationsand

$$x_p(t) = \begin{cases} \dfrac{1}{2}(1 + \cos 2\pi t), & |t| \le \dfrac{1}{2} \\[2mm] 0, & \text{otherwise} \end{cases}$$

and

$$x(t) = \sum_{n=-\infty}^{\infty} x_p(t - 2n) \tag{3-83}$$

The following MATLAB program (note that a function program is used to keep it compact) plots three pulses of this waveform and three of its derivatives. These plots are shown in Figure 3-15. Its third derivative contains impulses. Therefore, its amplitude spectrum approaches zero as $(n\omega_0)^{-3}$.

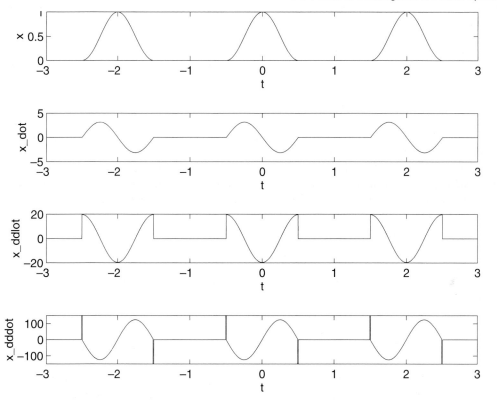

FIGURE 3-15. Plots of a raised cosine periodic waveform and its first three derivatives.

```
EDU»3ex13
%       Example 3-13 plots
%
t=-3:.001:3;
[x1,x1_dot,x1_ddot,x1_dddot]=rsd_cos(t+2);
[x2,x2_dot,x2_ddot,x2_dddot]=rsd_cos(t);
[x3,x3_dot,x3_ddot,x3_dddot]=rsd_cos(t-2);
x=x1+x2+x3;
x_dot=x1_dot+x2_dot+x3_dot;
x_ddot=x1_ddot+x2_ddot+x3_ddot;
x_dddot=x1_dddot+x2_dddot+x3_dddot;
subplot(4,1,1),plot(t,x),xlabel('t'),ylabel('x')
subplot(4,1,2),plot(t,x_dot),xlabel('t'),ylabel('x_dot')
subplot(4,1,3),plot(t,x_ddot),xlabel('t'),ylabel('x_ddlot')
subplot(4,1,4),plot(t,x_dddot),axis([-3 3 -150 150]), . . .
     xlabel('t'),ylabel('x_dddot')

%     Function to compute raised cosine and derivatives
%
function [x,x_dot,x_ddot,x_dddot]=rsd_cos(t)
x_cos=.5*(1+cos(2*pi*t));
x=x_cos.*pls_fn(t);
x_dot=-pi*sin(2*pi*t).*pls_fn(t);
x_ddot=-2*pi^2*cos(2*pi*t).*pls_fn(t);
```

```
x_dddot=4*pi^3*sin(2*pi*t).*pls_fn(t)+ . . .
    2*pi^2*(impls_fn(t+5,.01)-impls_fn(t-5,.01));
% Function to compute square pulse
%
Function y=pls_fn(t)
y=stp_fn(t+0.5)-stp_fn(t-0.5-eps)
```

3-10 Fourier Series and Signal Spaces

It is often useful to visualize signals as analogous to vectors in a generalized vector space. The Fourier series and the Fourier transform that we have just studied are tools for resolving power and energy signals, respectively, into such generalized vector spaces. We briefly consider this approach to Fourier representations, referred to as *generalized Fourier series.*

We begin our consideration of generalized Fourier series with a review of some concepts about ordinary vectors in three-dimensional space. We were probably first introduced to vectors as physical quantities whose specification involves both magnitude and direction and that obey the parallelogram law of addition. Later, we were perhaps shown that it was convenient to represent a vector, $\mathbf{A}$, in a Cartesian coordinate system in terms of a mutually perpendicular triad of unit vectors $\hat{\mathbf{i}}$, $\hat{\mathbf{j}}$, and $\hat{\mathbf{k}}$ along the x-, y-, and z-axes, respectively, as

$$\mathbf{A} = A_x\hat{\mathbf{i}} + A_y\hat{\mathbf{j}} + A_z\hat{\mathbf{k}} \tag{3-84}$$

The components A_x, A_y, and A_z can be expressed as

$$A_x = \mathbf{A} \cdot \hat{\mathbf{i}}, \qquad A_y = \mathbf{A} \cdot \hat{\mathbf{j}}, \qquad A_z = \mathbf{A} \cdot \hat{\mathbf{k}}$$

where the dot denotes the dot product of two vectors, which is the product of their magnitudes and the cosine of the angle between them. Since $\hat{\mathbf{i}}$, $\hat{\mathbf{j}}$, and $\hat{\mathbf{k}}$ are unit-length, mutually perpendicular vectors, the *xyz* components of $\mathbf{A}$ are easily obtained by simply projecting $\mathbf{A}$ onto the x-, y-, and z-axes, respectively.

We now consider representation of a finite-energy signal, $x(t)$, defined on a T-second interval $(t_0, t_0 + T)$ in terms of a set of preselected time functions, $\phi_1(t), \ldots, \phi_n(t)$. It is convenient to choose these functions with properties analogous to $\hat{\mathbf{i}}$, $\hat{\mathbf{j}}$, and $\hat{\mathbf{k}}$ of three-dimensional vector space. The mutually perpendicular property, referred to as *orthogonality,* is expressed as

$$\int_{t_0}^{t_0+T} \phi_m(t)\phi_n^*(t)\, dt = 0, \qquad m \neq n \tag{3-85}$$

where the conjugate suggests that complex-valued $\phi_n(t)$'s may be convenient in some cases. We further assume that the $\phi_n(t)$'s have been chosen such that

$$\int_{t_0}^{t_0+T} |\phi_n(t)|^2\, dt = 1 \tag{3-86}$$

In view of (3-86), the $\phi_n(t)$'s are said to be *normalized.* This assumption is invoked to simplify future equations.

We now attempt to approximate $x(t)$ as best we can by a series of the form

$$y(t) = \sum_{n=1}^{N} d_n\phi_n(t) \tag{3-87}$$

where the d_n's are constants to be chosen such that $y(t)$ represents $x(t)$ as closely as possible according to some criterion. It is convenient to measure the error in the integral-square sense, which is defined as

$$\text{integral-square error} = \epsilon_N = \int_{t_0}^{t_0+T} |x(t) - y(t)|^2\, dt \tag{3-88}$$

We wish to find $d_1, d_2, \ldots, d_N$ such that ϵ_N as expressed by (3-88) is a minimum. Substituting (3-87) into (3-88), we obtain

$$\epsilon_N = \int_T \left[x(t) - \sum_{n=1}^N d_n \phi_n(t) \right] \left[x^*(t) - \sum_{n=1}^N d_n^* \phi_n^*(t) \right] dt \tag{3-89}$$

where

$$\int_T (\cdot) \, dt \triangleq \int_{t_0}^{t_0 + T} (\cdot) \, dt$$

This can be expanded to yield

$$\epsilon_N = \int_T |x(t)|^2 \, dt - \sum_{n=1}^N \left[d_n^* \int_T x(t) \phi_n^*(t) \, dt + d_n \int_T x^*(t) \phi_n(t) \, dt \right] + \sum_{n=1}^N |d_n|^2 \tag{3-90}$$

which was obtained by making use of (3-85) and (3-86) after interchanging the orders of summation and integration.

It is convenient to add and subtract the quantity

$$\sum_{n=1}^N \left| \int_T x(t) \phi_n^*(t) \, dt \right|^2 \tag{3-91}$$

to (3-90) which, after rearrangement of terms, yields

$$\epsilon_N = \int_T |x(t)|^2 \, dt - \sum_{n=1}^N \left| \int_T x(t) \phi_n^*(t) \, dt \right|^2 + \sum_{n=1}^N \left| d_n - \int_T x(t) \phi_n^*(t) \, dt \right|^2 \tag{3-92}$$

To show the equivalence of (3-92) and (3-90), it is easiest to work backward from (3-92).

Now the first two terms on the right-hand side of (3-92) are independent of the coefficients d_n. The last summation of terms on the right-hand side is nonnegative and is added to the first two terms. Therefore, to minimize ϵ_N through choice of the d_n's, the best we can do is make each term of the last sum zero. That is, we choose the nth coefficient, d_n, such that

$$d_n = \int_{t_0}^{t_0 + T} x(t) \phi_n^*(t) dt, \qquad n = 1, 2, \ldots, N \tag{3-93}$$

This choice for $d_1, d_2, \ldots, d_N$ minimizes the integral-square error, ϵ_N. The resulting coefficients are called the *generalized Fourier coefficients*.

The minimum value for ϵ_N is

$$\epsilon_{N, \text{min}} = \int_{t_0}^{t_0 + T} |x(t)|^2 \, dt - \sum_{n=1}^N |d_n|^2 \tag{3-94}$$

It is natural to inquire as to the possibility of $\epsilon_{N, \text{min}}$ being zero. For special choices of the set of functions $\phi_1(t), \phi_2(t), \ldots, \phi_N(t)$, referred to as *complete sets* in the space of all integrable-square functions, it will be true that

$$\lim_{N \to \infty} \epsilon_{N, \text{min}} = 0 \tag{3-95}$$

for any signal that is square integrable, that is, for any signal for which

$$\int_{t_0}^{t_0 + T} |x(t)|^2 \, dt < \infty \tag{3-96}$$

In the sense that the integral-square error is zero, we may then write

$$x(t) = \sum_{n=1}^{\infty} d_n \phi_n(t) \tag{3-97}$$

For a complete set of orthogonal functions, $\phi_1(t)$, $\phi_2(t)$, $\ldots$, we obtain from (3-94) that

$$\int_{t_0}^{t_0+T} |x(t)|^2 \, dt = \sum_{n=1}^{\infty} |d_n|^2 \tag{3-98}$$

This is a generalized version of Parseval's theorem.

EXAMPLE 3-14

Given the set of functions shown in Figure 3-16. (a) Show that this is an orthogonal set and that each member of the set is normalized. (b) If $x(t) = \cos 2\pi t$, $0 \le t \le 1$, find $y(t)$ as expressed by (3-87) such that the integral-square error is minimized. (c) Same question as in part (b) but $x(t) = \sin 2\pi t$, $0 \le t \le 1$.

Solution:

(a) It is clear that each function is normalized according to (3-86). To show orthogonality, we calculate

$$\int_0^1 \phi_1(t)\phi_2(t) \, dt = \int_0^{0.5} (1)(1) \, dt + \int_{0.5}^1 (1)(-1) \, dt = 0 \tag{3-99}$$

Similarly, it can be shown that $\int_0^1 \phi_1(t)\phi_3(t) \, dt = 0$ and $\int_0^1 \phi_2(t)\phi_3(t) \, dt = 0$.

(b) The generalized Fourier coefficients are found according to (3-93). The coefficients are

$$d_1 = \int_0^1 (1) \cos 2\pi t \, dt$$

$$= \frac{\sin 2\pi t}{2\pi} \Big|_0^1$$

$$= 0 \tag{3-100}$$

$$d_2 = \int_0^{0.5} (1) \cos 2\pi t \, dt + \int_{0.5}^1 (-1) \cos 2\pi t \, dt$$

$$= \frac{1}{2\pi} \left[\sin 2\pi t \Big|_0^{0.5} - \sin 2\pi t \Big|_{0.5}^1 \right]$$

$$= 0 \tag{3-101}$$

$$d_3 = \int_0^{0.25} (1) \cos 2\pi t \, dt + \int_{0.25}^{0.75} (-1) \cos 2\pi t \, dt + \int_{0.75}^1 (1) \cos 2\pi t \, dt$$

$$= \frac{1}{2\pi} \left[\sin 2\pi t \Big|_0^{0.25} - \sin 2\pi t \Big|_{0.25}^{0.75} + \sin 2\pi t \Big|_{0.75}^1 \right] = \frac{2}{\pi} \tag{3-102}$$

Therefore,

$$y(t) = \frac{2}{\pi} \phi_3(t) \tag{3-103}$$

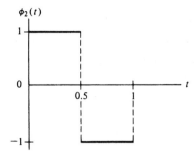

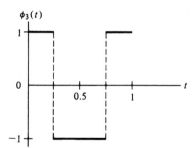

FIGURE 3-16. Orthogonal set of functions for Example 3-11.

The minimum integral-square error is

$$\epsilon_3 = \int_0^1 \cos^2 2\pi t \, dt - \left(\frac{2}{\pi}\right)^2 = \frac{1}{2} - \left(\frac{2}{\pi}\right)^2 \qquad (3\text{-}104)$$

(c) In a similar manner to the calculations carried out for part (b), it follows that for $x(t) = \sin 2\pi t$.

$$d_1 = 0$$
$$d_2 = \frac{2}{\pi} \qquad (3\text{-}105)$$
$$d_3 = 0$$

The minimum integral-square error is the same as for part (b). The student is advised to sketch $y(t)$ for each case and compare it with $x(t)$.

Note that the exponential Fourier series could have been derived as a generalized Fourier series with

$$\phi_n(t) = e^{jn\omega_0 t}, \qquad n = 0, \pm 1, \ldots \tag{3-106}$$

where the interval under consideration is $(t_0, t_0 + T_0)$ and

$$\omega_0 = 2\pi f_0 = \frac{2\pi}{T_0} \tag{3-107}$$

Thus, *partial sums of exponential (and trigonometric) Fourier series minimize the integral-square error between the series and the signal under consideration.*

Summary

In this chapter we have considered the representation of periodic signals by trigonometric sums and by rotating phasor sums whose terms have frequencies which are harmonically related. The former representation is called the trigonometric Fourier series, and the latter is referred to as an exponential Fourier series. Fourier series representations facilitate the plotting of spectra for periodic signals, which consist of discrete lines spaced at integer multiples of the fundamental frequency. Although we naturally think of the Fourier series as being a useful representation for periodic signals, an aperiodic signal may be represented by a Fourier series over a finite interval. We also returned to the idea of a frequency response function of a fixed, linear system in this chapter which is a complex function of frequency. A distortionless system has an output which is a scaled, delayed replica of the input signal; for such systems, the magnitude of the transfer function is a constant, and its argument is a linear function of frequency. The following are the major points covered in this chapter:

1. A *trigonometric Fourier series* has the form

$$x(t) = a_0 + \sum_{n=1}^{\infty} a_n \cos n\omega_0 t + \sum_{n=1}^{\infty} b_n \sin n\omega_0 t$$

   where

$$a_0 = \frac{1}{T_0} \int_{T_0} x(t)\, dt$$

$$a_n = \frac{2}{T_0} \int_{T_0} x(t) \cos n\omega_0 t\, dt, \qquad n \neq 0$$

   and

$$b_n = \frac{2}{T_0} \int_{T_0} x(t) \sin n\omega_0\, dt$$

   where $\omega_0 = 2\pi/T_0$, with T_0 being the period of the periodic signal or the interval of expansion if the series is used to approximate an aperiodic function over a finite interval.

2. At points of continuity of x, a Fourier series converges pointwise under fairly lenient restrictions. At points of jump discontinuity of x,, say at $t = t_0$, a Fourier series converges to the average of the left- and right-hand limits of $x(t)$ at t_0.

3. The Gibbs phenomenon of a Fourier series, to be treated more precisely in Chapter 4, refers to the tendency of the partial sums of a Fourier series to overshoot a jump discontinuity of a signal, no matter how many terms are included in the series. This is illustrated by Example 3-4 for a square wave.

4. A *complex exponential Fourier series* has the form

$$x(t) = \sum_{n=-\infty}^{\infty} X_n e^{jn\omega_0 t}$$

where

$$X_n = \frac{2}{T_0} \int_{T_0} x(t) e^{-jn\omega_0 t}\, dt$$

5. The coefficients for the trigonometric and exponential forms of the Fourier series are related by

$$X_0 = a_0$$
$$X_n = \tfrac{1}{2}(a_n - jb_n), \qquad n > 0$$

and

$$X_n = \tfrac{1}{2}(a_{-n} + jb_{-n}), \qquad n < 0$$

or

$$a_n = 2\, \mathrm{Re}\, X_n, \qquad n > 0$$

and

$$b_n = -2\, \mathrm{Im}\, X_n, \qquad n > 0$$

6. The following *symmetry properties* hold for the Fourier coefficients of a real signal:

$$X_n = X_{-n}^*$$
$$|X_n| = |X_{-n}|$$
$$\underline{/X_n} = -\underline{/X_{-n}}$$

If a signal is even, then
  a. The X_n's are real.
  b. All b_n's are zero.
On the other hand, if a signal is odd, then
  a. The X_n's are imaginary.
  b. All a_n's are zero.
If a signal is half-wave (odd) symmetrical, i.e., $x(t) = -x(t - T_0/2)$, then all even-indexed co-efficients are zero. If a signal has zero average value, then $X_0 = a_0 = 0$.

7. *Parseval's theorem* relates to the average power contained in a periodic signal and states that

$$P_{av} = \frac{1}{T_0} \int_{T_0} |x(t)|^2\, dt = \sum_{n=-\infty}^{\infty} |X_n|^2 = X_0^2 + 2\sum_{n=1}^{\infty} |X_n|^2$$

That is, the power can be found by averaging the square of the time-domain signal over a period, or by summing the powers in its rotating phasor components, or by summing the powers in its sinusoidal components plus the power in the dc component.

8. Line spectra for a signal are plotted from its Fourier series representation. The *two-sided amplitude spectrum* of a signal is obtained by plotting the magnitude of its complex exponential Fourier series coefficients versus frequency. Its *two-sided phase spectrum* is obtained by plotting the arguments of its complex exponential Fourier series coefficients versus frequency. The *single-sided amplitude spectrum* is obtained by doubling each line in the double-sided amplitude spectrum for $f > 0$ and leaving the line at $f = 0$ as it is. The *single-sided phase spectrum* is

obtained by keeping each line in the double-sided amplitude spectrum for $f \geq 0$ as it is. (Single-sided line spectra are zero for $f < 0$, as their name implies.)

9. The *frequency response function* of a system is given by the integral

$$H(\omega) = \int_{-\infty}^{\infty} h(t)e^{-j\omega t}\, dt$$

where $h(t)$ is its impulse response.

10. For any periodic input expressed in the form of a complex exponential Fourier series, the *steady-state output* of a fixed, linear system may be expressed as

$$y(t) = \sum_{n=-\infty}^{\infty} X_n H(n\omega_0)e^{jn\omega_0 t}$$

11. If $H(\omega)$ is represented in polar form as

$$H(\omega) = A(\omega)e^{j\theta(\omega)}$$

where $A(\omega)$ is called the *amplitude response* and $\theta(\omega)$ is called the *phase response* of the system, it follows that the steady-state output of a fixed, linear system can be expressed as

$$y(t) = X_0 H(0) + \sum_{n=1}^{\infty} 2|X_n| A(n\omega_0) \cos[n\omega_0 t + \theta_n + \theta(n\omega_0)]$$

12. For $h(t)$ real,

$$A(\omega) = A(-\omega)$$

and

$$\theta(\omega) = -\theta(-\omega)$$

13. A fixed, linear system that has no effect on its input except, possibly, for an amplitude scaling and delay which are the same for all input-frequency components is called *distortionless*. Such a system is defined by the input-output relationship $y(t) = Kx(t - \tau_0)$, where K is the amplitude scaling factor and τ_0 is the delay. Its amplitude and phase responses are given by

$$A(\omega) = K \quad \text{and} \quad \theta(\omega) = -\tau_0 \omega$$

14. Fixed, linear systems with nonconstant amplitude responses are said to introduce *amplitude distortion* into the input signal. Systems with phase responses that are not linear with frequency are said to introduce *phase distortion*.

15. If the kth derivative of a waveform is piecewise continuous such that its $(k + 1)$st derivative contains impulses, its Fourier spectrum decreases with frequency as the inverse $(k + 1)$st power.

16. An alternative way to develop the Fourier expansion of a signal is as a special case of its minimum integral-squared-error expansion over an interval of interest using a complete orthonormal set of functions over the interval. Such a development shows that when a Fourier series is truncated at any finite number of terms, the best representation of the signal of any equivalent length series is obtained in terms of minimum integral-squared error.

Further Reading

Fourier series and the Fourier transform are introduced in almost all beginning circuits texts, but their use as systems analysis tools are not fully exploited until the student becomes involved in more systems-oriented courses, such as communications and signal processing. Nevertheless, rather than cite references in these specialized areas, we give two that deal with these topics from a more mathematical viewpoint.

E. KREYSZIG, *Advanced Engineering Mathematics,* 6th ed. New York: Wiley, 1988 (Chapter 10).

A. PAPOULIS, *Signal Analysis.* New York: McGraw-Hill, 1977 (Chapter 3).

An older reference that provides a very readable treatment of Fourier analysis is long out of print. However, it is well worth exposure to if the student can obtain a copy. It is

E. A. GUILLEMIN, *The Mathematics of Circuit Analysis.* New York: Wiley, 1949 (Chapter 10).

Problems

Section 3-2

3-1. Plot the first through third partial sums of the series

$$x(t) = \frac{4}{\pi} (\cos \omega_0 t - \frac{1}{3} \cos 3\omega_0 t + \frac{1}{5} \cos 5\omega_0 t - \cdots)$$

Comment on its similarity to a square wave. Is it even or odd? Why is this series an alternating series, whereas the series of Example 3-1 was not? (*Hint:* In calculating the data for making the plot, note that symmetry can be used to save work.)

Section 3-3

3-2. Prove the relationships (3-12) and (3-13), and show all steps in deriving (3-16).

3-3. The uniqueness of the Fourier series means that if we can somehow find the Fourier series of a waveform, we are assured that there is no other waveform with that Fourier series, except for waveforms differing from the waveform under consideration only over an inconsequential set of values of the independent variable (this is referred to in mathematics as a set of measure zero). With this assistance, find the following trigonometric Fourier series of the signals without doing any integration:

(a) $x_1(t) = \cos^2(100\pi t)$;
(b) $x_2(t) = \exp(j200\pi t)$;
(c) $x_3(t) = \sin(2\pi t)\cos^2(10\pi t)$;
(d) $x_4(t) = \cos^3(20\pi t)[1 - \sin^2(10\pi t)]$

3-4. Obtain the trigonometric Fourier series of the square wave

$$x(t) = \begin{cases} A, & \dfrac{-T_0}{4} < t \le \dfrac{T_0}{4} \\[2mm] -A, & \dfrac{-T_0}{2} < t \le \dfrac{-T_0}{4} \quad \text{and} \quad \dfrac{T_0}{4} < t \le \dfrac{T_0}{2} \end{cases}$$

with $x(t) = x(t + T_0)$, all t. Why is it composed of only cosine terms?

3-5. (a) In Example 3-3, the violin string was viewed as one full period of a triagular waveform. Find the Fourier series representation if it is viewed as one-half of a period.
(b) In Figure 3-6, it is seen that the two points of maximum error occur at the ends of the string. With the representation found in part (a), what is the error at the ends of the string now?

Sections 3-4 to 3-6

3-6. Obtain the results for the exponential Fourier series examples given in Table 3-1 with the exception of number 3, which was worked in Example 3-5.

3-7. Suppose that differentiation of the periodic signal $x(t)$ results in a signal that has a Fourier series.
   (a) Integrate by parts the expression

$$X'_n = \frac{1}{T_0} \int_{T_0} \frac{dx}{dt} e^{-jn\omega_0 t} \, dt$$

   to show that

$$X'_n = (jn\omega_0)X_n$$

   That is, the Fourier series coefficients of the signal dx/dt are related to the Fourier series coefficients of $x(t)$ through multiplication by $jn\omega_0$.
   (b) From Table 3-1, waveform 5, obtain the Fourier series coefficients of an odd-symmetry square wave.

3-8. **(a)** Use Parseval's theorem expressed by (3-55a) and (3-55b) to find P_{av} for

$$x(t) = 2 \sin^2(2500\pi t) \cos(2 \times 10^4 \pi t)$$

   (b) If $x(t)$ is a signal that is transmitted through a telephone system which blocks dc and frequencies above 12 kHz, compute the ratio of received to transmitted power.

3-9. Consider the periodic pulse-train signal of Example 3-5. Find the ratio of the power in frequency components for which $|nf_0| \leq \tau^{-1}$ to total power if:
   (a) $T_0/\tau = 2$
   (b) $T_0/\tau = 4$
   (c) $T_0/\tau = 10$

3-10. Obtain the exponential Fourier series of the sawtooth waveform defined by

$$x(t) = At, \quad -\frac{T_0}{2} \leq t < \frac{T_0}{2}$$

$$x(t) = x(t + T_0), \quad \text{all } t$$

3-11. The complex exponential Fourier series of a signal over an interval $0 \leq t \leq T_0$ is

$$x(t) = \sum_{n=-\infty}^{\infty} \frac{1}{1 + j\pi n} e^{j(3\pi n t/2)}$$

   (a) Determine the numerical value of T_0.
   (b) What is the average value of $x(t)$ over the interval $(0, T_0)$?
   (c) Determine the amplitude of the third-harmonic component.
   (d) Determine the phase of the third-harmonic component.
   (e) Write down an expression for the third harmonic term in the Fourier series in terms of a cosine.

3-12. Represent the signal

$$x(t) = e^{-|t|}$$

over the interval $(-1, 1)$ using:
(a) The exponential Fourier series
(b) The trigonometric Fourier series
(c) Sketch the function to which either Fourier series converges

3-13. Obtain complex exponential Fourier series expansions for the following signals without doing any integration. Give the fundamental frequency for each case.
(a) $x_a(t) = \cos^2 20\pi t \sin 10\pi t$
(b) $x_b(t) = \sin^3 30\pi t + 2 \cos 25\pi t$
(c) $x_c(t) = \sin^2 40\pi t \cos^2 20\pi t + \sin 10\pi t \cos 5\pi t$
(*Hint:* Use trigonometric identities and Euler's theorem.)

3-14. Given that the complex exponential Fourier series coefficients of a signal $x(t)$ are $\ldots, X_{-1}, X_0, X_1, \ldots$. Find the Fourier series coefficients of $y(t)$ in terms of those for $x(t)$ if the two signals are related by:
(a) $y(t) = x(t - \tau)$, where τ is a constant;
(b) $y(t) = \exp(j2\pi f_0 t)x(t)$, where f_0 is a constant.

3-15. **(a)** Prove the relationships (3-32a) and (3-32b), which relate the coefficients of the exponential Fourier series to the trigonometric Fourier series.
(b) Show that the trigonometric Fourier series of a real, even signal consists only of cosine terms, and that of an odd signal consists only of sine terms.

3-16. Show the following symmetry properties for the complex exponential Fourier coefficients:
(a) X_n is real and an even function of n if $x(t)$ is real and even.
(b) X_n is imaginary and odd if $x(t)$ is real and odd.
(c) $X_n = 0$ for even n if $x(t) = -x(t \pm T_0/2)$ (i.e., has odd half-wave symmetry).
(*Hint:* (1) One way to do this is to write down the Fourier series for $x(t)$. Substitute $t + T_0/2$ for t to obtain a series for $x(t + T_0/2)$. What must be true of the coefficients in order for this new series to equal the original series for $x(t)$? (2) A second way to prove $X_n = 0$, n even, is to write down the integral for X_n and break the integral into one from $-T_0/2$ to 0 and 0 to $T_0/2$. Change variables in the one from $-T_0/2$ to 0 to make its limits 0 to $T_0/2$. Use the fact that $x(t - T_0/2) = -x(t)$ in the integrand. Consider n even and n odd separately.

3-17. Consider the waveforms in Figure P3-17 and their complex exponential Fourier series. For which ones are the following true: real coefficients; purely imaginary coefficients; even-indexed coefficients zero; $X_0 = 0$?
If true, check the appropriate box in the table.

3-18. Given the periodic waveform shown in Figure P3-18.
(a) What is the value of a_0 in the sine-cosine Fourier series? Why?
(b) What are the values of the b_m coefficients in the sine-cosine Fourier series? Why?
(c) Are the coefficients in the complex exponential Fourier series real, imaginary, or complex? Why?
(d) Does this waveform have half-wave odd symmetry?

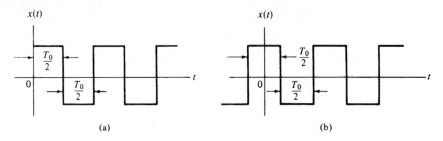

(a)

(b)

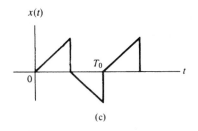

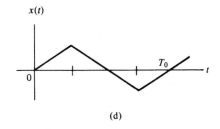

(c)

(d)

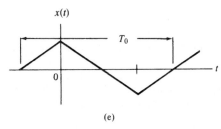

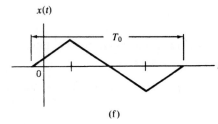

(e)

(f)

FIGURE P3-17

Property	Waveform					
	a	**b**	**c**	**d**	**e**	**f**
Purely real coefficients						
Purely imaginary coefficients						
Complex coefficients						
Even-indexed coefficients zero						
$X_0 = 0$						

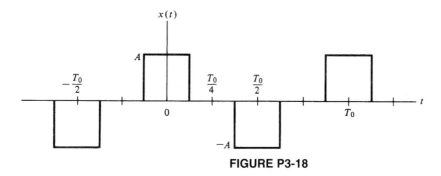

FIGURE P3-18

3-19. The phenomenon of "beats" when two slightly out of tune band instruments play the same nominal note at the same time may be seen by considering the signal

$$x(t) = \cos(\omega_0 t) + \cos[(\omega_0 + \Delta\omega)t]$$

Use trigonometric identies to rewrite this in the form

$$x(t) = A(t)\cos[\omega_0 t + \theta(t)]$$

That is, find explicit expressions for $A(t)$ and $\theta(t)$ in terms of $\Delta\omega$. Plot the waveform for the case of $\omega_0/2\pi = 1000$ hertz and $\Delta\omega/2\pi = 5$ hertz. Discuss how this would sound to the ear.

Section 3-7

3-20. Plot the two-sided amplitude and phase spectra for the full-wave rectified sinewave (waveform 2 of Table 3-1). Convert the two-sided plots to single-sided plots.

3-21. Plot and compare amplitude spectra for the triangular and square waves (waveforms 4 and 5 of Table 3-1). Which requires the most bandwidth for a given fidelity of reproduction?

3-22. Plot and label accurately Figure 3-11 for specific values of τ and T_0 as follows:
   (a) $\tau = 2$ ms; $T_0 = 8$ ms
   (b) $\tau = 1$ ms; $T_0 = 8$ ms
   (c) $\tau = 2$ ms; $T_0 = 16$ ms

Section 3-8

3-23. Show all the steps in the derivation of (3-73).

3-24. For the *RL* filter shown in Figure P3-24, obtain the steady-state output to a triangular waveform with zero average value if $R/L = \omega_0$, where ω_0 is the fundamental frequency of the waveform in rad/s.

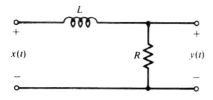

FIGURE P3.24

3-25. If the input to a system is $\cos 10\pi t + 2 \cos 20\pi t$, tell what kinds of distortion, if any, a system introduces if its output is:
- **(a)** $\cos(10\pi t - \pi/4) + 5 \cos(20\pi t - \pi/2)$
- **(b)** $\cos(10\pi t - \pi/4) + 2 \cos(20\pi t - \pi/4)$
- **(c)** $\cos(10\pi t - \pi/4) + 2 \cos(20\pi t - \pi/2)$
- **(d)** $2 \cos(10\pi t - \pi/4) + 4 (20\pi t - \pi/8)$

3-26. A function generator generates approximations of the following types of waveforms: (1) square wave; (2) triangular wave; (3) sine wave.

  The square wave is generated by a multivibrator circuit that has an asymmetry in switching between positive and negative values so that the waveform is actually given by

$$x(t) = \begin{cases} A, & |t| \le T_0/4 + \epsilon/2 \\ -A, & T_0/4 + \epsilon/2 < |t| \le T_0/2 \end{cases}$$

The triangular waveform is generated by integrating the square wave with an operational amplifier integrator as described in Example 1-2.

  The sine wave is obtained by filtering the square wave with a second-order filter with transfer function

$$H(\omega) = \frac{1}{1 + jQ[\omega/\omega_0 - \omega_0/\omega]}$$

where $\omega_0 = 2\pi/T_0$ and Q is a parameter called the filter quality factor.
- **(a)** Sketch the actual and ideal square-wave output. Determine the integral-square error between actual and ideal outputs for the square-wave mode. Divide the integral-square error by T_0 to determine the mean-square error. Plot it as a function of ϵ/T_0 for $A = 1$.
- **(b)** Obtain and sketch the triangular-mode output for an integrator constant $RC = 1$. Other than being inverted due to the operational amplifier circuit, how might one characterize its error from the ideal case?
- **(c)** For the sine-wave mode, define the percent harmonic distortion for the nth harmonic as

$$(HC)_n = \frac{P_{n\text{th har}}}{P_{\text{fund}}} \times 100$$

  - **(i)** Find $(HC)_n$ in terms of ϵ and Q. If $\epsilon/T_0 = 0.05$, what must Q be to make $(HC)_2 = 0.1\%$?
  - **(ii)** Find $(HC)_3$ in terms of ϵ and Q. What must Q be to make $(HC)_3 = 0.05\%$? ($\epsilon/T_0 = 0.05$)
  - **(iii)** Which is the most stringent condition, the one for $(HC)_2$ or the one for $(HC)_3$?

Section 3-9

3-27. Give the rate of convergence of the amplitude spectrum for the signals shown in Problem 3-17(c) and (d) to zero as frequency increases. Justify your answer.

3-28. Give the rate of convergence of the amplitude spectrum for a half-rectified sine wave to zero as frequency increases. Justify your answer.

Section 3-10

3-29. Show that (3-92) and (3-90) are equivalent.

3-30. Given the set of functions

$$f_1(t) = A_1 e^{-t}$$
$$f_2(t) = A_2 e^{-2t}$$
$$\vdots$$

defined on the interval $(0, \infty)$.
   (a) Find A_1 such that $f_1(t)$ is normalized to unity on $(0, \infty)$. Call this function $\phi_1(t)$.
   (b) Find B such that $\phi_1(t)$ and $f_2(t) - B \phi_1(t)$ are orthogonal on $(0, \infty)$. Normalize this new function and call it $\phi_2(t)$.
   (c) Do this for the third function, e^{-3t}. That is, choose C and D such that $f_3 - C\phi_2 - D\phi_1$ is orthogonal to both $\phi_1(t)$ and $\phi_2(t)$. Normalize it and call it $\phi_3(t)$. Comment on the feasibility of continuing this procedure.

3-31. What is the integral-square error in representing a square wave with zero average value by:
   (a) Its fundamental
   (b) Its fundamental plus third harmonic

Computer Exercises

3-1. Using the MATLAB application of Example 3-6 as a model, write a program to compute and plot the partial sums of each waveform given in Table 3-1. You might consider using MATLAB function programs to keep your program well organized.

3-2. Consider the use of a half-wave rectifier followed by an RC lowpass filter to implement a direct current (dc) power supply as shown in the following figure. Write a program to display the output waveform from the RC lowpass filter with a half-wave rectified sine wave at its input. Assume an operational amplifier isolation stage between rectifier and lowpass filter so that you can ignore loading of the rectifier by the filter.

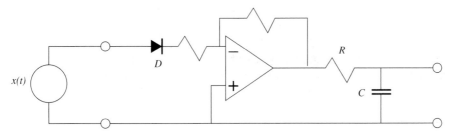

FIGURE P3-01

3-3. Rework Computer Exercise 3-2 for a full-wave rectified sine wave at the input.

3-4. Write a *simulink* program to rework Computer Exercise 3-3 without the isolation amplifier between the diode and the RC lowpass filter. Assume an ideal diode with zero forward resistance and infinite back resistance.

3-5. Rework Examples 3-10 and 3-11 for a triangular waveform at the RC lowpass filter input. Comment on the comparison of your results with those of Examples 3-10 and 3-11.

CHAPTER 4

The Fourier Transform and Its Applications

4-1 Introduction

The Fourier integral may be viewed as a limiting form of the Fourier series of a signal as the period goes to infinity. Thus, it is useful for pulse-type signals in providing spectral characterization and a system analysis tool. Since it converges for a relatively restricted class of signals, it finds limited use in the latter application. However, it also provides a bridge to a more general transform for system analysis, namely, the Laplace transform, which we take up in the next chapter. In this chapter, we provide a derivation of the Fourier transform and prove several theorems useful for deriving Fourier transform pairs. Several applications of the Fourier transform are also illustrated.

4-2 The Fourier Integral

Figure 3-11 showed that the spectral components of a pulse train became closer together as the period of the pulse train increased, while the shape of the spectrum remained unchanged if the pulse width stayed constant. We now use the insight gained in that exercise to develop a spectral representation for nonperiodic signals. Specifically, in the complex Fourier series sum and the integral for the coefficients of the Fourier series, we let the frequency spacing $f_0 = T_0^{-1}$ approach zero and the index n approach infinity such that the product nf_0 approaches a continuous frequency variable f. Repeating the exponential form of the Fourier series below for convenience,

$$x(t) = \sum_{n=-\infty}^{\infty} X_n e^{j2\pi nf_0 t} \tag{4-1}$$

$$X_n = \frac{1}{T_0} \int_{-T_0/2}^{T_0/2} x(t) e^{-j2\pi nf_0 t}\, dt \tag{4-2}$$

we note that the limiting process $f_0 = T_0^{-1} \to 0$ evidently implies that $X_n \to 0$ provided that the integral

$$\left| \int_{-\infty}^{\infty} x(t) e^{-j2\pi nf_0 t}\, dt \right| \le \int_{-\infty}^{\infty} |x(t)|\, dt \tag{4-3}$$

exists. Thus, before carrying out the limiting process, it is useful to write (4-1) and (4-2) in the forms

$$x(t) = \sum_{nf_0=-\infty}^{\infty} \frac{X_n}{f_0} e^{j2\pi nf_0 t}\, \Delta(nf_0) \tag{4-4}$$

$$\tilde{X}(nf_0) \triangleq \frac{X_n}{f_0} = \int_{-1/2f_0}^{1/2f_0} x(t) e^{-j2\pi nf_0 t}\, dt \tag{4-5}$$

respectively, where the symbol $\Delta(nf_0)$ means the increment in the variable nf_0 which is f_0. The limiting process then formally becomes

$$T_0 \to \infty$$

$$f_0 \to 0 \quad \text{and} \quad n \to \infty \quad \text{such that} \quad nf_0 \to f$$

$$\frac{1}{T_0} = \Delta(nf_0) \to df$$

$$\frac{X_n}{f_0} \to \tilde{X}(f)$$

which results in the sum in (4-4) becoming an integral over the continuous variable f, and the integral in (4-5) extending over the entire t-axis, $-\infty < t < \infty$. The resulting expressions are

$$x(t) = \int_{-\infty}^{\infty} X(f)e^{j2\pi ft}\, df \tag{4-6}$$

and

$$X(f) = \int_{-\infty}^{\infty} x(t)e^{-j2\pi ft}\, dt \tag{4-7}$$

where the tilde has been dropped from $X(f)$ to simplify notation.

These expressions define a *Fourier transform pair* for the signal $x(t)$, and are sometimes denoted by the notation $x(t) \leftrightarrow X(f)$, where the double-headed arrow means that $X(f)$ is obtained from $x(t)$ by applying (4-7) and $x(t)$ is obtained from $X(f)$ by applying (4-6). The integral (4-7) is referred to as the *Fourier transform* of $x(t)$, and is frequently denoted symbolically as $X(f) = \mathcal{F}[x(t)]$. Conversion back to the time domain by means of (4-6) is referred to as the *inverse Fourier transform* and is often denoted symbolically as $x(t) = \mathcal{F}^{-1}[X(f)]$.[†]

Several symmetry properties of the Fourier transform may be shown by writing $X(f)$ in terms of magnitude and phase as

$$X(f) = |X(f)|e^{j\theta(f)} \tag{4-8}$$

For example, assuming that $x(t)$ is real, it can be shown that[‡]

$$|X(f)| = |X(-f)| \quad \text{and} \quad \theta(f) = -\theta(-f) \tag{4-9}$$

That is, the magnitude is an even function of frequency and the argument, or phase, is an odd function of frequency. Plots of $|X(f)|$ and $\theta(f) = \underline{/X(f)}$ with f chosen as the abscissa are referred to as the amplitude and phase spectra of $x(t)$, respectively.[*] Thus the amplitude spectrum of a real, Fourier-transformable signal is even, and its phase spectrum is odd.

[†]Sufficient conditions for the convergence of the Fourier transform integral are that (1) the integral of $|x(t)|$ from $-\infty$ to ∞ exists, and (2) any discontinuities in $x(t)$ be finite. These being sufficient conditions means that signals exist which violate at least one of them and yet have Fourier transforms. An example is $(\sin \alpha t)/\alpha t$. For a discussion of other criteria that ensure convergence of the Fourier transform integral, see Bracewell (1978).

Later, we will include functions that do not possess Fourier transforms in the ordinary sense by generalization to Fourier transforms in the limit. Examples are the signals $x(t) = $ constant, all t, and $x(t) = \delta(t)$, the former violating condition (1) and the latter, condition (2).

[‡]For complex signals $|X(f)|$ need not be even, and $\theta(f)$ need not be odd.

[*]More properly, the amplitude spectrum should be referred to as the magnitude spectrum, but the former term is customary. Note that $|X(f)|$ is an amplitude, or magnitude, *density*. Its units are volts (or whatever the dimensions of $x(t)$ are) per hertz. This differs from the Fourier coefficients, which have the same dimensions as the original signal, $x(t)$.

Further properties which can be derived for $X(f)$ if $x(t)$ is real is that the *real part* of $X(f)$ is an *even* function of frequency and the *imaginary part* of $X(f)$ is an *odd* function of frequency.

The foregoing properties are proved by expanding $\exp(-j2\pi ft)$ in (4-7) as $\cos 2\pi ft - j \sin 2\pi ft$ by employing Euler's theorem and writing $X(f)$ as the sum of two integrals which are the real and imaginary parts of $X(f)$. Since the frequency dependence of the real part of $X(f)$ is through $\cos 2\pi ft$, it is therefore an even function of f. Similarly, since the frequency dependence of the imaginary part of $X(f)$ is through $\sin 2\pi ft$, we conclude that it is odd.

Further properties that can be proved (see Problem 4-4) about $X(f)$ are:

1. If $x(t)$ is even [i.e., $x(t) = x(-t)$], then $X(f)$ is a *real, even function of f*.
2. If $x(t)$ is odd [i.e., $x(t) = -x(-t)$], then $X(f)$ is an *imaginary, odd function of f*.

To illustrate the computation of Fourier transforms as well as their symmetry properties discussed above, we consider the following example.

EXAMPLE 4-1 _____

The Fourier transform of $x_a(t) = \Pi(t/2)$, which is sketched in Figure 4-1a, is real and even. On the other hand, the Fourier transform of $x_b(t) = \Pi(t + \frac{1}{2}) - \Pi(t - \frac{1}{2})$, which is the odd function sketched in Figure 4-1b, is imaginary and odd. To prove these statements, we derive their Fourier transforms in this example.

Since $\Pi(t/2) = 1$, $-1 \le t \le 1$, and is zero otherwise, the Fourier transform of $x_a(t)$ is given by

$$X_a(f) = \int_{-1}^{1} e^{-j2\pi ft}\, dt$$

$$= -\frac{e^{-j2\pi ft}}{j2\pi f}\Bigg|_{-1}^{1}$$

$$= 2\,\frac{\sin 2\pi f}{2\pi f}$$

$$= 2 \operatorname{sinc} 2f \tag{4-10}$$

which is clearly a real, even function of f.

From the sketch of $x_b(t)$ shown in Figure 4-1 we may write the integral for its Fourier transform as

$$X_b(f) = \int_{-1}^{0} e^{-j2\pi ft}\, dt - \int_{0}^{1} e^{-j2\pi ft}\, dt$$

$$= -\frac{1}{j2\pi f}\left[e^{-j2\pi ft}\Big|_{-1}^{0} - e^{-j2\pi ft}\Big|_{0}^{1} \right]$$

$$= \frac{j}{2\pi f}(2 - 2\cos 2\pi f)$$

$$= j2\,\frac{\sin^2 \pi f}{\pi f}$$

$$= j2\pi f \operatorname{sinc}^2 f \tag{4-11}$$

This is clearly an imaginary, odd function of f.

The amplitude and phase spectra of $x_a(t)$ and $x_b(t)$ are shown in Figure 4-2. The student should be convinced that these are correct, particularly the phase spectrum for $x_b(t)$.

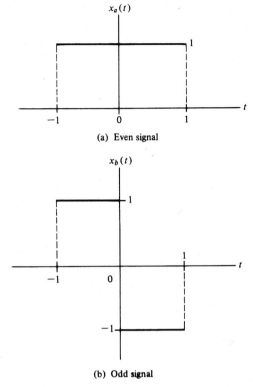

FIGURE 4-1. Even and odd pulse-type signals for illustration of the symmetry properties of Fourier transforms.

4-3 Energy Spectral Density

The energy of a signal can be expressed in the frequency domain by proceeding as follows. By definition of the normalized energy for a signal, given by (1-75), we write

$$E \triangleq \int_{-\infty}^{\infty} |x(t)|^2 \, dt$$

$$= \int_{-\infty}^{\infty} x^*(t) \left[\int_{-\infty}^{\infty} X(f) e^{j2\pi ft} \, df \right] dt \tag{4-12}$$

where $x(t)$ has been written in terms of its Fourier transform. Reversal of the orders of integration results in

$$E = \int_{-\infty}^{\infty} X(f) \left[\int_{-\infty}^{\infty} x^*(t) e^{j2\pi ft} \, dt \right] df$$

$$= \int_{-\infty}^{\infty} X(f) \left[\int_{-\infty}^{\infty} x(t) e^{-j2\pi ft} \, dt \right]^* df$$

$$= \int_{-\infty}^{\infty} X(f) X^*(f) \, df \tag{4-13}$$

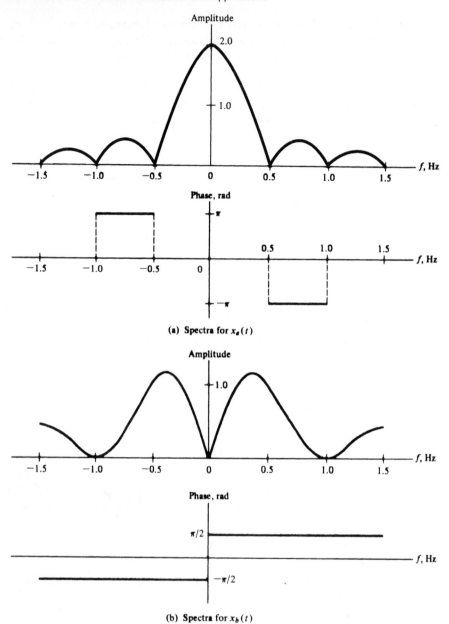

FIGURE 4-2.   Amplitude and phase spectra for the signals of Example 4-1.

or

$$E = \int_{-\infty}^{\infty} |x(t)|^2 \, dt = \int_{-\infty}^{\infty} |X(f)|^2 \, df \qquad (4\text{-}14)$$

This is referred to as *Parseval's theorem for Fourier transforms.*

The units of $|X(f)|^2$, assuming that $x(t)$ is a voltage, are (V-s)² or, since E represents energy on a per ohm basis, (W-s)/Hz = J/Hz. Thus $|X(f)|^2$ has the units of energy density with frequency, and we define the *energy spectral density,* first referred to in connection with (1-87), of a signal as

$$G(f) \triangleq |X(f)|^2 \qquad (4\text{-}15)$$

Integration of $G(f)$ over all frequencies from $-\infty$ to ∞ yields the total (normalized) energy contained in a signal. Similarly, integration of $G(f)$ over a finite range of frequencies gives the energy contained in the signal within the range of frequencies represented by the limits of integration.

EXAMPLE 4-2

The energy in the signal $x(t) = \exp(-\alpha t)u(t)$, $\alpha > 0$, has been shown to be $1/2\alpha$ in Example 1-10. The Fourier transform of this signal is (Problem 4-1)

$$X(f) = \frac{1}{\alpha + j2\pi f} \qquad (4\text{-}16)$$

and its energy spectral density is

$$G(f) = \frac{1}{\alpha^2 + (2\pi f)^2} \qquad (4\text{-}17)$$

The energy contained in this signal in the frequency range $-B < f < B$ is

$$E_B = \int_{-B}^{B} \frac{df}{\alpha^2 + (2\pi f)^2}$$

$$= \frac{1}{\pi\alpha} \int_{0}^{2\pi B/\alpha} \frac{dv}{1 + v^2}$$

$$= \frac{1}{\pi\alpha} \tan^{-1} \frac{2\pi B}{\alpha} \qquad (4\text{-}18)$$

where the fact that the integral is even has been used together with the substitution $v = 2\pi f/\alpha$. The student should plot E_B as a function of B and show that $\lim_{B\to\infty} E_B = 1/2\alpha$, in agreement with the result calculated in Example 1-10.

MATLAB *Application*

Student MATLAB has the capability to manipulate quantities symbolically. The symbolic manipulation capability is illustrated by the following program:

```
%       c4ex2 MATLAB application
%
x=sym('exp(-2*t)*Heaviside(t)') % Symbolically define signal
X=fourier(x)                     % Symbolically take the Fourier transform
Xf=subs(X,'2*pi*f','w')          % Change independent variable to f in
                                     hertz
Xf_conj=subs(Xf,'-i','i')        % Conjugate Fourier transform
GF=Xf*Xf_conj                    % Multiply the Fourier transform and its
                                     conjugate
ezplot(Gf)                       % Use the ezplot routine to plot energy
                                     spectrum
```

The result of running the program (each line above could be run directly from the command window) produces

```
EDU»c4ex2
x=
exp(-2*t)*Heaviside(t)
X=
1/(2+i*w)
Xf=
1/(2+2*i*pi*f)
Xf_conj=
1/(2-2*i*pi*f)
Gf=
1/(2+2*i*pi*f)/(2-2*i*pi*f)
```

The result of ezplot is shown in Figure 4-3.

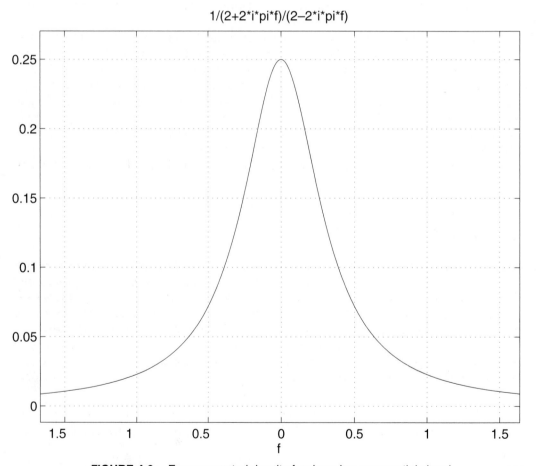

FIGURE 4-3. Energy spectral density for decaying exponential signal.

4-4 Fourier Transforms in the Limit

We now wish to add to our repertoire of Fourier transform pairs. In doing so, we will soon be faced with transforms of signals that are not absolutely integrable, as expressed by (4-3), or which have infinite discontinuities. An example of a signal that is not absolutely integrable is a sinusoid extending over all time from $t = -\infty$ to $t = \infty$. This signal we know has a discrete spectrum and yet cannot be Fourier-transformed through direct application of the defining integral (4-7). An example of a signal that has infinite discontinuities is the impulse function, which in the ordinary sense does not possess a Fourier transform. Rather than exclude such signals, which we know to be useful models, we broaden the definition of the Fourier transform to include Fourier transforms in the limit.

To define the Fourier transform of $\cos \omega_0 t$, $-\infty < t < \infty$, for example, we may first obtain the Fourier transform of the signal $\exp(-\alpha|t|) \cos \omega_0 t$, $\alpha > 0$, which is absolutely integrable.[†] The Fourier transform of $\cos \omega_0 t$ then is defined as the limit of the Fourier transform of $\exp(-\alpha|t|) \cos \omega_0 t$ as $\alpha \to 0$ since $\lim_{\alpha \to 0} \exp(-\alpha|t|) = 1$, $-\infty < t < \infty$. The Fourier transform of $\cos \omega_0 t$ will be given later.

As a second example of obtaining a Fourier transform in the limit, consider the approximation for the unit impulse function $\delta_\epsilon(t) = \Pi(t/2\epsilon)/2\epsilon$ discussed in Chapter 1. The Fourier transform of $\delta_\epsilon(t)$ exists for each finite value of ϵ and is, in fact, given by $\text{sinc } 2\epsilon f$. In the limit as $\epsilon \to 0$, $\delta_\epsilon(t)$ approaches a unit impulse and its Fourier transform approaches 1 and is then defined to be the Fourier transform of the unit impulse function.

As a final, somewhat more subtle, example of a Fourier transform in the limit, consider $x(t) = A$, a constant. It should be clear that the Fourier integral in this case does not exist. (Write it down and try to evaluate the limits after integration!) On the other hand, the Fourier transform of $A\Pi(t/T)$ does exist and is similar to $X_a(f)$ derived in Example 4-1. It is, in fact, $\mathcal{F}[A\Pi(t/T)] = AT \text{ sinc } fT$. If we now let $T \to \infty$ in order to obtain the Fourier transform of a constant, it is not clear what the limit of $AT \text{ sinc } fT$ is. You should sketch this Fourier transform to show that it is a damped oscillatory function with a main lobe centered at $f = 0$ which becomes very narrow and high for T large. Furthermore, $\int_{-\infty}^{\infty} AT \text{ sinc } fT \, df = A$ for any T. Thus we formally write that $\mathcal{F}[A] = A\delta(f)$.

Easier ways of deriving these transforms will be shown after some basic transforms and theorems are considered.

4-5 Fourier Transform Theorems

Several theorems involving operations on signals and the corresponding operations on their Fourier transforms are summarized in Table 4-1. These theorems are useful for obtaining additional Fourier transform pairs as well as in systems analysis. We now discuss these theorems, outlining proofs in some cases, and working examples to illustrate their application. To denote a Fourier transform pair we will sometimes use the notation $x(t) \leftrightarrow X(f)$, which means that $\mathcal{F}[x(t)] = X(f)$ and $\mathcal{F}^{-1}[X(f)] = x(t)$.

Linearity (Superposition) Theorem: Because the Fourier transform is an integral operation on a signal x(t), with x(t) appearing in the defining integral linearly, superposition is an obvious but extremely important property of the Fourier transform.

EXAMPLE 4-3 _____

Find the Fourier transform of $x(t) = \frac{1}{2}[x_a(t) - x_b(t)]$, where $x_a(t)$ and $x_b(t)$ are shown in Figure 4-1.

_____

[†]Without the use of some Fourier transform theorems to be given later, the derivation of this Fourier transform is too lengthy.

TABLE 4-1
Fourier Transform Theorems[a]

Name of Theorem		
1. Superposition (a_1 and a_2 arbitrary constants)	$a_1 x_1(t) + a_2 x_2(t)$	$a_1 X_1(f) + a_2 X_2(f)$
2. Time delay	$x(t - t_0)$	$X(f)e^{-j2\pi f t_0}$
3a. Scale change	$x(at)$	$\lvert a \rvert^{-1} X\left(\dfrac{f}{a}\right)$
b. Time reversal	$x(-t)$	$X(-f) = X * (f)$
4. Duality	$X(t)$	$x(-f)$
5a. Frequency translation	$x(t)e^{j\omega_0 t}$	$X(f - f_0)$
b. Modulation	$x(t) \cos \omega_0 t$	$\frac{1}{2}X(f - f_0) + \frac{1}{2}X(f + f_0)$
6. Differentiation	$\dfrac{d^n x(t)}{dt^n}$	$(j2\pi f)^n X(f)$
7. Integration	$\displaystyle\int_{-\infty}^{t} x(t')\,dt'$	$(j2\pi f)^{-1} X(f) + \frac{1}{2}X(0)\delta(f)$
8. Convolution	$\displaystyle\int_{-\infty}^{\infty} x_1(t - t')x_2(t')\,dt'$ $= \displaystyle\int_{-\infty}^{\infty} x_1(t')x_2(t - t')\,dt'$	$X_1(f)X_2(f)$
9. Multiplication	$x_1(t)x_2(t)$	$\displaystyle\int_{-\infty}^{\infty} X_1(f - f')X_2(f')\,df'$ $= \displaystyle\int_{-\infty}^{\infty} X_1(f')X_2(f - f')\,df'$

[a] $\omega_0 = 2\pi f_0$; $x(t)$ is assumed to be real in 3b.

Solution: This signal is a unit-high square pulse starting at $t = 0$ and ending at $t = 1$. Employing the superposition theorem and using the results of Example 4-1, we have

$$X(f) = X_a(f) - X_b(f) = \frac{1}{2}\left[2\text{sinc}(2f) - j2\pi f\,\text{sinc}^2(f)\right]$$

$$= \frac{1}{2}\left[2\frac{\sin 2\pi f}{2\pi f} - j2\pi f\frac{\sin^2 \pi f}{(\pi f)^2}\right]$$

$$= \frac{1}{2}\left[2\frac{\sin 2\pi f}{2\pi f} - j\frac{1 - \cos 2\pi f}{2\pi f}\right]$$

$$= \frac{1}{j2\pi f}[1 - \cos 2\pi f + j\sin 2\pi f]$$

$$= \frac{1}{j2\pi f}[1 - e^{-j2\pi f}]$$

$$= \frac{e^{j\pi f}}{\pi f}\frac{e^{j\pi f} - e^{-j\pi f}}{2j}$$

$$= \frac{\sin \pi f}{\pi f}e^{-j\pi f} = \text{sinc}(f)e^{-j\pi f} \qquad (4\text{-}19)$$

where Euler's theorem has been used to good advantage. Note that the final result is sinc(f) which, as will be seen later, is the Fourier transform of a unit-high square pulse centered on $t = 0$ and the factor $e^{-j\pi f}$, which is due to the delay of the pulse by $t = \frac{1}{2}$ to start it at $t = 0$ as will be shown in the next theorem.

Time-Delay Theorem: *The time-delay theorem follows by replacing x(t) in (4-7) with x(t − t_0) and making the change in variables t = t' + t_0, whereupon the factor* $\exp(-j2\pi ft_0)$ *is taken outside the integral. The remaining integral is then the Fourier transform of x(t) (note that what we call the variable of integration is immaterial).*

EXAMPLE 4-4

Obtain the Fourier transform of a unit-high square pulse 2 units wide starting at $t = 0$ and ending at $t = 2$.

Solution: This signal can be obtained from $x_a(t)$ of Example 4-1 by delaying it by $t_0 = 1$ unit of time. Thus

$$X(f) = 2 \operatorname{sinc} 2f\, e^{-j2\pi f} \tag{4-20}$$

Scale Change Theorem: *To prove this theorem, we first suppose that a > 0, which gives*

$$\mathscr{F}[x(at)] = \int_{-\infty}^{\infty} x(at)e^{-2\pi ft}\, dt$$

$$= \int_{-\infty}^{\infty} x(t')e^{-j2\pi ft'/a}\, \frac{dt'}{a}$$

$$= \frac{1}{a} X\left(\frac{f}{a}\right) \tag{4-21}$$

where the substitution t' = at has been made. Now consider a < 0 with at = −|a|t and the substitution t' = −|a|t = at made in the integral for $\mathscr{F}[x(at)]$. *This yields*

$$\mathscr{F}[x(at)] = \int_{-\infty}^{\infty} x(-|a|t)e^{-j2\pi ft}\, dt$$

$$= \int_{-\infty}^{\infty} x(t')e^{j2\pi ft'/|a|}\, \frac{dt'}{|a|}$$

$$= \frac{1}{|a|} X\left(\frac{-f}{|a|}\right)$$

$$= \frac{1}{|a|} X\left(\frac{f}{a}\right) \tag{4-22}$$

where the last step results because −|a| = a *for a < 0.*
   As a special case, we let $a = -1$, which results in the *time-reversal theorem*.

EXAMPLE 4-5 _____

An application of the scale change theorem is provided by considering the recording on magnetic tape of a signal at, say, 15 inches per second (ips) and playing it back at $7\frac{1}{2}$ ips. Assuming a flat frequency response for the recorder which is independent of playback speed and an original signal spectrum of

$$X_r(f) = \frac{1}{1 + (f/1000)^2} \tag{4-23}$$

we find the spectrum of the played back signal to be

$$X_{pb}(f) = \frac{2}{1 + (f/500)^2} \tag{4-24}$$

where $a = \frac{1}{2}$ in the scale change theorem. To see that $a = \frac{1}{2}$, consider an easily visualized signal such as a pulse 1 s in duration recorded at 15 ips. When played back at $7\frac{1}{2}$ ips it will be 2 s in duration. Thus $x_{pb}(t) = x_r(at)$ or $x_{pb}(2) = x_r(2a)$, where $t = 2$ corresponds to the trailing edge of the pulse on playback, which must correspond to the trailing edge on record. This requires that $2a = 1$ or $a = \frac{1}{2}$. The spectrum of the played-back signal must be narrower than that of the recorded signal (time is stretched out on playback, thus lowering the frequency content). The student should sketch $X_r(f)$ and $X_{pb}(f)$ and be convinced that they are indeed reasonable.

EXAMPLE 4-6 _____

As another application of the scale change theorem, consider the signal $x_a(t)$ of Example 4-1, for which we obtained the transform pair

$$x_a(t) = \Pi\left(\frac{t}{2}\right) \leftrightarrow 2 \text{ sinc } 2f \tag{4-25}$$

Instead of the Fourier transforms of a pulse of width 2, we wish to generalize this to a pulse of arbitrary width, say τ. Thus we let $a = 2/\tau$ to obtain the transform pair

$$x_a(at) = \Pi\left(\frac{t}{\tau}\right) \leftrightarrow \tau \text{ sinc } \tau f \tag{4-26}$$

Plots of this signal and the magnitude of its Fourier transform are shown in Figure 4-3a. Note that the signal duration and spectral width (or *bandwidth*) are inversely proportional. Even though the spectrum of $x(t)$ is infinite in extent in this case, we take some convenient measure as the bandwidth; in this case the width of the main lobe of the sinc function is taken as the bandwidth, which is $2/\tau$.

Duality Theorem: The duality theorem follows by virtue of the similarity of the direct and inverse Fourier transform relationships, the only difference in addition to the variable of integration being the sign in the exponent. The use of this theorem is illustrated in the following example.

EXAMPLE 4-7 _____

The duality theorem can be used to derive the Fourier transform pair

$$2W \text{ sinc } 2Wt \leftrightarrow \Pi\left(\frac{f}{2W}\right) \tag{4-27}$$

given the transform pair

$$x(t) = \Pi\left(\frac{t}{\tau}\right) \leftrightarrow \tau \operatorname{sinc} \tau f = X(f) \tag{4-28}$$

We do this by replacing f in $X(f)$ by t to get a new function of time, which is

$$X(t) = \tau \operatorname{sinc} \tau t$$

According to the duality theorem, this new time function has the Fourier transform

$$x(f) = \Pi\left(\frac{t}{\tau}\right)\bigg|_{t \to -f} = \Pi\left(\frac{-f}{\tau}\right) \tag{4-29}$$

Defining the new parameter $W = \tau/2$ to make the notation clearer, and noting that $\Pi(-f/\tau) = \Pi(f/\tau)$, we obtain the desired transform pair.

Plots of this signal and its spectrum as given in Figure 4-4 again illustrate the inverse relationship between signal duration and bandwidth. Since this signal is actually infinite in duration, we take a convenient measure for its practical duration, such as the width of its main lobe, which in this case is $1/W$.

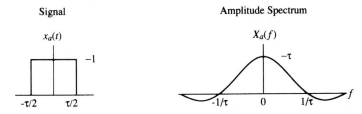

(a) Square-pulse signal

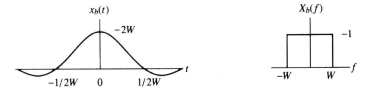

(b) Sinc-function signal

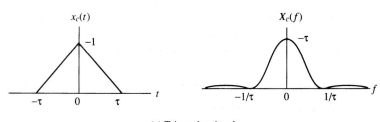

(c) Triangular signal

FIGURE 4-4. Various signals and their spectra.

Frequency Translation Theorem: This theorem is the dual of the time-delay theorem. It is proved by writing down the expression for the Fourier transform of $x(t)e^{j2\pi f_0 t}$, which is

$$\int_{-\infty}^{\infty} x(t)e^{j2\pi f_0 t}e^{-j2\pi ft}\, dt = \int_{-\infty}^{\infty} x(t)e^{-j2(f-f_0)t}\, dt \qquad (4\text{-}30)$$

We recognize the right-hand side as the Fourier transform of $x(t)$, with f replaced by $f - f_0$ and the theorem is proved. The modulation theorem follows from the frequency translation theorem by using Euler's theorem to write $\cos \omega_0 t = \frac{1}{2}e^{j\omega_0 t} + \frac{1}{2}e^{-j\omega_0 t}$ and then applying superposition.

EXAMPLE 4-8

The Fourier transform of the signal

$$x_1(t) = \Pi\!\left(\frac{t}{2}\right) \exp(\pm j20\pi t) \qquad (4\text{-}31)$$

is

$$X_1(f) = 2\, \text{sinc}[2(f \mp 10)] \qquad (4\text{-}32)$$

where the transform pair obtained in Example 4-1 for $x_a(t)$ has been used. From the modulation theorem, the Fourier transform of

$$x_2(t) = \Pi\!\left(\frac{t}{2}\right) \cos 20\pi t \qquad (4\text{-}33)$$

is

$$X_2(f) = \text{sinc}[2(f - 10)] + \text{sinc}[2(f + 10)] \qquad (4\text{-}34)$$

The student should sketch $X_2(f)$. Note that the modulation process shifts the spectrum of a signal to a new frequency. Figure 4-5 shows this signal and its spectrum.

MATLAB *Application*

The MATLAB program below provides a plot of the signal $x_1(t)$ and its Fourier transform. Note that sinc( ) is a MATLAB-defined function. The function program for the square pulse was given in Chapter 1.

```
EDU» c4ex8
%     Plot of signal and spectra of Example 4-8
%
t=-2:.005:2;                    % Define the t and f axes
f=-20:.005:20;
x=pls_fn(t/2).*cos(20*pi*t);    % Use the predefined pulse function to
                                % build up x(t) (page 136)
X=sinc(2*(f - 10))+sinc(2*(f+10));   % sinc() is a built-in MATLAB
                                     % function
subplot(2,1,1),plot(t,x),xlabel('t'),ylabel('x1(t)')
subplot(2,1,2),plot(f,X),xlabel('f'),ylabel('X1(f)')
```

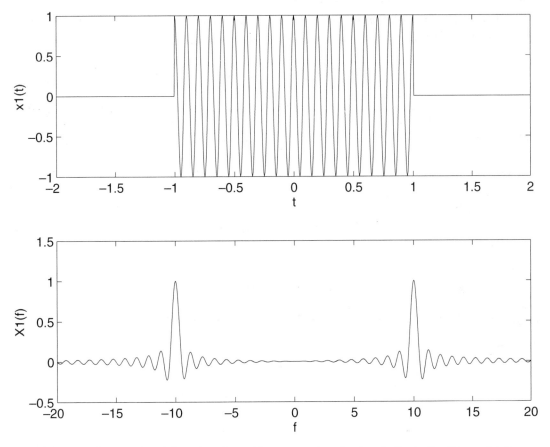

FIGURE 4-5. MATLAB plot of time-domain and frequency domain signals illustrating the modulation theorem. Top: Time-domain signal. Bottom: Fourier transform

EXAMPLE 4-9

Obtain the following transform pairs:

1. $A\delta(t) \leftrightarrow A$
2. $A\delta(t - t_0) \leftrightarrow Ae^{-j2\pi ft_0}$
3. $A \leftrightarrow A\delta(f)$
4. $Ae^{j2\pi f_0 t} \leftrightarrow A\delta(f - f_0)$
5. $A\cos 2\pi f_0 t \leftrightarrow A/2[\delta(f - f_0) + \delta(f + f_0)]$

The impulse function is not properly Fourier transformable. However, its transform is nevertheless useful. The same holds true for a constant. We could find their transforms using limiting operations, as discussed in Section 4-3. It is easier to prove the first transform pair by using the sifting property of the unit impulse:

$$\mathcal{F}[A\delta(t)] = A\int_{-\infty}^{\infty} \delta(t)e^{-j2\pi ft}\,dt = A \tag{4-35}$$

The second pair is obtained by applying the time-delay theorem to pair 1.

The third pair can be shown by using duality in conjunction with the first transform pair. Thus

$$X(t) = A \leftrightarrow A\delta(-f) = x(-f) = A\delta(f) \tag{4-36}$$

where we recall that the impulse function was defined as even in Chapter 1.

Transform pair 4 follows by applying the frequency translation theorem to pair 3, and pair 5 is obtained by applying the modulation theorem to pair 3.

EXAMPLE 4-10

A convenient signal for use in obtaining transforms of periodic signals is the signal

$$y_s(t) = \sum_{m=-\infty}^{\infty} \delta(t - mT_s) \tag{4-37}$$

It is a periodic train of impulses referred to as the *ideal sampling waveform.*

The Fourier transform of $y_s(t)$ can be obtained by noting that it is periodic and, in a formal sense, can be represented by the Fourier series

$$y_s(t) = \sum_{n=-\infty}^{\infty} Y_n e^{j2\pi n f_s t}, \qquad f_s = \frac{1}{T_s} \tag{4-38}$$

The Fourier coefficients are given by

$$Y_n = \frac{1}{T_s} \int_{-T_s/2}^{T_s/2} \delta(t) e^{-j2\pi n f_s t}\, dt = f_s \tag{4-39}$$

where the sifting property of the unit impulse function has been used. Therefore, $y_s(t)$ can be represented by the Fourier series

$$y_s(t) = f_s \sum_{n=-\infty}^{\infty} e^{j2\pi n f_s t} \tag{4-40}$$

Using the Fourier transform pair $e^{j2\pi f_0 t} \leftrightarrow \delta(f - f_0)$, we may take the Fourier transform of this Fourier series term-by-term, to obtain

$$Y_s(f) = f_s \sum_{n=-\infty}^{\infty} \mathcal{F}[e^{j2\pi n f_s t}]$$

$$= f_s \sum_{n=-\infty}^{\infty} \delta(f - nf_s) \tag{4-41}$$

Summarizing, we have obtained the transform pair

$$\sum_{m=-\infty}^{\infty} \delta(t - mT_s) \leftrightarrow f_s \sum_{n=-\infty}^{\infty} \delta(f - nf_s) \tag{4-42}$$

The transform pair (4-42) is useful in spectral representation of periodic signals by the Fourier transform, to be considered in Sections 4-7 and 4-11.

Differentiation and Integration Theorems: *To prove the differentiation theorem, consider*

$$\mathcal{F}\left[\frac{dx}{dt}\right] = \int_{-\infty}^{\infty} \frac{dx(t)}{dt} e^{-j2\pi f t}\, dt \tag{4-43}$$

Using integration by parts with $u = e^{-j2\pi ft}$ and $dv = (dx/dt)\, dt$, we obtain

$$\mathcal{F}\left[\frac{dx}{dt}\right] = x(t)e^{-j2\pi ft}\Big|_{-\infty}^{\infty} + j2\pi f \int_{-\infty}^{\infty} x(t)e^{-j2\pi ft}\, dt \tag{4-44}$$

But if $x(t)$ is absolutely integrable, $\lim_{t\to\pm\infty}|x(t)| = 0$. Therefore, $dx/dt \leftrightarrow j2\pi f X(f)$. Repeated application of integration by parts can be used to prove the theorem for n differentiations.

The integration theorem is proved in a manner similar to that for the differentiation theorem if $X(0) = \int_{-\infty}^{\infty} x(t)\, dt = 0$. If $X(0) \neq 0$, the proof of the integration theorem must be carried out using a transform-in-the-limit approach. The second term of the transform of pair 7, $\frac{1}{2}X(0)\delta(f)$, simply represents the spectrum of the constant $\int_{-\infty}^{\infty} x(t)\, dt$, which is the net area of $x(t)$.

Note that differentiation enhances the high-frequency content of a signal, which is reflected by the factor $j2\pi f$ in the transform of a derivative, while integration smooths out time fluctuations or suppresses the high-frequency content of a signal, as indicated by the $(j2\pi f)^{-1}$ in the transform of an integral.

EXAMPLE 4-11

A useful application of the differentiation theorem is that of obtaining Fourier transforms of piecewise linear signals such as the triangle signal. We generalize the triangle signal in this example to that of a trapezoidal signal and find its Fourier transform through application of the differentiation theorem. The trapezoidal signal is shown in Figure 4-6 along with its first two derivatives. The second derivative consists entirely of unit impulses, and may be written analystically as

$$\frac{d^2x(t)}{dt^2} = K[\delta(t + b) - \delta(t + a) - \delta(t - a) + \delta(t - b)] \tag{4-45}$$

where $K = A/(b - a)$. Applying item 2 in Example 4-9, we find the Fourier transform of the second derivative to be

$$\mathcal{F}\left[\frac{d^2x(t)}{dt^2}\right] = k[e^{j2\pi fb} - e^{j2\pi fa} - e^{-j2\pi fa} + e^{-j2\pi fb}] = 2K[\cos 2\pi fb - \cos 2\pi fa] \tag{4-46}$$

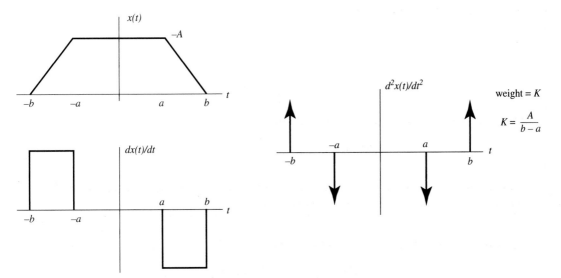

FIGURE 4-6. A trapezoidal pulse and its first two derivatives.

By the differentiation theorem, this transform equals $(j2\pi f)^2 X(f)$. Thus

$$X(f) = 2K \left[\frac{\cos 2\pi fb - \cos 2\pi fa}{(j2\pi f)^2} \right]$$

$$= K \left[b^2 \frac{\sin^2 \pi fb}{(\pi fb)^2} - a^2 \frac{\sin^2 \pi fa}{(\pi fa)^2} \right]$$

$$= K[b^2 \operatorname{sinc}^2(fb) - a^2 \operatorname{sinc}^2(fa)] \tag{4-47}$$

Another way to find the Fourier transform of a trapezoidal pulse is to write it as the difference of two triangle signals as

$$x(t) = B\Lambda(t/b) - (B - A)\Lambda(t/a) \tag{4-48}$$

where $B = Ab/(b - a)$. The unit-high triangle signal, $\Lambda(t/\tau)$, centered on $t = 0$ going from $t = -\tau$ to $t = \tau$ will be shown later to have the Fourier transform $\tau \operatorname{sinc}^2 f\tau$ [in fact this may be seen from (4-47) by setting $a = 0$ and $b = \tau$]. Applying this to (4-48) together with superposition gives

$$X(f) = B\tau b \operatorname{sinc}^2(fb) - (B - A)\tau a \operatorname{sinc}^2(fa) \tag{4-49}$$

which may be demonstrated to be the same result as (4-47). The student should also work out the special cases $a = 0$ and $a = b$.

MATLAB *Application*

We examine the effect of the pitch of the sides of the trapezoid on its spectrum. The following MATLAB program implements a plot of the trapezoid and the corresponding spectrum for several values of a with b fixed.

```
EDU»c4ex11
%       Spectrum of a trapezoidal pulse-Example 4-11
%
t_max=2;
f_max=2;
f=-f_max:.001:f_max;
t=-t_max:.001:t_max;
A=1;
b=1.5;
for k=1:4
k_odd=2*k-1;
k_even=2*k;
a=.5*(k-1)+.0000001;
aa=a;
if aa <=.000001
        aa=0;
end
B=A*b/(b-a);
x=B*trgl_fn(t/b)-(B-A)*trgl_fn(t/a);
K=A/(b-a);
X=K*(b^2*(sinc(f*b)).^2-a^2*(sinc(f*a)).^2);
subplot(4,2,k_even),plot(f,X),axis([-f_max f_max -.8 3.5]),...
        xlabel('f'),ylabel('X(f)')
subplot(4,2,k_odd),plot(t,x),axis([-t_max t_max 0 2]),...
        text(-1.9, 1.6,['a=',num2str(aa)]),xlabel('t'),ylabel('x(t)')
end
```

A plot of the pulse and its spectrum for four values of a are shown in Figure 4-7. Note that as the pulse approaches a triangle its spectrum is more concentrated about $f = 0$. For the limiting case of a square pulse, the spectrum has large lobes along side the main lobe. We infer from this that the smoother transitions of the trapezoid give better spectral containment within the main lobe of the spectrum.

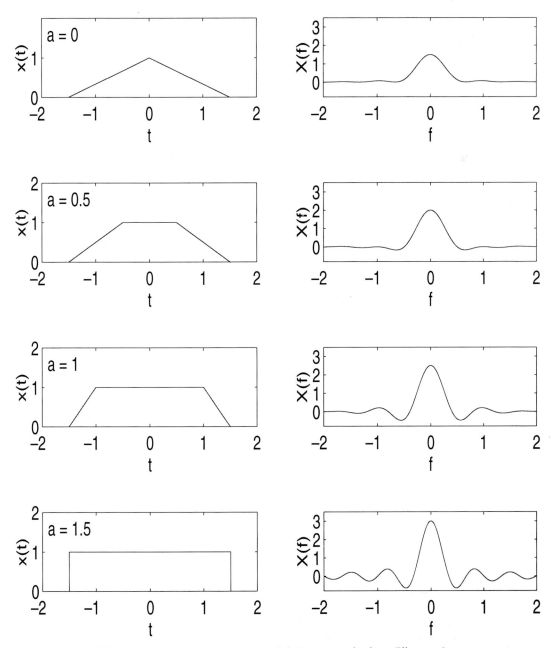

FIGURE 4-7. Trapezoidal pulses and their spectra for four different shapes.

EXAMPLE 4-12

As an application of the integration theorem, consider the Fourier transform of a unit step. Since $u(t) = \int_{-\infty}^{t} \delta(\lambda) \, d\lambda$ and $\int_{-\infty}^{\infty} \delta(t) \, dt = 1$, we obtain

$$\mathcal{F}[u(t)] = (j2\pi f)^{-1} + \tfrac{1}{2}\delta(f) \qquad (4\text{-}50)$$

where $x(t)$ in the integration theorem of Table 4-1 has been taken as $\delta(t)$. A signal that is related to the unit step is the signum function, defined as

$$\operatorname{sgn} t = 2u(t) - 1 = \begin{cases} 1, & t > 0 \\ -1, & t < 0 \end{cases} \qquad (4\text{-}51)$$

Using the result obtained above for the Fourier transform of a step, we find the Fourier transform of sgn t to be

$$\mathcal{F}[\operatorname{sgn} t] = (j\pi f)^{-1} \qquad (4\text{-}52)$$

By the duality theorem, it follows that

$$\mathcal{F}\left[\frac{1}{j\pi t}\right] = \operatorname{sgn}(-f) = -\operatorname{sgn} f \qquad (4\text{-}53)$$

or

$$\mathcal{F}\left[\frac{1}{\pi t}\right] = -j \operatorname{sgn} f \qquad (4\text{-}54)$$

which results because sgn f is odd. The last transform pair is useful in defining the Hilbert transform of a signal, which will be discussed in Example 4-14.

The Convolution Theorem: The proof of the convolution theorem follows by representing $x_2(t - t')$ in the second integral under Theorem 8 of Table 4-1 in terms of its inverse Fourier transform, which is

$$x_2(t - t') = \int_{-\infty}^{\infty} X_2(f)e^{j2\pi f(t-t')} \, df \qquad (4\text{-}55)$$

Substitution into the convolution integral gives

$$\begin{aligned} x_1 * x_2 &= \int_{-\infty}^{\infty} x_1(t')\left[\int_{-\infty}^{\infty} X_2(f)e^{j2\pi f(t-t')} \, df\right] dt' \\ &= \int_{-\infty}^{\infty} X_2(f)\left[\int_{-\infty}^{\infty} x_1(t')e^{-j2\pi ft'} \, dt'\right] e^{j2\pi ft} \, df \end{aligned} \qquad (4\text{-}56)$$

The last step results from reversing the order of the two integrations. Recognizing the bracketed term inside the integral as $X_1(f)$, the Fourier transform of $x_1(t)$, we have obtained

$$x_1 * x_2 = \int_{-\infty}^{\infty} X_1(f)X_2(f)e^{j2\pi ft} \, df \qquad (4\text{-}57)$$

which is the inverse Fourier transform of $X_1(f)X_2(f)$.

This theorem is useful in systems analysis applications as well as for deriving transform pairs. The latter application is illustrated by the following examples.

EXAMPLE 4-13 _____

Consider the transform of the signal that is the convolution of two rectangular pulses of equal width. The student can verify that

$$\Pi\left(\frac{t}{\tau}\right) * \Pi\left(\frac{t}{\tau}\right) = \tau\Lambda\left(\frac{t}{\tau}\right) \tag{4-58}$$

where $\Lambda(t/\tau)$ is the triangular signal. Therefore, according to the convolution theorem,

$$\mathcal{F}\left[\tau\Lambda\left(\frac{t}{\tau}\right)\right] = \left\{\mathcal{F}\left[\Pi\left(\frac{t}{\tau}\right)\right]\right\}^2$$

$$= \tau^2 \operatorname{sinc}^2 f\tau \tag{4-59}$$

EXAMPLE 4-14 _____

The Hilbert transform, $\hat{x}(t)$, of a signal, $x(t)$, is obtained by convolving $x(t)$ with $1/\pi t$. That is,

$$\hat{x}(t) = x(t) * \frac{1}{\pi t} = \frac{1}{\pi}\int_{-\infty}^{\infty}\frac{x(\lambda)}{t - \lambda}\,d\lambda \tag{4-60}$$

From Example 4-12 and the convolution theorem it follows that

$$\mathcal{F}[\hat{x}(t)] = \mathcal{F}\left[\frac{1}{\pi t}\right]\mathcal{F}[x(t)] = -j\operatorname{sgn}(f)X(f) \tag{4-61}$$

A Hilbert transform operation therefore multiplies all positive-frequency spectral components of a signal by $-j$, and all negative-frequency spectral components by j. The amplitude spectrum of the signal is left unchanged by the Hilbert transform operation and the phase is shifted by $\pi/2$ rad.

The Multiplication Theorem: *The proof of the multiplication theorem proceeds in a manner analogous to the proof of the convolution theorem. It is left to the student as a problem. Its application will be illustrated by an example.*

EXAMPLE 4-15 _____

We will use the multiplication theorem to obtain the Fourier transform of the cosinusoidal pulse given by

$$x(t) = A\Pi\left(\frac{t}{\tau}\right)\cos 2\pi f_0 t \tag{4-62}$$

The makeup of this waveform is illustrated in Figure 4-8c on the left-hand side; that is, the product of the waveforms shown in Figure 4-8a and b is the oscillatory pulse shown in Figure 4-8c. Shown on the right-hand side in each part of the figure are the Fourier transforms of the corresponding waveforms on the left. From the multiplication theorem, we know that the convolution of the spectra shown in Figure 4-8a and b is the spectrum of the cosinusoidal pulse of Figure 4-8c. In terms of equations, we have the transform pair

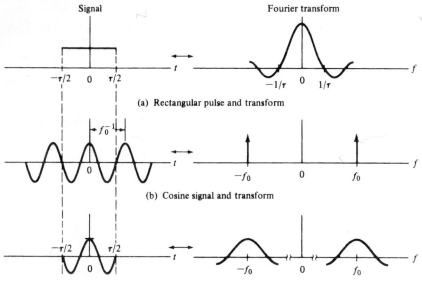

FIGURE 4-8. Application of the multiplication theorem in obtaining the Fourier transform of a finite-duration cosinusoidal signal.

$$A\Pi\left(\frac{t}{\tau}\right) \leftrightarrow A\tau \operatorname{sinc} f\tau \tag{4-63}$$

which was obtained previously. The Fourier transform of $\cos 2\pi f_0 t$ follows by using Euler's theorem to write

$$\mathscr{F}[\cos 2\pi f_0 t] = \mathscr{F}[\tfrac{1}{2}e^{j2\pi f_0 t} + \tfrac{1}{2}e^{-j2\pi f_0 t}]$$

$$= \tfrac{1}{2}\delta(f - f_0) + \tfrac{1}{2}\delta(f + f_0) \tag{4-64}$$

where the superposition theorem has been used together with the transform pair $e^{j2\pi f_0 t} \leftrightarrow \delta(f - f_0)$. The multiplication theorem for this example then states that

$$X(f) = \mathscr{F}\left[A\Pi\left(\frac{t}{\tau}\right)\cos 2\pi f_0 t\right]$$

$$= (A\tau \operatorname{sinc} f\tau) * [\tfrac{1}{2}\delta(f - f_0) + \tfrac{1}{2}\delta(f + f_0)]$$

$$= \frac{A\tau}{2}[\operatorname{sinc}(f - f_0)\tau + \operatorname{sinc}(f + f_0)\tau] \tag{4-65}$$

where use has been made of

$$\operatorname{sinc} f\tau * \delta(f \pm f_0) = \int_{-\infty}^{\infty} \operatorname{sinc} \lambda\tau\, \delta(\lambda - f \pm f_0)\, d\lambda$$

$$= \operatorname{sinc}(f \pm f_0)\tau \tag{4-66}$$

Note that the modulation theorem could have equally well been used to obtain this result.

TABLE 4-2
Fourier Transform Pairs

Pair Number	$x(t)$	$X(f)$	Comments on Derivation
1.	$\Pi\left(\dfrac{t}{\tau}\right)$	$\tau \operatorname{sinc} \tau f$	Direct evaluation
2.	$2W \operatorname{sinc} 2Wt$	$\Pi\left(\dfrac{f}{2W}\right)$	Duality with pair 1, Example 4-7
3.	$\Lambda\left(\dfrac{t}{\tau}\right)$	$\tau \operatorname{sinc}^2 \tau f$	Convolution using pair 1
4.	$\exp(-\alpha t)u(t),\ \alpha > 0$	$\dfrac{1}{\alpha + j2\pi f}$	Direct evaluation
5.	$t\exp(-\alpha t)u(t),\ \alpha > 0$	$\dfrac{1}{(\alpha + j2\pi f)^2}$	Differentiation of pair 4 with respect to α
6.	$\exp(-\alpha|t|),\ \alpha > 0$	$\dfrac{2\alpha}{\alpha^2 + (2\pi f)^2}$	Direct evaluation
7.	$e^{-\pi(t/\tau)^2}$	$\tau e^{-\pi(f\tau)^2}$	Direct evaluation
8.	$\delta(t)$	1	Example 4-9
9.	1	$\delta(f)$	Duality with pair 7
10.	$\delta(t - t_0)$	$\exp(-j2\pi f t_0)$	Shift and pair 7
11.	$\exp(j2\pi f_0 t)$	$\delta(f - f_0)$	Duality with pair 9
12.	$\cos 2\pi f_0 t$	$\frac{1}{2}\delta(f - f_0) + \frac{1}{2}\delta(f + f_0)$	Exponential representation of cos and sin and pair 10
13.	$\sin 2\pi f_0 t$	$\dfrac{1}{2j}\delta(f - f_0) - \dfrac{1}{2j}\delta(f + f_0)$	
14.	$u(t)$	$(j2\pi f)^{-1} + \frac{1}{2}\delta(f)$	Integration and pair 7
15.	$\operatorname{sgn} t$	$(j\pi f)^{-1}$	Pair 8 and pair 13 with superposition
16.	$\dfrac{1}{\pi t}$	$-j\operatorname{sgn}(f)$	Duality with pair 14
17.	$\hat{x}(t) = \dfrac{1}{\pi}\displaystyle\int_{-\infty}^{\infty}\dfrac{x(\lambda)}{t - \lambda}\,d\lambda$	$-j\operatorname{sgn}(f)X(f)$	Convolution and pair 15
18.	$\displaystyle\sum_{m=-\infty}^{\infty}\delta(t - mT_s)$	$f_s\displaystyle\sum_{m=-\infty}^{\infty}\delta(f - mf_s),$ $f_s = T_s^{-1}$	Example 4-10

The Fourier transform pairs derived in the preceding section are collected in Table 4-2 together with other pairs that are often useful. A column is provided that gives suggestions on the derivation of each pair.

4-6 System Analysis with the Fourier Transform

In Chapter 2 the superposition integral was developed as a systems analysis tool. Application of the convolution theorem of Fourier transforms, pair 8 of Table 4-1, to the superposition integral gives the result

$$Y(f) = H(f)X(f) \qquad (4\text{-}67)$$

where $X(f) = \mathscr{F}[x(t)]$, $Y(f) = \mathscr{F}[y(t)]$, and $H(f) = \mathscr{F}[h(t)]$. The latter is referred to as the *transfer function* of the system and is identical to the frequency-response function $H(\omega)$ defined by (3-70) but with

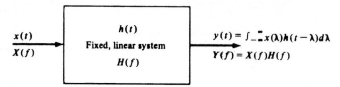

FIGURE 4-9.   Representation of a fixed, linear system in terms of its impulse response or, alternatively, its transfer function.

$f = \omega/2\pi$ as the independent variable. Either $h(t)$ or $H(f)$ are equally good characterizations of the system, since its output can be found in terms of either one as illustrated in Figure 4-9. In the frequency domain, the output of the system is the inverse Fourier transform of (4-67), which is

$$y(t) = \int_{-\infty}^{\infty} X(f)H(f)e^{j2\pi ft}\, df \tag{4-68}$$

Since $H(f)$ is, in general, a complex quantity, we write it as

$$H(f) = |H(f)|e^{j\underline{/H(f)}} \tag{4-69}$$

where $|H(f)|$ is the *amplitude-response* function and $\underline{/H(f)}$ the *phase-response* function of the network. If $H(f)$ is the Fourier transform of a real time function $h(t)$, which it is in all cases being considered in this book, it follows that[†]

$$|H(f)| = |H(-f)| \tag{4-70}$$

and

$$\underline{/H(f)} = -\underline{/H(-f)} \tag{4-71}$$

Since $H(f)$ is such an important system function, we illustrate its computation by the following example.

EXAMPLE 4-16

To obtain the transfer function of the *RC* filter shown in Figure 4-10, any of the methods illustrated in Figure 4-10 may be used. The first method (Fig. 4-10a) involves Fourier transformation of the governing differential equation (integrodifferential equation, in general), which for this system is

$$RC\frac{dy}{dt} + y(t) = x(t), \qquad -\infty < t < \infty \tag{4-72}$$

Assuming that the Fourier transform of each term exists, we may apply the superposition and differentiation theorems to obtain

$$(j2\pi fRC + 1)Y(f) = X(f) \tag{4-73}$$

or

$$H(f) = \frac{Y(f)}{X(f)} = \frac{1}{1 + j(f/f_3)} = \frac{1}{\sqrt{1 + (f/f_3)^2}}\, e^{-j\tan^{-1}(f/f_3)} \tag{4-74}$$

[†]It is useful to model some fixed, linear systems as having complex impulse responses. In such cases (4-70) and (4-71) do not hold.

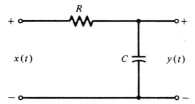

(a) Fourier-transforming the differential equation relating input
and output: $RC \dfrac{dy(t)}{dt} + y(t) = x(t)$

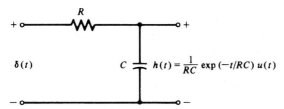

(b) Obtaining the impulse response of the system and finding its
Fourier transform

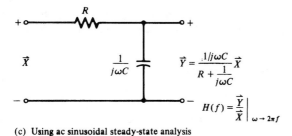

(c) Using ac sinusoidal steady-state analysis

FIGURE 4-10. Illustration of methods that can be used to obtain the transfer function of a system.

The frequency $f_3 = 1/2\pi RC$ is called the 3-dB or half-power frequency.[†] Equivalently, we can use Laplace transform theory as discussed in Chapter 5 and replace s by $j2\pi f$.

A second method is to obtain the impulse response and Fourier transform it to obtain $H(f)$. This method is illustrated in Figure 4-10b.

A third alternative is to use ac sinusoidal steady-state analysis as shown in Figure 4-10c and find the ratio of output to input phasors, $\vec{Y}/\vec{X}$.

The amplitude and phase responses of this system are obtained from (4-74). They are

$$|H(f)| = \left[1 + \left(\frac{f}{f_3}\right)^2\right]^{-1/2} \quad \text{and} \quad \underline{/H(f)} = -\tan^{-1}\frac{f}{f_3} \tag{4-75}$$

respectively. They are illustrated in Figure 4-11.

To end this example, we find the response of the system to the input

$$x(t) = Ae^{-\alpha t}u(t), \qquad \alpha > 0 \tag{4-76}$$

[†]The half-power frequency of a two-port system is defined as the frequency of an input sine wave which results in a steady-state sinusoidal output of amplitude $1/\sqrt{2}$ of the maximum possible amplitude of the output sinusoid. The power of the output sinusoid is then one-half of the maximum possible output power. The ratio of actual to maximum powers in decibels (dB) is therefore (P_{out}/P_{max}) dB $= 10 \log_{10} \frac{1}{2} = -3$ dB.

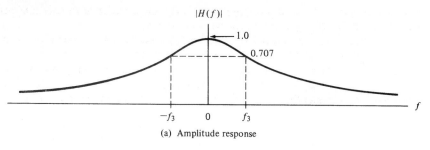

(a) Amplitude response

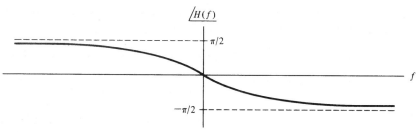

FIGURE 4-11. Amplitude and phase responses for a low-pass *RC* filter.

by using Fourier transform techniques. The Fourier transform of this input, from Table 4-2 is

$$X(f) = \frac{A}{\alpha + j2\pi f} \tag{4-77}$$

The Fourier transform of the output due to this input is

$$Y(f) = \frac{A/RC}{(\alpha + j2\pi f)(1/RC + j2\pi f)} \tag{4-78}$$

For $\alpha RC \neq 1$, this can be written as[†]

$$Y(f) = \frac{A}{\alpha RC - 1}\left[\frac{1}{1/RC + j2\pi f} - \frac{1}{\alpha + j2\pi f}\right] \tag{4-79}$$

The inverse Fourier transform of $Y(f)$ can be found with the aid of Table 4-2 and the superposition theorem. It is given by

$$y(t) = \frac{A}{\alpha RC - 1}\left[\exp\left(\frac{-t}{RC}\right) - \exp(-\alpha t)\right]u(t) \tag{4-80}$$

We can find the response for $\alpha RC = 1$ by using pair 5 of Table 4-2 or by taking the limit of the foregoing result as $\alpha RC \to 1$. In either case, the result is

$$y(t) = A\left(\frac{t}{RC}\right)\exp\left(\frac{-t}{RC}\right)u(t), \qquad \alpha = \frac{1}{RC} \tag{4-81}$$

The student should plot $y(t)$ for various combinations of α and RC and consider what their relationship must be for the output to resemble the input except for a scale change.

[†]Ways of obtaining partial-fraction expansions such as this one will be reviewed in Chapter 5.

As an example of how system characteristics influence the response of the system to a given input, we consider the *RC*-circuit response to a rectangular-pulse input.

EXAMPLE 4-17 _____

Again, we consider the system of Figure 4-10 but with the input

$$x(t) = A\Pi\left(\frac{t - T/2}{T}\right) = A[u(t) - u(t - T)] \tag{4-82}$$

Because the inverse Fourier transform relationship (4-68) is difficult to evaluate in this case, the output is found using time-domain techniques. From Example 2-16 the step response is

$$a_s(t) = (1 - e^{-t/RC})u(t) \tag{4-83}$$

Noting that the $x(t)$ consists of the difference of two steps and using superposition, we find the output to be

$$y(t) = \begin{cases} 0, & t < 0 \\ A(1 - e^{-t/RC}), & 0 \leq t \leq T \\ A[e^{-(t-T)/RC} - e^{-t/RC}], & t > T \end{cases} \tag{4-84}$$

This result is plotted in Figure 4-12 for several values of T/RC together with $|X(f)|$ and $|Y(f)|$. The parameter $T/RC = 2\pi f_3/T^{-1}$ is proportional to the ratio of the 3-dB frequency of the filter to the spectral width (T^{-1}) of the pulse. When the filter bandwidth is large compared with the spectral width of the input pulse, the input is essentially passed undistorted by the system. On the other hand, for $2\pi f_3/T^{-1} << 1$, the system distorts the input signal spectrum and the output does not resemble the input.

Since the energy spectral density of a signal is proportional to the magnitude of its Fourier transform squared, it follows that

$$G_y(f) = |H(f)|^2 G_x(f) \tag{4-85}$$

where $G_x(f)$ and $G_y(f)$ are the energy spectral densities of the system input and output, respectively. Note that the energy spectral density is an even function of frequency. The total energy in the output can be obtained by integrating (4-85) over all frequency. Since $G_y(f)$ is an even function of frequency, the integral can be carried out from $f = 0$ to $f = \infty$ and then doubling this result to get total energy. The energy within any bandwidth of frequencies can be obtained by integrating over that bandwidth. Once again, if the integration is over only positive frequencies, the result of the integration must be doubled.

4-7 Steady-State System Response to Sinusoidal Inputs by Means of the Fourier Transform

In Section 2-7 the frequency response function of a fixed, linear system was defined in terms of the attenuation and phase shift suffered by a sinusoid in passing through the system. We may use (4-67) to

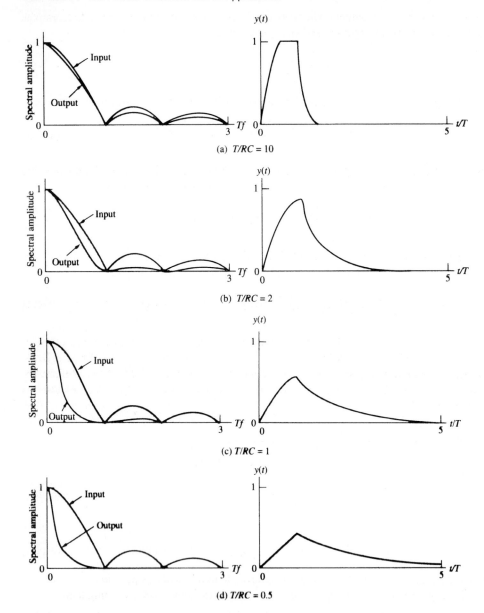

FIGURE 4-12. Input and output signals and spectra for a low-pass *RC* filter (only right side of spectra shown due to evenness).

relate the Fourier transforms of the input and output of a fixed, linear system in response to a periodic waveform. In particular, a periodic input signal $x(t)$ can be represented by its Fourier series as

$$x(t) = \sum_{n=-\infty}^{\infty} X_n e^{j2\pi n\omega_0 t} \qquad (4\text{-}86a)$$

Using the transform pair $e^{j2\pi n f_0 t} \leftrightarrow \delta(f - nf_0)$, this can be Fourier-transformed term-by-term to yield

$$X(f) = \sum_{n=-\infty}^{\infty} X_n \delta(f - nf_0) \qquad (4\text{-}86b)$$

When (4-86b) is multiplied by the system transfer function $H(f)$, and use is made of the relationship $H(f)\delta(f - nf_0) = H(nf_0)\delta(f - nf_0)$, we may express the Fourier transform of the output as

$$Y(f) = \sum_{n=-\infty}^{\infty} X_n H(nf_0)\delta(f - nf_0) \qquad (4\text{-}87)$$

Writing X_n and $H(nf_0)$ in terms of magnitude and phase as $|X_n| \exp j\underline{/X_n}$ and $|H(nf_0)| \exp j\underline{/H(nf_0)}$, respectively, (4-87) may be inverse-Fourier-transformed term-by-term to give the output

$$y(t) = \sum_{n=-\infty}^{\infty} |X_n| |H(nf_0)| \exp\left[j\left\{ 2\pi n f_0 t + \underline{/X_n} + \underline{/H(nf_0)} \right\} \right] \qquad (4\text{-}88)$$

To emphasize the significance of (4-88), note that the nth spectral component of the input, X_n, appears at the output with amplitude attenuated (or amplified) by the amplitude-response function $|H(nf_0)|$, and a phase which is the input phase-shifted by the system phase response $\underline{/H(nf_0)}$, both of which are evaluated *at the frequency of the particular spectral component* under consideration. Note that (4-88) is equivalent to (3-71) or (3-73).

EXAMPLE 4-18

Consider a system with amplitude- and phase-response functions given by

$$|H(f)| = K\Pi\left(\frac{f}{2B}\right) = \begin{cases} K, & |f| \le B \\ 0, & \text{otherwise} \end{cases} \qquad (4\text{-}89a)$$

and

$$\underline{/H(f)} = -2\pi t_0 f \qquad (4\text{-}89b)$$

respectively. A filter with this transfer function is referred to as an *ideal low-pass filter.*
Its output in response to $x(t) = A \cos(2\pi f_0 t + \theta_0)$ is obtained from (4-88) by noting that

$$X_1 = \tfrac{1}{2}Ae^{j\theta_0} = X_{-1}^* \qquad (4\text{-}90)$$

with all other X_n's = 0. Thus the output is

$$y(t) = \begin{cases} 0, & f_0 > B \\ KA \cos[2\pi f_0(t - t_0) + \theta_0], & f_0 \le B \end{cases} \qquad (4\text{-}91)$$

Thus an ideal low-pass filter completely rejects all spectral components with frequencies greater than some cutoff frequency B, and passes all input spectral components below this cutoff frequency except that their amplitudes are multiplied by a constant K and they are phase-shifted by an amount $-2\pi t_0 f_0$ or delayed in time by t_0.

Because ideal filters are used quite often in systems analysis, we consider the characteristics of three types of ideal filters in more detail in the following section.

4-8 Ideal Filters

It is often convenient to work with idealized filters having amplitude-response functions which are constant within the passband[†] and zero elsewhere. In general, three types of ideal filters, referred to as low-pass, high-pass, and bandpass filters, can be defined. A phase-shift function that is a linear function of frequency throughout the passband is assumed in each case. Figure 4-13 illustrates the frequency-response functions for each type of ideal filter.

The impulse response of any filter can be found by obtaining the inverse Fourier transform of its transfer function. For example, for the ideal low-pass filter, the impulse response can be found from[‡]

$$h_{LP}(t) = \int_{-\infty}^{\infty} H_{LP}(f)e^{j2\pi ft}\,df$$

$$= \int_{-B}^{B} Ke^{-j2\pi ft_0}e^{j2\pi ft}\,df$$

$$= \int_{-B}^{B} Ke^{j2\pi f(t-t_0)}\,df$$

$$= 2BK\,\mathrm{sinc}[2B(t - t_0)] \tag{4-92}$$

Since $\mathrm{sinc}[2B(t - t_0)]$ is nonzero for all t, except when $2B(t - t_0)$ takes on an integer value, it follows that $h_{LP}(t)$ is nonzero for $t < 0$. In other words, an ideal low-pass filter is *noncausal,* as are all types of ideal filters.

In fact, it can be shown that the ideal bandpass filter has impulse response

$$h_{BP}(t) = 2KB\,\mathrm{sinc}[B(t - t_0)]\cos[2\pi f_0(t - t_0)] \tag{4-93}$$

and that the ideal high-pass filter has impulse response

$$h_{HP}(t) = K\delta(t - t_0) - 2BK\,\mathrm{sinc}[2B(t - t_0)] \tag{4-94}$$

Both of these impulse responses are nonzero for $t < 0$ if t_0 is finite. Figure 4-14 illustrates ideal filter impulse responses. The noncausal nature of ideal filters is a consequence of their ideal attenuation characteristics in going from the passband to the stop band.

Several methods of approximating ideal filter frequency-response characteristics by means of causal filters are discussed in Appendix E. In general, the more closely a causal filter approximates a corresponding ideal filter, the more delay a signal suffers in passing through it.

4-9 Bandwidth and Rise Time

The rise time of a pulse is the amount of time that it takes in going from a prespecified minimum value, say 10% of the final value of the pulse, to a prespecified maximum value, say 90% of the final value of the pulse. For example, whereas the square pulse input to a low-pass RC filter has zero rise time, the

---

[†]The passband of a filter is defined as the frequency range where its amplitude response is greater than some arbitrarily chosen minimum value. Quite often, this minimum value is chosen as $1/\sqrt{2}$ of the maximum amplitude response.

[‡]B denotes filter bandwidth, and W signal bandwidth.

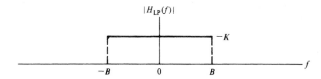

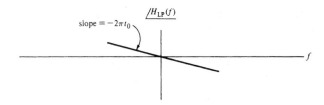

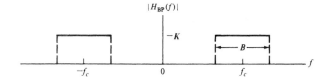

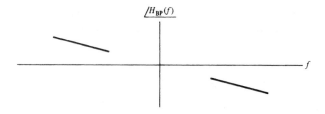

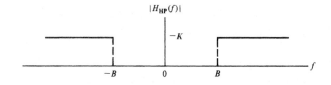

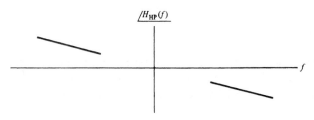

FIGURE 4-13.   Ideal filter amplitude- and phase-response functions.

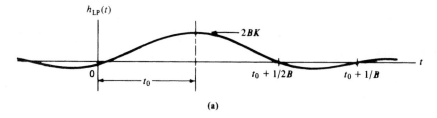

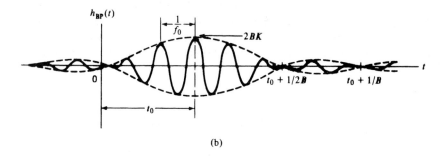

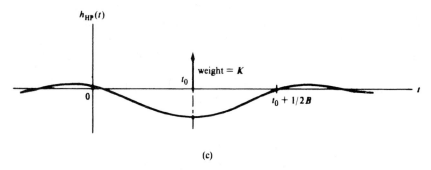

FIGURE 4-14. Impulse responses for ideal filters: (a) Low-pass; (b) bandpass; (c) high-pass.

filter output has a finite rise time as shown by the plots in Figure 4-12. The square-pulse response of this filter is given analytically by (4-84), where the width of the square-pulse input is T and its amplitude is A. The final value of the output pulse is also A. The prespecified minimum value of $0.1A$ (10% of the final value of the output pulse) occurs at time t_m, where

$$A(1 - e^{-t_m/RC}) = 0.1A \tag{4-95a}$$

or

$$\frac{t_m}{RC} = -\ln(0.9)$$

The prespecified maximum value of $0.9A$ (90% of the final value) occurs at time t_M, given by

$$A(1 - e^{-t_M/RC}) = 0.9A \tag{4-95b}$$

or

$$\frac{t_M}{RC} = -\ln(0.1)$$

The rise time is

$$T_R = t_M - t_m = RC[-\ln(0.1) + \ln(0.9)] \tag{4-95c}$$

In terms of the 3-dB bandwidth of the RC filter, which is $f_3 = (2\pi RC)^{-1}$, this becomes

$$T_R = \frac{-\ln(0.1) + \ln(0.9)}{2\pi f_3} = \frac{0.35}{f_3} \tag{4-96}$$

That is, for the lowpass filter the rise time is inversely proportional to the bandwidth. This inverse relationship is illustrated by Example 4-17 and, in particular, Figure 4-12. Although shown only for the particular case of a low-pass RC filter, this inverse relationship between rise time and bandwidth is true in general. In closing this section, note that the 10%–90% definition of rise time used here is only one of the many possible. No matter what the definition of rise time, the inverse relationship between rise time and bandwidth still holds.

*4-10 Window Functions and the Gibbs Phenomenon

In our consideration of trigonometric series at the beginning of Chapter 3, it was noted that the sum of such a series tended to overshoot the signal being approximated at a discontinuity. We can easily examine the reason for this behavior, referred to as the *Gibbs phenomenon,* with the aid of the Fourier transform. Consider a signal $x(t)$ with Fourier transform $X(f)$. We consider the effect of reconstructing $x(t)$ from only the low-pass part of its frequency spectrum. That is, we approximate the signal by

$$\tilde{x}(t) = \mathcal{F}^{-1}\left[X(f)\Pi\left(\frac{f}{2W}\right)\right] \tag{4-97}$$

where

$$\Pi\left(\frac{f}{2W}\right) = \begin{cases} 1, & |f| \le W \\ 0, & \text{otherwise} \end{cases} \tag{4-98}$$

According to the convolution theorem of Fourier transform theory,

$$\tilde{x}(t) = x(t) * \mathcal{F}^{-1}\left[\Pi\left(\frac{f}{2W}\right)\right]$$

$$= x(t) * (2W \, \text{sinc} \, 2Wt) \tag{4-99}$$

where the last step follows by virtue of the transform pair developed in Example 4-7. Recalling that convolution is a folding-product, sliding-integration process, it is now apparent why the overshoot phenomenon occurs at a discontinuity. Figure 4-15 illustrates (4-99) for $W = 2$ assuming that $x(t)$ is a unit-square pulse. In accordance with (4-99), as W increases, more and more of the frequency content of the rectangular pulse is used to obtain the approximation $\tilde{x}(t)$. Nevertheless, we see that a finite value of W means that the signal $x(t)$ is viewed through the *window function* $2W \, \text{sinc} \, 2Wt$ in the time domain.[†] As a result, anticipatory wiggles are present in $\tilde{x}(t)$ before the onset of the main pulse, and echoing wiggles remain after the discontinuity of the leading edge of $x(t)$ passes.

To put this discussion on a mathematical basis, we represent the square-pulse signal as

$$x(t) = u(t) - u(t - 1) \tag{4-100}$$

[†]Actually, it is often customary to refer to the multiplying function $\Pi(f/2W)$ as the window function.

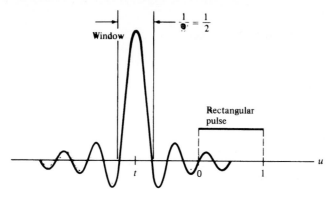

FIGURE 4-15. Convolution of the window function $2W$ sinc $2Wt$ with a rectangular pulse.

which results in

$$\tilde{x}(t) = [u(t) - u(t-1)] * 2W \text{ sinc } 2Wt \qquad (4\text{-}101)$$

Considering, for the moment, only the first term of (4-101), we write it as

$$\tilde{u}(t) = u(t) * 2W \text{ sinc } 2Wt = 2W \text{ sinc } 2Wt * u(t)$$

$$= 2W \int_{-\infty}^{\infty} \text{sinc } 2W\lambda \, u(t-\lambda) \, d\lambda$$

$$= 2W \int_{-\infty}^{t} \text{sinc } 2W\lambda \, d\lambda \qquad (4\text{-}102)$$

This integral cannot be evaluated in closed form but can be expressed in terms of the sine-integral function, Si(x), defined as

$$\text{Si}(x) = \int_{0}^{x} \frac{\sin u}{u} \, du \qquad (4\text{-}103)$$

which is available in tabular form.[†] It is useful to note that Si(x) is an even function and that Si(∞) = $\pi/2$. In terms of Si(x), $\tilde{u}(t)$ can be expressed as

$$\tilde{u}(t) = \begin{cases} \dfrac{1}{2} + \dfrac{1}{\pi} \text{Si}(2\pi Wt), & t > 0 \\[2mm] \dfrac{1}{2} - \dfrac{1}{\pi} \text{Si}(2\pi Wt), & t < 0 \end{cases} \qquad (4\text{-}104)$$

A plot of $\tilde{u}(t)$ versus Wt is shown in Figure 4-16a.

From (4-101) it follows that we may obtain $\tilde{x}(t)$ from $\tilde{x}(t)$ as

$$\tilde{x}(t) = \tilde{u}(t) - \tilde{u}(t-1) \qquad (4\text{-}105)$$

an operation that is easily carried out graphically; $\tilde{x}(t)$ is shown in Figures 4-16b and c for $W = 2$ and $W = 5$. The approximation of $\tilde{x}(t)$ to $x(t)$ becomes better as W increases relative to the pulse width,

[†]M. Abramowitz and A. Stegun, *Handbook of Mathematical Functions with Formulas, Tables, and Graphs* (New York: Dover, 1972), p. 243.

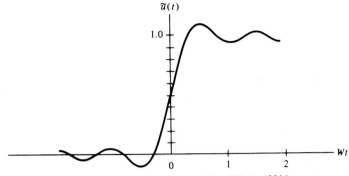

(a) A unit step viewed through the window $2W$ sinc $(2Wt)$

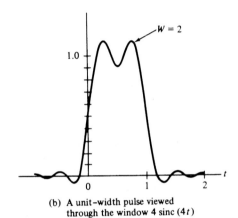

(b) A unit-width pulse viewed
through the window 4 sinc $(4t)$

FIGURE 4-16. Construction of windowed rectangular pulse.

except for the overshoot at the discontinuities at $t = 0$ and $t = 1$, which eventually approaches a value of 9% of the pulse height as W becomes large.

The question of what can be done to combat this behavior is answered by considering the possibility of reducing the sidelobes of the $2W$ sinc $2Wt$ window function, which was the inverse Fourier transform of $\Pi(f/2W)$. The function

$$A(f) = \begin{cases} 0.5 + 0.5 \cos \dfrac{\pi f}{W}, & |f| \le W \\[2mm] 0, & \text{otherwise} \end{cases} \tag{4-106}$$

provides a smooth transition from zero at $f = \pm W$ to its maximum value of unity at $f = 0$.[†] Its inverse Fourier transform is

$$a(t) = W[\text{sinc } 2Wt + \tfrac{1}{2} \text{sinc}(2Wt - 1) + \tfrac{1}{2} \text{sinc } (2Wt + 1)] \tag{4-107}$$

This function and its transform are shown in Figure 4-17, where it is seen that its sidelobes are much lower than the function $2W$ sinc $2Wt$.

[†]This function is called a *Hanning window.*

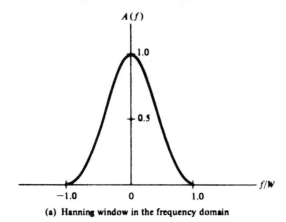

(a) Hanning window in the frequency domain

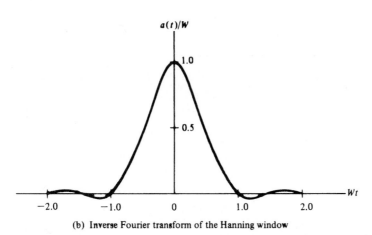

(b) Inverse Fourier transform of the Hanning window

FIGURE 4-17. Window with smoother transitions than the rectangular window.

Note that the inverse relationship between rise time and bandwidth discussed in the previous section is illustrated by Figure 4-16a. The waveform can be viewed as the response of an ideal low-pass filter to a unit step input. From this figure it is seen that the 10%–90% rise time is approximately $T_R = 0.5/W$. In this case W appears as the filter bandwidth rather than B because of the notation used in (4-97), from which we considered building up a square-pulse signal from a finite range of its frequency spectrum.

EXAMPLE 4-19 _____

To study the effect of window half width, W, the following MATLAB program cycles through four windows widths for both rectangular and Hanning windows. It produces the plots shown in Figure 4-18.

```
%       c4ex19
%
t_max=2;
t=-t_max+1:.01:t_max;
L=length(t);
for k=1:4                              % Loop to do 4 values of window width
k_even=2*k;
```

```
k_odd=2*k-1;
W=2+(k-1);
tp=[2*t(1):.01:2*t(L)];
w_r=2*W*sinc(2*W*t);                    % Inverse transform of rectangular
                                        % window
w_h=W*(sinc(2*W*t)+.5*sinc(2*W*t-1)+.5%sinc(2*W*t+1));  % Hanning window
x=pls_fn(t-.5);
x_tilde_r=.01*conv(w_r,x);              % Convolve rectangular window with
                                        % pulse
x_tilde_h=.01*conv(w_h,x);             % Convolve Hanning window with pulse
subplot(4,2,k_odd),plot(tp, x_tilde_r),xlabel('t'),
ylabel('wind.x(t)'),...
text(-.9,1,'rectan.'),text(-.9,.5,['W = ',num2str(W)]),
axis([-t_max+1 t_max -.2 1.5])
subplot(4,2,k_even),plot(tp, x_tilde_h),xlabel('t'),
ylabel('wind. x(t)'),...
text(-.9,1,'Hann.'),text(-.9,.5,['W = ',num2str(W)]),
axis([-t_max+1 t_max -.2 1.5])
end
```

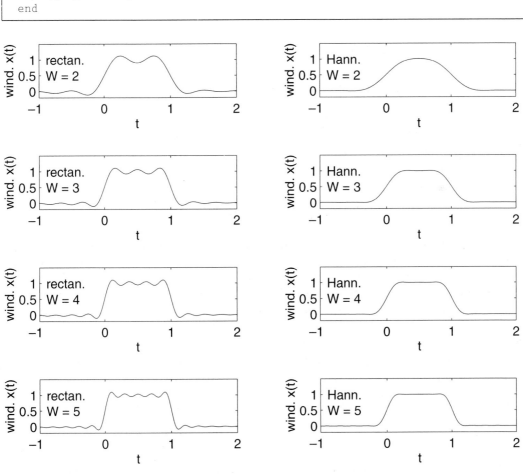

FIGURE 4-18.   Effect of windowing the spectrum of a rectangular pulse with rectangular (left) and Hanning (right) frequency windows for four windows half widths, W.

*4-11 Fourier Transforms of Periodic Signals

The Fourier transform of a periodic signal, in a strict mathematical sense, does not exist since periodic signals are not Fourier transformable. However, using the transform pairs derived in Example 4-9 for a constant and a phasor signal we could, in a formal sense, write down the Fourier transform of a periodic signal by Fourier-transforming its complex Fourier series term-by-term. Thus, for a periodic signal $x(t)$ with Fourier series $\sum_{n=-\infty}^{\infty} X_n e^{jn\omega_0 t}$, where $T_0 = 2\pi/\omega_0 = 1/f_0$ is the period, we have the transform pair

$$\sum_{n=-\infty}^{\infty} X_n e^{jn\omega_0 t} \leftrightarrow \sum_{n=-\infty}^{\infty} X_n \delta(f - nf_0) \tag{4-108}$$

Either representation in (4-108) contains the same information about $x(t)$, and either result can be used to plot the two-sided spectra of a signal. If the Fourier transform representation is used [right-hand side of (4-108)], the amplitude spectrum consists of impulses rather than lines.

A somewhat more useful form for the Fourier transform of a periodic signal than (4-108) is obtained by applying the convolution theorem and pair 17 of Table 4-2 for the ideal sampling wave. To obtain it, consider the result of convolving the ideal sampling waveform with a pulse-type signal $p(t)$ to obtain a new signal $x(t)$. If $p(t)$ is an energy signal of limited time extent such that $p(t) = 0$, $|t| \geq T/2 \leq T_s/2$, then $x(t)$ is a periodic power signal. This is apparent when one carries out the convolution with the aid of the sifting property of the unit impulse:

$$x(t) = \left[\sum_{m=-\infty}^{\infty} \delta(t - mT_s) \right] * p(t) = \sum_{m=-\infty}^{\infty} p(t - mT_s) \tag{4-109}$$

Applying the convolution theorem and the Fourier transform pair (4-42), the Fourier transform of $x(t)$ is

$$X(f) = \mathcal{F}\left[\sum_{m=-\infty}^{\infty} \delta(t - mT_s) \right] P(f)$$

$$= f_s P(f) \sum_{n=-\infty}^{\infty} \delta(f - nf_s)$$

$$= \sum_{n=-\infty}^{\infty} f_s P(nf_s) \delta(f - nf_s) \tag{4-110}$$

where $P(f) = \mathcal{F}[p(t)]$. Summarizing, we have obtained the Fourier transform pair

$$\sum_{m=-\infty}^{\infty} p(t - mT_s) \leftrightarrow \sum_{n=-\infty}^{\infty} f_s P(nf_s) \delta(f - nf_s) \tag{4-111}$$

The usefulness of (4-111) will be illustrated with an example.

EXAMPLE 4-20 _____

As an example of the use of (4-111), we obtain the spectrum of the signal shown in Figure 4-19 and of its periodic extension. We may write $x_1(t)$ as

$$x_1(t) = \Lambda(t) - \Pi(t - 1.5) \tag{4-112}$$

Using previously derived Fourier transform pairs for the triangle and the pulse, we obtain

$$X_1(f) = \text{sinc}^2 f - \text{sinc} f\, e^{-j3\pi f} \tag{4-113}$$

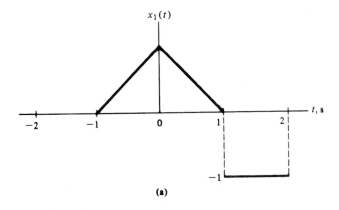

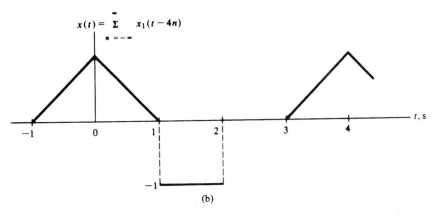

FIGURE 4-19. Pulse signal and its periodic extension.

Recalling that sinc $u = (\sin \pi u)/\pi u$, we see that the spectrum of $x_1(t)$ goes to zero as $1/f$. Checking Figure 4-16a, we see that one derivative is required to produce impulses from the square-pulse portion of $x_1(t)$, which implies the decrease as f^{-1} (see Section 3-11).

We proceed to find the Fourier series of the periodic waveform $x(t)$ by using (4-111). Since $x_1(t)$ is repeated every 4 s to produce $x(t)$, $f_s = 0.25$ in (4-111), and the Fourier transform of $x(t)$, using (4-113) in (4-111), is

$$X(f) = 0.25 \sum_{n=-\infty}^{\infty} [\text{sinc}^2 \, 0.25n - \text{sinc} \, 0.25n \, e^{-j0.75\pi n}] \, \delta(f - 0.25n) \qquad (4\text{-}114)$$

Recalling that $e^{j2\pi f_0 t} \leftrightarrow \delta(f - f_0)$, we may inverse-Fourier-transform (4-114) to obtain the exponential Fourier series of $x(t)$. The result is

$$x(t) = 0.25 \sum_{n=-\infty}^{\infty} (\text{sinc}^2 \, 0.25n - \text{sinc} \, 0.25n \, e^{-j0.75\pi n}) e^{j0.5\pi n t} \qquad (4\text{-}115)$$

Again, it is seen that the predominant term as $n \to \infty$ is sinc $0.25n \exp(-j0.75\pi n)$, which approaches zero as n^{-1}. The student should plot the amplitude and phase spectra of this signal.

*4-12 Applications of the Hilbert Transform

The Hilbert transform, $\hat{x}(t)$ of a signal $x(t)$, was introduced in Example 4-14, and its Fourier transform in terms of $x(t)$ was given as pair 16 of Table 4-2. We discuss two uses of the Hilbert transform in this section.

Analytic Signals

An analytic signal $z(t)$ is a complex-valued signal whose spectrum is single-sided (i.e., is nonzero only for $f > 0$ or $f < 0$). Because of this property of its spectrum, it follows that the real and imaginary parts of an analytic signal cannot be specified independently. It turns out that one is the Hilbert transform of the other. To show this, suppose that $x(t)$ and $y(t)$ are the real and imaginary parts of $z(t)$, respectively. Let $Z(f)$, $X(f)$, and $Y(f)$ be the Fourier transforms of $z(t)$, $x(t)$, and $y(t)$, respectively. It follows that

$$Z(f) = X(f) + jY(f) \tag{4-116}$$

being zero for $f < 0$, say, requires that

$$Y(f) = jX(f), \qquad f < 0 \tag{4-117}$$

[note that $X(f)$ and $Y(f)$ may be complex]. If we double the positive-frequency portion of $Z(f)$, it then follows that

$$Y(f) = -jX(f), \qquad f > 0 \tag{4-118}$$

so that

$$Y(f) = -j \operatorname{sgn}(f) X(f), \qquad \text{all } f \tag{4-119}$$

and

$$Z(f) = \begin{cases} 2X(f), & f > 0 \\ 0, & f < 0 \end{cases} \tag{4-120}$$

From (4-119) and pair 16 of Table 4-2, we see that

$$y(t) = \hat{x}(t) \qquad [Z(f) = 0, \ f < 0] \tag{4-121}$$

where $\hat{x}(t)$ is the Hilbert transform of $x(t)$. Had we required that $Z(f)$ be zero for $f > 0$ and nonzero for $f < 0$, it follows that the signs in front of the j in (4-117) and (4-118) would have been reversed, so that

$$y(t) = -\hat{x}(t) \qquad [Z(f) = 0, \ f > 0] \tag{4-122}$$

Analytic signals are of use in modulation theory applications, and in particular to describe single-side-band modulated signals mathematically.

Causality

A causal system is defined as one whose output does not anticipate the input. If a causal system is also linear *and* fixed, so that it can be characterized by an impulse response, $h(t)$, the causality of the system requires that

$$h(t) = 0, \qquad t < 0 \tag{4-123}$$

The Fourier transform of $h(t)$ is the transfer function $H(f)$ of the system and, from the discussion under analytic signals, it should be apparent that the real and imaginary parts of $H(f)$ cannot be independently specified, but are in fact Hilbert transforms of each other. To show this, consider $h(t)$ written in terms of its even and odd parts as

$$h(t) = h_e(t) + h_o(t) \tag{4-124}$$

where

$$h_e(t) = \tfrac{1}{2}[h(t) + h(-t)] \tag{4-125}$$

and

$$h_o(t) = \tfrac{1}{2}[h(t) - h(-t)] \tag{4-126}$$

In order that $h(t)$ be zero for $t < 0$, it follows that

$$h_o(t) = \begin{cases} h_e(t), & t > 0 \\ -h_e(t), & t < 0 \end{cases}$$

$$= \operatorname{sgn}(t)\, h_e(t) \tag{4-127}$$

Substitution of (4-127) into (4-124) results in

$$h(t) = h_e(t) + \operatorname{sgn}(t)\, h_e(t) \tag{4-128}$$

which, when Fourier-transformed with the aid of pair 14 of Table 4-2 and the multiplication theorem, allows the transfer function to be written as

$$H(f) = H_e(f) + \frac{1}{j\pi f} * H_e(f)$$

$$= H_e(f) - j\hat{H}_e(f) \tag{4-129}$$

where $\hat{H}_e(f)$ is the Hilbert transform of $H_e(f) = \mathcal{F}[h_e(t)]$.

*4-13 The Discrete Fourier Transform

The Fourier transform of a continuous-time signal, $x(t)$, as developed in Chapter 4, is

$$X(f) = \int_{-\infty}^{\infty} x(t)e^{-j2\pi ft}\, dt \tag{4-7}$$

In this section we consider the approximation of (4-7) by a discrete sum, and illustrate fast MATLAB programs, called fast Fourier transform (FFT) algorithms to compute this discrete sum efficiently on a computer. In Chapter 10, several FFT algorithms will be developed mathematically. To this end, we assume that $x(t)$ is nonzero only over the finite time interval $[0, T]$ and is bandlimited to W hertz. We saw earlier by means of several examples that it is not possible for a signal to be both strictly time limited and bandlimited. However, from a practical standpoint, we can define a bandwidth beyond which any strictly time limited signal has negligible energy, and vice versa. It is in this context that we take $x(t)$ to be time limited to T seconds and bandlimited to W hertz. Since $x(t)$ is effectively both time limited and bandlimited, we may represent it, to a good approximation, in terms of samples of $x(t)$.* This yields

$$x(t) \approx \sum_{n=0}^{N-1} x(n\Delta t)\delta(t - n\Delta t) \tag{4-130}$$

*See Chapter 7 for a derivation of the Sampling Theorem.

where $\Delta t = T/N$, with N being the total number of samples taken in $[0, T]$. In keeping with the band-limited nature of $x(t)$, it follows that $\Delta t \leq 1/2W$ in order to avoid significant aliasing. Substituting (4-130) into (4-7) and simplifying by interchanging orders of summation and integration, we find that

$$X(f) = \sum_{n=0}^{N-1} x(n\Delta t)e^{-j2\pi fn\,\Delta t} \tag{4-131}$$

This summation is the Fourier transform of the discrete-time signal represented by the sample values $\{x(n\Delta t)\}$ and is often expressed as a function of the variable $\omega = 2\pi f\,\Delta t = 2\pi r$, where r is the normalized frequency, f/f_s.

Since we are interested in digital computation of (4-131), we restrict f to the discrete set of values $\{0, 1/T, 2/T, \ldots, (N-1)/T\}$. Thus setting $f = k/T = k/(N\,\Delta t)$ in (4-131), where k takes on the discrete set of values $\{0, 1, 2, \ldots, N-1\}$, we obtain

$$X_k = \sum_{n=0}^{N-1} x_n e^{-j2\pi kn/N}, \qquad k = 0, 1, \ldots, N-1 \tag{4-132}$$

The explicit dependence of $x(n\,\Delta t)$ on Δt has been dropped, and both $X(f)$ and $x(t)$ are now replaced by the sequences $\{X_k\}$ and $\{x_n\}$, respectively. This is defined to be the *discrete Fourier transform* of the sequence $\{x_0 = x(0), x_1 = x(\Delta t), \ldots, x_{N-1} = x[(N-1)\,\Delta t]\}$. Because it was derived using a sampling approach, it should be clear that the sequence $\{X_k\}$ is periodic with period N. By replacing k by $k + N$ in (4-132), we see that this is indeed the case.

The original time-domain sequence $\{X_n\}$ is obtained from the sequence of frequency-domain samples $\{X_k\}$ by the inverse relationship

$$x_n = \frac{1}{N}\sum_{k=0}^{N-1} X_k e^{j2\pi nk/N}, \qquad n = 0, 1, \ldots, N-1 \tag{4-133}$$

Thus (4-133) defines the *inverse DFT* operation. We must show that (4-133) with (4-132) substituted produces an identity. Making this substitution, using l for the summation index in (4-132), and summing over k first, we obtain

$$x_n \stackrel{?}{=} \sum_{l=0}^{N-1} \frac{1}{N} x_l \sum_{k=0}^{N-1} e^{j(2\pi k/N)(n-l)} \tag{4-134}$$

where the $\stackrel{?}{=}$ means that the equality is being tested.

The inside sum over k can be written as

$$\sum_k \stackrel{\Delta}{=} \sum_{k=0}^{N-1} [e^{j2\pi(n-l)/N}]^k \tag{4-135}$$

Now the sum of a geometric series is

$$S_N \stackrel{\Delta}{=} \sum_{k=0}^{N-1} x^k = \frac{1 - x^N}{1 - x} \tag{4-136}$$

Letting $x = \exp[j2\pi(n-l)/N]$, we find that

$$\sum_N = \frac{1 - \exp[j2\pi(n-l)]}{1 - \exp[j2\pi(n-l)/N]} = 0, \qquad n \neq l \tag{4-137}$$

which follows because $\exp[j2\pi(n-l)] = 1$. If $n = l$, the result reduces to the indeterminate form $0/0$, so we must sum (4-135) directly for this special case. With $k = l$, the exponential in (4-135) is unity and we obtain

$$\sum_{N} = N, \qquad n = l \tag{4-138}$$

Summarizing, we have shown that

$$\sum_{N} = \sum_{k=0}^{N-1} e^{j(2\pi k/N)(n-l)} = \begin{cases} N, & n = l \\ 0, & n \neq l \end{cases} \triangleq N\delta_{nl} \tag{4-139}$$

where $\delta_{nl} = 1, n = 1$, and $\delta_{kl} = 0, n \neq 1$, is called the *Kronecker delta function*. Thus (4-134) becomes

$$x_n \overset{?}{=} \frac{1}{N} \sum_{l=0}^{N-1} N x_l \delta_{nl} = x_n \tag{4-140}$$

which shows that (4-132) and (4-133) are indeed inverse operations and therefore constitute a valid transform pair.

Before finding a fast computational technique for evaluating (4-132) and (4-133), we note that they are really equivalent operations. From (4-133) it follows that the complex conjugates of the time samples, x_n^*, are given by

$$x_n^* = \frac{1}{N} \sum_{k=0}^{N-1} X_k^* e^{-j(2\pi/N)nk}, \qquad n = 0, 1, \ldots, N - 1 \tag{4-141}$$

so that

$$N x_n^* = \sum_{k=0}^{N-1} X_k^* e^{-j(2\pi/N)nk}, \qquad n = 0, 1, \ldots, N - 1 \tag{4-142}$$

Comparison of (4-142) and (4-132) show that algorithms used to calculate the forward transform can also be used for the calculation of the inverse transform if the frequency samples X_k are conjugated prior to performing the computation. Since the result yields $N x_n^*$ rather than x_n, it must be conjugated and divided by N. Often, however, the time samples are real and the conjugation of the result can be omitted. Thus once we have developed a computationally efficient algorithm for calculating the sum given by (4-132), we can use the same algorithm for computing the inverse DFT.

EXAMPLE 4-21

To illustrate the FFT program in MATLAB consider the Fourier transformation of square pulses of various widths. The following MATLAB program does this for three different pulse widths. After FFT of a pulse, the inverse FFT is taken and this is also plotted to show that the FFT and inverse FFT are indeed inverse operations. The output of the program is given in Figure 4-20. Note that the shorter pulses have the higher frequency content and more overlap takes place in the middle of the spectral plot (high frequencies). Also note that the FFTs are not symmetric about 0, but the normally negative frequency portion is on the right of the plot.

```
%       Example 4-21 - illustration of the FFT and inverse FFT
%       by taking the FFT of a square pulse of various widths
%
T=4;
del_t=.2;
t=0:del_t:T;
L_t=length(t);
```

```
del_f=1/T;
f_max=(L_t-1)*del_f;
f=0:del_f:f_max;
for k=1:3
width=*0.6;
k_1=3*(k-1)+1;
k_2=3*k-1;
k_3=3*k;
x=pls_fn((t - width/2)/width);
X=fft(x);
X_inv=ifft(X);
subplot(3,3,k_1),stem(t,x),axis([0 T 0 1]),...
      xlabel('t'),ylabel('x(t)')
subplot(3,3,k_2),stem(f,abs(X)),axis([0 f_max 0 10]),...
      xlabel('f'),ylabel('X(f)')
subplot(3,3,k_3),stem(t,abs(X_inv)),axis([0 T 0 1]),...
      xlabel('t'),ylabel('X_inv(t)')
end
```

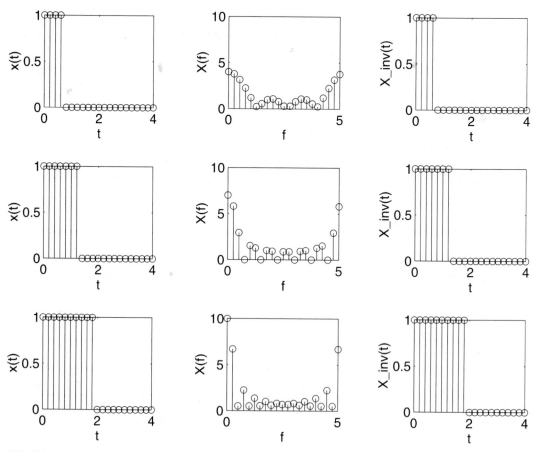

FIGURE 4-20. FFTs (middle column) of square pulses (left column) and inverse FFT of the FFT (right column).

Summary

In this chapter we have considered the representation of signals in terms of the *Fourier transform*. One class of signals to which the Fourier transform representation applies directly is energy signals. Power signals can also be represented in terms of Fourier transforms, but their spectra contain impulses representing finite power (infinite energy) at discrete frequencies. Fourier transforms of energy signals contain no impulsive components. The following are the major points made in this chapter.

1. A signal $x(t)$ and its *Fourier transform* $X(f)$ are related by

$$x(t) = \int_{-\infty}^{\infty} X(f)e^{j2\pi f}\, df$$

$$X(f) = \int_{-\infty}^{\infty} x(t)e^{-j2\pi f}\, dt$$

The former is the inverse relationship, and the latter is the direct relationship.

2. It is useful to represent the Fourier transform of a signal in terms of magnitude and argument or phase as

$$X(f) = |X(f)|e^{j\theta(f)}$$

where $|X(f)|$ plotted versus frequency is known as the *amplitude spectrum*, and $\theta(f)$ plotted versus frequency is called the *phase spectrum*.

3. If a signal is *real*, it follows that its *amplitude spectrum* is an *even* function of frequency and its *phase spectrum* is an *odd* function of frequency. Furthermore, *Fourier transforms of real, even signals are real, even functions of frequency,* while *Fourier transforms of real, odd signals are imaginary, odd functions of frequency.*

4. The *energy spectral density* of a signal is its magnitude-squared Fourier transform.

5. Several useful theorems pertaining to Fourier transforms of signals are summarized in Table 4-1.

6. Several useful Fourier transform pairs are given in Table 4-2. An important conclusion can be drawn from these transform pairs, namely that the *duration and bandwidth of pulse-type signals are inversely proportional.*

7. The Fourier transform, $Y(f)$, of the *output of a fixed, linear system* is related to its transfer function, $H(f)$, and the Fourier transform of the system input, $X(f)$, by

$$Y(f) = H(f)X(f)$$

By taking the magnitude squared of both sides of this equation, it follows that the *energy spectral density of input and output* are related by

$$G_y(f) = |H(f)|^2 G_x(f)$$

where

$$G_x(f) = |X(f)|^2 \quad \text{and} \quad G_y(f) = |Y(f)|^2$$

The *energy contained in the output within a band of frequencies* can be obtained by integrating $G_y(f)$ over the frequency band of interest.

8. The *transfer function* of a fixed, linear system can be obtained by applying any one of the following techniques:

   (a) Fourier-transforming the differential equation relating output to input;

   (b) Fourier-transforming the impulse response of the system; and

(c) Representing the lumped elements of the system in terms of their ac sinusoidal steady-state impedances and using ac circuit analysis.

These three methods are illustrated pictorially in Figure 4-10.

9. *Ideal filters* are filters passing all frequency components of the input within the passband with the same gain and the same time delay, and rejecting completely all frequency components of the input outside the passband. Usually, three types of ideal filters are defined; *low-pass, high-pass,* and *bandpass. Ideal filters are noncausal* since their impulse responses exist before the impulsive input is applied.

10. The *rise time* of a pulse can be defined in various ways. One common method is to define rise time as the time it takes for a pulse to go from 10% to 90% of its final value. It is true in general that a *pulse's bandwidth and rise time are inversely proportional.*

11. The *Gibb's phenomenon* refers to the property that the Fourier approximation to a pulse that undergoes an abrupt change which requires infinite frequency content tends to overshoot the abrupt change by about 9%. This overshoot phenomenon was shown to be directly attributable to the bandlimiting of its spectrum, which can be viewed in the time domain as a convolution with a sinc-function envelope.

12. An *analytic signal* is one whose spectrum is single sided (i.e., is nonzero only for $f \leq 0$ or for $f \geq 0$). Such a signal has real and imaginary parts which are Hilbert transforms of each other. The *Hilbert transform* of a signal is obtained by phase-shifting its positive-frequency spectral components by $-\pi/2$ rad and its negative-frequency components by $\pi/2$ rad.

13. It can be shown that a *causal system* has a transfer function whose real and imaginary parts are Hilbert transforms of each other.

14. The fast Fourier transform is introduced as a means for computing signal spectra efficiently.

Further Reading

In addition to the references given in Chapter 3, the first reference will provide a refreshing approach to Fourier transforms and their applications. The second reference gives extensive information on the FFT.

R. N. BRACEWELL, *The Fourier Transform and Its Applications,* 2nd ed. New York: McGraw-Hill, 1978.

W. W. SMITH and J. M. SMITH, *Handbook of Real-Time Fast Fourier Transforms,* Piscataway, NJ: IEEE Press, 1995.

Problems

Section 4-1

4-1. Obtain the Fourier transforms of the following signals ($\alpha > 0$):

(a) $x_a(t) = Ae^{-\alpha t}u(t)$;

(b) $x_b(t) = Ae^{\alpha t}u(-t)$;

(c) $x_c(t) = Ae^{-\alpha|t|}$;

(d) $x_d(t) = Ae^{-\alpha t}u(t) - Ae^{\alpha t}u(-t)$.

4-2. Plot and compare the amplitude and phase spectra of $x_a(t)$ and $x_b(t)$ in Problem 4-1.

4-3. Plot and compare the amplitude and phase spectra of $x_c(t)$ and $x_d(t)$ in Problem 4-1.

4-4. (a) Show that if $x(t)$ is real and even [i.e., $x(t) = x(-t)$], then the Fourier transform of $x(t)$ may be reduced to

$$X(f) = 2 \int_0^\infty x(t) \cos 2\pi f t \, dt$$

Therefore, conclude that $X(f)$ is real and even in this case.

(b) Show that for $x(t)$ real and odd [i.e., $x(t) = -x(-t)$], then

$$X(f) = -2j \int_0^\infty x(t) \sin 2\pi f t \, dt$$

Therefore, conclude that $X(f)$ is imaginary and odd if $x(t)$ is real and odd.

4-5. (a) Suppose we want the Fourier transform of a complex signal $x(t) = x_R(t) + j x_I(t)$, where $x_R(t)$ is the real part and $x_I(t)$ is the imaginary part. Show that $X(f)$ can be written as

$$X(f) = \int_{-\infty}^\infty x_R(t) \cos 2\pi f t \, dt + \int_{-\infty}^\infty x_I(t) \sin 2\pi f t \, dt$$

$$+ j \left[\int_{-\infty}^\infty x_I(t) \cos 2\pi f t \, dt - \int_{-\infty}^\infty x_R(t) \sin 2\pi f t \, dt \right]$$

(b) Provide an argument for the fact that if $x(t) = x(-t)$ (i.e., even), then the Fourier transform reduces to

$$X(f) = \int_{-\infty}^\infty x_R(t) \cos 2\pi f t \, dt + j \int_{-\infty}^\infty x_I(t) \cos 2\pi f t \, dt$$

$$= \int_{-\infty}^\infty x(t) \cos 2\pi f t \, dt$$

(c) Show that if $x(t) = -x(-t)$ (i.e., odd), then the Fourier transform reduces to

$$X(f) = \int_{-\infty}^\infty x_I(t) \sin 2\pi f t \, dt - j \int_{-\infty}^\infty x_R(t) \sin 2\pi f t \, dt$$

$$= -j \int_{-\infty}^\infty x(t) \sin 2\pi f t \, dt$$

4-6. Using the Fourier transform integral, find Fourier transforms of the following signals:
   (a) $x_a(t) = t \exp(-\alpha t) u(t)$, $\alpha > 0$;
   (b) $x_b(t) = t^2 u(t) u(1 - t)$;
   (c) $x_c(t) = \exp(-\alpha t) u(t) u(1 - t)$, $\alpha > 0$.

4-7. Compute and plot the spectrum of the pulse with raised half-cosine leading and trailing edges as shown.

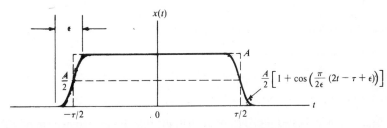

FIGURE P4-7

4-8. Obtain the Fourier transforms of the following signals:
   (a) $x_1(t) = \frac{1}{2}[\delta(t + 1) + \delta(t + \frac{1}{2}) + \delta(t - \frac{1}{2}) + \delta(t - 1)]$
   (b) $x_2(t) = \text{sinc}(t)\, u(t)$
   (c) $x_3(t) = \text{sinc}(t)\, \text{sgn}(t)$
   (d) $x_4(t) = e^{-|t|}u(t)$
   (e) $x_5(t) = e^{-|t|}\, \text{sgn}(t)$

4-9. If the Fourier transform of a real signal $x(t)$ is expressed in terms of its magnitude and phase as

$$X(f) = |X(f)| \exp[j\theta(f)]$$

show that

$$|X(f)| = |X(-f)|$$
$$\theta(f) = -\theta(-f)$$

(i.e., the magnitude is even and the phase is odd).

4-10. Given the following Fourier transforms, find the corresponding time-domain signals:
   (a) $X_1(f) = A\exp(-\alpha|f|)$;
   (b) $X_2(f) = A\exp(-\alpha|f| - j\pi\, \text{sgn}(f))$ where $\text{sgn}(f) = 1$ for $f > 0$ and $\text{sgn}(f) = -1, f < 0$;

Section 4-2

4-11. Obtain and plot the energy spectral densities of the signals given in Problem 4-1.

4-12. (a) Find the energy contained in the signal $x(t) = A\exp(-\alpha t)u(t), \alpha > 0$, for frequencies $|f| < \alpha/\pi$.
   (b) Same question as part (a) for frequencies $|f| < \alpha/2\pi$. What percent of the total energy is this?

Section 4-3

4-13. (a) Use the result of part (d), Problem 4-1, to find the Fourier transform of the signum function

$$\text{sgn } t \triangleq \begin{cases} 1, & t > 0 \\ -1, & t < 0 \end{cases}$$

   (b) Noting that the unit step $u(t)$ can be written in terms of the signum function as

$$u(t) = \frac{1}{2}(\text{sgn } t + 1)$$

   find its Fourier transform.

4-14. Using appropriate theorems, consider the Fourier transform of

$$x(t) = e^{-a|t|} \cos 2\pi f_0 t$$

in the limit as $a \to 0$ to get the Fourier transform of $\cos 2\pi f_0 t$. Be sure to justify the approach of certain normal functions to delta functions in the limit.

4-15. Consider the use of the transform pair

$$\tau^{-1}\Pi\left(\frac{t}{\tau}\right) \leftrightarrow \text{sinc } f\tau$$

to obtain the Fourier transform of $\delta(t)$. In what sense does $\tau^{-1}\Pi(t/\tau)$ approximate a unit impulse? What is $\lim_{\tau \to 0} \text{sinc } f\tau$?

Section 4-4

4-16. (a) Use the superposition and time delay theorems together with the pair $\Pi(t/\tau) \leftrightarrow \tau \, \text{sinc} \, f\tau$ to obtain Fourier transforms for the signals shown.

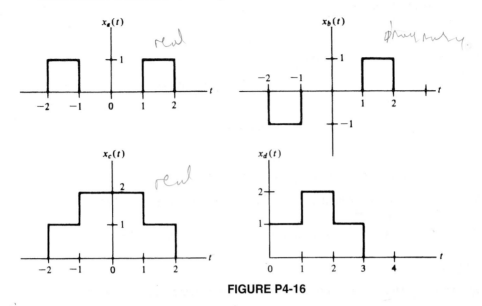

FIGURE P4-16

(b) Using symmetry, show that each transform is real, imaginary, or neither, as the case may be.

4-17. A signal with an approximately rectangular spectrum extending from 100 to 10,000 Hz is recorded on a tape recorder at 15 inches per second (ips). It is played back at $7\frac{1}{2}$ ips. Sketch the spectra for the recorded and played-back signals.

4-18. Use the duality theorem to find the signal whose transform is shown.

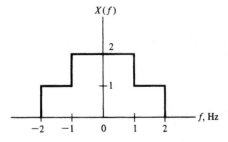

FIGURE P4-18

Hint: Work Problem 4-16 for $x_c(t)$ first.

4-19. If the signals of Problem 4-16 are each multiplied by the signal $\cos 20\pi t$, obtain and sketch the magnitudes of the Fourier transforms of the resulting signals.

4-20. Use the differentiation theorem and the transform pair $\delta(t - t_0) \leftrightarrow \exp(-j2\pi f t_0)$ to find the Fourier transforms of the signals shown. Which transform(s) should be real and even and which should be imaginary and odd?

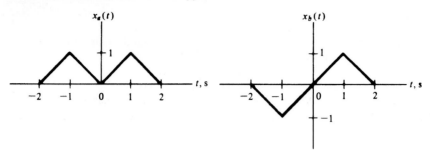

FIGURE P4-20

4-21. Given the signal

$$x_0(t) = (1 - t)u(t)u(1 - t)$$

Obtain its Fourier transform, and using it plus Fourier Transform theorems, find the Fourier transforms of the signal shown below:

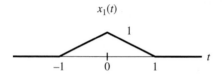

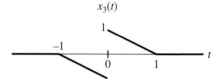

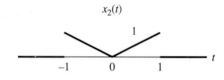

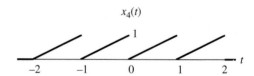

FIGURE P4-21

4-22. Obtain the inverse Fourier transform of

$$X(f) = 10\,\frac{\text{sinc } 2f}{3 + j2\pi f}$$

Hint: Use the convolution theorem.

4-23. **(a)** Prove the multiplication theorem.
(b) Use the multiplication theorem and the Fourier transform pair derived in Example 4-7 to inverse-Fourier-transform

$$X(f) = 15\,\Lambda\!\left(\frac{f}{3}\right)$$

4-24. Consider the rectangular, raised-cosine, and half-cosine pulses given, respectively, by

$$x_R(t) = A\Pi(t/\tau)$$

$$x_{RC}(t) = B[1 + \cos(2\pi t/\tau)]\Pi(t/\tau)$$

$$x_{HC}(t) = C \cos(\pi t/\tau) \, \Pi(t/\tau)$$

(a) Relate the constants A, B, and C to the respective energies of the pulses.

(b) Fixing the amplitudes of each so that all three have equal energies, compute and plot their spectra.

4-25. Use the convolution theorem of Fourier transforms to find the convolutions of the following signals (i.e., *do not* evaluate the convolution integral of Chapter 2):

(a) $x_1(t) = t \exp(-\alpha t) \, u(t)$ and $h_1(t) = \exp(-\beta t) \, u(t)$, α, $\beta > 0$;

(b) $x_2(t) = \exp(-\alpha t) \, u(t)$ and $h_2(t) = t\exp(-\beta t) \, u(t)$, α, $\beta > 0$;

(c) $x_3(t) = \exp(-\alpha t) \, u(t)$ and $h_3(t) = \exp(\beta t) \, u(-t)$, α, $\beta > 0$.

Section 4-6

4-26. (a) Obtain the transfer function of the RC high-pass filter shown.

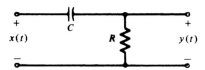

+
$x(t)$
C
R
+
$y(t)$

FIGURE P4-25

(b) Under what conditions will it perform as an ideal differentiator defined by the input-output relationship

$$y(t) = A \frac{dx}{dt}$$

Hint: A differentiator has transfer function $H_d(f) = j2\pi f$.

4-27. The input

$$x(t) = 4 \, \text{sinc}(2t)\cos^2(4\pi t)$$

is applied to a linear, time invariant system. Determine and plot the transfer function of the system, $H(f)$, such that its response will be

(a) $y(t) = 4 \, \text{sinc}(2t)$;

(b) $y(t) = 3 \, \text{sinc}(2t) \cos(8\pi t)$

4-28. Consider an ideal integrator for which input and output are related by

$$y(t) = A \int^{t} x(t') \, dt'$$

Use the integration theorem to obtain its transfer function. Under what conditions will the RC low-pass filter considered in Example 4-16 perform as an integrator?

4-29. Consider a fixed, linear system with amplitude and phase responses as shown.

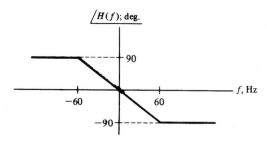

FIGURE P4-29

Obtain the output for the following inputs and tell what type of distortion, if any, results:
(a) $x(t) = \cos 20\pi t + \cos 60\pi t$
(b) $x(t) = \cos 20\pi t + \cos 140\pi t$
(c) $x(t) = \cos 20\pi t + \cos 220\pi t$
(d) $x(t) = \cos 130\pi t + \cos 220\pi t$

4-30. To improve the performance of a pulse-waveform radar in noise, several pulses are sent in sequence and the return from the target is filtered by a narrowband comb filter as illustrated in Figure P4-30. This problem is designed to illustrate the improvement that can be expected as a function of the number of pulses transmitted (and returned from the target).

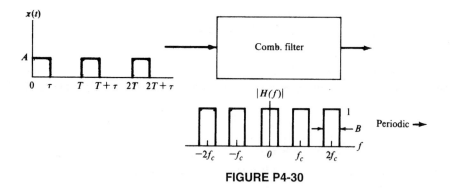

FIGURE P4-30

(a) If a single pulse is transmitted, and filtering from null-to-null of the main lobe of the signal spectrum is performed, what fraction of the signal energy is passed by the filter?
(b) If a sequence of two pulses separated by T seconds is transmitted, show that the spectrum of this waveform is given by

$$X(f) = A\tau \, \text{sinc}(f\tau) \, [1 + e^{-j2\pi fT}]e^{-j\pi f\tau}$$

$$= 2A\tau \, \text{sinc}(f\tau) \cos(\pi fT) \, e^{-j\pi f(T+\tau)}$$

Plot $|X(f)|^2 = G_x(f)$ and discuss how one might design a filter to pass only the peaks in the spectrum, thereby passing most of the signal energy and eliminating some of the background noise which is assumed to have uniform energy at all frequencies.
(c) Consider a train of N pulses of width τ and each separated by T seconds. Show that the resulting spectrum is given by

$$X(f) = A\tau \operatorname{sinc}(f\tau) \sum_{n=0}^{N-1} [e^{-j2\pi fT}]^n e^{-j\pi ft}$$

$$= A\tau \frac{\sin \pi fNT}{\sin \pi fT} \operatorname{sinc}(f\tau) \, e^{-j\pi f[(N-1)T+\tau]}$$

Plot the resulting energy spectral density for $N = 4$, and discuss how the filtering idea presented in part (b) appears to give better performance as N increases.

4-31. A Butterworth low-pass filter has the circuit diagram shown. Obtain and plot $|H(f)|$ and $\underline{/H(f)}$.

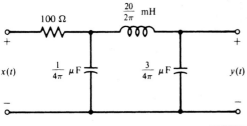

FIGURE P4-31

4-32. The Fourier transforms of two signals, $x(t)$ and $y(t)$, are defined as

$$X(f) = \cos(\pi f) \, \Pi(f)$$

and

$$Y(f) = X(f - f_0) + X(f + f_0)$$

(a) Find a closed-form expression for $x(t)$.
(b) Find a closed-form expression for $y(t)$.
(c) Design the system shown in the block diagram in terms of choosing the parameters A, f_1, and f_2 so that the output is $y(t)$.

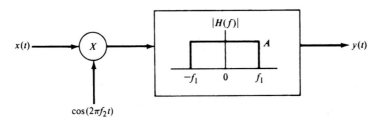

FIGURE P4-32

Section 4-7

4-33. The current source of the RLC circuit shown is a square wave of amplitude A and period T.
(a) Obtain the transfer function of the RLC filter. Plot its amplitude response. Put the latter in terms of the parameters $f_b = R/(2\pi L)$ and $f_0 = 1/2\pi\sqrt{LC}$.
(b) If $1/T = f_0$, obtain and plot, using Fourier transforms, the spectrum of the output. Give the weight of each line in the output spectrum.

(c) Obtain an expression giving the ratio of the powers in the fundamental component and nth harmonic of the output.

(d) If this arrangement is to be used as a sinusoidal source, design the filter (i.e., choose a value for f_b) such that the third harmonic is at least 30 dB down from the power in the fundamental. (*Note:* $x_{dB} = 20 \log_{10} x$ for x a voltage.)

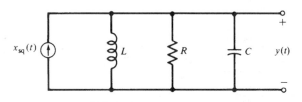

RLC circuit used to generate a sine wave

FIGURE P4-33

4-34. Show that (4-88) is equivalent to (3-73).

Section 4-8

4-35. Obtain the impulse response of the ideal high-pass filter with transfer function given by

$$H_{HP}(f) = H_0\left[1 - \Pi\left(\frac{f}{2B}\right)\right]e^{-j2\pi ft_0}$$

4-36. An ideal notch filter has the amplitude-response function shown. Assuming a linear phase response given by $\theta(f) = -2\pi t_0 f$, obtain and plot the impulse response without doing any integration.

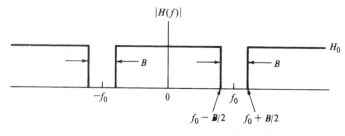

FIGURE P4-36

Section 4-9

4-37. Find the 10% to 90% rise time of the filter of Problem 4-31 in response to a unit step.

Section 4-10

4-38. Plot a figure similar to Figure 4-15b and c for $W = 10$ and $x(t) = u(t) - u(t - \frac{1}{2})$. Use MATLAB to do the plotting. Does overshoot ever go away in your estimation?

Section 4-11

4-39. Given the transform pair $\Lambda(t/\tau) \leftrightarrow \tau \operatorname{sinc}^2 f\tau$, obtain the Fourier transforms of the periodic triangular waveforms shown. Sketch the amplitude spectrum for each. Comment on the rapidity with which $|X_n| \to 0$ with $n \to \infty$.

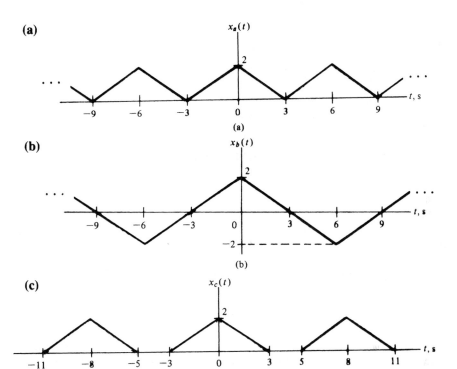

(a)

$x_a(t)$

(a)

(b)

$x_b(t)$

(b)

(c)

$x_c(t)$

(c)

FIGURE P4-39

4-40. Obtain the Fourier transform of the periodic raised-cosine pulse train

$$x(t) = \frac{1}{2} A \sum_{n=-\infty}^{\infty} \left[1 + \cos\left(\frac{2\pi(t - nT_0)}{\tau}\right) \right] \Pi\left(\frac{t - nT_0}{\tau}\right)$$

where $T_0 \geq \tau$. Sketch the waveform and the amplitude spectrum for the case $\tau = T_0$.

Section 4-12

4-41. Obtain Hilbert transforms of the following signals:
   (a) $\cos(2\pi f_0 t)$ (*Hint:* Use the fact that a Hilbert transform is a $-\pi/2$ radian phase shifter);
   (b) $\sin(2\pi f_0 t)$;
   (c) $\Pi(t/\tau)$(*Hint:* Use the convolution integral for this one).

Section 4-13

4-42. Use the fast Fourier transform of MATLAB to show that (4-106) and (4-107) are indeed transform pairs. Experiment with the number of points used.

Computer Exercises

4-1. Obtain the symbolic Fourier transforms of the following:
   (a) $x_1(t) = \exp(-t^2)$
   (b) $x_2(t) = t\exp(-t^2)$
   (c) $x_3(t) = (e^{-3t} + e^{-2t})u(t)$
   Check your results with Table 4-2 and by performing inverse symbolic Fourier transforms to show that the original signal results in each case.

4-2. (a) Compute and plot the amplitude and phase spectra of the signals given in Computer Exercise 4-1.
   (b) Compute and plot the energy spectra of the signals given in Computer Exercise 4-1.

4-3. Write a MATLAB program to verify the plots given in Figure 4-12.

4-4. From Table 4-2, we note that the Hilbert transform a signal is a convolution of it with $1/(\pi t)$. Write a MATLAB program using the conv function to obtain the Hilbert transforms of the following signals:
   (a) A square pulse;
   (b) A triangle;
   (c) The signal of Computer Exercise 4-1(a);
   (d) The signal of Computer Exercise 4-1(b).
   To avoid the $1/(\pi t)$ going to infinity (i.e., the largest number the computer can represent), add a small constant to t, such as eps.

4-5. Generalize the MATLAB program given in Example 4-19 to the following pulse shapes: (a) sawtooth; (b) trapezoidal; (c) raised cosine. Discuss the Gibb's phenomenon in the light of your results for bandlimiting these pulse shapes.

4-6. Experiment with using the FFT in MATLAB to find the spectra of the signals in Computer Exercise 4-1. What effect does length of the representation interval have? Spacing of samples of the time signal?

5

The Laplace Transform

5-1 Introduction

Systems analysis in the time domain involves the solution of differential equations or the evaluation of the superposition integral. Both techniques can result in tedious mathematical operations for relatively simple systems analysis problems. The Fourier transform provided an alternative approach wherein a differential equation relating the input and output of a system was transformed to an algebraic equation, the Fourier transform of the output was solved for, and the transform of the output was then inverse-Fourier-transformed to provide the system output as a function of time. Unfortunately, there are many signals of interest that arise in systems analysis problems for which Fourier transforms do not exist.

The Fourier transform technique for systems analysis provides a hint for a more general transform analysis procedure, however. We seek a transform that applies to a wider class of signals than the Fourier transform does. This can be accomplished by multiplying a signal $x(t)$ by an exponential convergence factor and Fourier-transforming the product. For signals that are zero for $t < 0$, an appropriate factor is $e^{-\sigma t}$, where σ is positive, resulting in the Fourier transform (with $\omega = 2\pi f$ as the frequency in rad/s)

$$X(\sigma + j\omega) = \int_0^\infty x(t)e^{-\sigma t}e^{-j\omega t}\, dt \tag{5-1}$$

With $s = \sigma + j\omega$, (5-1) can be written as

$$\mathcal{L}[x(t)] = X(s) = \int_0^\infty x(t)e^{-st}\, dt \tag{5-2}$$

where $\mathcal{L}[\,\cdot\,]$ denotes the operation of obtaining the *Laplace transform of $x(t)$* defined by (5-2). Since the lower limit of the integral is zero, (5-2) is referred to as the *single-sided Laplace transformation*.

Use of the Laplace transform for systems analysis results in several advantages. Among these are:

1. The solution of differential equations progresses systematically and involves only algebraic manipulations.
2. The total solution—particular integral and homogeneous solution (i.e., forced and transient responses)[†]—is obtained.
3. Initial conditions are automatically included in the solution of the equations.

[†]The particular integral, which emulates the forcing function or excitation of the system, is referred to as the *forced response*. That part of the total response which is influenced by characteristics of the system itself is referred to as the *transient response*. It is the nontrivial solution of the system differential equation with the excitation identically zero.

Not every signal possesses a Laplace transform. An example is exp[exp (t)], $t > 0$, which grows faster than exp$(-\sigma t)$ decays. Thus the Laplace transform integral (5-2) does not converge for this signal. We will discuss convergence properties of Laplace transforms more fully in the following section.

The operation that changes $X(s)$ back to $x(t)$ is referred to as the *inverse Laplace transformation,* and is symbolized by $\mathcal{L}^{-1}[\,\cdot\,]$. To obtain $x(t)$ in terms of $X(s) = X(\sigma + j\omega)$, we observe from the inverse Fourier transform of (5-1) that

$$x(t)e^{-\sigma t} = \mathcal{F}^{-1}[X(\sigma + j\omega)] = \frac{1}{2\pi} \int_{-\infty}^{\infty} X(\sigma + j\omega)e^{j\omega t}\, d\omega \qquad (5\text{-}3\text{a})$$

Multiplying both sides by $e^{\sigma t}$ and assuming σ constant, we obtain the *inversion integral,*

$$x(t) = \frac{1}{2\pi j} \int_{\sigma - j\infty}^{\sigma + j\infty} X(s)e^{st}\, ds \qquad (5\text{-}3\text{b})$$

where the change of variable $s = \sigma + j\omega$ and $ds = j\, d\omega$ has been used. With $\omega = \pm\infty$ in (5-3a) the limits for (5-3b) are clearly $\sigma \pm j\infty$.

Although not easy to show rigorously, (5-3b) may be generalized to a *complex inversion integral* where $s = \sigma + j\omega$ is a complex variable with a real part that may vary along with its imaginary part ω. It is discussed in more detail in Section 5-5, where the technique of *contour integration* is briefly examined for the purpose of evaluating inverse Laplace transforms.

For the most part, contour integration can be avoided when finding inverse Laplace transforms in systems analysis problems simply by making use of a table of Laplace transform pairs. We now begin the construction of such a table and illustrate the evaluation of (5-2) for several simple signals.

5-2 Examples of Evaluating Laplace Transforms

As an example of evaluating (5-2), let $x(t) = 1$. Then

$$X(s) = \int_0^\infty e^{-st}\, dt$$

$$= \frac{e^{-st}}{-s}\Big|_0^\infty = \frac{e^{-\sigma t}e^{-j\omega t}}{-s}\Big|_0^\infty$$

$$= -\frac{e^{-\sigma t}}{s}\,(\cos \omega t - j \sin \omega t)\Big|_0^\infty \qquad (5\text{-}4)$$

Clearly, unless $\sigma = \operatorname{Re}(s) > 0$, the limit as $t \to \infty$ does not exist since $\cos \omega t$ and $\sin \omega t$ oscillate with increasing t and $e^{-\sigma t}$ grows without bound if $\sigma < 0$. Requiring that $\operatorname{Re}(s) > 0$, we obtain

$$\mathcal{L}[1] = \frac{1}{s}, \qquad \operatorname{Re}(s) > 0 \qquad (5\text{-}5)$$

Because the lower limit of (5-2) is zero, values of $x(t)$ for $t < 0$ have no effect on $X(s)$. We see that $x(t) = 1$ and $x(t) = u(t)$ have the same single-sided Laplace transform. If we consider signals that are nonzero only for $t \geq 0$, this presents no problem.[†] This is reasonable for our purposes because all signals must start sometime. We may choose this starting time conveniently as $t = 0$.

[†]The *double-sided Laplace transform* can be used for signals that are nonzero for $t < 0$. An introduction to this transform is given in Section 5-6.

Some discussion is in order about the convergence of the Laplace transform integral. In obtaining the Laplace transform of $x(t) = 1$, it was necessary to restrict $\text{Re}(s) > 0$ in order for the integral to exist. Recall that an integral with one or both of its limits unbounded is referred to as an *improper integral*. An improper integral of the form

$$\int_0^\infty x(t)\, dt \triangleq \lim_{L \to \infty} \int_0^L x(t)\, dt \tag{5-6}$$

is said to be convergent if the limit on the right-hand side exists. The integral *converges absolutely* if and only if

$$\int_0^\infty |x(t)|\, dt < \infty \tag{5-7}$$

The Laplace transform integral of a signal $x(t)$ can be shown to converge absolutely for all

$$\sigma = \text{Re}(s) > c \qquad \text{if } \int_0^L |x(t)|\, dt \le K < \infty \tag{5-8}$$

for any positive real number L and some real constant K, and if

$$|x(t)| \le Ae^{ct}, \qquad t > L \tag{5-9}$$

where A and c are appropriately chosen real constants. Such a signal is said to be of exponential order. The smallest possible value of c is called the *abscissa of absolute convergence*. A proof of this theorem is given in Appendix C.

Applying this theorem to $x(t) = u(t)$, we see that $A = 1$ and $c = 0$ are appropriate choices, and the Laplace transform of $u(t)$ therefore converges absolutely for $\text{Re}(s) > c = 0$.

By applying the theorem to $x(t) = t^{-1/2}$ with $L \ge 1$, $A = 1$, and $c = 0$, we know that this signal also has a Laplace transform which, in fact, can be shown to be $(\pi/s)^{1/2}$ for $\text{Re}(s) > 0$. Note the importance of the role of L in this example to ensure convergence.

Consider next the Laplace transform of the signal

$$x(t) = e^{-\alpha t}u(t) \tag{5-10}$$

where α may be complex. By definition, the Laplace transform is

$$\mathcal{L}[e^{-\alpha t}u(t)] = \int_0^\infty e^{-(s+\alpha)t}\, dt$$

$$= -\frac{e^{-(\alpha+\sigma)t}e^{-j\omega t}}{s + \alpha}\bigg|_0^\infty \tag{5-11}$$

or

$$X(s) = \frac{1}{s + \alpha}, \qquad \text{Re}(s + \alpha) > 0 \quad \text{or} \quad \text{Re}(s) > -\text{Re}(\alpha) \tag{5-12}$$

The abscissa of absolute convergence, or simply abscissa of convergence, in this case is

$$c = -\text{Re}(\alpha) \tag{5-13}$$

As a final example, we obtain the Laplace transform of $\delta(t)$. By definition of the Laplace transform, it is

$$\mathscr{L}[\delta(t)] = \int_0^\infty \delta(t)e^{-st}\, dt \tag{5-14}$$

whereupon we are faced with a dilemma. Since the unit impulse function occurs at $t = 0$, do we integrate through half of it, none of it, or all of it? We will assume in this book that the lower limit on the Laplace transform is $t = 0^-$. Thus (5-14) evaluates to $\mathscr{L}[\delta(t)] = 1$. The value of s makes no difference; that is, the region of convergence is the entire s-plane.

A final word about convergence. The abscissas of convergence for the signals $u(t)$ and $e^{-\alpha t}u(t)$ are shown in Figure 5-1 together with the regions of absolute convergence of their respective Laplace transform integrals. For any value of s lying in the region of convergence, the respective Laplace transform is finite. It follows that any singularities of the Laplace transform of a signal must lie to the left of the abscissa of absolute convergence. The Laplace transforms of the unit step and decaying exponential signals, viewed as functions of the complex variable s, each have a single singularity that is referred to as

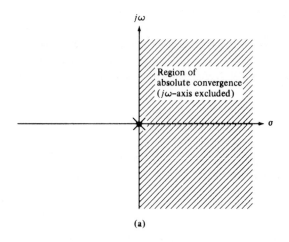

(a)

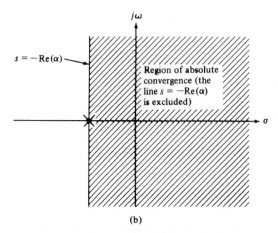

(b)

FIGURE 5-1. Regions of absolute convergence for the Laplace transforms of (a) $u(t)$ and (b) $e^{-\alpha t}u(t)$.

a *first-order pole.*[†] They are shown as ×'s in Figure 5-1 and do indeed lie outside the regions of absolute convergence of the Laplace transform integral. One potential point of confusion is the distinction between the region of absolute convergence of the Laplace transform integral of a signal $x(t)$ and the region in the complex s-plane where $X(s) = \mathcal{L}[x(t)]$ is a well-behaved function of s.[‡] The only place that $X(s)$ is undefined is at its singular points, which may be of the several different types defined in Appendix C. At every other point in the complex plane $X(s)$ is a well-behaved function of the complex variable s, both within and outside the region of absolute convergence of the Laplace transform integral.

The integral of (5-3) is carried out along any line to the right of the singularities of $X(s)$. If this line can be chosen as the $j\omega$-axis, *it then follows from (5-1) and (5-3) that the ordinary Fourier transform of $x(t)$ exists and can be obtained from $X(s)$ by substituting $s = j\omega$.*

The three Laplace transform pairs just derived are summarized in Table 5-1 together with their abscissas of absolute convergence. We will dispense with specifying the abscissas of absolute convergence of single-sided Laplace transforms since it can be shown that there is no ambiguity in the inverse single-sided Laplace transform, even with the region of convergence unspecified. This is not true of the double-sided Laplace transform for which it is necessary to specify the region of convergence together with the Laplace transform in order to determine uniquely the inverse Laplace transform. Since we are not considering the two-sided Laplace transform, specifications of the abscissas of absolute convergence are not necessary.

We now will extend Table 5-1, but to avoid the labor of carrying out integrations, we will give several Laplace transform theorems and use them to extend Table 5-1 later. Because the Fourier and Laplace transforms are both linear integral transforms, many of the theorems to be stated will be self-evident from our consideration of the Fourier transform.

5-3 Some Laplace Transform Theorems

Theorem 1: Linearity: *Since (5-2) is an integral in which $x(t)$ appears linearly, it follows, for two constants a_1 and a_2 which may be complex, that*

$$\mathcal{L}[a_1x_1(t) + a_2x_2(t)] = a_1X_1(s) + a_2X_2(s) \tag{5-15}$$

where $X_1(s) = \mathcal{L}[x_1(t)]$ and $X_2(t) = \mathcal{L}[x_2(t)]$. The linearity theorem will be made use of in the Laplace transformation of differential equations. We illustrate its use for extending Table 5-1 by an example.

TABLE 5-1
Table of Laplace Transforms

Signal	Laplace Transform	Abscissa of Convergence
$\delta(t)$	1	$-\infty$
1	$\dfrac{1}{s}$	0
$e^{-\alpha t}$	$\dfrac{1}{s + \alpha}$	$-\alpha$

[†]Appendix C gives a short summary of several definitions and theorems pertaining to functions of a complex variable. For our purposes here, a pole can be defined as a value for s for which $X(s) \to \infty$.

[‡]In terms of the terminology of functions of a complex variable, $X(s)$ is an *analytic function* of s. See Appendix C.

EXAMPLE 5-1

The Laplace transform of cos $\omega_0 t$ may be obtained by expressing it in exponential form, applying the linearity theorem, and using the transform pair (5-12) with $\alpha = \pm j\omega_0$, which gives

$$\mathscr{L}[\cos \omega_0 t] = \mathscr{L}[\tfrac{1}{2}e^{j\omega_0 t} + \tfrac{1}{2}e^{-j\omega_0 t}]$$

$$= \tfrac{1}{2}\mathscr{L}[e^{j\omega_0 t}] + \tfrac{1}{2}\mathscr{L}[e^{-j\omega_0 t}]$$

$$= \frac{1}{2}\frac{1}{s - j\omega_0} + \frac{1}{2}\frac{1}{s + j\omega_0}$$

$$= \frac{s}{s^2 + \omega_0^2} \tag{5-16}$$

In a similar manner, the Laplace transform of sin $\omega_0 t$ can be derived as follows:

$$\mathscr{L}[\sin \omega_0 t] = \mathscr{L}\left[\frac{1}{2j}e^{j\omega_0 t} - \frac{1}{2j}e^{-j\omega_0 t}\right]$$

$$= \frac{1}{2j}\mathscr{L}[e^{j\omega_0 t}] - \frac{1}{2j}\mathscr{L}[e^{-j\omega_0 t}]$$

$$= \frac{1}{2j}\frac{1}{s - j\omega_0} - \frac{1}{2j}\frac{1}{s + j\omega_0}$$

$$= \frac{\omega_0}{s^2 + \omega_0^2} \tag{5-17}$$

MATLAB Application

MATLAB may be used to find Laplace transforms and inverse Laplace transforms. For the signals in this example, the MATLAB command window entries are as follows:

```
EDU» x1 = sym('cos(omega_0*t)')
x1 =
cos(omega_0*t)
EDU» X1 = laplace(x1)
X1 =
s/(s^2+omega_0^2)
EDU» x2 = sym('sin(omega_0*t)')
x2 =
sin(omega_0*t)
EDU» X2 = laplace(x2)
X2 =
omega_0/(s^2+omega_0^2)
EDU» pretty(X2)
                omega_0
              -----------
               2         2
              s + omega_0
EDU» x2_p = ilaplace(X2)
x2_p =
sin(omega_0*t)
```

In the second last command line entry, the pretty function is used to make the MATLAB output look more like the actual equation. In the very last entry, the ilaplace function is used to show that the original time-domain signal is obtained by inverse Laplace transforming the s-domain result. Note that in all command line entries, a semi-colon could have been used to suppress the echo of the entered quantity. For example, to save space, we could have entered

```
EDU» x1 = sym('cos(omega_0*t)');
```

and this would have suppressed the echo

```
x1 =
cos(omega_0*t)
```

An alternative procedure to the command window entries would have been to write a program and store it in an M-file, then run it from the command window.

Theorem 2: Transforms of Derivatives: *From (5-2) the Laplace transform of dx(t)/dt is*

$$\mathcal{L}\left[\frac{dx(t)}{dt}\right] = \int_0^\infty \frac{dx(t)}{dt} e^{-st}\, dt \tag{5-18}$$

which may be integrated by parts by letting $u = e^{-st}$, and $dv = dx(t)$ in the equation

$$\int_a^b u\, dv = uv\Big|_a^b - \int_a^b v\, du \tag{5-19}$$

Then $du = -se^{-st}\, dt$ and $v = x(t)$, so that (5-18) becomes

$$\mathcal{L}\left[\frac{dx(t)}{dt}\right] = e^{-st}x(t)\Big|_0^\infty + s\int_0^\infty x(t)e^{-st}\, dt$$

$$= sX(s) - x(0^-) \tag{5-20}$$

provided that $\lim_{t\to\infty} x(t)e^{-st} = 0$, where we recall that we have agreed to use 0^- for the lower limit. To find $\mathcal{L}[d^2x(t)/dt^2]$, we write

$$\frac{d^2x(t)}{dt^2} = \frac{d}{dt}\left[\frac{dx(t)}{dt}\right] \tag{5-21a}$$

which results in

$$\mathcal{L}\left[\frac{d^2x(t)}{dt^2}\right] = s\mathcal{L}\left[\frac{dx(t)}{dt}\right] - \frac{dx}{dt}\Big|_{t=0^-}$$

$$= s[sX(s) - x(0^-)] - x^{(1)}(0^-)$$

$$= s^2X(s) - sx(0^-) - x^{(1)}(0^-) \tag{5-21b}$$

Using induction, we may show that

$$\mathcal{L}\left[\frac{d^nx(t)}{dt^n}\right] = s^nX(s) - s^{n-1}x(0^-) - \cdots - x^{(n-1)}(0^-) \tag{5-21c}$$

where $x^{(n)}(0^-)$ denotes the nth derivative of x(t) evaluated at $t = 0^-$. The use of (5-20) will be illustrated with an example.

EXAMPLE 5-2 _____

Consider the circuit shown in Figure 5-2, where the switch is switched from ① to ② at $t = 0$. Find the current through the inductor as a function of time.

Solution: The inductor current obeys the differential equation

$$\frac{di(t)}{dt} + 2i(t) = \begin{cases} 4, & t \le 0 \\ 0, & t > 0 \end{cases} \qquad (5\text{-}22)$$

Taking the Laplace transform of both sides starting at $t = 0^-$, we obtain

$$sI(s) - i(0^-) + 2I(s) = 0 \qquad (5\text{-}23)$$

To proceed further, we require $i(0^-)$. Assuming that the circuit was in steady state for $t < 0$, we see from the circuit diagram that

$$i(0^-) = \frac{4}{2} = 2 \qquad (5\text{-}24)$$

Thus

$$I(s)(s + 2) - 2 = 0 \qquad (5\text{-}25a)$$

or

$$I(s) = \frac{2}{s + 2} \qquad (5\text{-}25b)$$

Using (5-12), the current through the inductor is

$$i(t) = \begin{cases} 2e^{-2t}, & t > 0 \\ 2, & t \le 0 \end{cases} \qquad (5\text{-}26)$$

Theorem 3: Laplace Transform of an Integral: *The Laplace transform of*

$$y(t) = \int_{-\infty}^{t} x(\lambda)\, d\lambda \qquad (5\text{-}27)$$

is

$$\mathscr{L}\!\left[\int_{-\infty}^{t} x(\lambda)\, d\lambda\right] = \frac{X(s)}{s} + \frac{y(0^-)}{s} \qquad (5\text{-}28)$$

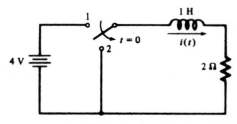

FIGURE 5-2. Circuit for the illustration of Laplace transform solution techniques.

where y(0⁻) is shorthand notation for

$$y(0^-) = \int_{-\infty}^{t} x(\lambda) \, d\lambda \bigg|_{t=0^-}$$

This theorem is proved by using the formula for integration by parts. With

$$u = \int_{-\infty}^{t} x(\lambda) \, d\lambda, \qquad du = x(t) \, dt \tag{5-29a}$$

and

$$dv = e^{-st} \, dt, \qquad v = \frac{-e^{-st}}{s} \tag{5-29b}$$

the Laplace transform of (5-27) becomes

$$\mathcal{L}\left[\int_{-\infty}^{t} x(\lambda) \, d\lambda \right] = -\frac{e^{-st}}{s} \int_{-\infty}^{t} x(\lambda) \, d\lambda \bigg|_{0^-}^{\infty} + \frac{1}{s} \int_{0^-}^{\infty} x(t) e^{-st} \, dt \tag{5-30}$$

which gives (5-28) when the limits are substituted provided that

$$\lim_{t \to \infty} e^{-st} \int_{-\infty}^{t} x(\lambda) \, d\lambda = 0 \tag{5-31}$$

EXAMPLE 5-3

The use of the integration theorem will be illustrated by finding an expression for the Laplace transform of the current in the circuit of Figure 5-3. Kirchhoff's voltage law results in the loop equation

$$L\frac{di(t)}{dt} + Ri(t) + \frac{1}{C} \int_{-\infty}^{t} i(\lambda) \, d\lambda = x(t) \tag{5-32}$$

where the voltage-current relationships for each element have been substituted. Application of the linearity, differentiation, and integration theorems yields

$$LsI(s) + RI(s) + \frac{I(s)}{sC} + \frac{v_c(0^-)}{s} = X(s) \tag{5-33}$$

where $i(0^-) = 0$ because the switch is open prior to $t = 0$ and

$$\frac{i^{(-1)}(0^-)}{C} = \frac{1}{C} \int_{-\infty}^{0^-} i(\lambda) \, d\lambda = \frac{q_c(0^-)}{C} = v_c(0^-) \tag{5-34}$$

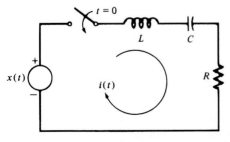

FIGURE 5-3. Circuit for illustration of the integration theorem.

is the voltage across the capacitor at $t = 0^-$. Solving for $I(s)$, we obtain

$$I(s) = \frac{sX(s) - v_c(0^-)}{L[s^2 + (R/L)s + 1/LC]} \tag{5-35}$$

The inversion of $I(s)$ in response to particular inputs will be postponed until the next section.

Theorem 4: Complex Frequency Shift (s-Shift) Theorem: *The Laplace transform of*

$$y(t) = x(t)e^{-\alpha t} \tag{5-36}$$

is

$$Y(s) = X(s + \alpha) \tag{5-37}$$

where $X(s) = \mathcal{L}[x(t)]$.

The theorem is proved by substituting (5-36) into the definition of the Laplace transform and noting that the integral is the Laplace transform of $x(t)$ with the variable $s + \alpha$. The proof is left to the problems at the end of the chapter.

The s-shift theorem is useful for extending the table of Laplace transforms. For example, application of the s-shift theorem to (5-16) shows that

$$\mathcal{L}[e^{-\alpha t} \cos \omega_0 t] = \frac{s + \alpha}{(s + \alpha)^2 + \omega_0^2} \tag{5-38}$$

while application to (5-17) results in the Laplace transform pair

$$\mathcal{L}[e^{-\alpha t} \sin \omega_0 t] = \frac{\omega_0}{(s + \alpha)^2 + \omega_0^2} \tag{5-39}$$

EXAMPLE 5-4

Consider the function of s,

$$X(s) = \frac{s + 8}{s^2 + 6s + 13} \tag{5-40}$$

To find the corresponding function of time, we write it as

$$X(s) = \frac{s + 8}{(s + 3)^2 + 4}$$

$$= \frac{s + 3}{(s + 3)^2 + 2^2} + \frac{5}{(s + 3)^2 + 2^2} \tag{5-41}$$

Applying the transform pairs (5-38) and (5-39), we obtain

$$x(t) = e^{-3t}(\cos 2t + \tfrac{5}{2} \sin 2t), \qquad t > 0 \tag{5-42}$$

MATLAB Application

We check the result of this example using MATLAB symbolic manipulation. This time we use a program stored in the M-file c5ex4.m and run it with echo on. The result is:

```
EDU» echo on
EDU» c5ex4
%      MATLAB solution of the inverse Laplace transform
%      for Example 5-4
%
X = sym('(s+8)/(s^2+6*s+13)');
x = ilaplace(X);
pretty(X)
                        s + 8
                     -----------

                        2
                     s + 6 s + 13
pretty(x)
              5/2 exp(-3 t) sin(2 t) + exp(-3t) cos(2 t)
```

Note that the same result is obtained as obtained analytically on this fairly simple example. In other cases that are more complex, MATLAB may give the answer in a different form. The pretty function has been employed here to make the equations look more like the analytical result.

Theorem 5: Delay Theorem: *If the Laplace transform of x(t)u(t) is X(s), then*

$$\mathscr{L}[x(t-t_0)u(t-t_0)] = e^{-st_0}X(s), \qquad t_0 > 0 \tag{5-43}$$

The proof follows easily by using the definition of the Laplace transform (5-2) with $x(t - t_0)u(t - t_0)$ substituted. Since $u(t-t_0) = 0$ for $t < t_0$, we obtain

$$\mathscr{L}[x(t - t_0)u(t - t_0)] = \int_{t_0}^{\infty} x(t - t_0)e^{-st}\, dt \tag{5-44}$$

Letting $t' = t - t_0$ in the integrand, this becomes

$$\mathscr{L}[x(t - t_0)u(t - t_0)] = \int_0^{\infty} x(t')e^{-s(t' + t_0)}\, dt' = X(s)e^{-st_0} \tag{5-45}$$

which proves the theorem.

We note that the step function $u(t - t_0)$ is necessary in (5-43) to give the proper lower limit on the Laplace transform. Equation (5-43) *does not hold if $t_0 < 0$ since the single-sided Laplace transform will not include the portion of $x(t - t_0)u(t-t_0)$, that exists for $t < 0$.*

EXAMPLE 5-5

The Laplace transform of a square wave beginning at $t = 0$ is

$$\mathscr{L}[x_{sq}(t)] = \mathscr{L}\left[u(t) - 2u\left(t - \frac{T_0}{2}\right) + 2u(t - T_0) - \cdots\right]$$

$$= \frac{1}{s}(1 - 2e^{-sT_0/2} + 2e^{-sT_0} - \cdots) \tag{5-46}$$

Using the series

$$\frac{1}{1 + x} = 1 - x + x^2 - x^3 + \dots, \qquad |x| < 1 \tag{5-47}$$

we obtain

$$\mathcal{L}[x_{sq}(t)] = \frac{1}{s}\left(\frac{2}{1 + e^{-sT_0/2}} - 1\right)$$

$$= \frac{1}{s}\frac{1 - e^{-sT_0/2}}{1 + e^{-sT_0/2}} \tag{5-48}$$

where $x_{sq}(t)$ is a square wave with amplitudes ± 1 and period T_0.

MATLAB Application

In this application, we illustrate the MATLAB function `Heaviside`, which is the symbolic equivalent to our `stp_fn`, and the factor operator in addition to those symbolic operations illustrated previously. A MATLAB M-file program that computes the first three terms of (5-46) and the symbolic Laplace transform of this three term series is given below (command window appearance with echo on):

```
» echo on
» c5ex5
%       Illustration of the symbolic toolbox of MATLAB
%       to implement the delay theorem
%
syms t s
x0 = sym('Heaviside(t-0)');
x1 = sym('Heaviside(t-1)');
x2 = sym('Heaviside(t-2)');
x3 = sym('Heaviside(t-3)');
xt = x0+x1+x2+x3;
X = laplace(xt);
factor(X)
ans =
(exp(-3*s)+exp(-2*s)+1+exp(-s))/s
pretty(X)

        exp(-3 s)    exp(-2 s)          exp(-s)
        ---------  + ---------  + 1/s + -------
            s            s                 s
```

Note that the factor operation pulls out the 1/s, but does not reorder the terms with the 1 at the end which we would do mentally. Note also that the pretty operation keeps the same order as the program output and the factor function. The author could not discover a way to use a for loop to form the symbolic series. Perhaps the student can discover a way to accomplish this.

Theorem 6: Laplace Transform of the Convolution of Two Signals: *Given two signals, $x_1(t)$ and $x_2(t)$, which are zero for $t < 0$, their convolution is*

$$y(t) \triangleq x_1(t) * x_2(t) = \int_0^t x_1(\lambda)x_2(t - \lambda)\, d\lambda = \int_0^\infty x_1(\lambda)x_2(t - \lambda)\, d\lambda \tag{5-49a}$$

with Laplace transform

$$Y(s) = X_1(s)X_2(s) \tag{5-49b}$$

The last equation of (5-49a) follows by virtue of $x_2(t) = 0$, $t < 0$, or $x_2(t-\lambda) = 0$, $\lambda > t$. The proof is as follows: By definition of the single-sided Laplace transform, we have

$$Y(s) = \mathcal{L}[y(t)] = \int_0^\infty \left[\int_0^\infty x_1(\lambda)x_2(t - \lambda)\, d\lambda \right] e^{-st}\, dt \qquad (5\text{-}50)$$

If we let $\eta = t-\lambda$, (5-50) becomes

$$Y(s) = \int_0^\infty x_1(\lambda) \left[\int_0^\infty x_2(\eta)e^{-s\eta}\, d\eta \right] e^{-s\lambda}\, d\lambda$$

$$= X_1(s)X_2(s) \qquad (5\text{-}51)$$

We recognize the inner integral as the Laplace transform of $x_2(t)$ and the outer integral as the Laplace transform of $x_1(t)$. Thus, the Laplace transform of the convolution of two signals is the product of their respective Laplace transforms, and the region of absolute convergence consists of at least the intersection of the regions for $X_1(s)$ and $X_2(s)$.

Theorem 7: Laplace Transform of a Product: *The Laplace transform of a product of two signals can be expressed as an integral in the complex plane. We will not use it in this chapter, but it is given in the table of Laplace transform theorems at the end of this section for reference.*

Theorem 8: Initial Value Theorem: The Laplace transform of the derivative of a signal has been shown to be

$$\mathcal{L}\left[\frac{dx}{dt}\right] \triangleq \int_0^\infty \frac{dx}{dt} e^{-st}\, dt = sX(s) - x(0^-) \qquad (5\text{-}52)$$

If $x(t)$ is continuous at $t = 0$, dx/dt does not contain an impulse at $t = 0$. Therefore, as $s \to \infty$ with σ greater than the abscissa of convergence, the integral vanishes, giving

$$\lim_{s \to \infty} sX(s) = x(0^-) = x(0^+) \qquad (5\text{-}53)$$

If $x(t)$ is discontinuous at $t = 0$, then dx/dt contains an impulse $[x(0^+) - x(0^-)]\delta(t)$, and

$$\lim_{s \to \infty} \int_{0^-}^\infty \frac{dx}{dt} e^{-st}\, dt = x(0^+) - x(0^-) \qquad (5\text{-}54)$$

Thus

$$x(0^+) - x(0^-) = \lim_{s \to \infty} sX(s) - x(0^-) \qquad (5\text{-}55a)$$

or

$$\lim_{s \to \infty} sX(s) = x(0^+) \qquad (5\text{-}55b)$$

provided that $x(0^+)$ exists. If the lower limit on the Laplace transform had been taken as 0^+, the impulse would not have been included and we would have again obtained (5-55b). Thus $\lim_{s \to 0} sX(s)$, provided that it exists, always gives the initial value of $x(t)$ as $t \to 0^+$ regardless of the lower limit used on the Laplace transform integral.

EXAMPLE 5-6 _____

The initial value of $\exp(-\alpha t) \cos \omega_0 t\, u(t)$ is given by

$$\lim_{s \to \infty} s\, \frac{s + \alpha}{(s + \alpha)^2 + \omega_0^2} = 1 \qquad (5\text{-}56)$$

where the transform pair given by (5-38) has been used.

The initial value of $\exp(-\alpha t) \sin \omega_0 t\, u(t)$ is given by

$$\lim_{s \to \infty} s \frac{\omega_0}{(s + \alpha)^2 + \omega_0^2} = 0 \tag{5-57}$$

where the transform pair (5-39) has been used.

The initial value of the time function whose Laplace transform is $s/(s + 10)$ cannot be found since $\lim_{s \to \infty} s^2/(s + 10)$ does not exist. The reason is that

$$\frac{s}{s + 10} = 1 - \frac{10}{s + 10} \tag{5-58}$$

has the inverse Laplace transform $\delta(t) - 10e^{-10t}u(t)$, which has an impulse at $t = 0$.

Theorem 9: Final Value Theorem: *If $x(t)$ and $dx(t)/dt$ are Laplace transformable, then*

$$\lim_{t \to \infty} x(t) = \lim_{s \to 0} sX(s) \tag{5-59}$$

provided that $\lim_{t \to \infty} x(t)$ exists, which is the case if $sX(s)$ has no poles on the $j\omega$-axis or in the right-half plane. The proof of this theorem is left to the problems.

EXAMPLE 5-7

The final value of a unit step is given by

$$\lim_{s \to 0} s \frac{1}{s} = 1 \tag{5-60}$$

The final value of $\exp(-\alpha t)u(t)$, $\alpha > 0$, is obtained from

$$\lim_{s \to 0} s \left(\frac{1}{s + \alpha} \right) = 0 \tag{5-61}$$

The final value of $\exp(\alpha t)u(t)$, $\alpha > 0$, is infinite. Observing that the pole of its Laplace transform is in the right half plane, we see that the final value theorem should not be applied.

As a final example, the final value of $\cos(\omega_0 t)u(t)$ doesn't exist (this signal oscillates between 1 and -1 as $t \to \infty$). If we attempt to apply the final value theorem, we get

$$\lim_{t \to \infty} s \frac{s}{s^2 + \omega_0^2} = 0 \tag{5-62}$$

which is wrong. However, the final value theorem does not apply to signals having Laplace transforms with poles on the $j\omega$-axis.

Theorem 10: The Scaling Theorem: *The Laplace transform of $x(at)$, where a is a positive constant, is $a^{-1}X(s/a)$. This is the generalization of the scale change theorem of Fourier transforms to the complex s-domain. Note that since we are dealing with the single-sided Laplace transform, a must be positive. If $a = -1$, for example, a time function to be Laplace transformed would be reversed and exist for t negative. The resulting negative-time portion would then not be included in the Laplace transform integral.*

EXAMPLE 5-8

In this example we use the symbolic toolbox of MATLAB to illustrate the scaling theorem. The program is run from an M-file with echo on to display the following in the command window. The comment statements make it self-explanatory.

```
» echo on
» c5ex8
%       Example 5-8 illustrates the scaling theorem
%
syms t
x = sym('exp(-alpha*t)*cos(omega_0*t)          % Define a signal
   *Heaviside(t)');
X =laplace(x);                                  % Find its Laplace transform
pretty(X)                                       % Display it for reference
                    s + alpha
              -------------------

                     2         2
              (s + alpha) + omega_0
X2 = subs(X, 's/2', 's');                       % Scale: substitute
                                                % s/2 for s

X3 = X2/2;                                       % Scale: multiply by 1/2
pretty(X3)                                       % Display for reference
                    1/2 s + alpha
          1/2  ----------------------

                       2          2
              (1/2 s + alpha) + omega_0
x3 = ilaplace(X3);                              % Inverse LT scaled LT
pretty(x3)                                      % Display: result is x(t)
                                                % with t replaced by 2t
              exp(- 2 alpha t) cos(2 omega_0 t)
```

The theorems and transform pairs that have been developed so far are collected for easy reference in Tables 5-2 and 5-3, respectively.

TABLE 5-2
Laplace Transform Theorems

Name	Operation in Time Domain	Operation in Frequency Domain
1. Linearity	$a_1 x_1(t) + a_2 x_2(t)$	$a_1 X_1(s) + a_2 X_2(s)$
2. Differentiation	$\dfrac{d^n x(t)}{dt^n}$	$s^n X(s) - s^{n-1} x(0^-) - \cdots - x^{(n-1)}(0^-)$
3. Integration	$\displaystyle\int_{-\infty}^{t} x(\lambda)\, d\lambda$	$\dfrac{X(s)}{s} + \dfrac{x^{(-1)}(0^-)}{s}$
4. s-shift	$x(t) \exp(-\alpha t)$	$X(s + \alpha)$
5. Delay	$x(t - t_0) u(t - t_0)$	$X(s) \exp(-s t_0)$
6. Convolution	$x_1(t) * x_2(t) = \displaystyle\int_0^{\infty} x_1(\lambda) x_2(t - \lambda)\, d\lambda$	$X_1(s) X_2(s)$
7. Product	$x_1(t) x_2(t)$	$\dfrac{1}{2\pi j} \displaystyle\int_{c-j\infty}^{c+j\infty} X_1(s - \lambda) X_2(\lambda)\, d\lambda$
8. Initial value (provided limits exist)	$\displaystyle\lim_{t \to 0^+} x(t)$	$\displaystyle\lim_{s \to \infty} s X(s)$
9. Final value (provided limits exist)	$\displaystyle\lim_{t \to \infty} x(t)$	$\displaystyle\lim_{s \to 0} s X(s)$
10. Time scaling	$x(at), \quad a > 0$	$a^{-1} X\!\left(\dfrac{s}{a}\right)$

TABLE 5-3
Extended Table of Single-Sided Laplace Transforms

Signal	Laplace Transform	Comments on Derivation
1. $\delta^{(n)}(t)$	s^n	Direct evaluation with aid of (1-66)
2. 1 or $u(t)$	$\dfrac{1}{s}$	Direct evaluation
3. $\dfrac{t^n \exp(-\alpha t)u(t)}{n!}$	$\dfrac{1}{(s + \alpha)^{n+1}}$	Differentiation applied to pair 3, Table 5-1
4. $\cos \omega_0 t\, u(t)$	$\dfrac{s}{s^2 + \omega_0^2}$	Example 5-1
5. $\sin \omega_0 t\, u(t)$	$\dfrac{\omega_0}{s^2 + \omega_0^2}$	Example 5-1
6. $\exp(-\alpha t) \cos \omega_0 t\, u(t)$	$\dfrac{s + \alpha}{(s + \alpha)^2 + \omega_0^2}$	s-shift and pair 4
7. $\exp(-\alpha t) \sin \omega_0 t\, u(t)$	$\dfrac{\omega_0}{(s + \alpha)^2 + \omega_0^2}$	s-shift and pair 5
8. Square wave: $u(t) - 2u\left(t - \dfrac{T_0}{2}\right) + 2u(t - T_0) - \cdots$	$\dfrac{1}{s} \dfrac{1 - e^{-sT_0/2}}{1 + e^{-sT_0/2}}$	Example 5-5
9. $(\sin \omega_0 t - \omega_0 t \cos \omega_0 t)u(t)$	$\dfrac{2\omega_0^3}{(s^2 + \omega_0^2)^2}$	Example 5-12, pair 5, and convolution
10. $(\omega_0 t \sin \omega_0 t)u(t)$	$\dfrac{2\omega_0^2 s}{(s^2 + \omega_0^2)^2}$	Pair 4 and convolution
11. $\omega_0 t \exp(-\alpha t) \sin \omega_0 t\, u(t)$	$\dfrac{2\omega_0^2(s + \alpha)}{[(s + \alpha)^2 + \omega_0^2]^2}$	s-shift and pair 10
12. $\exp(-\alpha t)(\sin \omega_0 t - \omega_0 t \cos \omega_0 t)u(t)$	$\dfrac{2\omega_0^3}{[(s + \alpha)^2 + \omega_0^2]^2}$	s-shift and pair 9

5-4 Inversion of Rational Functions

The examples of Section 5-3 illustrate that functions involving the ratio of two polynomials in s, called rational functions of s, are commonly occurring Laplace transforms. Indeed, this will be the form of the Laplace transform obtained when considering any fixed, linear, lumped system with a forcing function that is a power of t, exponential in t, sinusoidal, or a combination of these.

The techniques that we shall develop for inversion of rational functions of s are valid only for *proper rational functions,* that is, functions for which the numerator polynomial is of degree less than the degree of the denominator polynomial. For cases where this is not true, it is very simple to obtain a proper rational function through use of long division. For example, consider the nonproper rational function of s:

$$Z(s) = \frac{s + 2}{s + 1} \qquad (5\text{-}63)$$

This can be written as

$$Z(s) = 1 + \frac{1}{s+1} \tag{5-64}$$

by use of long division. We may easily apply the Laplace transform pairs of Table 5-3 to obtain its inverse Laplace transform. The important point is that the second term of (5-64) is a proper rational function.

To illustrate techniques for the inverse Laplace transformation of proper rational functions, we consider (5-35) for several special cases. To simplify the notation, we consider the voltage across the resistor in Figure 5-3 as the system output $y(t)$, and denote its Laplace transform as $Y(s)$. Also, let

$$2\zeta\omega_n = \frac{R}{L} \tag{5-65}$$

and

$$\omega_n^2 = \frac{1}{LC} \tag{5-66}$$

Thus (5-35) becomes

$$Y(s) = \frac{2\zeta\omega_n[sX(s) - v_c(0^-)]}{s^2 + 2\zeta\omega_n s + \omega_n^2} \tag{5-67}$$

We refer to ζ and ω_n as the damping ratio and natural frequency, respectively. The roots of the denominator of (5-67) are given by

$$s_{1,2} = [-\zeta \pm \sqrt{\zeta^2 - 1}]\omega_n \tag{5-68}$$

Thus, if $\zeta > 1$, the roots are real and distinct; if $\zeta < 1$, the roots are complex conjugates; and if $\zeta = 1$, they are real and equal.

EXAMPLE 5-9 SIMPLE FACTORS

Let the input to the circuit shown in Figure 5-3 be a unit step function having Laplace transform $X(s) = 1/s$. Also, assume that $v_c(0^-) = 0$, $\omega_n^2 = 16$, and $2\zeta\omega_n = 10$. Thus (5-67) becomes

$$Y(s) = \frac{10}{s^2 + 10s + 16}$$

$$= \frac{10}{(s+2)(s+8)} \tag{5-69}$$

Checking Table 5-3, we see that the required form for (5-69) does not appear. However, (5-69) may be expanded in partial fractions as

$$\frac{10}{(s+2)(s+8)} = \frac{A}{s+2} + \frac{B}{s+8} \tag{5-70}$$

We have three routes we can follow in determining the unknown coefficients, A and B.

1. *Common Denominator.* Placing each factor on the right-hand side of (5-70) over a common denominator results in

$$10 = (s+8)A + (s+2)B \tag{5-71a}$$

or

$$10 = (A + B)s + (8A + 2B) \tag{5-71b}$$

Setting coefficients of like powers of s equal on either side of this equation, we obtain the simultaneous equations

$$A + B = 0 \tag{5-72a}$$

$$8A + 2B = 10 \tag{5-72b}$$

The first of these equations gives $A = -B$, which, when substituted into the second, yields

$$-8B + 2B = 10 \tag{5-73}$$

or $B = -\frac{5}{3}$, and $A = \frac{5}{3}$. Thus

$$Y(s) = \frac{5}{3}\left(\frac{1}{s + 2} - \frac{1}{s + 8}\right) \tag{5-74}$$

Using the linearity theorem and transform pair 3, we find $y(t)$ to be

$$y(t) = \frac{5}{3}(e^{-2t} - e^{-8t})u(t) \tag{5-75}$$

2. *Substituting Specific Values of s.* Since (5-70) is an identity for any s, we may obtain two simultaneous equations for A and B by substituting two convenient values of s (neither of which are equal to either root of the denominator). For example, substituting $s = 0$ and $s = 2$ in (5-70), we obtain the two equations

$$\frac{10}{(2)(8)} = \frac{A}{2} + \frac{B}{8} \tag{5-76a}$$

and

$$\frac{10}{(4)(10)} = \frac{A}{4} + \frac{B}{10} \tag{5-76b}$$

or

$$4A + B = 5 \tag{5-77a}$$

$$5A + 2B = 5 \tag{5-77b}$$

which give the same values of A and B as before.

3. *Heaviside's Expansion Theorem.* The coefficients in (5-70) may be found by yet another technique, known as Heaviside's expansion theorem.[†] This procedure can be justified, in the case of (5-70), by noting that multiplication of both sides by $s + 2$ results in the equation

$$\frac{10}{s + 8} = A + B\frac{s + 2}{s + 8} \tag{5-78}$$

Since (5-78) holds for all values of s, we can obtain an equation for A by letting $s = -2$, which eliminates the second term on the right-hand side. The result is

$$A = \tfrac{10}{6} = \tfrac{5}{3} \tag{5-79}$$

[†]Named after Oliver Heaviside (1850–1925), an English engineer, who originated *operational calculus.*

as before. Similarly, B can be found by multiplying both sides of (5-70) by $s + 8$ and setting $s = -8$. This results in

$$\frac{10}{s + 2}\bigg|_{s=-8} = A\,\frac{s + 8}{s + 2}\bigg|_{s=-8} + B \tag{5-80}$$

or $B = 10/(-6) = -5/3$, as before. This works if roots are not repeated.

We have just seen how to expand a rational function of s, which contains only *simple factors* in the denominator, in partial fractions. That is, the procedures used in Example 5-9 can be used as long as the factors in the denominator are different or are not raised to a power. Furthermore, the degree of the numerator must be less than the degree of the denominator; that is, it must be a proper rational function of s. If not, long division is used to obtain a proper rational function.

We now wish to consider examples involving more complicated factors than simple ones in the denominator.

EXAMPLE 5-10 IMAGINARY ROOTS

As our next example, suppose that the input to the circuit shown in Figure 5-3 is $x(t) = (-0.5\cos t + 2.5\sin t)u(t)$, with Laplace transform

$$X(s) = -\frac{0.5s}{s^2 + 1} + \frac{2.5}{s^2 + 1} = \frac{-0.5s + 2.5}{s^2 + 1} \tag{5-81}$$

and that $v_c(0^-) = -2v$. All other parameters are assumed to be the same as in Example 5-9. Then the Laplace transform of the output is

$$Y(s) = \frac{10[(-0.5s^2 + 2.5s)/(s^2 + 1) + 2]}{s^2 + 10s + 16}$$

$$= \frac{15s^2 + 25s + 20}{(s^2 + 1)(s + 2)(s + 8)} \tag{5-82}$$

Again, we have several approaches at our disposal for obtaining the partial-fraction expansion of (5-82). For example, the factor $s^2 + 1$ in the denominator can be factored as $(s + j)(s - j)$ and Heaviside's expansion theorem used as in part (3) of Example 5-9. Using such a procedure, we would represent (5-82) as

$$Y(s) = \frac{A_1}{s + j} + \frac{A_2}{s - j} + \frac{A_3}{s + 2} + \frac{A_4}{s + 8} \tag{5-83}$$

Using Heaviside's expansion formula, A_3 and A_4 are found to be

$$A_3 = (s + 2)Y(s)\big|_{s=-2}$$

$$= \frac{15s^2 + 25s + 20}{(s^2 + 1)(s + 8)}\bigg|_{s=-2} = 1 \tag{5-84}$$

and

$$A_4 = (s + 8)Y(s)\big|_{s=-8}$$

$$= \frac{15s^2 + 25s + 20}{(s^2 + 1)(s + 2)}\bigg|_{s=-8} = -2 \tag{5-85}$$

respectively. A_1 and A_2 could be found similarly. However, by examining the first two terms in (5-83) further, we can simplify things to some extent. Putting the first two terms over a common denominator, we obtain

$$\frac{A_1}{s+j} + \frac{A_2}{s-j} = \frac{(A_1 + A_2)s + j(A_2 - A_1)}{s^2 + 1} \tag{5-86}$$

The numerator of (5-86) must be a real function of s. The only way that this can be true is for $A_1 = A_2^*$, so that $A_1 + A_2^* = 2\,\text{Re}(A_1) \triangleq B_1$ and $j(A_1 - A_2) = -2\,\text{Im}(A_1) \triangleq B_2$. Thus (5-83) can be written as

$$Y(s) = \frac{B_1 s + B_2}{s^2 + 1} + \frac{1}{s+2} - \frac{2}{s+8} \tag{5-87}$$

The unknown constants B_1 and B_2 in the first term on the right-hand side can be found in two ways.

1. Placing the right-hand side of (5-87) over a common denominator and setting the result equal to (5-82) results in the identity

$$(B_1 s + B_2)(s+2)(s+8) + (s^2+1)(s+8) - 2(s^2+1)(s+2) = 15s^2 + 25s + 20 \tag{5-88}$$

Multiplying factors, collecting like powers in s, and setting coefficients of like powers of s equal, we obtain

$$\left.\begin{array}{lll}
B_1 + 1 - 2 = 0 & \text{or} & B_1 = 1 \\
10B_1 + B_2 + 8 - 4 = 15 & \text{or} & 10B_1 + B_2 = 11 \\
16B_1 + 10B_2 + 1 - 2 = 25 & \text{or} & 16B_1 + 10B_2 = 26 \\
16B_2 + 8 - 4 = 20 & \text{or} & B_2 = 1
\end{array}\right\} \tag{5-89}$$

We note that although there are only two unknowns and four equations, the four equations are consistent; the second and third provide checks on the first and the fourth. Substituting these results in (5-87), $Y(s)$ becomes

$$Y(s) = \frac{s+1}{s^2+1} + \frac{1}{s+2} - \frac{2}{s+8} \tag{5-90}$$

Using Table 5-3, the output time function is

$$y(t) = [\cos t + \sin t + \exp(-2t) - 2\exp(-8t)]u(t) \tag{5-91}$$

The same result would have been obtained by finding A_1 and A_2 in (5-83) using Heaviside's expansion theorem. The student should show this.

2. Another approach that can be used to find the first term in (5-87) is to equate (5-87) and (5-82). Solving for $(B_1 s + B_2)/(s^2 + 1)$ results in

$$\begin{aligned}
\frac{B_1 s + B_2}{s^2 + 1} &= \frac{15s^2 + 25s + 20}{(s^2+1)(s+2)(s+8)} - \frac{1}{s+2} + \frac{2}{s+8} \\
&= \frac{15s^2 + 25s + 20 - (s^2+1)(s+8) + 2(s^2+1)(s+2)}{(s^2+1)(s+2)(s+8)} \\
&= \frac{s^3 + 11s^2 + 26s + 16}{(s^2+1)(s+2)(s+8)}
\end{aligned} \tag{5-92}$$

This looks worse than our initial expression, but if (5-87) holds, and we know that it does, then $(s + 2)(s + 8) = s^2 + 10s + 16$ must be a factor of the numerator. Carrying out the long division on (5-92), we find this to be the case, and

$$\frac{B_1s + B_2}{s^2 + 1} = \frac{s + 1}{s^2 + 1} \tag{5-93}$$

as before.

MATLAB Application

MATLAB may be used to find partial fraction expansions by means of the residue function. The numerator and demoninator polynomials are first specified by entering their coefficients as row vectors with the first element being the coefficient of the highest power. Both numerator and denominator vectors must have the same number of elements. If, for example, the numerator polynomial is of degree less than that of the denominator, 0s will have to be entered in the appropriate positions to make both the same dimension. The command window entries for obtaining the partial fraction expansion of $Y(s)$ in this example are:

```
EDU» num = [0 0 15 25 20];
EDU» den = [1 10 17 10 16];
EDU»[R, P, k]=residue(num, den)
R =
 -2.0000
 1.0000 + 0.0000i
 0.5000 - 0.5000i
 0.5000 + 0.5000i
P =
 -8.0000
 -2.0000
 0.0000 + 1.0000i
 0.0000 - 1.0000i
k =
 []
```

Note the two leading 0s in num to make it the same size as den. Also note that the factors of the denominator polynomial had to be first multiplied out before its vector representation could be entered in the command window. This may be done by means of the conv function which is really a polynomial multiply. First the three denominator factors are defined by means of their coefficients in a row vector, with the coefficient of the highest power of s first. Since the highest power is two in a denominator factor, all must be defined as second-order polynomials. Then the conv function is applied twice, in this case by concatenation. The MATLAB command window statements are given below:

```
EDU» D1 = [1 0 1];
EDU» D2 = [0 1 2];
EDU» D3 = [0 1 8];
EDU» den = conv(D1, conv(D2,D3))
den =
 0 0 1 10 17 10 16
```

Reading from right to left, with the right-most number being the coefficient of s^0, we read the denominator polynominal as

$$s^4 + 10s^3 + 17s^2 + 10s + 16$$

The outputs of the residue function are three arrays. The first, R in this case, gives the residues or the partial fraction expansion coefficients. The second, P, gives the denominator polynomial roots. The third, or k, gives the remainder, which in this case is empty because there is no remainder. Thus, the result of the residue operation is equivalent to the equation

$$Y(s) = -\frac{2}{s + 8} + \frac{1}{s + 2} + \frac{0.5 - j}{s + j} + \frac{0.5 + j}{s + j}$$

This may be shown to be equivalent to (5-90) by combining the last two terms over a common denominator to give the first term of (5-90).

EXAMPLE 5-11 REPEATED LINEAR FACTORS

In this example we consider the case of repeated linear factors. Let the forcing function for the circuit shown in Figure 5-3 be $x(t) = \exp(-2t)u(t)$, for which the Laplace transform is

$$X(s) = \frac{1}{s + 2} \tag{5-94}$$

Assume that $v_c(0^-) = 0$ and that the other parameters are the same as in Example 5-8. Thus

$$Y(s) = \frac{10s}{(s + 2)^2(s + 8)} \tag{5-95}$$

The partial-fraction expression of this rational function must be of the form

$$Y(s) = \frac{A_1}{s + 8} + \frac{A_2}{s + 2} + \frac{A_3}{(s + 2)^2} \tag{5-96}$$

where A_1 and A_3 can be found using the Heaviside technique. We obtain

$$A_1 = (s + 8)Y(s)\big|_{s=-8} = -\frac{20}{9} \tag{5-97a}$$

and

$$A_3 = (s + 2)^2 Y(s)\big|_{s=-2} = -\frac{10}{3} \tag{5-97b}$$

The same approach will not work for A_2 because multiplication by $(s + 2)$ removes only one of the $(s + 2)$ factors in the denominator of (5-95). Substitution of $s = -2$ then gives an undefined result for A_2. We could use the procedure employed in the preceding example and subtract $A_1/(s + 8)$ and $A_3/(s + 2)^2$ from $Y(s)$. However, it is easier to note that

$$\frac{d}{ds}[(s + 2)^2 Y(s)]\bigg|_{s=-2} = \frac{d}{ds}\frac{A_1(s + 2)^2}{s + 8}\bigg|_{s=-2} + \frac{d}{ds}(s + 2)A_2\bigg|_{s=-2} = A_2 \tag{5-98}$$

or

$$A_2 = \frac{d}{ds}\left(\frac{10s}{s + 8}\right)\bigg|_{s=-2} = \frac{10(s + 8) - 10s}{(s + 8)^2}\bigg|_{s=-2} = \frac{20}{9} \tag{5-99}$$

Thus

$$Y(s) = \frac{20}{9} \left[-\frac{1}{s+8} + \frac{1}{s+2} - \frac{\frac{3}{2}}{(s+2)^2} \right] \tag{5-100}$$

Using Table 5-3, we find the inverse Laplace transform to be

$$y(t) = \frac{20}{9}[-\exp(-8t) + \exp(-2t) - \frac{3}{2}t\exp(-2t)]u(t) \tag{5-101}$$

EXAMPLE 5-12 REPEATED LINEAR FACTORS

We consider next the generalization of the method for finding the partial-fraction expansion for repeated linear factors developed in Example 5-10. Suppose that the forcing function for the circuit of Figure 5-3 is now $x(t) = t\exp(-2t)u(t)$, with all other circuit parameters the same as in Example 5-10. Thus

$$X(s) = \frac{1}{(s+2)^2} \tag{5-102}$$

and

$$Y(s) = \frac{10s}{(s+2)^3(s+8)} \tag{5-103}$$

The partial-fraction expansion of $Y(s)$ is of the form

$$Y(s) = \frac{A_1}{s+2} + \frac{A_2}{(s+2)^2} + \frac{A_3}{(s+2)^3} + \frac{B}{s+8} \tag{5-104}$$

Using Heaviside's expansion formula, we obtain

$$B = (s+8)Y(s)\big|_{s=-8} = \frac{10}{27} \tag{5-105}$$

$$A_3 = (s+2)^3 Y(s)\big|_{s=-2} = -\frac{10}{3} \tag{5-106}$$

Using the differentiation technique of Example 5-10, we find that

$$\begin{aligned}
A_2 &= \frac{d}{ds}[(s+2)^3 Y(s)]\bigg|_{s=-2} \\
&= \frac{d}{ds}\left(\frac{10s}{s+8}\right)_{s=-2} \\
&= \frac{80}{(s+8)^2}\bigg|_{s=-2} \\
&= \frac{20}{9}
\end{aligned} \tag{5-107}$$

To find A_1, we note that multiplication of both sides of (5-104) by $(s+2)^3$ gives

$$(s+2)^3 Y(s) = (s+2)^2 A_1 + (s+2)A_2 + A_3 + \frac{(s+2)^3 B}{s+8} \tag{5-108}$$

Differentiation twice with respect to s gives

$$\frac{d^2}{ds^2}[(s+2)^3 Y(s)] = \frac{d^2}{ds^2}\left[\frac{(s+2)^3}{s+8}B\right] + 2A_1 + 0 + 0 \tag{5-109}$$

Setting $s = -2$, we obtain

$$2A_1 = \frac{d^2}{ds^2}\left[\frac{10s}{s+8}\right]_{s=-2}$$

$$= \frac{d}{ds}\left[\frac{80}{(s+8)^2}\right]_{s=-2}$$

$$= -\frac{160}{(s+8)^3}\bigg|_{s=-2}$$

$$= -\frac{20}{27} \tag{5-110}$$

Thus $A_1 = -\frac{10}{27}$ and

$$Y(s) = \frac{10}{27}\left(\frac{1}{s+8} - \frac{1}{s+2}\right) + \frac{20}{9}\frac{1}{(s+2)^2} - \frac{10}{3}\frac{1}{(s+2)^3} \tag{5-111}$$

Using Table 5-3, we find $y(t)$ to be

$$y(t) = [\tfrac{10}{27}(e^{-8t} - e^{-2t}) - \tfrac{5}{3}t(t - \tfrac{4}{3})e^{-2t}]u(t) \tag{5-112}$$

MATLAB *Application*

The command window entries to find the partial fraction expansion of $Y(s)$ in Example 5-12 are:

```
EDU» num = [0 0 0 10 0];
EDU» den = [1 14 60 104 64];
EDU» [R, P, k] = residue(num, den)
R=
   0.3704
  -0.3704
   2.2222
  -3.3333
P=
  -8.0000
  -2.0000
  -2.0000
  -2.0000
k=
   []
```

The first two elements in the R matrix are $10/27 = 0.3704$ and $-10/27 = -0.3704$, corresponding to the partial fraction expansion coefficients for the denominator factors $(s+8)$ and $(s+2)$, respectively. The third element is $20/9 = 2.2222$ corresponding to the partial fraction expansion coefficient for the denominator factor $(s+2)^2$, and the last element of the R matrix is $-10/3 = -3.3333$ corresponding to the partial fraction expansion coefficient for the denominator factor $(s+2)^3$. These values match those obtained in the analytical solution.

Based on the preceding two examples, we may deduce a general result for expanding rational functions involving repeated linear factors. If the rational function is of the form

$$Y(s) = \frac{P(s)}{(s + \alpha)^n Q(s)} \tag{5-113}$$

then its partial-fraction expansion is of the form

$$Y(s) = \frac{A_1}{(s + \alpha)} + \frac{A_2}{(s + \alpha)^2} + \cdots + \frac{A_n}{(s + \alpha)^n} + \frac{R(s)}{Q(s)} \tag{5-114}$$

where $P(s)$, $Q(s)$, and $R(s)$ are polynomials in s and the factors of $P(s)$ and $Q(s)$ are distinct from $s + \alpha$. The mth coefficient, A_m, is given by

$$A_m = \frac{1}{(n - m)!} \frac{d^{(n-m)}}{ds^{(n-m)}} [(s + \alpha)^n Y(s)]_{s=-\alpha} \tag{5-115}$$

To continue with the use of partial-fraction expansions in obtaining inverse Laplace transforms, we consider two examples which are really special cases of previous examples, but which are conveniently carried out in special ways.

EXAMPLE 5-13 COMPLEX-CONJUGATE FACTORS

In this example we consider the partial fraction expansion of a function having complex-conjugate roots. Consider the rational function of s

$$Y(s) = \frac{2s^2 + 6s + 6}{(s + 2)(s^2 + 2s + 2)}$$

$$= \frac{2s^2 + 6s + 6}{(s + 2)[(s + 1)^2 + 1]} \tag{5-116}$$

The quadratic factor in the denominator can be factored as

$$(s + 1)^2 + 1 = (s + 1 + j)(s + 1 - j) \tag{5-117}$$

and Heaviside's expansion theorem could be used as in Example 5-9 since all factors are distinct. However, it is easier to keep both of the complex-conjugate factors expressed by (5-117) together and expand (5-116) as

$$Y(s) = \frac{A}{s + 2} + \frac{Bs + C}{s^2 + 2s + 2} \tag{5-118}$$

This allows the inverse Laplace transform to be found easily with the help of pairs 3, 6, and 7 in Table 5-3.

As before, we may calculate A by using Heaviside's theorem:

$$A = (s + 2)Y(s)\big|_{s=-2}$$

$$= \frac{2s^2 + 6s + 6}{s^2 + 2s + 2}\bigg|_{s=-2} = 1 \tag{5-119}$$

The coefficients B and C may be obtained by substituting specific values of s in (5-118), or by placing both terms on the right-hand side over a common denominator and equating coefficients of like powers of s in the numerator. Or, one could subtract $1/(s + 2)$ from $Y(s)$ to produce the second term on the right-hand side of (5-118). We will use the first technique. First, letting $s = 0$ in (5-118), we obtain

$$Y(0) = \frac{A}{2} + \frac{C}{2} \tag{5-120a}$$

or

$$C = 2Y(0) - A = \frac{2(6)}{(4)} - 1$$

$$= 2 \tag{5-120b}$$

To find B, we multiply both sides of (5-118) by s and let $s \to \infty$. This gives

$$\lim_{s \to \infty} sY(s) = A + B \tag{5-121a}$$

or

$$B = \lim_{s \to \infty} sY(s) - A$$

$$= 2 - 1 = 1 \tag{5-121b}$$

Thus (5-118) becomes

$$Y(s) = \frac{1}{s + 2} + \frac{s + 2}{s^2 + 2s + 2}$$

$$= \frac{1}{s + 2} + \frac{s + 1}{(s + 1)^2 + 1} + \frac{1}{(s + 1)^2 + 1} \tag{5-122}$$

Application of transform pairs 3, 6, and 7 in Table 5-3 then yields

$$y(t) = [e^{-2t} + e^{-t}(\cos t + \sin t)]u(t) \tag{5-123}$$

EXAMPLE 5-14 REPEATED QUADRATIC FACTORS

Consider the rational function

$$Y(s) = \frac{s^4 + 5s^3 + 12s^2 + 7s + 15}{(s + 2)(s^2 + 1)^2} \tag{5-124}$$

The second factor in the denominator could be expanded as $(s + j)^2(s - j)^2$ and (5-115) used to treat the repeated roots. However, it is more convenient to keep the complex conjugate factors in the denominator together and expand (5-124) as

$$Y(s) = \frac{A_1}{s + 2} + \frac{B_1 s + C_1}{s^2 + 1} + \frac{B_2 s + C_2}{(s^2 + 1)^2} \tag{5-125}$$

Using Heaviside's technique, we find A_1 to be

$$A_1 = (s + 2)Y(s)\big|_{s=-2} = 1 \tag{5-126}$$

We can obtain the remaining coefficients by putting the right-hand side of (5-125) over a common denominator and equating the resulting numerator to the numerator of (5-124). This results in the identity

$$A_1(s^2 + 1)^2 + (B_1s + C_1)(s + 2)(s^2 + 1) + (B_2s + C_2)(s + 2)$$
$$= s^4 + 5s^3 + 12s^2 + 7s + 15 \qquad (5\text{-}127)$$

Multiplying the factors on the left-hand side together and collecting like powers of s, we obtain

$$(A_1 + B_1)s^4 + (C_1 + 2B_1)s^3 + (2A_1 + B_1 + 2C_1 + B_2)s^2 + (2B_1 + C_1 + 2B_2 + C_2)s$$
$$+ (A_1 + 2C_1 + 2C_2) = s^4 + 5s^3 + 12s^2 + 7s + 15 \qquad (5\text{-}128)$$

The coefficients of like powers of s on either side of (5-128) must be equal since (5-128) is an identity. This results in the equations (since A_1 is known only four equations are needed)

$$A_1 + B_1 = 1 \qquad (5\text{-}129\text{a})$$

$$C_1 + 2B_1 = 5 \qquad (5\text{-}129\text{b})$$

$$2A_1 + B_1 + 2C_1 + B_2 = 12 \qquad (5\text{-}129\text{c})$$

$$A_1 + 2C_1 + 2C_2 = 15 \qquad (5\text{-}129\text{d})$$

Since $A_1 = 1$, the first of these equations results in

$$B_1 = 1 - A_1 = 0 \qquad (5\text{-}130)$$

This immediately gives

$$C_1 = 5 \qquad (5\text{-}131)$$

from (5-129b). Thus (5-129c) and (5-129d) become

$$2 + 10 + B_2 = 12 \qquad \text{or} \qquad B_2 = 0 \qquad (5\text{-}132\text{a})$$

and

$$1 + 10 + 2C_2 = 15 \qquad \text{or} \qquad C_2 = 2 \qquad (5\text{-}132\text{b})$$

respectively. The partial-fraction expansion for $Y(s)$, as given by (5-125), is

$$Y(s) = \frac{1}{s + 2} + \frac{5}{s^2 + 1} + \frac{2}{(s^2 + 1)^2} \qquad (5\text{-}133)$$

This can be inverse-Laplace-transformed by using the theorems and transform pairs given earlier. The first two terms are immediately inverse-transformed by using pairs 3 and 5, respectively, of Table 5-3. The last term can be handled by using the convolution theorem on pair 5 of Table 5-3. Thus

$$\mathcal{L}\{[\sin \omega_0 t\, u(t)] * [\sin \omega_0 t\, u(t)]\} = \frac{\omega_0^2}{(s^2 + \omega_0^2)^2} \qquad (5\text{-}134)$$

The left-hand side, using trigonometric identities, is

$$\int_0^t \sin \omega_0 \lambda \, \sin \omega_0(t - \lambda) \, d\lambda = \frac{1}{2}\int_0^t \cos(2\omega_0\lambda - \omega_0 t) \, d\lambda - \frac{1}{2}\int_0^t \cos \omega_0 t \, d\lambda$$

$$= \frac{1}{2\omega_0} \sin \omega_0 t - \frac{1}{2}t \cos \omega_0 t \qquad (5\text{-}135)$$

Therefore we have the transform pair

$$\sin \omega_0 t - \omega_0 t \cos \omega_0 t \leftrightarrow \frac{2\omega_0^3}{(s^2 + \omega_0^2)^2} \qquad (5\text{-}136)$$

which is pair 9 of Table 5-3. The inverse Laplace transform of (5-133) can be written as

$$y(t) = (e^{-2t} + 5 \sin t + \sin t - t \cos t)u(t)$$

$$= (e^{-2t} + 6 \sin t - t \cos t)u(t) \tag{5-137}$$

MATLAB Application

We test the capability of MATLAB to find the correct inverse Laplace transform in this more complex example. The command window entries for doing so are as follows:

```
EDU» Y=sym('(s^4+5*s^3+12*s^2+7*s+15)/((s+2)*(s^2+1)^2)')
Y =
(s^4+5*s^3+12*s^2+7*s+15)/((s+2)*(s^2+1)^2)
EDU»pretty(Y)
          4      3       2
        s + 5 s + 12 s + 7 s + 15
        - - - - - - - - - - - - - - - - - - - - - - - -
                    2       2
              (s + 2) (s + 1)
EDU» y = ilaplace(Y)
y =
exp(-2*t)+6*sin(t)-t*cos(t)
EDU» pretty(y)
          exp(-2 t) + 6 sin(t) - t cos(t)
```

This corresponds to (5-137), which provides both a test for MATLAB and a verification of our analytical result.

*5-5 The Inversion Integral and Its Use in Obtaining Inverse Laplace Transforms

In the introductory discussion on the Laplace transform given at the beginning of the chapter, we derived the inversion integral (5-3) in a logical way by making use of the inverse Fourier transform. In that derivation σ was assumed constant. It is much more useful to allow s to be a *complex variable* where both σ and ω may vary. The inversion integral in this case takes the form

$$\frac{1}{2\pi j} \int_{c-j\infty}^{c+j\infty} X(s)e^{st}\, da = \begin{cases} 0, & t < 0 \\ \frac{1}{2}x(0^+), & t = 0 \\ \frac{1}{2}[x(t^+) + x(t^-)], & t > 0 \end{cases} \tag{5-138}$$

where $c \geq 0$, $c > \sigma_c$, with σ_c an abscissa of convergence of $X(s)$.

To evaluate (5-138) we may make use of the *residue theorem* of the theory of complex variables.[†] The residue theorem pertains to functions of a complex variable $F(s)$ that contain a finite number of isolated singular points in the complex s-plane. Near such a singular point, say $s = s_0$, the function can be represented as a series of the form (referred to as a *Laurent series*)

[†]See Appendix C for a short summary of some pertinent definitions and theorems of complex variable theory.

$$F(s) = \sum_{n=-\infty}^{\infty} B_n(s - s_0)^n \tag{5-139}$$

The B_n's are expansion coefficients which can be found by various means. The coefficient B_{-1} is referred to as the *residue* of $F(s)$ at the singular point $s = s_0$.

With this introduction, we may now state the residue theorem.

Residue Theorem: *Let C be a closed path in the complex s-plane within and on which a function $F(s)$ is analytic[†] except for the isolated singular points $s_1, s_2, \ldots, s_n$ with residues $K_1, K_2, \ldots, K_n$, respectively. Then*

$$\oint_C F(s)\, ds = 2\pi j(K_1 + K_2 + \cdots + K_n) \tag{5-140}$$

where $\oint_C (\cdot)\, ds$ is understood to mean the line integral in a counterclockwise sense along C and is referred to as a contour integral.

How do we make use of the residue theorem to evaluate the inverse Laplace transform of a function of a complex variable, say $X(s)$? Equation (5-138) states that we are to evaluate the integral of $X(s)e^{st}$ along a line parallel to the $j\omega$-axis from $-j\infty$ to $j\infty$. We can apply the residue theorem by choosing a semicircular path in the complex plane as shown in Figure 5-4. Letting $F(s)$ in the residue theorem (5-140) be $X(s)e^{st}$, we obtain

$$\frac{1}{2\pi j} \oint_{C_1 + C_2} X(s)e^{st}\, ds = K_1 + K_2 + \cdots + K_n \tag{5-141}$$

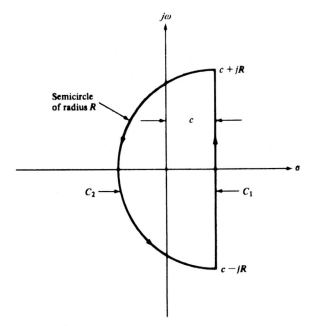

FIGURE 5-4. Closed contour in the complex plane that is appropriate for evaluating the inverse Laplace transform integral.

[†]Analyticity refers to the differentiability of a function of a complex variable, but is much stronger than requiring only that the function be differentiable at a point (see Appendix C).

where $K_1, K_2, \ldots, K_n$ are the residues of $X(s)e^{st}$, and $C_1 + C_2$ means the closed path consisting of the subpaths C_1 and C_2. Further, noting that the integral over the closed path $C_1 + C_2$ can be written as the sum of the integrals over the separate paths C_1 and C_2, and that the integral over C_1 is an integral between the limits $c - jR$ to $c + jR$, we obtain

$$\frac{1}{2\pi j}\int_{c-jR}^{c+jR} X(s)e^{st}\,ds + \frac{1}{2\pi j}\int_{C_2} X(s)e^{st}\,ds = \sum_{m=1}^{n} K_n \tag{5-142}$$

It can be shown that for $t > 0$ the second integral on the left-hand side contributes nothing to the result provided that $X(s) \to 0$ uniformly as $R \to \infty$. (By uniformly it is meant that the limit is approached at the same rate for all angles of s within the range defined by C_2.) For example, if $X(s)$ is the ratio of two polynomials, it is sufficient for the degree of the denominator to exceed that of the numerator by one or more. Given this restriction on $X(s)$, it follows that the left-hand side of (5-142) yields the inverse Laplace transform, as defined by (5-138), in the limit as $R \to \infty$. That is, for $t > 0$,

$$x(t) = \sum \text{residues of } X(s)e^{st} \text{ at the finite singularities of } X(s) \tag{5-143}$$

EXAMPLE 5-14

As an example, consider the function of s given by (5-40) which was inverse-transformed in Example 5-4. The function $X(s)$ has two singularities which are first-order poles, given by

$$s_{1,2} = -3 \pm j2 \tag{5-144}$$

Using the Heaviside expansion technique, we write

$$X(s) \triangleq \frac{s+8}{(s+3)^2 + 4} = \frac{A_1}{s+3+j2} + \frac{A_2}{s+3-j2} \tag{5-145}$$

where

$$A_1 = (s + 3 + j2)X(s)\big|_{s=-3-j2}$$
$$= \frac{5 - j2}{-j4} = \frac{1}{2} + j\frac{5}{4} \tag{5-146a}$$

and

$$A_2 = (s + 3 - j2)X(s)\big|_{s=-3+j2}$$
$$= \frac{1}{2} - j\frac{5}{4} \tag{5-146b}$$

To evaluate the integral, we require the residues of

$$X(s)e^{st} = \frac{s+8}{(s+3)^2 + 4}e^{st} = \frac{A_1 e^{st}}{s+3+j2} + \frac{A_2 e^{st}}{s+3-j2} \tag{5-147}$$

at the poles of $X(s)$. That is, we want the coefficient K_{-1} of the expansion

$$\frac{A_1 e^{st}}{s+3+j2} = \sum_{n=-\infty}^{\infty} K_n(s+3+j2)^n \tag{5-148a}$$

and the coefficient K'_{-1} of the expansion

$$\frac{A_2 e^{st}}{s+3-j2} = \sum_{n=-\infty}^{\infty} K'_n(s+3-j2)^n \tag{5-148b}$$

Now since $e^{st} = \sum_{n=0}^{\infty} s^n t^n/n!$, it is apparent that e^{st} has no singularities in the finite s-plane because no negative-power s-terms are present in its Laurent series expansion. It follows, therefore, that in this case we can find the coefficients K_{-1} and K'_{-1} simply by applying the Heaviside technique. That is,

$$K_{-1} = (s + 3 + j2)X(s)e^{st}\big|_{s=-3-j2}$$

$$= \frac{5 - j2}{-j4} e^{(-3-j2)t}$$

$$= \left(\frac{1}{2} + j\frac{5}{4}\right)e^{-(3+j2)t} \tag{5-149a}$$

and

$$K'_{-1} = (s + 3 - j2)X(s)e^{st}\big|_{s=-3+j2}$$

$$= \left(\frac{1}{2} - j\frac{5}{4}\right)e^{-(3-j2)t} \tag{5-149b}$$

[Compare with (5-146).] From (5-143) we obtain, for $t > 0$,

$$x(t) = \left(\frac{1}{2} + j\frac{5}{4}\right)e^{-(3+j2)t} + \left(\frac{1}{2} - j\frac{5}{4}\right)e^{-(3-j2)t}$$

$$= e^{-3t}\left(\frac{e^{j2t} + e^{-j2t}}{2} + \frac{5}{2}\frac{e^{j2t} - e^{-j2t}}{2j}\right)$$

$$= e^{-3t}(\cos 2t + \tfrac{5}{2}\sin (2t)), \qquad t > 0 \tag{5-150}$$

which is the same result as obtained in Example 5-4.

Generally, the inversion integral would not be used for a problem such as this. It is easier to use partial-fraction expansion and a table of Laplace transform pairs.

*5-6 The Double-Sided Laplace Transform

In the discussion leading to (5-1) it was assumed that the signal $x(t)$ was zero for $t < 0$. This resulted in the single-sided Laplace transform that was perfectly adequate for our purposes. Removing this restriction on $x(t)$ and following the same reasoning that resulted in (5-1), we obtain the *double-sided Laplace transform* of a signal $x(t)$, which is

$$X(s) = \int_{-\infty}^{\infty} x(t)e^{-st}\, dt \tag{5-151}$$

The inversion integral remains the same as before with an added restriction on choosing the path of integration, which will be discussed shortly. The integral (5-151) converges absolutely if

$$\int_{-\infty}^{\infty} |x(t)e^{-st}|\, dt < \infty \tag{5-152}$$

But

$$|x(t)e^{-st}| = |x(t)|e^{-\sigma t} \tag{5-153}$$

and the double-sided Laplace transform integral therefore converges absolutely if

$$\int_{-\infty}^{\infty} |x(t)| e^{-\sigma t} \, dt < \infty \tag{5-154}$$

A sufficient condition on $x(t)$ for (5-154) to hold is that there exists a real positive number A so that, for some real b and c, $|x(t)|$ is bounded by

$$|x(t)| \leq \begin{cases} Ae^{ct}, & t > 0 \\ Ae^{bt}, & t < 0 \end{cases} \tag{5-155}$$

Then (5-151) is absolutely convergent for

$$c < \sigma < b \tag{5-156}$$

When using contour integration to evaluate the inverse double-sided Laplace transform, the path of integration is chosen within this convergence strip and closed by a semicircular arc to the left for $t > 0$ and to the right for $t < 0$. To show (5-156), we break the integral (5-151) up into two parts as

$$X(s) = \int_{-\infty}^{0} x(t) e^{-st} \, dt + \int_{0}^{\infty} x(t) e^{-st} \, dt \tag{5-157}$$

and use (5-155) to bound $|X(s)|$ by

$$|X(s)| \leq \int_{-\infty}^{0} Ae^{(b-s)t} \, dt + \int_{0}^{\infty} Ae^{(c-s)t} \, dt$$

$$\leq A \left[\frac{e^{(b-s)t}}{b-s} \Big|_{-\infty}^{0} + \frac{e^{(c-s)t}}{c-s} \Big|_{0}^{\infty} \right] \tag{5-158}$$

The first integral will converge for $\mathrm{Re}(s) < b$ (the lower limit yields zero when substituted), and the second integral will converge for $\mathrm{Re}(s) > c$ (the upper limit yields zero when substituted). It follows that the singularities of $X(s)$ due to the *positive-time* half of $x(t)$ lie to the *left* of the line $s = c$, while the singularities of $X(s)$ due to the *negative-time* half of $x(t)$ lie to the *right* of the line $s = b$.

In contrast to the single-sided Laplace transform, the region of convergence must be specified together with a two-sided Laplace transform in order to identify the corresponding inverse transform uniquely. To illustrate this, consider the two-sided Laplace transforms of the signals

$$x_1(t) = e^{\alpha t} u(t) \tag{5-159}$$

and

$$x_2(t) = -e^{\alpha t} u(-t) \tag{5-160}$$

Both have the two-sided Laplace transform $1/(s-\alpha)$. However, their regions of convergence are different. In particular,

$$X_1(s) = \frac{1}{s-\alpha} \qquad \text{for } \sigma > \alpha \tag{5-161}$$

and

$$X_2(s) = \frac{1}{s-\alpha} \qquad \text{for } \sigma < \alpha \tag{5-162}$$

where $X_1(s)$ and $X_2(s)$ are the Laplace transforms of $x_1(t)$ and $x_2(t)$, respectively.

We may obtain the two-sided Laplace transform of a signal through use of the single-sided Laplace transforms by separating the signal into its positive- and negative-time components. In particular, assume that $x(t)$ is defined for $-\infty < t < \infty$. We may represent it as

$$x(t) = x_1(t) + x_2(t) \tag{5-163}$$

where

$$x_1(t) = x(t)u(t) \tag{5-164a}$$

and

$$x_2(t) = x(t)u(-t) \tag{5-164b}$$

The two-sided Laplace transform of $x(t)$ is

$$\mathcal{L}_d[x(t)] = \int_{-\infty}^{\infty} x(t)e^{-st}\,dt$$

$$= \int_{-\infty}^{0} x_2(t)e^{-st}\,dt + \int_{0}^{\infty} x_1(t)e^{-st}\,dt \tag{5-165}$$

where $\mathcal{L}_d[\,\cdot\,]$ denotes the double-sided Laplace transform. Changing variables in the first integral to $t' = -t$, we obtain

$$\mathcal{L}_d[x(t)] = \int_{0}^{\infty} x_2(-t')e^{st'}\,dt' + \int_{0}^{\infty} x_1(t)e^{-st}\,dt$$

$$= X_2(-s) + X_1(s) \tag{5-166}$$

where

$$X_1(s) = \mathcal{L}_s[x_1(t)] \tag{5-167a}$$

and

$$X_2(s) = \mathcal{L}_s[x_2(-t)] \tag{5-167b}$$

The notation $\mathcal{L}_s[\,\cdot\,]$ denotes the single-sided Laplace transform. The signal $x_2(-t)$ is, of course, the reflection (mirror image) of the negative-time portion of $x(t)$ about the $t = 0$ axis. Equation (5-166) tells us to take the single-sided Laplace transform of this signal, replace s by $-s$ in the result, and add this new function of s to the single-sided Laplace transform of $x_1(t)$, which is the positive-time portion of $x(t)$.

EXAMPLE 5-15

Find the two-sided Laplace transform of

$$x(t) = e^{-3t}u(t) + e^{2t}u(-t) \tag{5-168}$$

Solution: The signals $x_1(t)$ and $x_2(-t)$ are

$$x_1(t) = e^{-3t}u(t) \tag{5-169a}$$

and

$$x_2(-t) = e^{-2t}u(t) \tag{5-169b}$$

Their single-sided Laplace transforms are

$$X_1(s) = \frac{1}{s+3}, \qquad \sigma > -3 \tag{5-170}$$

and

$$X_2(s) = \frac{1}{s + 2}, \qquad \sigma > -2 \tag{5-171}$$

Replacing s by $-s$ in $X_2(s)$, we obtain

$$X_2(-s) = \frac{1}{-s + 2}, \qquad \text{Re}(-s) = -\sigma > -2 \quad \text{or} \quad \sigma < 2 \tag{5-172}$$

Therefore, from (5-166), we obtain

$$X(s) = \frac{1}{s + 3} + \frac{1}{-s + 2} = \frac{-5}{s^2 + s - 6}, \qquad -3 < \sigma < 2 \tag{5-173}$$

EXAMPLE 5-16

Find the inverse double-sided Laplace transform of

$$X(s) = \frac{2s}{(s + 3)(s + 1)}, \qquad -3 < \sigma < -1 \tag{5-174}$$

Solution: The partial-fraction expansion of $X(s)$ is

$$X(s) = \frac{3}{s + 3} - \frac{1}{s + 1}, \qquad -3 < \sigma < -1 \tag{5-175}$$

Since the pole at $s = -3$ lies to the *left* of the convergence region, the term $3/(s + 3)$ corresponds to the Laplace transform of the *positive-time* portion of $\mathcal{L}^{-1}[X(s)]$. Similarly, since the pole at $s = -1$ lies to the *right* of the convergence region, the term $-1/(s + 1)$ corresponds to the Laplace transform of the *negative-time* portion of $\mathcal{L}^{-1}[X(s)]$. Using (5-166), (5-167a), and (5-167b), we obtain

$$x_1(t) = \mathcal{L}_s^{-1}\left[\frac{3}{s + 3}\right] = 3e^{-3t}u(t) \tag{5-176a}$$

and

$$x_2(-t) = \mathcal{L}_s^{-1}\left[-\frac{1}{-s + 1}\right]$$

$$= e^t u(t) \tag{5-176b}$$

Putting these results together, we obtain

$$x(t) = x_1(t) + x_2(t) = 3e^{-3t}u(t) + e^{-t}u(-t) \tag{5-177}$$

Summary

In this chapter, the basic tool for the analysis of lumped, fixed, linear systems has been introduced, namely, the *Laplace transform.* Use of the Laplace transform for system analysis is advantageous in that it converts the governing differential equation for such systems to an algebraic equation, and initial conditions can be included automatically as a part of the solution. It allows transform techniques to be applied to a wider class of signals than is possible with the Fourier transform—that is, signals of exponential class (such signals grow no more rapidly with time than exponentially). The following are the major points included in this chapter.

1. The *single-sided Laplace transform* is defined by the integral

$$\mathscr{L}[x(t)] = X(s) = \int_0^\infty x(t)e^{-st}\, dt$$

   where $X(s)$ is called the Laplace transform of $x(t)$. In this book, the lower limit in the integral is taken as $t = 0^-$. Thus, the Laplace transform of an impulse is unity, and initial conditions, when solving differential equations, are taken as initial conditions at $t = 0^-$.

2. To get $x(t)$ back from a given $X(s)$, tables of Laplace transforms are used along with partial-fraction expansion techniques. The *inversion integral* is seldom used. However, for completeness, it is given by

$$x(t) = \frac{1}{2\pi j} \int_{\sigma-j\infty}^{\sigma+j\infty} X(s)e^{st}\, ds$$

   This integral is usually carried out as a contour integral in the complex plane by using Cauchy's integral theorem (a brief summary of complex variable theory is given in Appendix C).

3. Important *Laplace transform theorems* are summarized in Table 5-2. They are useful in generating new Laplace transforms from simpler ones and for obtaining inverse Laplace transforms.

4. A short table of *Laplace transform pairs* is given in Table 5-3. Much more complete tables are available, and the reader is referred to the references at the end of this chapter for them.

5. The method of *partial-fraction expansion* allows one to obtain the inverse Laplace transform of rational functions of s. The rational function must be first converted to a *proper rational function* by long division to obtain a function in which the numerator polynomial is at least one degree less than the denominator polynomial. For *simple roots* of the denominator polynomial, $D(s)$, the expansion is

$$\frac{N(s)}{D(s)} = \frac{A_1}{s - s_1} + \frac{A_2}{s - s_2} + \cdots + \frac{A_n}{s - s_n}$$

where

$$A_k = \left[\frac{N(s)}{D(s)} (s - s_k) \right]_{s=s_k}$$

   It is almost always convenient to combine complex conjugate pole-pair terms into a single term and inverse-Laplace-transform the combined term.
   For *repeated roots*, the partial-fraction expansion is of the form

$$\frac{P(s)}{(s + \alpha)^n Q(s)} = \frac{A_1}{(s + \alpha)} + \frac{A_2}{(s + \alpha)^2} + \cdots + \frac{A_n}{(s + \alpha)^n} + \frac{R(s)}{Q(s)}$$

where

$$A_m = \frac{1}{(n - m)!} \frac{d^{(n-m)}}{ds^{(n-m)}} [(s + \alpha)^n Y(s)]_{s=-\alpha}$$

6. The *double-sided Laplace transform* can be used for signals that are nonzero for $t < 0$. It is the same integral as the single-sided Laplace transform except that the lower limit on the integral

is $-\infty$. The function of s denoting a double-sided Laplace transform does not correspond to a unique inverse transform. Rather, both the function of s and the region of convergence of the Laplace transform must be specified to define a unique inverse transform.

Further Reading

Almost all circuits books provide an introductory treatment of the Laplace transform. The following engineering mathematics text includes a chapter on the single-sided Laplace transform and its application to solving ordinary, linear, constant-coefficient differential equations as well as its application to solving partial differential equations:
E. KREYSZIG, *Advanced Engineering Mathematics*, 6th ed. New York: Wiley, 1988 (Chapter 5 and Section 11.13).
A classic text on the Laplace transform, long since out of print, is given below partly for the sake of nostalgia. However, if the student can find a copy, it contains a very complete set of tables of Laplace transforms:
M. F. GARDNER and J. L. BARNES, *Transients in Linear Systems*. New York: Wiley, 1942 (Appendix A).
Assuming that the above reference is not available, the following book contains a fairly complete listing of Laplace transform pairs:
M. ABRAMOWITZ and I. STEGUN, eds., *Handbook of Mathematical Functions*. New York: Dover, 1972.
The following book provides a treatise on the two-sided Laplace transform:
B. VAN DER POL and H. BREMMER, *Operational Calculus: Based on the Two-Sided Laplace Integral*, 2nd ed.
    London: Cambridge University Press, 1955.

PROBLEMS

Sections 5-1 and 5-2

5-1. Obtain the Laplace transforms of the following signals:
    (a) $(1-e^{-2t})u(t)$
    (b) $(e^{-2t}-e^{-10t})u(t)$
    (c) $u(t)-u(t-10)$
    (d) $\delta(t)-\delta(t-10)$

5-2. Sketch each signal given and find its Laplace transform. Give the region of convergence in each case.
    (a) $x_1(t) = \delta(t)-\delta(t-1) + \delta(t-2)$
    (b) $x_2(t) = u(t) + e^{-3t}u(t)$
    (c) $x_3(t) = e^{-4t}u(t)-e^{-4(t-1)}u(t-1)$

5-3. Find the signals whose single-sided Laplace transforms are the following:
    (a) $X_1(s) = 1/s + 1/(s + 10)$
    (b) $X_2(s) = 1 + 1/s + 1/(s + 2)$
    (c) $X_3(s) = 1/(s + 5)-1/(s + 10)$

5-4. Which of the following signals have Laplace transforms only, and which have both Fourier and Laplace transforms? Why?
    (a) $e^{-10t}u(t)$
    (b) $e^{10t}u(t)$
    (c) $e^{-10|t|}$
    (d) $r(t)$, where $r(t)$ is the unit ramp
    (e) $te^{-10t}u(t)$

Section 5-3

5-5. Obtain the Laplace transforms of the following signals:
(a) $\cos 200\pi t$
(b) $\sin 200\pi t$
(c) $\sqrt{2}\cos(200\pi t - \pi/4)$. [*Hint:* Relate this signal to those given in parts (a) and (b).]
(d) If (c) is written as $\sqrt{2}\cos[200\pi(t - 1/800)]$ can time delay be applied? Why or why not?

5-6. Obtain the Laplace transform of the triangular signal $x(t) = \Lambda(t-1)$ by using the differentiation theorem, the time-delay theorem, and expressing dx/dt in terms of unit steps.

5-7. Solve the following differential equations by means of the Laplace transform:
(a) $\dfrac{d^2x(t)}{dt^2} + 6\dfrac{dx(t)}{dt} + 5x(t) = e^{-7t}u(t)$

with $x(0) = 0$ and $\left[\dfrac{dx(t)}{dt}\right]_{t=0} = 0$

(b) $\dfrac{dx(t)}{dt} + 3x(t) + 2y(t) = u(t)$ and $\dfrac{dy(t)}{dt} - x(t) = 0$

with $x(0) = y(0) = 0$.

Solve for $x(t)$ and $y(t)$. [*Hint:* Note that it is not necessary to obtain equations in terms of $x(t)$ and $y(t)$ alone. You may solve for the Laplace transforms, $X(s)$ and $Y(s)$, directly and inverse-transform them.]

5-8. Solve for the current in the circuit shown if the switch changes as indicated by the arrow at $t = 0$. Assume that the switch has been in position 1 since $t = -\infty$.

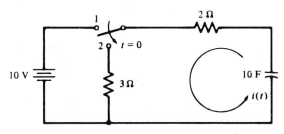

FIGURE P5-8

5-9. Obtain the voltage across the resistor as a function of time for $t > 0$. Assume that $i(0) = v_c(0) = 0$.

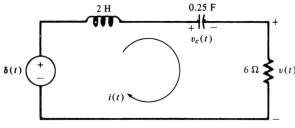

FIGURE P5-9

5-10. Find the inverse Laplace transforms of the functions of s given below;

(a) $\dfrac{s + 10}{s^2 + 8s + 20}$

(b) $\dfrac{s + 3}{s^2 + 4s + 5}$

(c) $\dfrac{s}{s^2 + 6s + 18}$

(d) $\dfrac{10}{s^2 + 10s + 34}$

(*Hint:* Use the s-shift theorem by completing the square in the denominator.)

5-11. Generalize the results of Example 5-5 to the case where we are given a time-limited pulse signal, $p(t)$, which is 0 for $t < 0$ and $t > T_0$. Let its Laplace transform be $P(s)$. For the signal

$$x(t) = \sum_{n=0}^{\infty} p(t - nT_0)$$

Obtain a closed form expression for the Laplace transform of $x(t)$ in terms of $P(s)$.

(b) Obtain the Laplace transform for the case where $p(t) = \Lambda(t/\tau - 1)$ for $\tau \le T_0$.

(c) Obtain the Laplace transform of a half-rectified sine wave.

(d) Obtain the Laplace transform of a full-rectified sine wave.

5-12. (a) Find the Laplace transform of the signal shown in the following figure, using appropriate theorems:

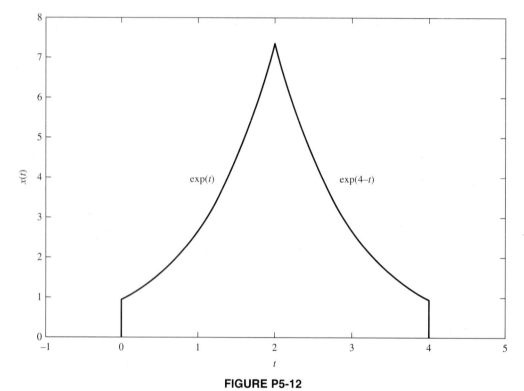

FIGURE P5-12

(b) Find the Laplace transform of

$$y(t) = \sum_{n=0}^{\infty} x(t - 4n)$$

where $x(t)$ is the signal given in part (a). Express in closed form.

5-13. Find the initial and final values, if they exist, of the signals with Laplace transforms given below.

(a) $\dfrac{s + 10}{s^2 + 3s + 2}$ (b) $\dfrac{5}{s^3 + s^2 + 9s + 9}$ (*Hint*: $s + 1$ is a factor)

(c) $\dfrac{s^2 + 5s + 7}{s^2 + 3s + 2}$ (d) $\dfrac{s + 3}{s^2 + 2s}$

5-14. A signal has Laplace transform

$$X(s) = \frac{s + 2}{s^2 + 4s + 5}$$

Find Laplace transforms, $Y(s)$, of the following signals. In each case, tell what Laplace transform theorems you used to find the signal.

(a) $y_1(t) = x(2t-1)u(2t-1)$
(b) $y_2(t) = tx(t)$ (*Hint*: See Problem 5-25b)
(c) $y_3(t) = e^{-3t}x(t)$
(d) $y_4(t) = x(t) * x(t)$
(e) $y_5(t) = dx/dt$
(f) $y_6(t) = 2x(t/4) + 3x(5t)$
(g) $y_7(t) = x(t)\cos(7t)$

5-15. Use the convolution theorem of Laplace transforms to find

$$y(t) = x_1(t) * x_2(t)$$

where $x_1(t)$ and $x_2(t)$ are given below:
(a) $x_1(t) = e^{-2t}u(t)$ and $x_2(t) = u(t-5)$
(b) $x_1(t) = \cos(5t)\,u(t)$ and $x_2(t) = \sin(3t)\,u(t)$
(c) $x_1(t) = e^{-2t}u(t)$ and $x_2(t) = \cos(5t)\,u(t)$
(d) $x_1(t) = \sin(3t)\,u(t)$ and $x_2(t) = u(t-5)$

Section 5-4

5-16. Obtain the inverse Laplace transforms of the functions of s given in Problem 5-13. Do the initial and final values found in Problem 5-13 agree with the results obtained in this problem?

5-17. (a) Referring to Example 2-21, use Laplace transforms to solve (2-130) for $\Omega(t)$ assuming all initial conditions are 0. To put the final result in a neat form, define the auxilary parameters

$$A = \frac{K}{LJ}; \quad 2\zeta\omega_n = \frac{LB + JR}{LJ}; \quad \omega_n^2 = \frac{BR + K^2}{LJ}$$

(b) Plot versus t for the parameter values given in Figures 2-22 and 2-23 to verify the simulation curves shown in Figure 2-23.

5-18. Obtain the inverse Laplace transform of

(a) $X(s) = \dfrac{7s^3 + 20s^2 + 33s + 82}{(s^2 + 4)(s + 2)(s + 3)}$

(b) $X(s) = \dfrac{2s^3 + 9s^2 + 22s + 23}{[(s + 1)^2 + 4](s + 1)(s + 3)}$

5-19. Obtain the inverse Laplace transform of

$$X(s) = \frac{7s^2 + 15s + 10}{(s + 1)^2(s + 3)}$$

5-20. Obtain the inverse Laplace transform of

$$X(s) = \frac{s^2(s + 9)}{(s + 3)^3(s + 1)}$$

5-21. Obtain the inverse Laplace transform of

(a) $X(s) = \dfrac{s^4 + 8s^2 + s + 17}{(s^2 + 4)^2(s + 1)}$
    (b) $X(s) = \dfrac{s^2 + 3}{[(s + 1)^2 + 4]^2(s + 1)}$

5-22. Derive pairs 1 and 3 of Table 5-3.

5-23. (a) Derive pair 10 of Table 5-3. Use the derivation of (5-136) as a guide.
(b) Derive pairs 11 and 12 of Table 5-3.

5-24. One way to define the *effective delay* t_0 and *duration* τ of a signal $x(t)$, where $x(t) = 0$, $t < 0$, is

$$t_0 = \frac{\int_0^\infty tx(t)\, dt}{\int_0^\infty x(t)\, dt}$$

and

$$\tau^2 = \frac{\int_0^\infty (t - t_0)^2 x(t)\, dt}{\int_0^\infty x(t)\, dt}$$

(a) If $X(s)$ is the Laplace transform of $x(t)$, show that

$$t_0 = \frac{X'(0)}{X(0)}$$

and

$$\tau^2 = \frac{X''(0)}{X(0)} - \left[\frac{X'(0)}{X(0)}\right]^2$$

where

$$X'(s) = \frac{dX(s)}{ds}$$

and

$$X''(s) = \frac{d^2 X(s)}{ds^2}$$

(b) Suppose that $x(t) = x_1(t) * x_2(t)$ and $X(s) = X_1(s)X_2(s)$. Let t_0, t_{01}, t_{02}, and τ, τ_1, τ_2 be the delay and duration times of $x(t)$, $x_1(t)$, and $x_2(t)$, respectively. Show that $t_0 = t_{01} + t_{02}$ and $\tau^2 = \tau_1^2 + \tau_2^2$.

(c) Apply the above theory to the case where

$$x_1(t) = Ae^{-\alpha t}u(t) \quad \text{and} \quad x_1(t) = Be^{-\beta t}u(t)$$

where A, B, α, and β are positive constants.

5-25. Derive the following theorems concerning the Laplace transform:
   (a) $\mathcal{L}[x(t/a)] = aX(as), a > 0$
   (b) $\mathcal{L}[tx(t)] = -dX(s)/ds$
   (c) Generalize part (b) to $\mathcal{L}[t^n x(t)]$.
   (d) Use part (c) to derive pair 3, Table 5-3.

5-26. Determine the inverse Laplace transforms of the following functions of s. Use the MATLAB symbolic toolbox to check your answers in those cases where you can.
   (a) $X_1(s) = \dfrac{s^2 + 2s + 9}{s(s^2 + 9)}$
   (b) $X_2(s) = \dfrac{2s^4 - 3s^3 + 36s^2 - 26s + 162}{s(s^2 + 9)^2}$
   (c) $X_3(s) = \dfrac{18}{s^2(s^2 + 9)}$
   (d) $X_4(s) = \dfrac{1}{(s + 3)(1 + e^{-2s})}$
   (e) $X_5(s) = \dfrac{2}{s + se^{-3s}}$

5-27. In the circuits shown, the capacitor is initially charged to V volts and the switch is closed at $t = 0$. Obtain the current, $i(t)$, for $t \geq 0$ using the Laplace transform.
   (a)   **(b)**

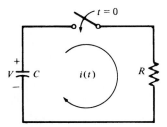

FIGURE P5-27a

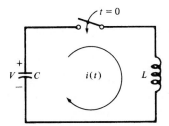

FIGURE P5-27b

5-28. In the circuits shown, the switch is moved in accordance with the arrow at $t = 0$, having been in the top position for a long time. Solve for the current, $i(t)$, for $t \geq 0$ using the Laplace transform.
   (a)

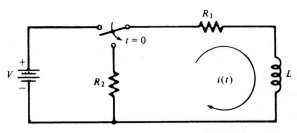

FIGURE P5-28a

(b)

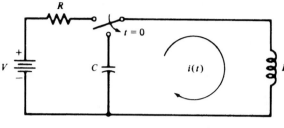

FIGURE P5-28b

5-29. Find the inverse Laplace transform of

$$X(s) = 1/(s + 1)^2$$

the following two ways:
(a) Use the convolution theorem of Laplace transforms.
(b) Find the inverse Laplace transform of $1/s^2$ using the integration theorem, and then apply the s-shift theorem.

5-30. Find inverse Laplace transforms of the following functions of s. Tell what Laplace transform theorems you used in each case.

(a) $X_1(s) = \dfrac{s + 5}{s^2 + 10s + 34}$

(b) $X_2(s) = \dfrac{1 - e^{-2s}}{s + 4}$

(c) $X_3(s) = \dfrac{1}{(s + 1)^3}$

5-31. Compute the initial and final values of the signals whose Laplace transforms are given in Problem 5-30. Compare these results with the limits as $t \to 0$ and $t \to \infty$ computed from the actual inverse Laplace transforms.

5-32. Find the inverse Laplace transforms of the functions of s given below:

(a) $X_1(s) = \dfrac{1}{s(s + 1)(s^2 + 1)}$

(b) $X_2(s) = \dfrac{1}{s(s + 1)^3}$

5-33. Find the double-sided Laplace transforms of the following signals. Sketch each signal.
(a) $x_1(t) = 2e^{-5t}u(t) + 2e^{3t}u(-t)$
(b) $x_2(t) = u(-t) + e^{-t}u(t)$
(c) $x_3(t) = \cos(-5t)\,u(-t) + e^{-t}u(t)$

5-34. Find the signals whose double-sided Laplace transforms are given:

(a) $X_1(s) = \dfrac{1}{(s + 5)(s + 1)}$, $-5 < \sigma < -1$

(b) $X_2(s) = \dfrac{2}{(s^2 + 1)(s + 1)}$, $-1 < \sigma < 0$

5-35. Consider the *RLC* circuit of Example 5-3 with $L = 1$ H, $R/L = 2$ s, and $(LC)^{-1} = 2$ s^{-2}. Let the initial voltage on the capacitor be zero. Assume a square-wave input of amplitudes ± 1 and period T_0. Show that the current can be written in series form as

$$i(t) = e^{-t}\sin(t)\,u(t) + 2\sum_{n=1}^{\infty}(-1)^n e^{-(t-nT_0/2)}\sin\left(t - \frac{nT_0}{2}\right)u\left(t - \frac{nT_0}{2}\right)$$

COMPUTER EXERCISES

5-1. Given the following polynomials in s, find the inverse Laplace transforms, $y_1(t)$, $y_2(t)$, $y_3(t)$, and $y_4(t)$. In symbolic MATLAB, $\text{Dirac}(n,t)$ denotes the nth derivative of the unit impulse function.

$$X_1(s) = s^3 + 6s^2 + 11s + 6$$
$$X_2(s) = [s^2 + 4]^2$$
$$X_3(s) = s^2 + 4s + 4$$

(a) $Y_1(s) = \dfrac{X_1(s)}{X_2(s)}$

(b) $Y_2(s) = \dfrac{X_1(s)}{X_3(s)}$

(c) $Y_3(s) = \dfrac{1}{X_1(s)X_2(s)}$

(d) $Y_4(s) = \dfrac{X_1(s)}{X_2(s)X_3(s)}$

5-2. Use the function of MATLAB to find the signals which are the convolutions of the following signals, where $y_1(t)$, $y_2(t)$, $y_3(t)$, and $y_4(t)$ correspond to the Laplace transformation asked for in Computer Exercise 5-1:
(a) $y_1(t)$ and $y_2(t)$;
(b) $y_3(t)$ and $y_4(t)$;
(c) $y_2(t)$ and $y_4(t)$;
(d) $y_1(t)$ and $y_4(t)$.

5-3. Use the symbolic toolbox in MATLAB to find Laplace transforms of the following signals. Check your results by performing inverse Laplace transforms to show that the original signals result.
(a) $x_1(t) = \cos(\omega_0 t)e^{-\alpha t}u(t)$
(b) $x_2(t) = t^2\cos(\omega_0 t)e^{-\alpha t}u(t)$
(c) $x_3(t) = x_1(t) + x_2(t)$

5-4. Use the MATLAB residue function to find partial fraction exansions for the rational functions of Computer Exercise 5-1. Use the symbolic capability of MATLAB to form the denominator in (c) and (d) without having to carry out the multiplication by hand.

5-5. Write a MATLAB program to plot the waveform of Problem 5-35 for several values of T_0. Comment on the effect of the square wave period in comparison with the circuit time constant (2 seconds).

5-6. A method of deriving difference equations that approximate continuous-time systems is known as Tustin's substitution method. We let $z = e^{sT}$ represent an advance by T, where T is the step size in the difference equation. The solution for s is approximated by

$$s = \frac{1}{T}\ln(z) \cong \frac{2}{T}\left(\frac{z-1}{z+1}\right)$$

where the approximation is obtained by truncating the Laurent series expansion of $\ln(z)$ (see Appendix C for a definition of the Laurent series). In Chapter 8, we will see the Tustin substitution again, where it will be used in the analysis of digital filters. The Tustin substitution is also called the bilinear z-transform. Integration, or s^{-1}, is then approximated by

$$s^{-1} \cong \frac{T}{2}\left(\frac{z+1}{z-1}\right)$$

$$= \frac{T}{2}\frac{1+z^{-1}}{1-z^{-1}}$$

This approximation is then substituted into the Laplace transform of a response (system output) to obtain the form

$$\frac{Y(z)}{X(z)} = \frac{A_n + A_{n-1}z^{-1} + \cdots + A_0 z^{-n}}{B_m + B_{m-1}z^{-1} + \cdots + B_0 z^{-m}}$$

Multiplying through by the denominator of the last expression gives

$$[B_m + B_{m-1}z^{-1} + \cdots + B_0 z^{-m}]Y(z) = [A_n + A_{n-1}z^{-1} + \cdots + A_0 z^{-n}]X(z)$$

Interpreting z^{-k} as a delay by kT, we write the difference equation as

$$[B_m y(kT) + B_{m-1}y[(k-1)T] + \cdots + B_0 y[(k-m)T]$$
$$= [A_n x(kT) + A_{n-1}x[(k-1)T] + \cdots + A_0 x[(k-n)T]$$

When rearranged, the system output at step kT is given in terms of the present and past input samples and the past output samples as

$$y(kT) = \frac{1}{B_M}\left[\sum_{i=0}^{\infty} A_i x[(k-i)T] - \sum_{j=1}^{m} B_{m-j}y[(k-j)T]\right]$$

In Chapter 8 we will see that this difference equation is a general algorithm for determining the output of a linear, time-invariant, discrete-time system given the input samples and past output samples.

Use this method to obtain the output of a system whose Laplace transformed output is

$$\frac{Y(s)}{X(s)} = \frac{1}{s + 1/\tau_0}$$

if the input is a unit step [i.e., $x(kT) = 1$, $k \geq 0$, and is zero for $k < 0$]. Compare with the analytically computed output for various step sizes, T. Assume $y(kT) = 0$ for $k < 0$.
(Reference: See Smith, listed in the references in Chapter 2, Appendix C.)

CHAPTER 6

Applications of the Laplace Transform

6-1 Introduction

In Chapter 5 we developed the Laplace transform and several of its properties and showed how to obtain inverse Laplace transforms using partial-fraction expansion. In this chapter we make use of these tools to analyze electrical networks. We introduce the idea of the Laplace-transformed network, a convenient framework for solution of problems involving both initial conditions and sources.

The latter portion of the chapter contains several topics of interest in the analysis of more general systems. The concept of a transfer function is discussed in some detail, and the Routh test for determining stability is presented. Bode plots are introduced as a means of conveniently displaying the frequency response of a system. Finally, the block diagram is presented as a convenient means of representing a system as a combination of mathematical and schematic models particularly useful for more complicated systems, and operational amplifier models are discussed.

6-2 Network Analysis Using the Laplace Transform

Laplace Transform Equivalent Circuit Elements

In analyzing the circuit of Figure 5-3 (repeated in Figure 6-1) we first wrote down the differential equation expressing Kirchhoff's voltage law and performed the Laplace transform of this equation for various forcing functions, $x(t)$, to find the Laplace transform of the voltage across the resistance R. This approach provided the rational functions for the inverse Laplace transform used in Examples 5-9 through 5-12. There is another approach that could have been taken. We now use the circuit of Figure 6-1 to explain this approach.

Each circuit element in Figure 6-1 is first replaced by its Laplace-transformed equivalent, including initial conditions. To see how this is done, consider the capacitor. Its voltage-current relationship is

$$v(t) = \frac{1}{C} \int_{-\infty}^{t} i(\lambda)\, d\lambda \tag{6-1}$$

where $v(t)$ is the voltage across it with reference direction in the same sense as the reference direction for the current $i(t)$ through it (across and through variables)—see Appendix B. We may rewrite (6-1) as

$$v(t) = \frac{1}{C} \int_{0^-}^{t} i(\lambda)\, d\lambda + v(0^-) \tag{6-2}$$

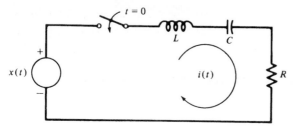

FIGURE 6-1. Series *RLC* circuit.

where

$$v(0^-) = \frac{1}{C} \int_{-\infty}^{0^-} i(\lambda) \, d\lambda = \frac{q(0^-)}{C} \tag{6-3}$$

is the initial voltage across the capacitor. The Laplace transform of (6-2) is

$$V(s) = \frac{1}{sC} I(s) + \frac{v(0^-)}{s} \tag{6-4}$$

which suggests the Laplace-transformed KVL equivalent circuit shown in Figure 6-2a. The capacitor is represented by an *impedance* of value $1/sC$, and the initial condition is represented by a voltage generator of value $v_c(0^-)/s$.

Alternatively, (6-1) can be differentiated to produce the relationship

$$i(t) = C\frac{dv}{dt} \tag{6-5}$$

which can be Laplace-transformed with the aid of the differentiation theorem of Table 5-2. This results in

$$I(s) = sCV(s) - Cv(0^-) \tag{6-6}$$

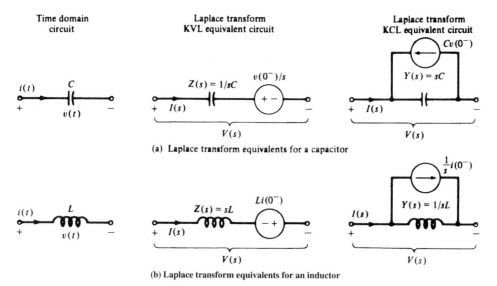

(a) Laplace transform equivalents for a capacitor

(b) Laplace transform equivalents for an inductor

FIGURE 6-2. Laplace-transformed equivalent-circuit elements, which include initial conditions.

which can be interpreted as the Laplace-transformed KCL equivalent circuit shown in Figure 6-2a. Equation (6-6) shows that $I(s)$ can be viewed as being composed of the difference of two currents—that through the capacitor, which is the product of the *admittance* sC and $V(s)$, minus that due to the initial condition, which is $Cv(0^-)$.

Consider next an inductor for which the voltage-current relationship may be written as

$$v(t) = L\frac{di}{dt} \tag{6-7}$$

with the Laplace-transformed equivalent

$$V(s) = sLI(s) - Li(0^-) \tag{6-8}$$

which can be rearranged to yield

$$I(s) = \frac{1}{sL}V(s) + \frac{1}{s}i(0^-) \tag{6-9}$$

These equations suggest the equivalent circuits shown in Figure 6-2b.

Since the voltage-current characteristic of a resistor is

$$v(t) = Ri(t) \tag{6-10}$$

its Laplace-transformed equivalent is

$$V(s) = RI(s) \tag{6-11}$$

Thus the Laplace-transformed equivalent circuit for a resistor is simply a resistor with current $I(s)$ through it and voltage $V(s)$ across it.

We now consider an example that illustrates the application of the models just developed to circuit analysis.

EXAMPLE 6-1

We again use the by now familiar circuit of Figure 6-1 as an example. Suppose that

$$x(t) = (-0.5\cos t + 2.5\sin t)u(t) \tag{6-12}$$

as in Example 5-10. This forcing function has the Laplace transform

$$X(s) = -\frac{0.5s}{s^2 + 1} + \frac{2.5}{s^2 + 1}$$

$$= \frac{-0.5s + 2.5}{s^2 + 1} \tag{6-13}$$

Other parameter values, which were used in Example 5-10, are

$$i_L(0^-) = 0\text{ A}$$

$$v_c(0^-) = -2\text{ V}$$

$$\omega_n^2 \triangleq \frac{1}{LC} = 16\text{ s}^{-2}$$

and

$$2\zeta\omega_n = \frac{R}{L} = 10\text{ s}^{-1} \tag{6-14}$$

For simplicity, we now assume that $R = 1\ \Omega$. This means that the loop current and the voltage across the resistor, which was the output variable considered previously, are numerically the same. Since $R/L = 10$, we have $L = 1/10$ H with $R = 1\ \Omega$. Thus $C = 1/16L = \frac{5}{8}$ F.

Using these circuit parameter values and the assumed $X(s)$ as well as the initial conditions $v_c(0^-) = -2$ V and $i_L(0^-) = 0$ A, we may now draw the Laplace-transformed equivalent circuit as shown in Figure 6-3. Writing Kirchhoff's voltage law around the loop in terms of Laplace-transformed voltages and current yields

$$\frac{-0.5s + 2.5}{s^2 + 1} = \left(\frac{s}{10} + \frac{8}{5s} + 1\right)I(s) - \frac{2}{s} \tag{6-15}$$

Solving for $I(s)$, we find that the Laplace transform of the current is

$$I(s) = \frac{(-0.5s + 2.5)/(s^2 + 1) + 2/s}{s/10 + 8/5s + 1} \tag{6-16}$$

Clearing fractions in the numerator and denominator, we obtain

$$I(s) = \frac{15s^2 + 25s + 20}{(s^2 + 1)(s^2 + 10s + 16)} \tag{6-17}$$

which is equivalent to (5-82), the result for Example 5-10.

MATLAB Application

A symbolic toolbox MATLAB program to solve for $I(s)$, inverse Laplace transform $I(s)$, and display the current in the time domain is as follows:

```
»echo on
»c6ex1
%       MATLAB symbolic toolbox solution for Example 6-1
%
V = sym('(-0.5*s+2.5)/(s^2+1)+2/s');           \
Z = sym('s/10+8/(5*s)+1');
I = V/Z

I =

((-.5*s+2.5)/(s^2+1)+2/s)/(1/10*s+8/5/s+1)

i = ilaplace(I)

i =

-2.*exp(-8.*t)+exp(-2.*t)+cos(t)+sin(t)

»
```

Although we do not get the same rational form for $I(s)$ as in Example 6-1, the result is nevertheless the same. A few trials with the symbolic toolboxes collect and simple functions does not seem to help. The partial fraction expansion may be displayed by concatenating the `diff` and `int` operations. Note that the inverse Laplace transform may be identified on a term-by-term basis with the partial fraction expansion.

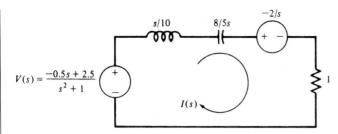

FIGURE 6-3.   Laplace-transformed equivalent circuit for Example 6-1.

Mutual Inductance

Circuits involving mutual inductance may be handled by using an equivalent circuit for the coupled coils called a T-equivalent. Consider the transformer model shown diagrammatically in Figure 6-4. With the reference directions for the terminal voltages and currents as shown in Figure 6-4, it is defined by the voltage-current relationships

$$v_1 = L_1 \frac{di_1}{dt} + M \frac{di_2}{dt} \qquad (6\text{-}18a)$$

$$v_2 = M \frac{di_1}{dt} + L_2 \frac{di_2}{dt} \qquad (6\text{-}18b)$$

The parameters L_1 and L_2 are referred to as the *self-inductances* and the parameter M is called the *mutual inductance*. Whereas self-inductances are always positive quantities, mutual inductance may be positive or negative, depending on the polarity marks, usually indicated by dots, used on the transformer circuit diagram. The value of M will be positive if each current reference is directed toward or away

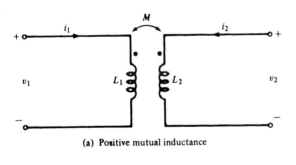

(a) Positive mutual inductance

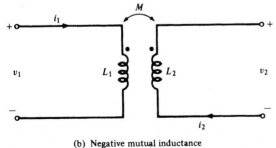

(b) Negative mutual inductance

FIGURE 6-4.   Schematic representation of a transformer.

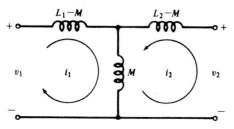

FIGURE 6-5.   T-equivalent circuit for a transformer.

from the polarity marks on the schematic diagram, whereas M is negative if one current reference is directed away and one current reference is directed toward the corresponding dot-marked terminal. This is determined by how the transformer is wound. Thus Figure 6-4a shows a case where M is positive, and Figure 6-4b illustrates a case for M negative.

An equivalent representation of the transformer defined by (6-18a) and (6-18b) may be obtained by rewriting them as

$$v_1 = (L_1 - M)\frac{di_1}{dt} + M\frac{d}{dt}(i_1 + i_2) \tag{6-19a}$$

and

$$v_2 = M\frac{d}{dt}(i_1 + i_2) + (L_2 - M)\frac{di_2}{dt} \tag{6-19b}$$

These equations suggest the T-equivalent circuit shown in Figure 6-5. This equivalent circuit may be used to replace a transformer in a circuit by equivalent inductances and the analysis of the circuit then proceeds as in the case of an ordinary circuit containing only R, L, and C elements. The next example illustrates the use of the T-equivalent model shown in Figure 6-5.

EXAMPLE 6-2

The coils of a large horseshoe type of electromagnet used in a magnetohydrodynamic generator are represented by the circuit model shown in Figure 6-6. (a) Find the current provided by the power supply after the switch is closed; (b) a lug drops off one of the coils, accidentally causing a break in the circuit as indicated by the dots. Find the voltage induced in L_1 by the current discontinuity in L_2.

Solution

(a) We replace the circuit of Figure 6-6 with the Laplace-transformed equivalent circuit of Figure 6-6, which employs the equivalent T-representation of the transformer. Since $i_1(0^-) = i_2(0^-) = 0$, there are no initial condition generators. Writing loop equations, we obtain

$$\frac{100}{s} = (5 + 10s)I_1(s) - (5 + 5s)I_2(s) \tag{6-20a}$$

$$0 = -(5 + 5s)I_1(s) + (10 + 10s)I_2(s) \tag{6-20b}$$

Multiplying the first equation by 2 and adding, we obtain

$$\frac{200}{s} = (5 + 15s)I_1(s) \tag{6-21a}$$

or

$$I_1(s) = \frac{40}{3s(s + \frac{1}{3})}$$

$$= \frac{A}{s} + \frac{B}{s + \frac{1}{3}} \qquad\qquad (6\text{-}21b)$$

where $A = 40$ and $B = -40$. Thus $i_1(t) = 40(1 - e^{-t/3})u(t)$.

(b) For the second part of the problem, we let the time at which the lug drops off be $t = 0$. The initial conditions for $i_1(t)$ and $i_2(t)$ are the steady-state values that occur when the switch has been closed for a long time before the lug separates. They are $i_1(0^-) = 40$ A and $i_2(0^-) = 20$ A. The $t = 0^-$ inductor currents for the equivalent circuit of Figure 6-6 are then easily found to be 20 A for each of the inductors $L_1 - M$ and $L_2 - M$, and 40 A for M. With initial-condition generators included for these inductor currents, the equivalent circuit after the lug drops off is as shown in Figure 6-7. The loop equation for $I_1(s)$ is now

$$\frac{100}{s} + 300 = (5 + 10s)I_1(s) \qquad\qquad (6\text{-}22)$$

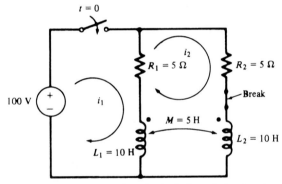

(a) Time domain equivalent circuit of a large electromagnet

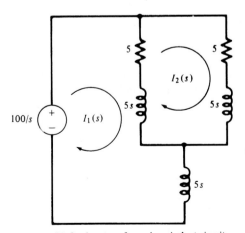

(b) Laplace-transformed equivalent circuit

FIGURE 6-6. Circuits for Example 6-2.

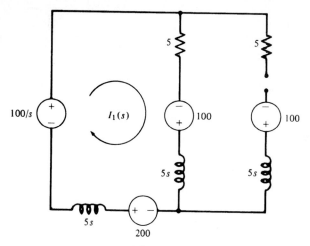

FIGURE 6-7. Equivalent circuit for Example 6-2(b).

Solving for $I_1(s)$, we obtain

$$I_1(s) = \frac{30(s + \frac{1}{3})}{s(s + \frac{1}{2})}$$

(6-23)

The Laplace transform of the voltage induced in L_1 is

$$V_{L_1}(s) = \frac{30(s + \frac{1}{3})}{s(s + \frac{1}{2})}(10s) - 300$$

$$= \frac{300(s + \frac{1}{3})}{(s + \frac{1}{2})} - 300$$

$$= -\frac{50}{s + \frac{1}{2}}$$

(6-24)

Thus the induced voltage is

$$V_{L_1}(t) = -50e^{-t/2}u(t)$$

(6-25)

MATLAB Application

We use the symbolic capability of MATLAB to find the voltage across inductor L_2 due to the break and subsequent abrupt change of current in inductor L_1. Since $i_2(t)$ is 0 for time after the break $(t > 0)$, we have

$$v_{L_2}(t) = M\frac{di_1(t)}{dt}$$

or, Laplace transforming,

$$V_{L_2}(s) = sMI_1(s) - MI_1(0^-) = 5s\frac{30(s + 1/2)}{s(s + 1/2)} - 5(40)$$

The following MATLAB program then finds the voltage across inductor 2:

```
» echo on
» c6ex2
%       MATLAB symbolic toolbox solution for voltage vL2 in
%       Example 5-2 after open circuit occurs
%
VL2 = sym('5*s*30*(s+1/3)/(s*(s+1/2))-200');
VL2_P = diff(int(VL2))

VL2_P =

-50-150/(6*s+3)

vL2 = ilaplace(VL2)

vL2 =

-50*Dirac(t)-25*exp(-1/2*t)
```

Note the unit impulse function of weight 50 in the inductor voltage. Also note the use of the `diff` and `int` operations to carry out the partial fraction expansion.

Network Theorems in Terms of the Laplace Transform

The student will recall a number of theorems, such as Thevenin's and Norton's theorems, from introductory circuit theory, that are useful for simplified analysis of electrical circuits. This section will review those theorems and put them in terms of the Laplace transform variable, s. Recall, also, that a circuit can be described in terms of the number of ports of interest. If one node pair is of interest, say, as an input node pair, we refer to the network as a one-port network. If two node pairs are of interest, say, one pair as an input and the other as an output, we refer to the circuit as a two-port network.

Equivalence and Network Reduction. Two one-port networks are equivalent if their voltage-current characteristics at their respective input ports are the same. Consequences of this definition are the familiar series and parallel equivalents of passive networks. In particular, if a one-port circuit is composed of n passive subcircuits in series, the ith having impedance $Z_i(s)$, the impedance looking into the one-port is

$$Z(s) = Z_1(s) + Z_2(s) + \cdots + Z_n(s) \tag{6-26}$$

On the other hand, if a one-port network is composed of n passive subcircuits in parallel, the ith having admittance $Y_i(s)$, the admittance looking into the one-port is

$$Y(s) = Y_1(s) + Y_2(s) + \cdots + Y_n(s) \tag{6-27}$$

Thevenin's Theorem: Insofar as a load is concerned, any one-port network consisting of R, L, and C passive elements and sources (either independent or controlled) can be replaced by a series combination consisting of an ideal voltage source, $V_T(s)$, and a series impedance, $Z_T(s)$, where V_T is the open-circuit voltage, V_{oc}, of the one-port and $Z_T(s)$ can be found by one of the following methods:

(1) $Z_T(s) = V_{oc}/I_{sc}$, where I_{sc} is the short-circuit current of the one-port.
(2) If no controlled sources are present in the network, $Z_T(s) = Z_{eq}(s)$, where $Z_{eq}(s)$ is the equivalent impedance looking back into the one-port with all (independent) sources set to zero (voltage sources replaced by short circuits and current sources replaced by open circuits).
(3) $Z_T(s) = V_{test}/I$ or $Z_T(s) = V/I_{test}$, where V_{test} and I_{test} are conveniently chosen (say, unity-valued) test voltage or current sources, respectively, with I the resultant current due to V_{test} and V the voltage across the terminals of I_{test}, and all independent sources are set to zero in the network.

To illustrate the above methods of finding the Thevenin equivalent circuit, consider the following example.

EXAMPLE 6-3

Find the Thevenin equivalent circuit of the network shown in Figure 6-8a using the above three methods.

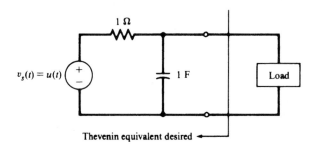

(a) Circuit for which Thevenin equivalent is desired (zero initial conditions)

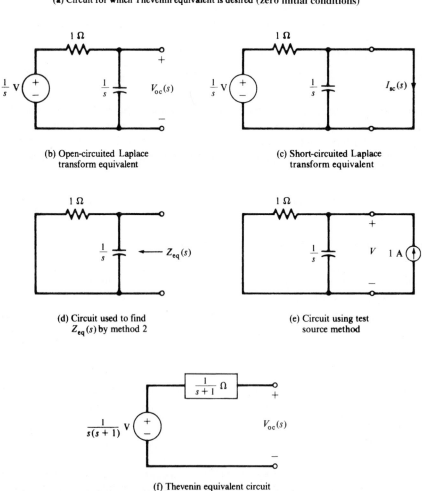

(b) Open-circuited Laplace transform equivalent

(c) Short-circuited Laplace transform equivalent

(d) Circuit used to find $Z_{eq}(s)$ by method 2

(e) Circuit using test source method

(f) Thevenin equivalent circuit

FIGURE 6-8. Circuits pertinent to finding the Thevenin equivalent circuit for Example 6-3.

Solution: Using method 1, we replace the circuit with the Laplace transform equivalent to the left of the nodes *a-b*. Two cases must be considered, one with the nodes *a-b* open-circuited and one with them short-circuited. These are shown in Figures 6-8b and c, respectively. From Figure 6-8b and voltage division, the open-circuit voltage is

$$V_{oc}(s) = \frac{1/s}{1 + 1/s}\left(\frac{1}{s}\right) = \frac{1}{s(s + 1)} \tag{6-28}$$

Using Figure 6-8c and Ohm's law, the short-circuit current is

$$I_{sc}(s) = \frac{1/s}{1} = \frac{1}{s} \tag{6-29}$$

The Thevenin equivalent impedance is

$$Z_T(s) = \frac{V_{oc}(s)}{I_{sc}(s)} = \frac{1/s(s + 1)}{1/s} = \frac{1}{s + 1} \tag{6-30}$$

The Thevenin equivalent circuit is shown in Figure 6-8f.

Using method 2, we replace the voltage source by a short circuit as shown in Figure 6-8d. The equivalent impedance seen looking into the node pair *a-b* is

$$Z_{eq}(s) = \frac{1}{s} \| 1 = \frac{1}{s + 1} \tag{6-31}$$

$I_{...}(s)$ has already been found, and the Thevenin equivalent circuit is the same as obtained using method 1. (The notation ‖ means the parallel combination of two impedances.)

To apply method 3, we apply a 1-A test current source to the node pair *a-b* as shown in Figure 6-8e. Since independent sources are to be set to zero, the voltage source is replaced by a short circuit. The voltage between nodes *a-b* is given by

$$V(s) = \frac{1 \text{ A}}{Y_{eq}(s)} = \frac{1}{s + 1} \text{ V} \tag{6-32}$$

and $Z_{eq}(s) = V/I_{test}$ is the same as found previously. The open-circuit voltage is the same as found before and the Thevenin equivalent circuit is as shown in Figure 6-8f.

Norton's Theorem: *As far as a load is concerned, any one-port of R, L, and C passive components and energy sources (independent or controlled) can be replaced by a parallel combination of an ideal current source, $I_N(s)$, and an admittance, $Y_N(s)$, where $I_N(s)$ is the short-circuit current of the one-port and $Y_N(s)$ can be found by one of the following methods:*

(1) $Y_N(s) = I_{sc}(s)/V_{oc}(s)$.
(2) *If no dependent sources are present, $Y_N(s) = Y_{eq}(s)$, where $Y_{eq}(s)$ is the equivalent admittance looking into the node pair of interest with all (independent) sources set to zero.*
(3) $Y_N(s) = I_{test}/V$ *or* $Y_N(s) = I/V_{test}$, *where I_{test} and V_{test} are defined as before. All independent sources are set to zero.*

Source Transformations. Recall that it is sometimes useful to convert from Thevenin to Norton equivalent circuits to simplify complex circuits. This method of source transformation is summarized in Figure 6-9.

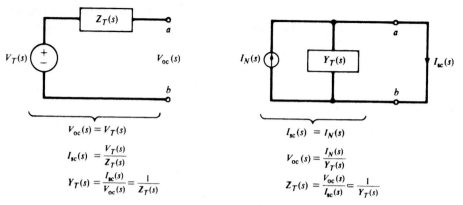

$$V_{oc}(s) = V_T(s)$$

$$I_{sc}(s) = \frac{V_T(s)}{Z_T(s)}$$

$$Y_T(s) = \frac{I_{sc}(s)}{V_{oc}(s)} = \frac{1}{Z_T(s)}$$

$$I_{sc}(s) = I_N(s)$$

$$V_{oc}(s) = \frac{I_N(s)}{Y_T(s)}$$

$$Z_T(s) = \frac{V_{oc}(s)}{I_{sc}(s)} = \frac{1}{Y_T(s)}$$

FIGURE 6-9. Source transformations from Thevenin to Norton equivalent circuits.

The necessity of having the test source method of finding a Thevenin or Norton equivalent circuit is illustrated by the following example.

EXAMPLE 6-4

Consider the circuit shown in Figure 6-10a. Find its Norton equivalent circuit.

Solution: The circuit is shown in Figure 6-10b with its terminals short-circuited. Writing a node equation at V, we obtain

$$I + 2I = I_c \quad \text{or} \quad 3I = I_c \tag{6-33}$$

But

$$I = \frac{1/s - V}{1} \quad \text{and} \quad I_c = sV \tag{6-34}$$

Substituting these results into (6-33) and solving for V results in

$$V(s) = \frac{3}{s(s + 3)} \tag{6-35}$$

Using (6-34) to find I, we obtain

$$I = \frac{1}{s + 3} \tag{6-36}$$

Since $I_{sc}(s) = -2I$, we have

$$I_{sc}(s) = -\frac{2}{s + 3} \tag{6-37}$$

To obtain the equivalent admittance looking into the node pair a-b, we use the test source method with a voltage source $V_{test}(s)$ applied between a-b with the independent voltage source replaced by a short circuit and the dependent current source left alone. Note that the voltages and currents within the circuit are now different from before, so primes have been used to denote this. By Kirchhoff's current law, written at the node marked V', we obtain

$$3I' = I_c \tag{6-38a}$$

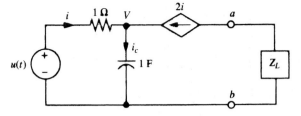

(a) Circuit for which Norton equivalent is to be found

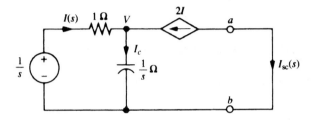

(b) Laplace-transformed equivalent with short-circuited terminals

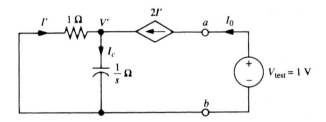

(c) Circuit with test voltage source applied to output terminals and
independent sources set to zero

FIGURE 6-10. Circuits pertinent to finding Norton equivalent circuit using the test source method.

or

$$3(-V'/1) = sV' \qquad (6\text{-}38\text{b})$$

or

$$(s + 3)V' = 0 \qquad (6\text{-}38\text{c})$$

This implies that $V' = 0$ so that $I' = 0$. Hence,

$$I_0 = 2I' = 0 \qquad (6\text{-}39)$$

The Norton equivalent admittance is

$$Y_N(s) = \frac{I_0(s)}{V_{\text{test}}(s)} = 0 \qquad (6\text{-}40)$$

The Norton equivalent circuit therefore consists of an ideal current source, alone, of value given by (6-37).

6-3 Loop and Node Analyses of Circuits by Means of the Laplace Transform

Recall that the governing equations for a planar, fixed, linear network consisting of inductances, capacitances, resistances, ideal voltage sources, and ideal current sources can be written in two forms. These two forms are obtained by writing Kirchhoff's voltage law around properly chosen loops (loop analysis), and by using Kirchhoff's current law at a properly chosen set of nodes (nodal analysis). These two techniques will now be reviewed, and the Laplace transform used to simplify their application.

Loop Analysis

For a network consisting of N_b branches and N_v nodes it follows that $N_b - (N_v - 1)$ independent equations could be written by employing KVL around $N_b - (N_v - 1)$ properly chosen loops. For planar networks, one such proper choice for these loops is the meshes or "windows" of the network. In this case, the equations have the form (dependent sources excluded for the moment):

$$Z_{11}(D)i_1(t) + Z_{12}(D)i_2(t) + \cdots + Z_{1n}(D)i_n(t) = e_1(t)$$

$$Z_{21}(D)i_1(t) + Z_{22}(D)i_2(t) + \cdots + Z_{2n}(D)i_n(t) = e_2(t)$$

$$\vdots$$

$$Z_{n1}(D)i_1(t) + Z_{n2}(D)i_2(t) + \cdots + Z_{nn}(D)i_n(t) = e_n(t)$$

(6-41)

where $n = N_b - (N_v - 1)$. For a clockwise reference direction choice of the mesh currents, the Z_{ii}'s have the form

$$Z_{ii}(D) = L_{ii}D + R_{ii} + S_{ii}D^{-1}$$

(6-42)

where L_{ii} = total series inductance around ith mesh
$\quad R_{ii}$ = total series resistance around ith mesh
$\quad S_{ii}$ = sum of all reciprocal capacitances around ith mesh $(S = C^{-1})$

Also,

$$Z_{ij}(D) = -(L_{ij}D + R_{ij} + S_{ij}D^{-1})$$

(6-43)

where L_{ij} = sum of all inductances common to meshes i and j
$\quad R_{ij}$ = sum of all resistances common to meshes i and j
$\quad S_{ij}$ = sum of all reciprocal capacitances common to meshes i and j
$\quad D = d/dt$
$\quad D^{-1} = \int_{-\infty}^{t} (\cdot) \, d\lambda$
$\quad i_j(t)$ = jth mesh current
$\quad e_j(t)$ = algebraic sum of the voltage sources around mesh j

Taking the Laplace transform of (6-41), we obtain

$$\mathbf{Z}(s)\mathbf{I}(s) = \mathbf{E}(s) + \mathbf{V}_i(s)$$

(6-44)

where $\mathbf{I}(s)$ is a column matrix of Laplace-transformed loop currents, $\mathbf{E}(s)$ is the Laplace-transformed source voltage matrix, $\mathbf{V}_i(s)$ is a Laplace-transformed initial condition generator matrix, and the matrix

$$\mathbf{Z}(s) = [Z_{ij}(s)]$$

(6-45)

has elements given by

$$Z_{ii}(s) = sL_{ii} + R_{ii} + \frac{S_{ii}}{s}$$

$$Z_{ij}(s) = -\left(sL_{ij} + R_{ij} + \frac{S_{ij}}{s}\right) \tag{6-46}$$

$\mathbf{Z}(s)$ is called the *loop impedance matrix.* If no dependent sources are present, it is symmetric. The *loop current matrix,* $\mathbf{I}(s)$, in (6-44) is defined as

$$\mathbf{I}(s) = \begin{bmatrix} \mathscr{L}[i_1(t)] \\ \mathscr{L}[i_2(t)] \\ \vdots \\ \mathscr{L}[i_n(t)] \end{bmatrix} \triangleq \mathscr{L} \begin{bmatrix} i_1(t) \\ i_2(t) \\ \vdots \\ i_n(t) \end{bmatrix} \tag{6-47}$$

and the *source voltage matrix* $\mathbf{E}(s)$ is defined as

$$\mathbf{E}(s) = \mathscr{L} \begin{bmatrix} e_1(t) \\ e_2(t) \\ \vdots \\ e_n(t) \end{bmatrix} \tag{6-48}$$

The second term on the right-hand side of (6-44) is the *initial-condition matrix.* By using the equivalent circuits in the second column of Figure 6-2, the initial conditions can be easily included with the source voltage matrix. Just how this is accomplished will be illustrated by an example.

It is emphasized that the above equations only hold if there are no *dependent sources* present in the network. With dependent sources present one must go through writing the KVL equations for each mesh from the beginning, the end result being that some elements of $\mathbf{Z}(s)$ will have additional terms present due to the dependent sources. Thus $\mathbf{Z}(s)$ is not symmetric if dependent sources are present.

To illustrate the use of (6-44) through (6-48), we now consider an example.

EXAMPLE 6-5 _____

Consider the circuit of Figure 6-11 with an initial current through the inductor of $i_c(0^-) = i_0$ and an initial voltage across the capacitor of $v_c(0^-) = v_0$. (a) First, assume that $v_c(t)$ is an independent source. We may write down the governing equations on a loop basis in the time domain immediately and Laplace-transform them, or we may work with the Laplace-transformed equivalent circuit. We choose the latter course and can immediately write down the matrix equations as

$$\begin{bmatrix} R_1 + R_2 + sL & -R_2 & 0 \\ -R_2 & R_2 + \dfrac{1}{sC} & -\dfrac{1}{sC} \\ 0 & -\dfrac{1}{sC} & R_3 + R_4 + \dfrac{1}{sC} \end{bmatrix} \begin{bmatrix} I_1(s) \\ I_2(s) \\ I_3(s) \end{bmatrix} = \begin{bmatrix} Li_0 + V_s(s) \\ \dfrac{-v_0}{s} \\ \dfrac{v_0}{s} + V_c(s) \end{bmatrix} \tag{6-49}$$

Note that the loop impedance matrix is symmetric. The diagonal elements are the sums of the impedances around each loop. The off-diagonal elements have magnitudes equal to the impedance common to the loops involved and are preceded by a minus sign; for example, R_2 is common to loops 1 and 2 so it appears in the matrix as the element Z_{12} with a minus sign because $i_1(t)$ and $i_2(t)$ were

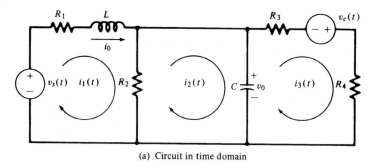

(a) Circuit in time domain

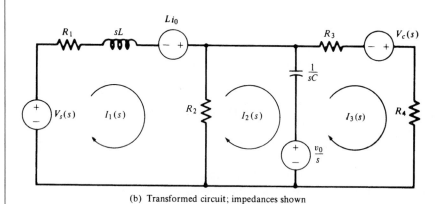

(b) Transformed circuit; impedances shown

FIGURE 6-11. Circuits for the illustration of loop analysis.

chosen to have opposite reference directions through R_2. (b) Now assume that $v_c(t)$ is a dependent source; in particular, let

$$v_c(t) = Ki_1(t) \tag{6-50}$$

where K is a constant of proportionality. By writing down the loop equations and rearranging, the student should show that the element $Z_{13}(s)$ of the loop impedance matrix in (6-49) becomes $-K$; that is, now

$$\mathbf{Z}(s) = \begin{bmatrix} R_1 + R_2 + sL & -R_2 & 0 \\ -R_2 & R_2 + \dfrac{1}{sC} & -\dfrac{1}{sC} \\ -K & -\dfrac{1}{sC} & R_3 + R_4 + \dfrac{1}{sC} \end{bmatrix} \tag{6-51}$$

Hence the presence of the dependent source means that $\mathbf{Z}(s)$ is no longer symmetric.

MATLAB Application

The following MATLAB program computes the loop currents for Example 6-5 for an initial conductor current of 2 A, and initial capacitor voltage of 0 V, all resistances equal to 1 ohm, the capacitance value equal to 0.5 F, and an inductance value of 1 H.

```
» echo on
» c6ex5
%       Solution of Example 6-5 for currents using MATLAB
%       symbolic toolbox
%
syms t s
Z = sym('[2+s, -1, 0; -1, 1+2/s,-2/s; 0, -2/s, 2+2/s]'); % Define
                                                          % impedance
                                                          % matrix
vs = sym('cos(2*t)')

vs =

cos(2*t)

Vs = laplace(vs)

Vs =

s/(s^2+4)

vc = sym('2*cos(2*t)')

vc =

2*cos(2*t)

Vc = laplace(vc)

Vc =

2*s/(s^2+4)

E = sym('[s/(s^2+4); 0; 2*s/(s^2+4)]');   % Define LT independent source
                                          % matrix
Vi = sym('[2; 0; 0]');                    % Define LT initial condition
                                          % matrix
E_plus_Vi = E+Vi                          % Form LT of source plus IC
                                          % matrix

E_plus_Vi =

[s/(s^2+4)+2]
[        0]
[2*2/(s^2+4)]

Z_inv = inv(Z);                           % Obtain inverse of impedance
                                          % matrix
I = Z_inv*E_plus_Vi                       % Obtain LT current matrix and
                                          % print

I =

[    (s+3)/(s^2+4*s+5)*(s/(s^2+4)+2)+2/(s^2+4*s+5)*s/(s^2+4)]
[(s+1)/(s^2+4*s+5)*(s/(s^2+4)+2)+2*(s+2)/(s^2+4*s+5)*s/(s^2+4)]
[1/(s^2+4*s+5)*(s/(s^2+4)+2)+(s^2+3*s+4)/(s^2+4*s+5)*s/(s^2+4)]
```

Strange as it may seem `ilaplace` cannot handle a matrix. Furthermore, using tricks to reduce the matrix to one row won't work either. The problem seems to be in size. The I matrix has size 3×45 and each row has size 1×45. Yet, if the size of the string expression making up each row is determined, it is of different size (1×36 for row 1). Thus, the only method the author found for determining the time responses for the currents was to copy each expression from the command window and put it in the argument of `ilaplace` in single quotation marks as shown in the following for $i_1(t)$:

```
EDU»i1 = ilaplace(sym('(13*s+7*s^2+24+2*s^3)/(5+4*s+s^2)/(s^2+4)'))

i1 =

77/65*exp(-2*t)*sin(t)+109/65*exp(-2*t)*cos(t)+38/65*sin(2*t)+21/65*
cos(2*t)

EDU»pretty(i1)


 77                        109                       38              21
 ----exp(-2 t) sin(t)  +  ----exp(-2 t) cos(t)  +  ----sin(2 t)  +  ----cos(2 t)
 65                        65                        65              65
```

Node Analysis

For a network consisting of N_b branches and N_v nodes, it follows that $N_v - 1$ independent equations can be written by writing KCL at $N_v - 1$ properly chosen nodes (see Appendix B). If the node for which we do not write a KCL equation is the same one that we choose as a reference node (i.e., "ground"), the equations have the following symmetric form:

$$Y_{11}(D)v_1(t) + Y_{12}(D)v_2(t) + \cdots + Y_{1n}(D)v_n(t) = i_1(t)$$
$$Y_{21}(D)v_1(t) + Y_{22}(D)v_2(t) + \cdots + Y_{2n}(D)v_n(t) = i_2(t) \tag{6-52}$$
$$\vdots$$
$$Y_{n1}(D)v_1(t) + Y_{n2}(D)v_2(t) + \cdots + Y_{nn}(D)v_n(t) = i_n(t)$$

where $Y_{ii}(D) = C_{ii}D + G_{ii} + \Gamma_{ii}D^{-1}$

$\quad C_{ii}$ = sum of all capacitances between node i and all other nodes
$\quad G_{ii}$ = sum of all conductances between node i and all other nodes $\qquad$ (6-53)
$\quad \Gamma_{ii}$ = sum of all reciprocal inductances between node i and all other
$\qquad$ nodes ($\Gamma = L^{-1}$)

$$D = \frac{d}{dt}, \qquad D^{-1} = \int_{-\infty}^{t} (\cdot)\, d\lambda$$

Also,

$$Y_{ij}(D) = -(C_{ij}D + G_{ij} + \Gamma_{ij}D^{-1}) \tag{6-54}$$

where G_{ij} = sum of all conductances between nodes i and j
$\quad C_{ij}$ = sum of all capacitances between nodes i and j
$\quad \Gamma_{ij}$ = sum of all reciprocal inductances between nodes i and j
$\quad v_j(t)$ = voltage drop between node j and the reference node
$\quad i_j(t)$ = total source current flowing into node j

Taking the Laplace transform of (6-52), we obtain

$$\mathbf{Y}(s)\mathbf{V}(s) = \mathbf{I}(s) + \mathbf{J}(s) \tag{6-55}$$

where $\mathbf{Y}(s)$ is called the *node admittance matrix* and has elements given by

$$Y_{ii}(s) = sC_{ii} + G_{ii} + \frac{\Gamma_{ii}}{s} \tag{6-56}$$

$$Y_{ij}(s) = -\left(sC_{ij} + G_{ij} + \frac{\Gamma_{ij}}{s}\right) \tag{6-57}$$

If no *dependent sources* are present, $\mathbf{Y}(s)$ is a symmetric matrix. The other matrices in (6-55) are defined as

$$\mathbf{V}(s) \triangleq \mathcal{L} \begin{bmatrix} v_1(t) \\ v_2(t) \\ \vdots \\ v_n(t) \end{bmatrix} \qquad (n = N_v - 1) \tag{6-58}$$

$$\mathbf{I}(s) \triangleq \mathcal{L} \begin{bmatrix} i_1(t) \\ i_2(t) \\ \vdots \\ i_n(t) \end{bmatrix} \qquad (n = N_v - 1) \tag{6-59}$$

and $\mathbf{J}(s)$ is the *initial-condition matrix* which results when the derivatives and integrals of (6-52) are Laplace-transformed. It can be included directly with the *source current matrix,* $\mathbf{I}(s)$, by using the Laplace-transformed equivalent circuits of Figure 6-2 for capacitive and inductive elements with nonzero initial conditions.

EXAMPLE 6-6

We consider the same circuit as Figure 6-11, except that the two voltage sources and the series resistors are replaced by their Norton equivalents.[†] The resulting circuit is shown in Figure 6-12. Working with the Laplace-transformed equivalent circuit of Figure 6-12b, where initial conditions have been replaced by equivalent-current generators, we may immediately write down the Laplace-transformed matrix equations on a nodal basis. They are

$$\begin{bmatrix} \dfrac{1}{R_1} + \dfrac{1}{sL} & -\dfrac{1}{sL} & 0 \\[2ex] -\dfrac{1}{sL} & \dfrac{1}{R_2} + \dfrac{1}{R_3} + sC + \dfrac{1}{sL} & -\dfrac{1}{R_3} \\[2ex] 0 & -\dfrac{1}{R_3} & \dfrac{1}{R_3} + \dfrac{1}{R_4} \end{bmatrix} \begin{bmatrix} V_1(s) \\[2ex] V_2(s) \\[2ex] V_3(s) \end{bmatrix} = \begin{bmatrix} \dfrac{V_s(s)}{R_1} - \dfrac{i_0}{s} \\[2ex] \dfrac{i_0}{s} + Cv_0 - \dfrac{V_c(s)}{R_3} \\[2ex] \dfrac{V_c(s)}{R_3} \end{bmatrix} \tag{6-60}$$

We note that the admittance matrix, $\mathbf{Y}(s)$, is symmetric and can be written down directly from the Laplace-transformed circuit diagram by inspection. Also, by using the initial-condition generators

---

[†]This step need not be carried out, but in that case the equilibrium equations cannot be written down by inspection in matrix form.

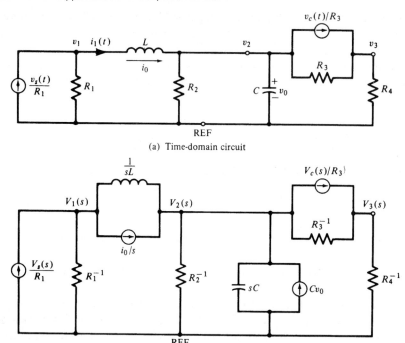

(a) Time-domain circuit

(b) Laplace-transformed circuit including initial-condition generators: admittances shown

FIGURE 6-12.   Circuit for Example 6-6.

of Figure 6-2, the initial conditions are included automatically with the source current matrix. Note also that a convenient way to handle voltage sources in nodal analysis is to convert them to Norton equivalents.

The student should write down the KCL equations for this circuit assuming that $v_c(t)$ is a dependent source; for example, $V_c(s) = \alpha R_3 I_1(s)$. The admittance matrix is no longer symmetric in this case.

6-4 Transfer Functions

Definition of a Transfer Function

The concept of a *transfer function* is of great importance in system studies. It provides a method of completely specifying the behavior of a system subjected to arbitrary inputs. We define the *transfer function* of a fixed, linear system as the ratio of the Laplace transform of the system output to the Laplace transform of the system input when all initial conditions are zero. If we denote $y(t)$ as the output and $x(t)$ as the input, we may express the transfer function, $H(s)$, as

$$H(s) = \frac{Y(s)}{X(s)} \bigg|_{\text{all initial conditions zero}} \tag{6-61}$$

This relation must hold true for any input. Suppose that $x(t) = \delta(t)$. Then it is obvious that $H(s)$ is the Laplace transform of the unit impulse response of the system since $X(s) = 1$. There are some things that we should note about it:

1. The transfer function is independent of the particular input and is a property of the circuit only.
2. The transfer function is obtained for the case of zero initial conditions.
3. The transfer function for a lumped, linear, fixed circuit is a rational function of s.
4. We obtain the frequency response of a system from $H(s)$ by setting $s = j\omega = j2\pi f$ to obtain $H(j2\pi f)$. The magnitude of $H(j2\pi f)$ is then the amplitude response function and its argument is the phase response function.

Properties of Transfer Function for Linear, Lumped Stable Systems

If the system is lumped, in addition to being linear and fixed, the transfer function may be written as

$$H(s) = \frac{b_m s^m + b_{m-1} s^{m-1} + \cdots + b_0}{a_n s^n + a_{n-1} s^{n-1} + \cdots + a_0} \triangleq \frac{N(s)}{D(s)} \tag{6-62}$$

where the a_n's and b_m's are the coefficients of the differential equation that describes the system:

$$a_n \frac{d^n y(t)}{dt^n} + a_{n-1} \frac{d^{n-1} y(t)}{dt^{n-1}} + \cdots + a_0 y(t) = b_m \frac{d^m x(t)}{dt^m} + b_{m-1} \frac{d^{m-1} x(t)}{dt^{m-1}} + \cdots + b_0 x(t) \tag{6-63}$$

The a_n's and b_m's are real and nonnegative because they result from real system components such as resistors, capacitors, and inductors (springs, masses, and dampers if we are dealing with mechanical systems). Thus the numerator and denominator polynomials, $N(s)$ and $D(s)$, have real coefficients. This implies that their roots are either real or occur in complex conjugate pairs. The roots of $D(s)$ are referred to as the *poles* of the transfer function $H(s)$ and the roots of $N(s)$ are referred to as the *zeros* of $H(s)$.

Since $H(s)$ is the Laplace transform of the impulse response, it follows that the roots of $N(s)$ and $D(s)$ must satisfy certain conditions if $h(t)$ represents a response from a stable system.

First, requiring the system to be bounded-input, bounded-output (BIBO) stable, we note *that the degree of $N(s)$ must be less than or equal to the degree of $D(s)$*, for if this were not the case, long division of $D(s)$ into $N(s)$ results in the following form for $H(s)$:

$$H(s) = C_k s^k + C_{k-1} s^{k-1} + \ldots + C_1 s + C_0 + \frac{N'(s)}{D(s)} \tag{6-64}$$

Multiplication of $H(s)$ by the Laplace transform of a bounded input signal (e.g., a unit step) and inverse Laplace transforming will then result in a response that is not bounded. For example, if the term $C_1 s$ alone is present, which would result if $N(s)$ were of degree one more than $D(s)$, the output in response to a unit step would contain the term $\mathcal{L}^{-1}[(C_1 s)(1/s)] = C_1 \delta(t)$. Also, all roots of $D(s)$—the poles of $H(s)$—must have real parts which are negative; *that is, the poles must lie in the left half of the s-plane.*[†] The reason for this is apparent if one considers the inverse Laplace transform of a partial-fraction expansion of $Y(s) = X(s)H(s)$. Assume that $H(s)$ contains the simple complex conjugate pair of poles $s_1 = -\alpha + j\omega_0$ and $s_2 = -\alpha - j\omega_0$, giving the denominator quadratic factor $(s + \alpha)^2 + \omega_0^2$. Incomplete partial-fraction expansion of $Y(s)$ then gives

$$Y(s) = \frac{As + B}{(s + \alpha)^2 + \omega_0^2} + Y_1(s) \tag{6-65}$$

The first term of $Y(s)$, according to Table 5-3, corresponds to terms in $y(t)$ of the form

$$\exp(-\alpha t) \cos \omega_0 t \, u(t)$$

[†]The poles of $H(s)$ give rise to response terms which are due to *natural frequencies*.

and

$$\exp(-\alpha t) \sin \omega_0 t\, u(t)$$

Since $s_{1,2} = -\alpha \pm j\omega_0$ corresponds to the complex conjugate pole locations, we see that α must be positive (poles in the left-half plane) if $\exp(-\alpha t)$ is to decrease with increasing t. Otherwise, the output will be unbounded regardless of whether the input is bounded or not. Similar considerations of pole factors of the form $[(s + \alpha)^2 + \omega_0^2]^n$ in $H(s)$ result in time-domain terms of the form

$$At^{n-1} \exp(-\alpha t) \cos \omega_0 t\, u(t)$$

or

$$At^{n-1} \exp(-\alpha t) \sin \omega_0 t\, u(t)$$

Clearly, if $\alpha > 0$, a bounded response results. That is, poles in the left-half s-plane need not be simple.

If $\alpha = 0$, a bounded output will not result, even if $n = 1$, when the system is forced with a sinusoid of frequency ω_0 (a bounded input). This can be seen since $Y(s)$ will then have a denominator factor of $(s^2 + \omega_0^2)^2$. This will give a term in the partial-fraction expansion of the form $A/(s^2 + \omega_0^2)^2$, which Table 5-3 indicates will give an unbounded term in $y(t)$. Therefore, BIBO stability requires strictly left-half-plane poles.

With these considerations on permissible pole locations for a stable system, we are naturally curious about restrictions, if any, on zero locations. With no condition other than BIBO stability placed on a transfer function, there are no restrictions on zero locations. If we impose an additional condition on the transfer function, defined as *minimum phase,* zeros must be located in the left-half plane also. A minimum-phase transfer function has the property that the absolute value of its phase is the smallest possible of all transfer functions with the same poles.

Components of System Response

If the system differential equation, given by (6-63), is Laplace-transformed term-by-term through the use of the differentiation theorem of Laplace transforms, the result is

$$a_n[s^n Y(s) - s^{n-1}y(0) - s^{n-2}\dot{y}(0) - \cdots - sy^{(n-2)}(0) - y^{(n-1)}(0)]$$
$$+ a_{n-1}[s^{n-1}Y(s) - s^{n-2}y(0) - s^{n-3}\dot{y}(0) - \cdots - y^{(n-2)}(0)]$$
$$+ a_{n-2}[s^{n-2}Y(s) - s^{n-3}y(0) - s^{n-4}\dot{y}(0) - \cdots - y^{(n-3)}(0)] + \cdots$$
$$+ a_1[sY(s) - y(0)] + a_0 Y(s)$$
$$= [b_m s^m + b_{m-1}s^{m-1} + \cdots + b_0]X(s) \tag{6-66}$$

where $Y(s) = \mathcal{L}[y(t)]$, $X(s) = \mathcal{L}[x(t)]$, and $y^{(n)}(0) = d^n y/dt^n|_{t=0}$. It has been assumed that the derivatives on $x(t)$ at $t = 0$ are zero. Equation (6-66) can be written

$$D(s)Y(s) - C(s) = N(s)X(s) \tag{6-67}$$

or

$$Y(s) = \frac{C(s)}{D(s)} + H(s)X(s) \tag{6-68}$$

where

$$N(s) = \sum_{j=0}^{m} b_j s^j \tag{6-69a}$$

$$D(s) = \sum_{j=0}^{n} a_j s^j \tag{6-69b}$$

$$H(s) = \frac{N(s)}{D(s)} \tag{6-69c}$$

and

$$C(s) = \sum_{i=1}^{n} a_i s^{i-1} y(0) + \sum_{i=2}^{n} a_i s^{i-2} \dot{y}(0) + \cdots + [a_n s + a_{n-1}] y^{(n-2)}(0) + a_n y^{(n-1)}(0) \tag{6-69d}$$

The inverse Laplace transform of (6-68) gives the total response of the system including that due to the initial conditions as well as to the forcing function. It is referred to as the *initial-state response (ISR)*, and is composed of the two terms

$$y_{zir}(t) = \mathcal{L}^{-1}[C(s)/D(s)] \tag{6-70}$$

and

$$y_{zsr}(t) = \mathcal{L}^{-1}[H(s)X(s)] \tag{6-71}$$

The first results from the initial conditions present in the system alone, and is called the *zero-input response (ZIR)*. The second is due to the forcing function or system input, assuming zero initial conditions, and is referred to as the *zero-state response (ZSR)*.

As mentioned in Chapter 5, the response of a system can be divided into the transient and forced responses. The former is due to the natural response of the circuit and approaches zero as $t \to \infty$. The latter is independent of the natural response and depends only on the form of the input. We have just seen that the system response can also be decomposed into the ZIR and ZSR. Clarification of these two viewpoints is provided by the following example.

EXAMPLE 6-7

Consider the *RC* circuit of Figure 6-13a. Instead of the output being taken across the capacitor, as previously, we take it across the resistor. Find the output for the input

$$v_s(t) = 5 \cos 2t \, u(t) \tag{6-72}$$

and an initial capacitor voltage of v_0 with $RC = 1$ s. Express it as (a) the sum of the forced and transient responses, and (b) as the sum of the ZIR and ZSR.

Solution: The Laplace-transformed equivalent circuit is shown in Figure 6-13b. The Laplace transform of the loop current is

$$I(s) = \frac{V_s(s) - v_0/s}{R + 1/sC} \tag{6-73}$$

and the Laplace transform of the output (the voltage across the resistor) is

$$Y(s) = \frac{s}{s + 1/RC} V_s(s) - \frac{v_0}{s} \frac{s}{s + 1/RC}$$

$$= H(s)\left[V_s(s) - \frac{v_0}{s} \right] \tag{6-74}$$

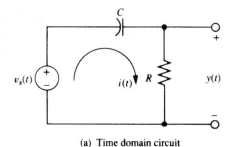

(a) Time domain circuit

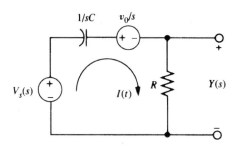

(b) Laplace-transformed equivalent circuit

FIGURE 6-13. A circuit used to illustrate the concepts of transient plus forced responses and ZIR plus ZSR.

where

$$H(s) = \frac{s}{s + 1/RC}$$

Clearly, the first term is due to the input, and the second term is due to the initial condition. For the given input and the given value for RC, the Laplace transform of the output is

$$Y(s) = \frac{5s^2}{(s + 1)(s^2 + 4)} - \frac{v_0}{s + 1}$$

$$= \frac{1}{s + 1} + 4\left[\frac{s}{s^2 + 4} - \frac{1}{2}\frac{2}{s^2 + 4}\right] - \frac{v_0}{s + 1} \tag{6-75}$$

Inverse-Laplace-transforming this expression, we find the output to be

$$y(t) = \left\{e^{-t} + 4\left[\cos 2t - \frac{1}{2}\sin 2t\right]\right\}u(t) - v_0 e^{-t}u(t) \tag{6-76a}$$

If the initial condition is zero, the last term disappears. Hence, the last term is the ZIR. The terms inside the braces disappear if the input is zero. Therefore the terms inside the braces constitute the ZSR. The output may be rewritten as

$$y(t) = \left\{(1 - v_0)e^{-t} + 4\left[\cos 2t - \frac{1}{2}\sin 2t\right]\right\}u(t) \tag{6-76b}$$

We can now identify the transient and steady-state responses as corresponding to the exponential and sinusoidal terms, respectively, for the former goes away as $t \to \infty$. Note in (6-75) that the pole of the transfer function appears both in the term due to the input and in the term due to the initial condition. This produces the response $e^{-t}u(t)$ in both the ZSR and ZIR. In other words, both the input and the initial condition excite the natural frequencies of the system. When the output is viewed as the sum of the transient and forced responses, terms dependent on the natural frequencies appear only in the transient portion.

MATLAB *Application*

The output of a system may be simulated using the function `lsim` from the signal processing tool box of MATLAB (a subset of this toolbox is included in *The Student Edition of MATLAB*) as shown in the following program listing. Plots of the output (voltage across resistor) obtained by simulation and that computed from the analytical formula (6-65b) are given in Figure 6-14. Similarly, the impulse and step responses may be obtained by using the functions `impulse` and `step`.

```
EDU» echo on
EDU» c6ex7
%      lsim plot for Example 6-7
%
v0 = -2;                       % Define IC value
t = 0:.05:40;                  % Define time variable
num = [1 0];
den = [1 1];                   % Define H(s) numerator and denominator
                               % polynomials
v = 5*cos(2*t)-v0;             % Define input in time domain; include IC
y_sim = lsim(num,den,v,t);     % Use lsim() to get output (without y =
                               % lsim(), plot is done)
```

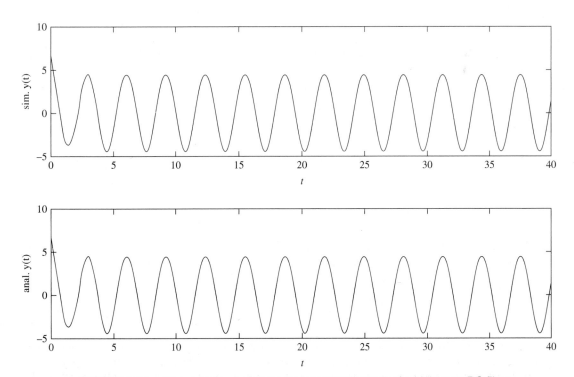

FIGURE 6-14.   Simulated and analytically computed output of a high pass RC filter.

```
y_anal = (1-v0)*exp(-t)+4*(cos(2*t)-.5*sin(2*t)); % Analytical output for
                                                   % comparison
% Plot simulated and analytically computed output
subplot(2,1,1),plot(t,y_sim),xlabel('t'),ylabel('sim. y(t)')
subplot(2,1,2),plot(t,y_anal),xlabel('t'),ylabel('anal. y(t)')
```

Asymptotic and Marginal Stability

A system is said to be *asymptotically stable* if $y_{zir}(t) \to 0$ as $t \to \infty$ for *all* initial conditions, $y(0)$, $\dot{y}(0)$, $y^{(2)}(0), \ldots, y^{(n-1)}(0)$. That is, the system response due to any initial stored energy in the system dies out to zero as time goes to infinity.

The system is marginally stable if

$$\left| y_{zir}(t) \right| \le M, \qquad \text{all } t \ge 0 \text{ and all initial conditions} \tag{6-77}$$

where M is a finite positive constant. The system is unstable if the magnitude of the ZIR grows without bound for some values of the initial conditions.

It can be shown that if a system is asymptotically stable (also referred to as internal stability), it must be BIBO stable (also referred to as external stability). The converse is not necessarily true (see Kailath, pp. 175 ff).

6-5 Stability and the Routh Array

The concept of stability is extremely important in system studies. This is true for the simple reason that stability is, in almost all cases, the minimum condition that a system must meet in order to behave acceptably. On the other hand, it does not follow that a system will exhibit acceptable behavior just because it is stable.

There are two concepts of stability in common use. The first notion of stability involves the behavior of the system when it is subjected to general inputs. The concept is quite simple. If the system output is bounded for all time for every bounded input, the system is said to be bounded-input, bounded-output (BIBO) stable, as discussed in Section 6-4.

The second concept of stability involves the behavior of the system after it is subjected to small disturbances of short duration. Let us assume that the system is in some equilibrium condition and that some disturbance causes it to be moved from its operating point to some other operating point. This second operating point is assumed to be not much different from the original one. Now suppose that the disturbance is removed, and we observe the system behavior after this time. If the system output always remains bounded and eventually tends toward the original operating point, the system is said to be *stable*. If the system output grows without bound, the system is said to be *unstable*. If the output remains bounded but does not tend toward the original operating point, the system is said to be *marginally stable*. For example, a marginally stable system may exhibit a sustained oscillation or a constant output. These concepts are illustrated in Figure 6-15, where the horizontal position of the ball represents the output. As is readily seen, this concept of stability is quite different from that of BIBO stability. However, for lumped, linear, time-invariant systems it can be shown that the two concepts of stability are equivalent; that is, one implies the other, assuming that the system is controllable and observable (see Section 7-8).

It is well known that a necessary and sufficient condition for the stability of a linear, time-invariant system is that the system poles all lie in the left-half s-plane. The system poles are simply the roots of the characteristic polynomial, which is the denominator of the overall system transfer function. The va-

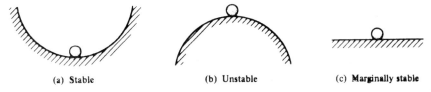

(a) Stable   (b) Unstable   (c) Marginally stable

FIGURE 6-15. Illustration of stability concepts.

lidity of the stability condition is readily established. The system response will contain terms of the form $e^{p_i t}$, where the p_i are the system poles. If any of the poles have positive real parts, the response will clearly increase without bound. If the real part is zero, the response will, at best, oscillate with constant amplitude or maintain a nonzero dc value. Thus the response for a small initial disturbance does not decay to zero. On the other hand, if all poles have negative real parts (i.e., are in the left-half s-plane) all terms in the response ultimately decay to zero.

We have established that the determination of system stability (for linear, time-invariant systems) is equivalent to determining whether or not all poles lie in the left-half s-plane (l.h.p.). How can we do this? One approach would be to factor the characteristic polynomial. This would tell us the exact root locations. However, this involves factoring an nth-degree polynomial in s, which is no easy task unless we resort to computer-aided methods. Furthermore, the exact root locations really give more information than is required—we only need to know if all the roots are in the l.h.p. Thus we turn to methods other than factoring the characteristic polynomial.

A necessary, but not sufficient, condition for a polynomial to have strictly l.h.p. roots is that all its coefficients be strictly positive (i.e., greater than zero). The reason for this is fairly easy to see. Let us write the polynomial in factored form and then multiply it out so as to examine the form of the coefficients. We have

$$p(s) = (s + p_1)(s + p_2)(s + p_3) \cdots (s + p_n)$$

$$= s^n + (p_1 + p_2 + \ldots + p_n)s^{n-1} + \left(\begin{array}{c}\text{sum of products of } p\text{'s} \\ \text{taken two at a time}\end{array}\right)s^{n-1}$$

$$+ \cdots + \left(\begin{array}{c}\text{sum of products of } p\text{'s} \\ \text{taken } n - 1 \text{ at a time}\end{array}\right)s + p_1 p_2 \cdots p_n \tag{6-78}$$

From the way in which the coefficients are constructed it is clear that if a negative or zero coefficient is present, it must arise due to one or more of the p_i having negative or zero real part. Since the p_i are the negatives of the roots, this means that at least one root lies in the r.h.p. or on the imaginary axis. Thus the system is not stable. Unfortunately, having all the coefficients positive does not guarantee stability. However, this condition is sometimes useful in that it allows us to immediately recognize as unstable a system whose characteristic polynomial contains negative coefficients.

A stability test that is both necessary and sufficient is clearly desirable. Such a test is provided by the *Routh criterion*. It is sometimes called the Routh-Hurwitz criterion because Hurwitz developed an equivalent test formulated in terms of determinants. The Routh criterion makes use of an array of coefficients generated from the coefficients of the characteristic polynomial. When the array is complete, the number of sign changes in the first column is equal to the number of r.h.p. roots. For a stable system there must be no such roots (i.e., no first-column sign changes).

The Routh array is constructed in a straightforward manner from the coefficients of the characteristic polynomial, $D(s)$. Consider a polynomial

$$a_n s^n + a_{n-1} s^{n-1} + a_{n-2} s^{n-2} + \cdots + a_1 s + a_0 \tag{6-79}$$

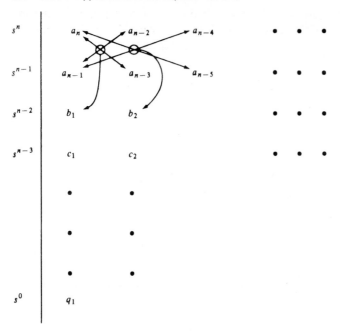

FIGURE 6-16. Formation of Routh array.

The Routh array is formed as shown in Figure 6-17. The first two rows are obtained directly from the polynomial coefficients. The coefficients of the third row are then generated from those of the first two rows as follows:

$$b_1 = \frac{a_{n-1}a_{n-2} - a_n a_{n-3}}{a_{n-1}}, \qquad b_2 = \frac{a_{n-1}a_{n-4} - a_n a_{n-5}}{a_{n-1}} \tag{6-80}$$

and so on, until there are no more a terms from which to compute coefficients. The multiplication is indicated by arrows in Figure 6-16. Once the s^{n-2} row is obtained, it is used with the coefficients of the s^{n-1} row to compute the coefficients of the next (s^{n-3}) row. This is done in exactly the same manner; that is,

$$c_1 = \frac{b_1 a_{n-3} - b_2 a_{n-1}}{b_1}, \qquad c_2 = \frac{b_1 a_{n-5} - b_3 a_{n-1}}{b_1} \tag{6-81}$$

and so on, until this row stops. Succeeding rows are computed in the same way until the array terminates with only one element in the s^0 row. It is important to label the power of s corresponding to each row so that we know when the array terminates. Finally, when the array is complete, the elements of the first column are examined for sign changes. The number of sign changes is equal to the number of r.h.p. roots of the polynomial from whose coefficients the array was generated. The process sounds a good deal more complicated than it actually is. The following examples will illustrate the procedure.

EXAMPLE 6-8

Consider the following polynomial:

$$p(s) = s^3 + 14s^2 + 41s - 56 \tag{6-82}$$

Find the number of r.h.p. roots.

We can readily see that there is at least one root in the r.h.p. because of the negative coefficient, but let us apply the Routh test and see how many r.h.p. roots there are. The Routh array is found to be

$$
\begin{array}{c|cc}
s^3 & 1 & 41 \\
s^2 & 14 & -56 \\
s & \dfrac{41(14) + 56}{14} = 45 & \\
s^0 & -56 &
\end{array}
$$

We see that there is one sign change in the first column, so there is one r.h.p. root. The result can be checked for this simple case by factoring the polynomial as

$$p(s) = (s - 1)(s + 7)(s + 8) \tag{6-83}$$

Clearly, the r.h.p. root is at $s = 1$.

EXAMPLE 6-9

Find the number of r.h.p. roots of

$$p(s) = s^4 + 5s^3 + s^2 + 10s + 1 \tag{6-84}$$

The Routh array is

$$
\begin{array}{c|ccc}
s^4 & 1 & 1 & 1 \\
s^3 & 5 & 10 & \\
s^2 & -1 & 1 & \\
s & 15 & & \\
s^0 & 1 & &
\end{array}
$$

Here we have two sign changes ($+5$ to -1 and -1 to $+15$), so there are two r.h.p. roots.

There are two cases in which development of the Routh array requires slight modification of the procedure discussed. Both pertain to zero entries in the array. The first case is one in which the first element of a row is zero, but there is at least one nonzero element in the row. The zero element causes difficulty because we must divide by it in order to compute the elements of the next row. To avoid the difficulty, we replace the zero element by a small positive number ϵ and continue. When the array is completed we let ϵ tend toward zero and examine the sign changes in the first column. This technique is illustrated in Example 6-10.

EXAMPLE 6-10

Determine the stability of a system whose characteristic polynomial is

$$p(s) = s^4 + s^3 + s^2 + s + 3 \tag{6-85}$$

The Routh array is begun as usual, but we find a zero first element in the s^2 row with another element in that row nonzero. So we replace the zero by ϵ and continue as shown.

s^4	1	1	3
s^3	1	1	
s^2	$\cancel{0}^\epsilon$	3	
s	$\dfrac{\epsilon - 3}{\epsilon}$		
s^0	3		

We then let ϵ tend toward zero. All terms in the first column remain positive except the term in the s row, which becomes very large and negative and leads to two sign changes. Hence the system is unstable since it has two r.h.p. poles.

The second case requiring special treatment is one in which an entire row of the array is zero. This situation arises whenever some or all of the roots are arranged symmetrically about the origin. The zero row is always one corresponding to an odd power of s. The preceding row contains coefficients that may be used to form a polynomial that is even in s. This polynomial is known as the *auxiliary polynomial*, and it contains the roots that cause the zero row. In addition, the degree of the auxiliary polynomial is equal to the number of roots that occur in symmetric pairs. To continue with the array, we differentiate the auxiliary polynomial and use the resulting coefficients in place of the row of zeros. The remaining portion of the array is computed as usual. The procedure is best illustrated by an example.

EXAMPLE 6-11

Find the number of r.h.p. roots of

$$p(s) = s^7 + 3s^6 + 3s^5 + s^4 + s^3 + 3s^2 + 3s + 1 \tag{6-86}$$

The Routh array is formed as shown.

s^7	1	3	1	3	
s^6	3	1	3	1	
s^5	$\dfrac{8}{3}$	0	$\dfrac{8}{3}$		
s^4	1	0	1		$\leftarrow$ auxiliary polynomial $s^4 + 1$
s^3	$\cancel{0}4$	0			$\leftarrow \dfrac{d}{ds}(s^4 + 1) = 4s^3$
s^2	$\cancel{0}^\epsilon$	1			
s	$-\dfrac{4}{\epsilon}$				
s^0	1				

We see that two r.h.p. roots are predicted. The roots that cause the zero row are contained in the auxiliary polynomial and are found to be $\pm \sqrt{2}/2 \pm j\sqrt{2}/2$. We may write $p(s)$ as

$$p(s) = (s + 1)^3 (s^4 + 1) \tag{6-87}$$

so that there are in fact two r.h.p. roots—the two contained in $(s^4 + 1)$.

A common application of the Routh criterion is that of determining the range of some system parameter for which the system is stable, that is, for which the characteristic polynomial has no r.h.p. roots. Such a situation arises frequently in the study of control systems, where the variable parameter is most frequently an amplifier gain. The procedure is illustrated in Example 6-12.

EXAMPLE 6-12

Find the range of K for which

$$s^3 + 3s^2 + 3s + (1 + K) = 0 \qquad (6\text{-}88)$$

has no r.h.p. roots. The Routh array is

s^3	1	3
s^2	3	$1 + K$
s	$\dfrac{8 - K}{3}$	
s^0	$1 + K$	

For the polynomial to have no r.h.p. roots, there must be no sign changes in the first column, that is,

$$\frac{8 - K}{3} > 0, \qquad 1 + K > 0 \qquad (6\text{-}89)$$

Then the condition of K is

$$-1 < K < 8 \qquad (6\text{-}90)$$

in order to have no r.h.p. roots.

6-6 Frequency Response and Bode Plots

Frequently, it is desirable to display graphically the frequency response of a system as discussed in Chapters 3 and 4. Such graphical displays are useful in control system analysis and design and in filter design. The frequency response is a sinusoidal steady-state concept and is obtained from the system transfer function by letting $s = j\omega$. Usually, we are interested in plotting both the magnitude and the angle of the transfer function versus the angular frequency ω. It is conventional to plot the magnitude in decibels (dB) and use a log scale for ω. (The magnitude of a function H in decibels, denoted $|H|_{dB}$, is given by $|H|_{dB} = 20 \log_{10}|H|$.) There are two reasons for using this choice of scales. First, since rather large ranges of frequency and magnitude may be involved, the log scales may help to compress the data so that they can be displayed on a conventionally sized page. Second, the nature of the transfer function is such that the choice of a decibel scale leads to computational simplicity.

Recall that for a linear, lumped, time-invariant system, the transfer function is a rational function of s; that is,

$$H(s) = \frac{P(s)}{Q(s)} \qquad (6\text{-}91)$$

where P and Q are polynomials in s. Therefore, $H(s)$ can be written as a product of factors of the following forms:

1. Constant factor, K.
2. Poles or zeros at the origin, $s^{\pm N}$.
3. Real poles or zeros, $(Ts + 1)^{\pm N}$.
4. Complex-conjugate poles or zeros, $(T^2s^2 + 2\zeta Ts + 1)^{\pm N}$, where ζ is the damping ratio and $0 < \zeta < 1$.

Since

$$|H|_{dB} = 20 \log_{10}|H| \tag{6-92}$$

plotting the magnitude of H in decibels allows us to plot the factors individually and then sum the results to obtain the complete magnitude plot. A frequency-response plot obtained in this way is known as a *Bode plot,* after H. W. Bode, who invented and used such plots to analyze feedback amplifiers in connection with his work at Bell Laboratories.

The phase plot is also obtained by summing the plots of the phase of the individual factors. This follows directly from the fact that the angle of the product of two complex numbers is the sum of the angles of the two numbers. Since we can obtain both the magnitude and phase plots for the complete transfer function by summing the plots for the individual factors, we now turn to developing the Bode plot for each of the factors of interest.

1. *Constant Factor, K.* For the constant factor K, we have

$$|K|_{dB} = 20 \log_{10}|K| \tag{6-93}$$

which is just a constant. Therefore, the magnitude plot is just a horizontal line. The angle is zero or $\pm 180°$, depending on whether K is positive or negative, respectively. Therefore, the phase curve is a horizontal line.

2. *Poles or Zeros at the Origin, $s^{\pm N}$.* In this case

$$H(j\omega) = (j\omega)^{\pm N}$$

so

$$|H|_{dB} = 20 \log_{10} \omega^{\pm N} = \pm 20N \log_{10} \omega \tag{6-94}$$

Since a log scale is used for ω, (6-94) describes a family of straight lines that intersect the ω-axis (0 dB) at $\omega = 1$ and which have slope $\pm 20N$ dB/decade. (A decade is a factor of 10 in ω, for which $\log_{10} \omega$ changes by 1 unit.) This family of lines is shown in Figure 6-17.

The angle of H is given by

$$\underline{/H} = \pm N90° \tag{6-95}$$

so poles or zeros at the origin simply introduce a constant phase.

3. *Real Poles or Zeros, $(Ts + 1)^{\pm N}$.* Here we have amplitude and phase curves which are not straight lines; however, they are asymptotic to straight lines. Hence our strategy will be to develop these asymptotes and make corrections as necessary to obtain the actual curves. For the magnitude we obtain

$$\begin{aligned} |H|_{dB} &= 20 \log_{10}(\sqrt{(\omega T)^2 + 1})^{\pm N} \\ &= \pm 20N \log_{10} \sqrt{(\omega T)^2 + 1} \\ &= \pm 10N \log_{10}(\omega^2 T^2 + 1) \end{aligned} \tag{6-96}$$

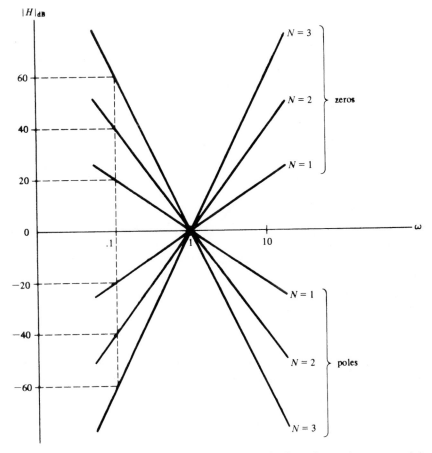

FIGURE 6-17. Magnitude plot for poles and zeros at origin.

When ω is very small (i.e., $\omega << 1/T$), (6-96) becomes

$$|H|_{dB} \simeq \pm 10N \log_{10} 1 = 0 \quad \text{dB} \tag{6-97a}$$

Therefore, the low-frequency asymptote is 0 dB. When $\omega >> 1/T$ we obtain

$$|H|_{dB} = \pm 20N \log_{10} \omega T = \pm 20N \left(\log_{10} \omega - \log_{10} \frac{1}{T} \right) \tag{6-97b}$$

which is the equation of the high-frequency asymptote. It describes a straight line of slope $\pm 20N$ dB/decade which intersects 0 dB when $\omega = 1/T$. The actual curve lies almost on the asymptotes except within about one-half decade either side of $\omega = 1/T$, which is called the *corner frequency*. In this region, the magnitude varies smoothly. At the corner frequency

$$|H|_{dB} = \pm 10N \log_{10} 2 \simeq \pm 3N \quad \text{dB} \tag{6-98}$$

This gives us enough information to sketch the magnitude curve quite closely. A sketch is shown for a real pole factor of order N in Figure 6-18. A zero factor would have a magnitude plot which is the mirror image through the ω-axis of Figure 6-19.

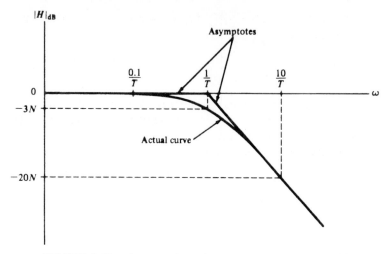

FIGURE 6-18. Asymptotic and actual magnitude curves for a real pole factor of order N.

The phase curve is obtained from

$$\underline{/H} = \pm N \tan^{-1} \omega T \qquad (6\text{-}99)$$

At low frequencies the angle of H is about $0°$, whereas at high frequencies it approaches $\pm N90°$. At the corner frequency the angle is $\pm N45°$. The phase curve for a pole of order N is shown in Figure 6-19 together with an approximation consisting of straight lines. Note that the sloped-line asymptote starts one decade before the corner frequency and ends a decade after it. The slope is $-N45°$/decade. For a zero factor, the curve would be the mirror image through the ω-axis of Figure 6-19.

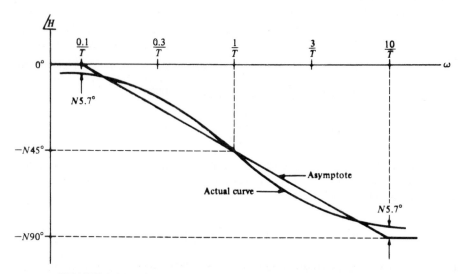

FIGURE 6-19. Asymptotic and actual phase curves for a real pole factor of order N.

4. *Complex-Conjugate Poles or Zeros* $(T^2s^2 + 2\zeta Ts + 1)^{\pm N}$. For simplicity we consider only the case of a single pair of complex conjugate poles,

$$H(s) = \frac{1}{T^2s^2 + 2\zeta Ts + 1} \tag{6-100}$$

If the poles are repeated N times, all ordinates on the curves we generate are simply multiplied by N. Their basic shape remains unchanged. Of course, if we have zeros instead of poles, curves are mirror images through the ω-axis of the curves corresponding to poles.

From (6-100) we may write the magnitude of H in decibels as

$$|H|_{dB} = -20 \log_{10}[(1 - T^2\omega^2)^2 + 4\zeta^2T^2\omega^2]^{1/2}$$

$$= -10 \log_{10}[(1 - T^2\omega^2)^2 + 4\zeta^2T^2\omega^2] \tag{6-101}$$

Again we consider the asymptotic cases. For very small ω ($\omega << 1/T$) we obtain

$$|H|_{dB} \simeq -10 \log_{10}1 = 0 \quad dB \tag{6-102}$$

Therefore, the curve is asymptotic to the ω-axis at low frequencies. For $\omega >> 1/T$ we may write (6-101) as

$$|H|_{dB} \simeq -10 \log_{10}(\omega^4T^4 + 4\zeta^2T^2\omega^2)$$

$$\simeq -10 \log_{10} \omega^4T^4 = -40 \log_{10} \omega T$$

$$= -40\left(\log_{10} \omega - \log_{10}\frac{1}{T}\right) \tag{6-103}$$

Note that we have neglected the ω^2 term compared to the ω^4 term because ω is large. Equation (6-103) describes a straight line of slope -40 dB/decade. The line intersects the ω-axis at the corner frequency $\omega = 1/T$. The actual magnitude curve is asymptotic to this line at high frequencies. In the general vicinity of the corner frequency the actual curve does not lie on the asymptotes but varies smoothly. However, the manner in which it varies depends on the damping ratio. For $\zeta < \sqrt{2}/2 = 0.707$, the curve exhibits a hump. The magnitude of this resonant peak (in decibels) is given by $20 \log_{10}(1/2\zeta \sqrt{1 - \zeta^2})$. The frequency at which this maximum occurs, which is called the *resonant frequency,* is $\omega_p = (1/T)\sqrt{1 - 2\zeta^2}$. These values give enough information to sketch the magnitude curve. Figure 6-20 shows the asymptotic curve and actual magnitude curves for several values of the damping ratio.

The phase curve for the complex pole factor may be obtained from

$$\underline{/H} = -\tan^{-1}\frac{2\zeta T\omega}{1 - \omega^2 T^2} \tag{6-104}$$

However, if we wish to use the principal value of the inverse tangent, (6-104) is valid only for $0 < \omega < 1/T$. For $\omega > 1/T$ the curve lies below $-90°$ and approaches $-180°$ for large ω since two poles are present. An expression valid for $\omega > 1/T$ is

$$\underline{/H} = -180° + \tan^{-1}\frac{2\zeta T\omega}{\omega^2 T^2 - 1} \tag{6-105}$$

Equations (6-104) and (6-105) may be used to generate phase curves. Note that these curves depend on the damping ratio. We may still use a straight-line approximation beginning at $\omega = 0.1/T$ and ending

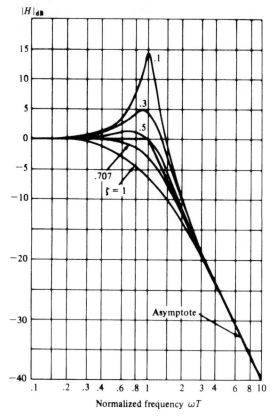

FIGURE 6-20. Asymptotic and actual magnitude curves for a complex pole factor for various ζ.

at $\omega = 10/T$ with slope $-90°$/decade, but the error between the approximation and the actual curve increases as ζ is made small.[†] Figure 6-21 shows the asymptotic curves and actual phase curves for several values of ζ.

Now that we know how to obtain Bode plots for the individual factors which may appear in a transfer function, we are in a position to construct a Bode plot for almost any rational transfer function. The usual strategy is to plot the asymptotes for the magnitude and phase curves of the individual factors and then add to obtain the asymptotic curves for the complete transfer function. Finally, the corrected curves are sketched. We close this section with a detailed example illustrating the complete procedure.

EXAMPLE 6-13 _____

Construct a Bode plot for the transfer function

$$H(s) = \frac{100(s + 2)}{s(s^2 + 5s + 25)} \tag{6-106}$$

_____
[†]The asymptotic approximation given here is common in the study of control systems. Other straightline approximations would be more accurate and are sometimes encountered.

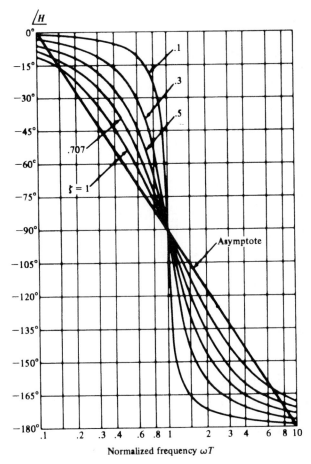

FIGURE 6-21. Asymptotic and actual phase curves for a complex pole factor for various ζ.

First we must convert this to the standard form in which the factors appear as $sT + 1$ and $s^2T^2 + 2\zeta Ts + 1$. Factoring 2 from the numerator and 25 from the denominator yields

$$H(s) = \frac{8(0.5s + 1)}{s[(0.2)^2s^2 + 2(0.5)(0.2)s + 1]} \tag{6-107}$$

We can recognize the following factors:

1. Constant factor, $K = 8$ or 18.06 dB.
2. Simple zero factor $(0.5s + 1)$, corner frequency $\omega = 2$.
3. Simple pole factor at origin, $1/s$.
4. Complex pole factor $(0.2)^2s^2 + 2(0.5)(0.2)s + 1$, corner frequency $\omega = 5$, damping ratio $\zeta = 0.5$.

The asymptotes for these factors are obtained as previously outlined and are shown in Figures 6-22 and 6-23. Note that the constant factor introduces no contribution to the phase curve, while the pole at the origin yields −90° regardless of frequency. The complete asymptotic magnitude and phase curves are shown as dashed lines. They are obtained by adding the individual curves. Finally, the

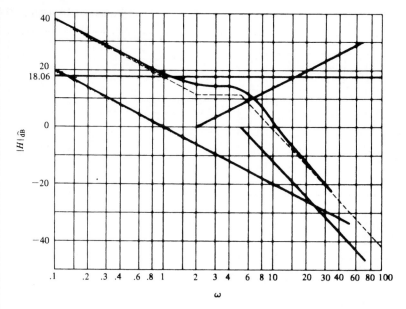

FIGURE 6-22. Asymptotic and actual magnitude curves for Example 6-13.

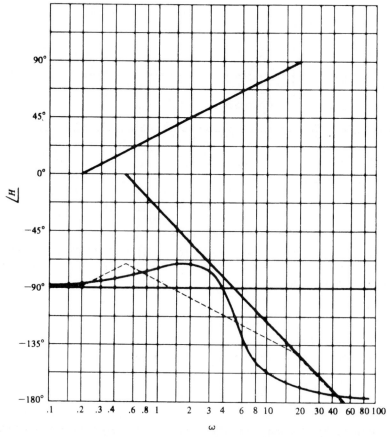

FIGURE 6-23. Asymptotic and actual phase curves for Example 6-13.

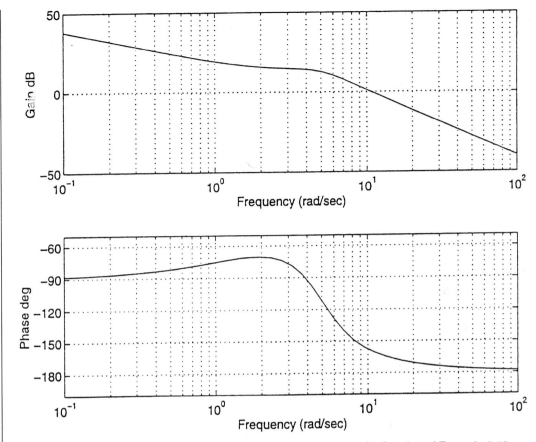

FIGURE 6-24. Output of the MATLAB function bode for the transfer function of Example 6-13.

corrected curves are the solid smooth curves shown. These are constructed by using the results shown in Figures 6-17 through 6-21 and adding the various corrections to the final asymptotic curve.

MATLAB Application

MATLAB may be used to plot Bode plots using the function `bode`. Since only three statements are required to plot the Bode plots for Example 6-13, we do them directly in the command window as shown. Plots of the output of `bode` (magnitude and phase) are shown in Figure 6-24, which shows that the same results are obtained as when the plotting is done by hand.

```
EDU» num = [0 0 100 200];
EDU» den = [1 5 25 0];
EDU» bode(num, den)
```

6-7 Block Diagrams

We now discuss a special form of system model that is very useful in the analysis of large systems—the block diagram. The block diagram is a combination of mathematical and schematic models of the

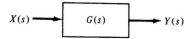

FIGURE 6-25. Basic element of block diagram.

system. As such, it not only gives the equations of each of the parts of the system, but also shows how these parts are interconnected. Therefore, the block diagram offers considerable insight into the system structure and provides all necessary information to determine the system equations or transfer function.

The basic element of block diagrams is the single block shown in Figure 6-25. Signal transmission is assumed possible only in the direction of the arrows. In Figure 6-25, $X(s)$ is the Laplace transform of the input, $Y(s)$ is the Laplace transform of the output, and $G(s)$ is the transfer function discussed in Section 6-4. Accordingly, the output is given by

$$Y(s) = G(s)X(s) \tag{6-108}$$

and this relation is assumed to hold *regardless of what is connected to the output of the block.* Such an assumption is equivalent to neglecting all loading effects. Obviously, there will be cases in which this assumption is not valid. In such cases, we must develop the transfer function under the conditions of system operation (i.e., we must take into account the effects of loading). Once we do this, we may then proceed as though loading were not present. From this point on, then, we will assume that the effects of loading have already been taken into account in the development of the transfer function and need not be considered further.

The simplest interconnection of blocks is the cascade connection shown in Figure 6-26. The transfer function can be found easily. Note that

$$Y(s) = G_2(s)W(s), \qquad W(s) = G_1(s)X(s) \tag{6-109}$$

Hence

$$Y(s) = G_2(s)G_1(s)X(s) \tag{6-110}$$

or

$$G(s) = \frac{Y(s)}{X(s)} = G_1(s)G_2(s) \tag{6-111}$$

The result is readily extended to several blocks in cascade: The transfer function of blocks in cascade is the product of the transfer functions of the individual blocks. Rules such as this become very important in simplifying complicated block diagrams, as we shall see shortly. Before doing this, however, we need to define one further element of the block diagram—the summer. A summer with several inputs is shown in Figure 6-27. The output is just the sum of the inputs, each with the sign indicated; that is, for the case shown

$$Y(s) = X_1(s) - X_2(s) - X_3(s) + X_4(s) \tag{6-112}$$

Summers are usually used to model the comparison function so essential to the operation of a feedback control system, as well as to indicate points where disturbances or other inputs may enter the system.

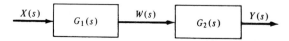

FIGURE 6-26. Cascade connection of blocks.

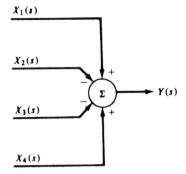

FIGURE 6-27. Symbol for summer.

When the system block diagram is very complicated it becomes tedious to write equations, as we did for the cascade connection, and solve them for the overall system transfer function. It is helpful to have some way of systematically reducing the block diagram to some form we can recognize. The most common simple form is the single-loop feedback system of Figure 6-28. We can obtain the system transfer function easily. First note that

$$C(s) = G(s)E(s) \tag{6-113}$$

Furthermore,

$$E(s) = R(s) - H(s)C(s) \tag{6-114}$$

Substitution of (6-113) into (6-114) yields

$$C(s) = G(s)R(s) - G(s)H(s)C(s) \tag{6-115}$$

or

$$C(s)[1 + G(s)H(s)] = G(s)R(s) \tag{6-116}$$

Rearranging, we obtain the transfer function as

$$T(s) = \frac{C(s)}{R(s)} = \frac{G(s)}{1 + G(s)H(s)} \tag{6-117}$$

Thus the single-loop system may be replaced by a single block whose transfer function is given by (6-117). Several relations useful for simplification of block diagrams are given in Table 6-1. It is easy to verify these relations by writing down the algebraic relations for each case and showing that the same transfer functions result.

When we reduce a block diagram we make use of the relations shown in Table 6-1 to combine blocks in cascade or parallel, moving summing points, pick off points, and simplify loops until only one loop

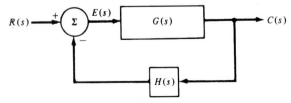

FIGURE 6-28. Single-loop feedback system.

TABLE 6-1
Relations for Block Diagram Reduction

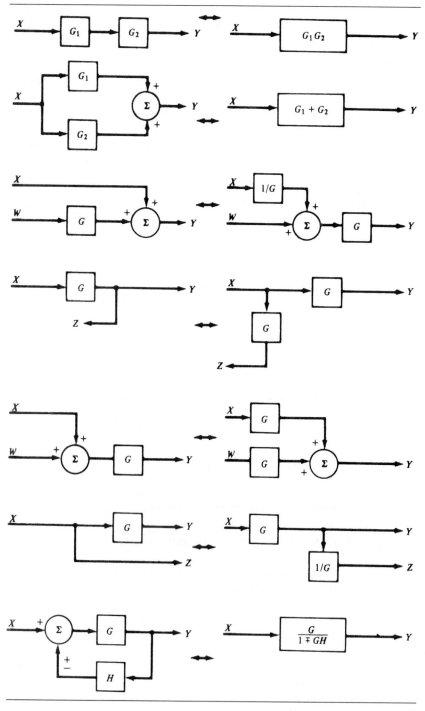

is left. At that point we can simply write down the answer using the last entry of the table. Although it sounds complicated, the procedure is really very straightforward. It is best discussed with a particular example in mind.

EXAMPLE 6-14

Find $T(s) = Y(s)/X(s)$ for the block diagram of Figure 6-29.

First, we replace the cascade connection of G_2 and G_4 by its equivalent block with transfer function G_2G_4. This new block is then in parallel with G_3, so we perform a further simplification, yielding a block with transfer function $G_3 + G_2G_4$. At this stage the system appears as shown in Figure 6-30. Next we replace the inner loop consisting of G_1 and H_1 by its equivalent transfer function $G_1/(1 + G_1H_1)$ and further simplify by combining this block with $G_3 + G_2G_4$. Now the system is reduced to the single loop shown in Figure 6-31. We can immediately write the transfer function of this single-loop system as

$$T(s) = \frac{Y(s)}{X(s)} = \frac{G_1(G_3 + G_2G_4)/(1 + G_1H_1)}{1 + H_2G_1(G_3 + G_2G_4)/(1 + G_1H_1)} \tag{6-118}$$

The result may be simplified somewhat to yield

$$T(s) = \frac{G_1(G_3 + G_2G_4)}{1 + G_1H_1 + H_2G_1(G_3 + G_2G_4)} \tag{6-119}$$

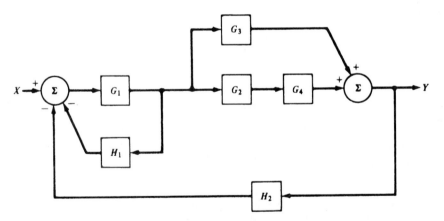

FIGURE 6-29. Block diagram to be reduced.

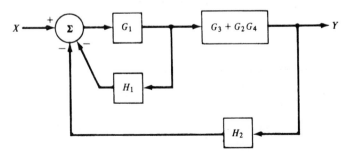

FIGURE 6-30. Partially simplified block diagram.

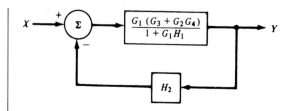

FIGURE 6-31. System after reduction to single-loop form.

Although this example has not used all the equivalencies of Table 6-1, it does indicate the general procedure to be followed in reducing block diagrams.

We close this section with a perhaps simpler, but considerably more realistic, example.

EXAMPLE 6-15

The armature-controlled dc servomotor is a common element in position control systems. We wish to find the transfer function between the input voltage (armature voltage E_a) and the output position (angular shaft position θ).

First, we must define some parameters of the motor and obtain equations governing its behavior. Once we have done this we can construct a block diagram and then determine the transfer function. This is an approach commonly employed in the development of models for physical systems. We assume that the motor has inertia J and friction B. Then the torque T produced by the motor and the shaft position θ are related by

$$J\ddot{\theta} + B\dot{\theta} = T \tag{6-120}$$

Laplace-transforming with zero initial conditions and rearranging, we obtain

$$\frac{\theta(s)}{T(s)} = \frac{1}{s(sJ + B)} \tag{6-121}$$

as the transfer function between torque and shaft position. In an armature-controlled dc motor the torque produced is proportional to the armature current,

$$T = K_m I_a \tag{6-122}$$

where K_m is called the *motor constant* and depends on the particular motor being used. To obtain an expression for the armature current, two factors must be considered. First, the armature contains resistance and inductance R_a and L_a, respectively. Thus its impedance is

$$Z_a(s) = sL_a + R_a \tag{6-123}$$

Second, a *back emf* (electromotive force) appears in the armature circuit as a result of the speed of the motor:

$$v_b(s) = K_B \dot{\theta}(t) \tag{6-124a}$$

or

$$V_b(s) = K_B s \theta(s) \tag{6-124b}$$

where K_B is a constant of proportionality. This back emf is in opposition to the applied voltage E_a, so the voltage appearing across the armature impedance is

$$V_a(s) = E_a(s) - V_b(s) \tag{6-125}$$

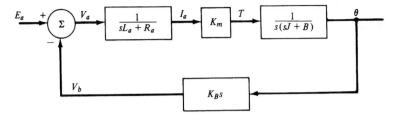

FIGURE 6-32. Block diagram of armature-controlled dc motor.

The armature current is then given by

$$I_a(s) = \frac{V_a(s)}{sL_a + R_a} \tag{6-126}$$

Equations (6-121) through (6-126) define a linear model of the armature-controlled dc motor. A block diagram representing these equations is shown in Figure 6-32. This is a single-loop negative-feedback system with $G(s)$ given by the product of the forward-loop transfer functions; that is,

$$G(s) = \frac{K_m}{s(sL_a + R_a)(sJ + B)} \tag{6-127}$$

and

$$H(s) = K_B s \tag{6-128}$$

Therefore,

$$\frac{\theta(s)}{E_a(s)} = \frac{G}{1 + GH} = \frac{K_m/[s(sL_a + R_a)(sJ + B)]}{1 + K_m K_B/[sL_a + R_a)(sJ + B)]} \tag{6-129}$$

or

$$\frac{\theta(s)}{E_a(s)} = \frac{K_m}{s[(sL_a + R_a)(sJ + B) + K_m K_B]} \tag{6-130}$$

It is perhaps worth pointing out that in most small motors L_a may be neglected, so that

$$\frac{\theta(s)}{E_a(s)} = \frac{K_m}{s[sJR_a + (R_a B + K_m K_B)]} \tag{6-131}$$

By grouping the various parameters in an obvious way we could also obtain the following representation as an approximation that is valid for many small motors:

$$\frac{\theta(s)}{E_a(s)} = \frac{K'_m}{s(s\tau_m + 1)} \tag{6-132}$$

In Example 6-12 we showed how to use the Routh criterion to determine the range of a system parameter in order to ensure stability. We are now in a position to see how such a problem might come about.

EXAMPLE 6-16 _____

The system shown in Figure 6-33 is a block diagram of a position-control system which, if it is stable, will follow a ramp angular position input in steady state. The motor is modeled by

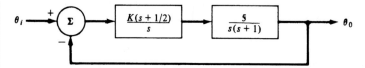

FIGURE 6-33. Position-control system.

(6-132) with $K'_m = 5$ and $\tau_m = 1$ s. The controller includes a variable gain K. We wish to know the range of K, if any, which yields a stable system.

Straightforward block diagram reduction yields the system transfer function as (show this)

$$\frac{\theta_o(s)}{\theta_i(s)} = \frac{5K(s + \frac{1}{2})}{s^3 + s^2 + 5Ks + 2.5K} \tag{6-133}$$

The characteristic polynomial is just

$$s^3 + s^2 + 5Ks + 2.5K \tag{6-134}$$

and must have no r.h.p. roots if the system is to be stable. The Routh array is

s^3	1	$5K$
s^2	1	$2.5K$
s	$2.5K$	
s^0	1	

For stability, there must be no sign changes in the first column; therefore, any $K > 0$ will yield a stable system. In actual design of such a system it would be necessary to choose a value of K that led to a suitable transient response. Such topics are discussed in detail in courses concerning control systems.

*6-8 Operational Amplifiers as Elements in Feedback Circuits

As pointed out by the first two examples in Chapter 1, operational amplifiers are convenient elements in analog signal-processing circuits. We will not go into the details of operational amplifier electrical characteristics here (see Siebert for more details), but we will instead look at a frequency-dependent model for operational amplifiers that is convenient for system analysis.

In Chapter 1, and in most introductory circuits courses, the operational amplifier is represented as an infinite-gain amplifier with an infinite input impedance at both its inverting and noninverting inputs. In linear signal-processing applications, an operational amplifier is typically used with a feedback connection from its output to the inverting input to ensure stability. In reality, the operational amplifier does not have the ideal characteristic of infinite input impedance and infinite gain. We keep the assumption of infinite input impedance here, but account for a frequency-dependent gain by the operational amplifier model shown in Figure 6-34. A simple model for the frequency-dependent gain is

$$K(s) = \frac{A_0}{1 + s/\omega_p} \tag{6-135}$$

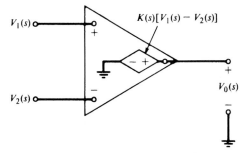

$$K(s)[V_1(s) - V_2(s)]$$

FIGURE 6-34. Frequency-dependent model for an operational amplifier.

where typical values for the parameters A_0, referred to as the low-frequency gain, and ω_p, called the open-loop bandwidth, are

$$A_0 = 2 \times 10^5$$

and

$$\omega_p = 40 \text{ rad/s} \quad (6 \text{ Hz})$$

A somewhat more complex model for $K(s)$ involving three poles is

$$K(s) = \frac{A_0}{(1 + s/\omega_{p1})(1 + s/\omega_{p2})^2} \tag{6-136}$$

Typical parameter values in this model for two popular operational amplifiers are

$$A_0 = 10^5; \quad \omega_{p1} = 20\pi \ \text{rad/s}; \quad \omega_{p2} = 2\pi \times 10^6 \ \text{rad/s} \quad \text{(type 741)}$$
$$A_0 = 10^5; \quad \omega_{p1} = 200\pi \ \text{rad/s}; \quad \omega_{p2} = 2\pi \times 10^6 \ \text{rad/s} \quad \text{(type 748)}$$

EXAMPLE 6-17

Consider the feedback operational amplifier circuit shown in Figure 6-35. Find the transfer function of this circuit.

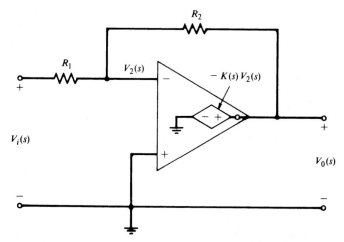

FIGURE 6-35. Feedback operational amplifier circuit pertaining to Example 6-17.

Solution: Writing a KCL equation at the node marked V_2, we obtain

$$\frac{V_o - V_2}{R_2} + \frac{V_i - V_2}{R_1} = 0 \qquad (6\text{-}137)$$

The defining relation of the operational amplifier, however, is

$$V_o(s) = K(s)(V_1 - V_2) = -K(s)V_2(s) \qquad (6\text{-}138)$$

since the noninverting input is grounded. Substitution of (6-138) into (6-136) and solving for the ratio of $V_o(s)$ to $V_i(s)$ gives

$$H(s) = \frac{V_o(s)}{V_i(s)} = -\frac{R_2}{R_1(1 + 1/K) + R_2/K} \qquad (6\text{-}139)$$

Note that for $K \rightarrow \infty$, the usual expression for the gain of this amplifier circuit results, namely, $-R_2/R_1$. Replacing K by the frequency-dependent model given by (6-135), however, we obtain the frequency-dependent transfer function

$$H(s) = -\frac{A_0/(1 + s/\omega_p)}{1 + R_1/R_2[1 + A_0/(1 + s/\omega_p)]}$$

$$= -\frac{A_0\omega_p\beta(R_2/R_1)}{s + (1 + A_0\beta)\omega_p} \qquad (6\text{-}140)$$

where

$$\beta = \frac{R_1}{R_1 + R_2} \qquad (6\text{-}141)$$

The low-frequency gain of the amplifier is obtained by setting $s = 0$, and is given by

$$H(0) = -\frac{A_0\beta(R_2/R_1)}{1 + A_0\beta} \qquad (6\text{-}142)$$

The half-power cutoff frequency is

$$\omega_3 = (1 + A_0\beta)\omega_p \qquad (6\text{-}143)$$

It is interesting to note that the product of the low-frequency gain and half-power frequency is nearly constant if $R_2 \geq 10R_1$ (i.e., the feedback circuit is serving as an amplifier). In particular, from (6-141), (6-142), and (6-143) we have

$$|H(0)|\omega_3 = A_0\omega_p \frac{R_2}{R_1 + R_2} \approx A_0\omega_p, \qquad R_2 > R_1 \qquad (6\text{-}144)$$

One can use this fact to get a rough estimate of the half-power frequency for a given low-frequency gain, and vice versa, for a particular feedback configuration.

Summary

In this chapter we have treated a number of important applications of the Laplace transform. In particular, we have used it to solve circuit examples of systems, introduced and discussed the properties of the transfer function, illustrated the use of Bode plots to obtain the frequency response of a lumped parameter system, introduced the use of the Routh array to determine whether the roots of a polynomial are in the left-half plane, and considered block diagram algebra. The following are the main points made in this chapter.

1. Initial conditions are easily included in Laplace transform analysis by *initial-condition genera-tors* in the *Laplace-transformed equivalent circuit.* Laplace-transformed equivalents for induc-tors and capacitors are given in Figure 6-2.

2. *Mutual inductance* between two coils is easily included in Laplace transform analysis by the T-equivalent circuit for a transformer as illustrated in Figure 6-5.

3. All the *network theorems* familiar from earlier circuits are applicable in the Laplace-transform do-main as well. These include *Thevenin's theorem, Norton's theorem,* and *source transformations.*

4. *Loop* and *node analyses* by means of the Laplace-transformed equivalent circuit can be done systematically on a loop or node basis. The pertinent equations for the loop basis are (6-44)–(6-48), while the applicable equations for node analysis are (6-55)–(6-59).

5. The *transfer function* for a fixed, linear system is redefined in this chapter as the ratio of the Laplace transform of the output to the Laplace transform of the input with all initial conditions zero. The definitions of earlier chapters are obtained by setting $s = j\omega = j2\pi f$ if the Fourier transform converges on the $j\omega$-axis.

6. For a *linear, lumped stable system,* the transfer function is a ratio of polynomials in s with the following properties:

   (a) *Numerator and denominator polynomials have real, nonnegative coefficients.*

   (b) *The degree of the numerator polynomial must be less than or equal to the degree of the de-nominator polynomial.*

   (c) *For BIBO* (recall from Chapter 2 that BIBO means bounded-input, bounded-output) *stabil-ity, the roots of the denominator polynomial* (i.e., the poles of the transfer function) *must be in the left-half s-plane for a causal system.*

   (d) *The roots of the numerator polynomial* (i.e., the zeros of the transfer function) *need not lie in the left-half s-plane unless the transfer function is required to be minimum phase.*

7. The response of a fixed, linear system may be written as the sum of the *zero-input response* (ZIR) and *zero-state response* (ZSR). The former results from the initial conditions present in the sys-tem alone, and the latter results from the input with zero initial conditions.

8. The response of a fixed, linear system may be written as the sum of the *transient* and *steady-state responses.* The former is due to the natural response of the circuit and dies out as $t \to \infty$, and the latter is independent of the natural response and depends only on the form of the input. These responses can also be identified as the *homogeneous solution* to the governing differen-tial equation and the *particular integral,* respectively.

9. A system is *asymptotically stable* if the ZIR approaches zero as time goes to infinity. It is *mar-ginally stable* if the ZIR is bounded as time goes to infinity. If a system is asymptotically stable, it is also BIBO stable.

10. The *Routh array* can be used to determine if the roots of a polynomial lie in the left-half plane, and can therefore be used to test a transfer function of a system for stability.

11. A *Bode plot* is a convenient means of obtaining the frequency response corresponding to a trans-fer function $H(s)$ (i.e., its magnitude and phase for $s = j\omega$).

12. *Block diagram algebra* allows one to simplify fixed, linear systems consisting of several sub-systems, including parallel, cascade, and feedback configurations.

13. The chapter closes with a discussion of operational amplifiers as elements in feedback circuits; some practical models for the frequency dependence of such amplifiers are introduced.

Further Reading

In addition to the circuit analysis texts cited in previous chapters, the following one provides a sound treatment with many good examples:

L. S. BOBROW, *Elementary Linear Circuit Analysis,* 2nd ed. New York: Holt, Rinehart, and Winston, 1987.

The following text on systems analysis provides several applications involving operational amplifiers and electromechanical circuits:

W. M. SIEBERT, *Circuits, Signals, and Systems.* New York: McGraw-Hill, 1986.

The following books provide a coverage of many control topics, including stability and tests for stability, block diagrams, and Bode plots:

K. OGATA, *Modern Control Engineering.* Englewood Cliffs, N.J.: Prentice-Hall, 1997.

T. KAILATH, *Linear Systems.* Englewood Cliffs, N.J.: Prentice-Hall, 1980.

PROBLEMS

Sections 6-2 and 6-3

6-1. Sketch the Laplace-transformed equivalents, including initial-condition generators, of the circuits shown. Choose initial-condition generators appropriate for writing loop equations.

(a)

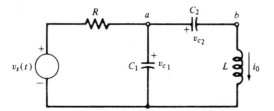

FIGURE P6-1a

(b)

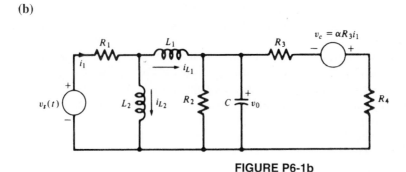

FIGURE P6-1b

6-2. Write the Laplace-transformed loop equations for the circuits of Problem 6-1 by inspection. Use matrix notation. Include initial conditions.

6-3. Sketch Laplace-transformed equivalent circuits for Problem 6-1 using initial-condition generators appropriate for writing node equations.

6-4. Write by inspection the Laplace-transformed node equations for the circuits of Problem 6-1. Use matrix notation. Include initial conditions.

6-5. Find the Laplace transform Thevenin equivalent circuit for the circuit of Problem 6-1a between the terminals a and b. Assume zero initial conditions.

6-6. Find $v_0(t)$ for the circuit shown. All initial conditions are zero.

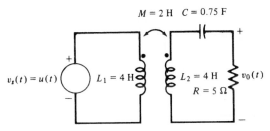

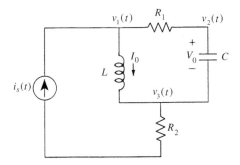

FIGURE P6-6

6-7. (a) Write node voltage (Kirchhoff's current law) equations by inspection for the circuit shown.

FIGURE P6-7

(b) Write mesh equations for this circuit. Note that one of the unknowns will be the voltage across the current source. However, the current source completely specifies the left-hand mesh current.

6-8. If more than two coupled coils are present in the same circuit, it is necessary to show polarity markings with more than one set of symbols (e.g., dots and triangles). An example of such a circuit is shown.

(a) Assuming mutual inductances of M_{12}, M_{13}, and M_{23} between the coils with self-inductances L_1, L_2 and L_3, respectively, write the governing differential equations on a loop basis for this circuit.

(b) Laplace transform the differential equations found in part (a), assuming zero initial conditions.

(c) Discuss the feasibility of replacing the coupled coils with equivalent T-circuits.

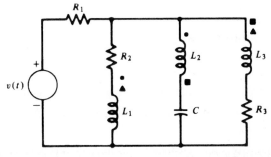

FIGURE P6-8

6-9. Repeat Example 6-6, assuming that $V_c(s) = \alpha R_3 I_1(s)$.

6-10. Find the Thevenin and Norton equivalents between a and b of the circuits shown (assume 0 initial conditions if not shown).

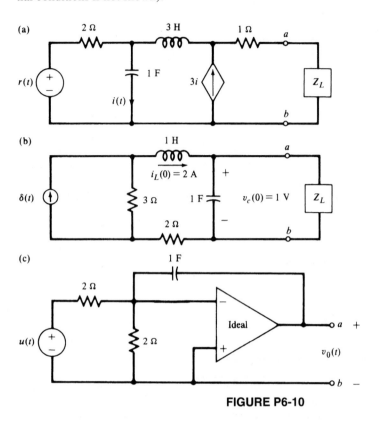

FIGURE P6-10

6-11. Use source transformations to find the load current in the circuit shown.

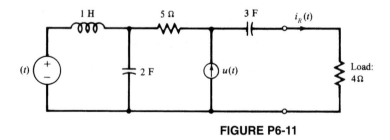

FIGURE P6-11

Section 6-4

6-12. Show that (6-62) results by Laplace-transforming (6-63), assuming zero initial conditions, and taking the ratio of $Y(s)$ to $X(s)$, where $Y(s)$ is the Laplace transform of the output and $X(s)$ is the Laplace transform of the input.

6-13. Which of the following rational functions of s are satisfactory for transfer functions of BIBO stable fixed, linear systems? Tell why or why not in each case. Then check your answers using

MATLAB to compute the response to (a) a step; (b) a sine wave. Note that the latter does not prove you are right, but will merely increase your confidence.

(a) $H_1(s) = \dfrac{s}{s^2 + 4}$;

(b) $H_2(s) = \dfrac{s + 1}{(s + 1)^2 + 9}$;

(c) $H_3(s) = \dfrac{s^2 + 1}{s^2 + 2s + 1}$;

(d) $H_4(s) = \dfrac{s^3 + 1}{s^2 + 2s + 5}$;

(e) $H_5(s) = \dfrac{s + 5}{(s - 1)(s + 2)(s + 3)}$

6-14. Given the following circuit with an initial current through the inductor of I_0 and a voltage step of magnitude A as input. Find the output voltage, $v_0(t)$, and write it in the two forms of (a) steady-state plus transient response, and (b) zero-input plus zero-state responses. Identify all components clearly.

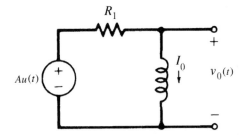

FIGURE P6-14

6-15. A linear, time-invariant two-port network is represented in terms of the currents and voltages at its ports in the block diagrams shown in Figure P6-15. The *zero-state response* equations of the two-port can be expressed in the form

$$I_1(s) = Y_{11}(s)V_1(s) + Y_{12}(s)V_2(s)$$

$$I_2(s) = Y_{21}(s)V_1(s) + Y_{22}(s)V_2(s)$$

where $Y_{11}(s)$, $Y_{12}(s)$, $Y_{21}(s)$, and $Y_{22}(s)$ can be defined in terms of various conditions on the terminals. For example,

$$Y_{11}(s) = \left.\dfrac{I_1(s)}{V_1(s)}\right|_{V_2(s)=0}$$

is called the short-circuit driving-point admittance because it refers to the input port and is obtained by short-circuiting the output port [$V_2(s) = 0$]. Also,

$$Y_{12}(s) = \left.\dfrac{I_1(s)}{V_2(s)}\right|_{V_1(s)=0}$$

is referred to as the short-circuit transfer admittance because it refers to a transfer quantity between input and output ports and is obtained by short-circuiting the input port. Suppose that the

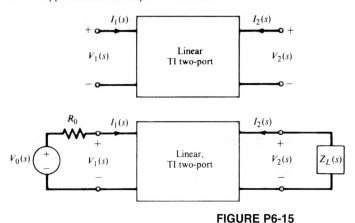

FIGURE P6-15

input is driven by a voltage source with source resistance R_0 as shown in the second figure. Show that the voltage transfer function is

$$H(s) = \frac{V_2(s)}{V_0(s)} = \frac{-Y_{21}(s)/R_0}{[1/R_0 + Y_{11}(s)][1/Z_L(s) + Y_{22}(s)] - Y_{12}(s)Y_{21}(s)}$$

6-16. (a) Derive the short-circuit driving-point and transfer admittances for the twin-T network shown in Figure P6-16 (see Problem 6-15 for a definition of these admittances).

(b) Use the result of Problem 6-15 to show that the voltage transfer function of the twin-T network is given by

$$H(s) = \frac{s^2 + 1/(RC)^2}{s^2 + 4s/RC + 1/(RC)^2}$$

Provide a pole-zero plot for this transfer function.

(c) Show that $|H(j\omega)| = 0.707|H(0)|$ for $\omega RC = \sqrt{5} \pm 2$. Noting these values, the location of the zero found in part (b), and the limiting values for $|H(j\omega)|$ as $\omega \to 0$ and $\omega \to \infty$, plot $|H(j\omega)|$ versus $\log_{10}(\omega RC)$.

(d) It is desired to use this twin-T network as a notch filter to eliminate 60-Hz power-line interference. If $C = 1\ \mu F$, calculate the value of R necessary to place the notch at 60 Hz.

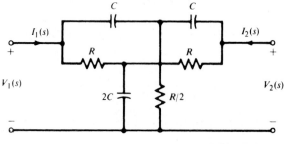

FIGURE P6-16

6-17. (a) Obtain the voltage transfer function of the operational amplifier filter shown.

(b) If $C_1 = 0.0022\ \mu F$, $C_2 = 330$ pF, $C_3 = 0.0056\ \mu F$, and $R = 10$ kΩ, find the location of the poles of this transfer function. (*Hint:* One of them must be real.)

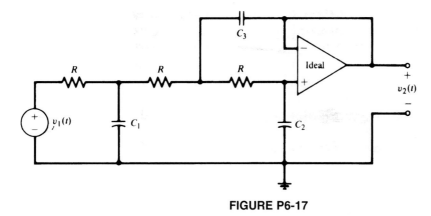

FIGURE P6-17

6-18. (a) For the circuit shown in Figure P6-18a demonstrate that under *zero-state response* conditions the transfer function is

$$H(s) = \frac{V_2(s)}{V_1(s)} = \frac{\omega_0^2/\beta}{s^2 + 2\alpha s + \omega_0^2}$$

where

$$\omega_0^2 = \frac{1}{(R_1 C_1 R_2 C_2)}$$

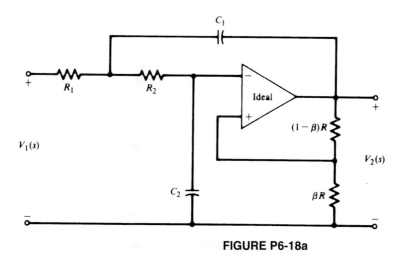

FIGURE P6-18a

and

$$2\alpha = \frac{1}{R_1 C_1} + \frac{1}{R_2 C_1} + \frac{1}{R_2 C_2}\left(1 - \frac{1}{\beta}\right)$$

This circuit, referred to as a Sallen-Key circuit, has the feature that the magnitude and angle of the pole locations can be adjusted independently.

(b) Noting that the poles of a second-order Butterworth filter are located at the s-locations $s_{1,2}$

$$s_{1,2} = -\frac{1}{\sqrt{2}}(1 \pm j)\omega_3$$

where ω_3 is the 3-dB frequency in rad/s, choose a set of circuit values and β so that the circuit realizes a second-order Butterworth filter with 3-dB frequency equal to 10 kHz.

(c) Choose the parameters of two cascaded Sallen-Key filter circuits so that a fourth-order Butterworth filter with 3-dB frequency of 10 kHz is realized. A fourth-order Butterworth filter has the pole locations as shown in Figure P6-18b.

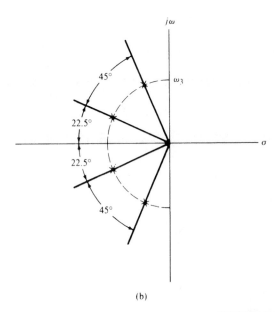

(b)

FIGURE P6-18b

Section 6-5

6-19. Determine the number of r.h.p. roots of the following polynomials.

(a) $s^3 + 6s^2 + 3s + 2$

(b) $s^4 + 3s^3 + 12s^2 + 12s + 36$

(c) $s^4 + s^3 - s - 1$

(d) $4s^5 + s^4 + 4s^3 + s^2 + 15s + 10$

6-20. Find the range, if any, of the parameter K for stability.

(a) $s^3 + 7s^2 + 11s + 5 + K$ (c) $s^3 + (2K + 1)s^2 + (K + 1)^2s + K^2 + 1$

(b) $s^4 + 2s^3 + 2s^2 + s + K$

6-21. Find the condition on the constants a, b, and c for the polynomial shown to have no r.h.p. roots.

$$s^3 + as^2 + bs + c$$

6-22. Using the Routh criterion and the models for the type 741 and 748 operational amplifiers given in connection with (6-135) and (6-136), discuss the stability of these operational amplifiers when used in the voltage-follower circuit in Figure P6-22.

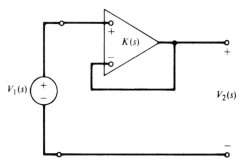

FIGURE P6-22

6-23. Using the models of the type 741 and 748 operational amplifiers given in connection with (6-141) and (6-142), apply the Routh criterion to discuss the stability of the noninverting amplifier shown. Why is the voltage-follower circuit of Problem 6-22 a more severe test of stability for these operational amplifiers? (*Note:* One of the time constants of the type 741 is intentionally made larger to avoid the difficulty explored here and in Problem 6-22.)

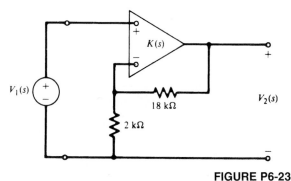

FIGURE P6-23

Section 6-6

6-24. Assume that the rational functions of s in Problem 5-13 are transfer functions. Construct amplitude and phase Bode plots (both asymptotic and corrected). You may check your results with MATLAB.

6-25. Construct a Bode plot for the transfer function

$$G(s) = \frac{250(s + 2)}{s(s^2 + 12s + 100)}$$

6-26. Construct a Bode plot for the transfer function

$$G(s) = \frac{1000(s^2 + s + 1)}{(s + 6)(s + 10)^2}$$

6-27. Show that the complex pole factor given by (6-103) will have no peak in the magnitude curve if $\zeta > \sqrt{2}/2 = 0.707$.

6-28. Given the form of the open-loop gain of an operational amplifier of (6-135) and (6-136):

    (a) Sketch a Bode plot for the open-loop gain using the type 741 parameters.

    (b) Sketch a Bode plot for the open loop gain using the type 748 parameters. Dimension fully.

Section 6-7

6-29. Obtain $Y(s)$ as a function of $X_1(s)$ and $X_2(s)$ for the system shown.

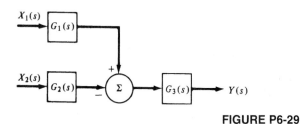

FIGURE P6-29

6-30. Find the transfer function $Y(s)/R(s)$.

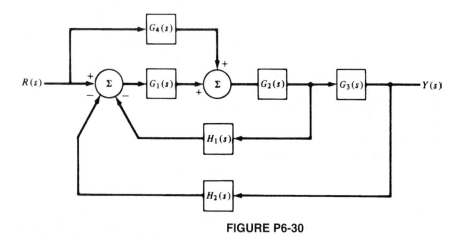

FIGURE P6-30

6-31. In the following system, let $G(s) = 6/(s + 1)(s + 3)$, $R(s) = a/s$, and $D(s) = b/s$, with a and b constants. Show that if K is made large, then $\lim_{t\to\infty} y(t) \simeq a$; that is, the disturbance has little effect on the system output in steady state.

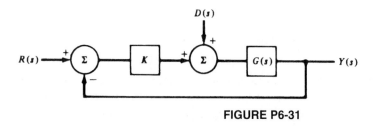

FIGURE P6-31

6-32. Given the feedback system shown.

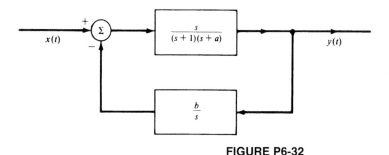

FIGURE P6-32

(a) Determine a and b such that the overall transfer function is

$$H(s) = \frac{s}{(s + 4)(s + 5)}$$

(b) If $a = 3$, what is the range of values for b for which the system is stable?

(c) For the choices of a and b determined in part (a), obtain the time response of the output if the input is a unit ramp.

6-33. The linear feedback system shown is designed to perform as a tracking system. Ideally the error signal, $\epsilon(t)$, is desired to be identically zero, or in other words it is desired that $x(t) = y(t)$.

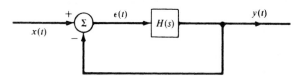

FIGURE P6-33

(a) If $H(s) = 1/s$ and the input is a unit ramp function, obtain the steady-state tracking error.

(b) If $H(s) = 1/s^2$ and the input is a unit ramp function, what is the nature of the steady-state tracking error?

(c) Let $H(s) = A(s)/s^2$ and choose $A(s)$ to avoid the problems uncovered in parts (a) and (b). That is, choose $A(s)$ so that $\epsilon(t) \rightarrow 0$ for a unit ramp input. Sketch $x(t)$, $y(t)$, and $\epsilon(t)$ for your choice of $A(s)$.

6-34. Figure P6-34 is a simplified diagram of an RC oscillator.

(a) With the switch open, show that

$$H(s) = \frac{V_0(s)}{V_1(s)} = \frac{K(RCs)}{(RCs)^2 + 3(RCs) + 1}$$

(b) With the switch closed, the circuit will be marginally stable if the values of s for which the open-loop gain $H(s) = 1$ lie on the $j\omega$-axis. Equivalently, these are the values of s for which

$$(RCs)^2 + (3 - K)(RCs) + 1 = 0$$

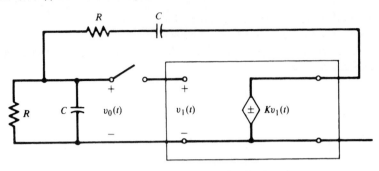

FIGURE P6-34

In practice, the value of K is chosen so that the poles lie just to the right of the $j\omega$-axis. Design an oscillator that will tune over the range of 10 to 10,000 Hz in decades by continuously varying R and changing C in steps. Try to use reasonable values for R and C (e.g., kΩ and μF).

Problems Extending Text Material

6-35. (a) Obtain the transfer function of the two-stage RC-coupled amplifier shown.

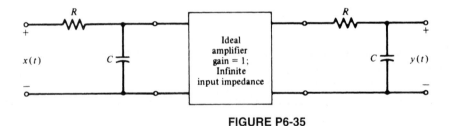

FIGURE P6-35

(b) Generalize the result of part (a) to n stages and show that

$$H_n(s) = (1 + RCs)^{-n}$$

where $H_n(s)$ is the transfer function of n RC stages coupled as shown.

(c) Show that the impulse response of the n-stage RC-coupled amplifier is

$$h_n(t) = \left[\frac{1}{RC}e^{-t/RC}u(t)\right] * \cdots * \left[\frac{1}{RC}e^{-t/RC}u(t)\right]$$

(d) Using the results of Problem 5-24, show that the effective delay and duration for the impulse response of the n-stage RC-coupled amplifier are given by $t_0 = nRC$ and $\tau = \sqrt{n}\ RC$, respectively.

6-36. *Singular cases with initial conditions.* If the following situations arise in circuits, we have the so-called "singular" cases, resulting in discontinuous capacitance voltages or inductor currents:
1. Loops consisting entirely of capacitors or capacitors and voltage generators.
2. Junctions of inductors and current generators.

Obtain $i_1(t)$ and $i_2(t)$ in the circuit shown, using two methods:

(a) Employ $t = 0^-$ initial conditions in the Laplace transform.

(b) Invoke KVL and conservation of charge through $t = 0$ around each loop. Show that both procedures give the same result.

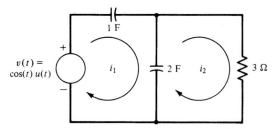

FIGURE P6-36

Both capacitors discharged at $t = 0$.

Answer:

$$i_2(t) = \tfrac{1}{82}(\tfrac{1}{9}e^{-t/9} + 9\cos t - \sin t)u(t)$$

$$i_1(t) = \tfrac{2}{3}\delta(t) + \tfrac{1}{82}(\tfrac{1}{27}e^{-t/9} + 3\cos t - 55\sin t)u(t)$$

Note that $v_{c_1}(0^+) = \tfrac{2}{3}$ V; $v_{c_2}(0^+) = \tfrac{1}{3}$ V.

6-37. *Singular example for inductor currents.* Referring to the discussion in Problem 6-36, obtain $v_1(t)$ and $v_2(t)$ in the circuit shown by using two methods:

(a) Employ $t = 0^-$ initial conditions in the Laplace transform.

(b) Invoke KCL and require conservation of flux linkages $[Li(t)]$ through $t = 0$ at each node.

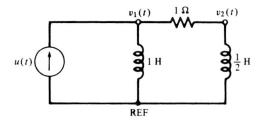

FIGURE P6-37

Both inductors have zero current flowing at $t = 0^-$.

Answer:

$$v_1(t) = \tfrac{1}{3}\delta(t) + \tfrac{4}{9}e^{-2t/3}u(t)$$

$$v_2(t) = \tfrac{1}{3}\delta(t) - \tfrac{2}{9}e^{-2t/3}u(t)$$

Note that the voltage impulses are required to cause the currents to change from 0 to $i_{L_1}(0^+) = \tfrac{1}{3}$ A and $i_{L_2}(0^+) = \tfrac{2}{3}$ A.

6-38. *Duality.* For planar networks, the following rules allow us to find a dual of a network; that is, the loop equations for one network are the same as the node equations for the other network except for the symbols:

1. Assign a node for the dual network to be constructed to each mesh of the given network plus an additional reference node.
2. For each *inductor* of L henries in the original network, assign a *capacitor* of L farads in the dual. For each *capacitor* of C farads in the original network, assign an *inductor* of C henries in the dual. For each *resistance* of R ohms in the original network, assign a *conductance* of R mhos in the dual. For each *current (voltage)* generator in the original network, assign a *voltage (current)* generator in the dual of the same numerical value.
3. The mesh (node) equations for the original network will now be identical to the node (mesh) equations of the dual network with the following replacements:

$$\text{node equations} \leftrightarrow \text{mesh equations}$$
$$v \leftrightarrow i$$
$$L \leftrightarrow C$$
$$R \leftrightarrow G$$
$$\text{open circuit} \leftrightarrow \text{short circuit}$$

Obtain the duals of the networks given in Problems 6-36 and 6-37. Show that they have equivalent equilibrium equations on a mesh and node basis.

6-39. One way to synthesize a circuit from a driving-point impedance is to use long division to produce a continued fraction for $Z(s)$. For example, show that the ladder structure illustrated in Figure P6-39 has a driving-point impedance of the form

$$Z(s) = Z_1(s) + \cfrac{1}{Y_2(s) + \cfrac{1}{Z_3(s) + \cfrac{1}{Y_4(s) + \cfrac{1}{Z_5(s) + \cfrac{1}{Y_6(s)}}}}}$$

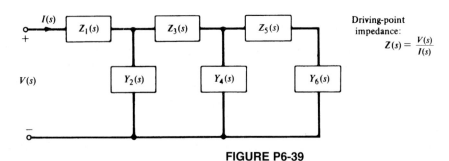

FIGURE P6-39

Computer Exercises

6-1. Assume the following parameters for Problem 6-1(a): $R = 1$ ohm; $C_1 = 1/6$ farad; $C_2 = 5/3$ farad; $L = 3/5$ henries. Use the symbolic toolbox of MATLAB to solve for the time-domain current in the left-hand mesh if the input voltage is a unit step, $v_{C1} = 0$, $v_{C1} = 0.5$ volts, and $i_0 = 2$ amperes.

6-2. A fixed linear system has transfer function

$$H(s) = \frac{s^2 + 3}{s^3 + 3s^2 + 3s + 1}$$

and a response given by

$$y(t) = \left[\left(\frac{1}{13} t^2 + \frac{11}{169} t - \frac{217}{4394} \right) e^{-t} + \left(\frac{865}{4394} \sin(5t) + \frac{217}{4394} \cos(5t) \right) e^{-2t} \right] u(t)$$

Using the symbolic toolbox of MATLAB, find the Laplace transform of the input assuming zero initial conditions.

6-3. Using the symbolic toolbox of MATLAB and following the MATLAB application, solve Example 6-5 for the time domain mesh currents for meshes 2 and 3.

6-4. (a) Using the signals and systems toolbox features of *The Student Edition of* MATLAB, numerically compute and plot the output of the system with numerator and denominator polynomials defined by

$$\text{num} = [0 \ 0 \ 3 \ 1] \quad \text{and} \quad \text{den} = [1 \ 3 \ 3 \ 1]$$

and input

$$x(t) = e^{-2t}\cos(5t)u(t)$$

(**b**) Compute and plot the step response of the previously defined system.
(**c**) Compute and plot the impulse response of the previously defined system.
(**d**) Check your results analytically by using the residue function of MATLAB to do a partial fraction expansion and thereby facilitate an inverse Laplace transform for the three cases (a)–(c).

6-5. Given the following transfer functions:

(i) $H_1(s) = \dfrac{s + 4}{s^2 + 5s + 6}$ (ii) $H_2(s) = \dfrac{s^2 + 3s + 4}{s^3 + 6s^2 + 11s + 6}$

(iii) $H_3(s) = \dfrac{s^3 + 2s^2 + 3s + 2}{s^4 + 10s^3 + 35s^2 + 50s + 24}$

(**a**) Find and plot the step response of each system;
(**b**) Find and plot the impulse response of each system;
(**c**) Provide Bode plots for each system.

6-6. Design (numerator and denominator polynomials) and provide Bode plots for the following filters (use "Help" in *The Student Edition of* MATLAB to learn how to insert the parameters).
(**a**) A 4th-order highpass Butterworth with cutoff frequency of 5 hertz (MATLAB function `butter`);
(**b**) A 3rd-order bandpass Chebyshev I with 0.5 dB ripple in the passband going from 4 to 9 hertz (MATLAB function `cheby1`);
(**c**) A 5th-order bandstop Chebyshev II; stopband from 2 to 6 hertz and ripple 50 dB down (MATLAB function `cheby2`).

CHAPTER 7

State-Variable Techniques

7-1 Introduction

So far the techniques of system analysis we have discussed have been applied primarily to systems with only a single input and a single output. Although such systems are often good models for individual parts of a system, they are rarely sufficient for modeling the overall system. Many engineering systems of interest have several inputs and several outputs. Although such a system may be modeled as an inter-connection of single-input, single-output systems, it is not always advantageous to do so. As we will see, the state-variable method provides a convenient formulation procedure. It allows us to handle systems with many inputs and outputs within precisely the same notational framework that we will use for single-input, single-output systems.

The state-variable formulation has other advantages as well. In our previous discussion we have been concerned with transfer functions between the system input and output. No information concerning the internal behavior of the system is readily available using this formulation. On the other hand, the state-variable formulation allows us to determine the internal behavior of the system easily while still giving the input-output information we desire. Furthermore, the state-variable formulation is often the most efficient form from the standpoint of computer simulation of the system. This characteristic alone leads us to prefer the state-variable formulation for highly complex systems.

In this chapter we develop the concepts and formulation of the state-variable approach to system analysis. We solve the state equations using both time-domain and Laplace transform techniques and examine some properties of the solution. Finally, we show how the state-variable method may be applied to circuit analysis.

7-2 State-Variable Concepts

There are several concepts which, if understood, will make our further discussion both easier and more meaningful. The most fundamental concept is that of the state of a system. Instead of giving a detailed mathematical definition of the state of a system, we prefer to follow a less formal approach that we believe yields more insight. We may say that the state of a system at time t_0 includes the minimum information necessary to specify completely the condition of the system at time t_0 and allow determination of all system outputs at time $t > t_0$ when inputs up to time t are specified. In short, the state of a system at time t_0 tells us everything that is important about the system at time t_0. In a linear, time-invariant electric network, for example, knowledge of all capacitor voltages and inductor currents uniquely specifies the network condition at any particular time. Furthermore, this information and knowledge of the input allow us to find the condition of the network at any future time. Thus for this example the values of the capacitor voltages and inductor currents at some time t_0 constitute the state of the system at time t_0.

The state of a system at time t_0 is simply the set of values, at time t_0, of an appropriately chosen set of variables. These variables are called the *state variables.* For the electric network example above, the state variables are thus the capacitor voltages and inductor currents. The number of state variables is equal to the order of the system and is usually just the number of energy storage elements. The choice of state variables is not unique. In fact, there are infinitely many choices for any given system, but not all are equally convenient. Usually, state variables are chosen so that they correspond to physically measurable quantities or lead to particularly simplified calculations.

It is often advantageous to think of an n-dimensional space in which each coordinate is defined by one of the state variables $x_1, x_2, \ldots, x_n$, where n is the order of the system. This space is called the *state space.* The *state vector* is an n-vector $\mathbf{x}$ whose elements are the state variables. Then at any time t the state vector defines a point in the state space (i.e., the state of the system at that time). As time progresses and the system state changes, a set of points will be defined. This set of points, the locus of the tip of the state vector as time progresses, is called a *trajectory* of the system.

The concepts and terminology of this section are used throughout the remainder of this chapter. The student may find it helpful to refer to this section from time to time.

7-3 Form of the State Equations

An nth-order *linear differential system* with m inputs and p outputs may always be represented by n first-order differential equations and p output equations in the following manner[†]:

$$\dot{x}_1 = a_{11}x_1 + a_{12}x_2 + \cdots + a_{1n}x_n + b_{11}u_1 + b_{12}u_2 + \cdots + b_{1m}u_m$$
$$\dot{x}_2 = a_{21}x_1 + a_{22}x_2 + \cdots + a_{2n}x_n + b_{21}u_1 + b_{22}u_2 + \cdots + b_{2m}u_m$$
$$\vdots \qquad\qquad\qquad\qquad\qquad\qquad\qquad\qquad\qquad\qquad (7\text{-}1)$$
$$\dot{x}_n = a_{n1}x_1 + a_{n2}x_2 + \cdots + a_{nn}x_n + b_{n1}u_1 + b_{n2}u_2 + \cdots + b_{nm}u_m$$

and

$$y_1 = c_{11}x_1 + c_{12}x_2 + \cdots + c_{1n}x_n + d_{11}u_1 + d_{12}u_2 + \cdots + d_{1m}u_m$$
$$\vdots \qquad\qquad\qquad\qquad\qquad\qquad\qquad\qquad\qquad\qquad (7\text{-}2)$$
$$y_p = c_{p1}x_1 + c_{p2}x_2 + \cdots + c_{pn}x_n + d_{p1}u_1 + d_{p2}u_2 + \cdots + d_{pm}u_m$$

where the dot indicates time differentiation ($\dot{x} = dx/dt$); the u_i, $i = 1, 2, \ldots, m$, are the system inputs, and the y_i, $i = 1, 2, \ldots, p$, are the system outputs. The x_i, $i = 1, 2, \ldots, n$, are called the state variables. Equations (7-1) are often called the *state equations,* while (7-2) comprise the *output equations.* Together, the $n + p$ equations constitute a state-equation model for the system. Many books call these the normal-form equations. In the most general case the a's, b's, c's, and d's may be functions of time. However, the solution of such a set of time-varying state equations is exceedingly difficult and must almost always be carried out with the aid of a computer. For this reason we restrict our attention to time-invariant systems, ones for which all the coefficients are constant. In this case we can obtain solutions without too much difficulty, as we shall see in the following section.

At this point it is reasonable to ask two questions. First, what are the advantages of this method, if any? Second, how do we find the state equations for a system? We consider the second question in a later section. For now, we concentrate on providing answers to the first question. The most obvious advantage of the state-variable representation is that it allows us to treat the case of multiple inputs and

---

[†]For convenience we have omitted the explicit time dependence. We will follow this approach unless understanding would be enhanced by including (t).

multiple outputs. This is something that can be done only with considerable difficulty using the other analysis techniques we have discussed. Of course, the single-input, single-output system is included as a special case of (7-1) and (7-2). For this situation $m = p = 1$.

There are other advantages of the state model as well. The model is given in the time domain, and it is straightforward to obtain a simulation diagram for the equations. This is extremely useful if we wish to use computer simulation methods to study the system. Furthermore, an extremely compact matrix notation can be used for the state model. With the laws of matrix algebra, it becomes much less cumbersome to maniplate the equations.

Let us turn our attention to the development of simulation diagrams for state equations. For brevity we consider a two-input, one-output second-order system given by

$$\dot{x}_1 = a_{11}x_1 + a_{12}x_2 + b_{11}u_1 + b_{12}u_2$$
$$\dot{x}_2 = a_{21}x_1 + a_{22}x_2 + b_{21}u_1 + b_{22}u_2 \qquad (7\text{-}3)$$
$$y = c_1x_1 + c_2x_2 + d_1u_1 + d_2u_2$$

It is evident that if we knew the states, we could find the output, for the inputs and all coefficients are assumed known. Therefore, consider only the first two equations. If we knew $\dot{x}_1$ and $\dot{x}_2$ we could obtain x_1 and x_2 by simple integration. Hence $\dot{x}_1$ and $\dot{x}_2$ should be the inputs of two integrators. The corresponding integrator outputs are x_1 and x_2. This leaves only the problem of obtaining $\dot{x}_1$ and $\dot{x}_2$ for use as inputs to the integrators. In fact, this is already specified by the state equations: $\dot{x}_1$ and $\dot{x}_2$ are simply the appropriate linear combinations of x_1, x_2, u_1, and u_2 specified by the first two equations of (7-3). Once we do this we simply use the inputs u_1 and u_2 and the integrator outputs x_1 and x_2 to form the system output y. The completed diagram is shown in Figure 7-1. Those familiar with analog-computer simulation techniques will recognize this as essentially an analog computer program for (7-3). Although the development has been for a relatively simple system, the extension to more complicated systems is obvious. One need only employ more integrators and provide the block diagram elements to reflect the mathematical statements of the state model. Of course, the diagram becomes cluttered and more complicated, but this is to be expected—so are the equations.

An examination of the simulation diagram reveals the significance of the various constants in the state model. The a_{ij} essentially specify the internal system connection and behavior, even in the absence of inputs, but with nonzero initial values of the states. The b_{ij} specify the manner in which inputs affect the internal behavior of the system. The c_{ij} specify the manner in which the internal workings of the system affect the output, while the d_{ij} represent direct transmission paths from input to output. If the state variables are chosen to be physically measurable variables of interest, considerable insight into system behavior can be obtained from the simulation diagram.

It is perhaps worth noting that the state equations can be obtained directly from a simulation diagram. All that is required is to write an expression for the output of each summer in terms of its inputs. In Figure 7-1 the inputs to the summer whose output is $\dot{x}_1$ are $b_{11}u_1$, $b_{12}u_2$, $a_{11}x_1$, and $a_{12}x_2$. Summing these yields the first of equations (7-3). Similar equations at the other summers yield the remaining two equations.

It should be abundantly clear that manipulations of (7-1) and (7-2) will present unnecessary difficulty unless some compact notation can be developed. Fortunately, such a notation is already available through the use of vectors and matrices. Let us define vectors

$$\mathbf{x} = \begin{bmatrix} x_1 \\ x_2 \\ \vdots \\ x_n \end{bmatrix}, \quad \mathbf{u} = \begin{bmatrix} u_1 \\ u_2 \\ \vdots \\ u_m \end{bmatrix}, \quad \mathbf{y} = \begin{bmatrix} y_1 \\ y_2 \\ \vdots \\ y_p \end{bmatrix} \qquad (7\text{-}4)$$

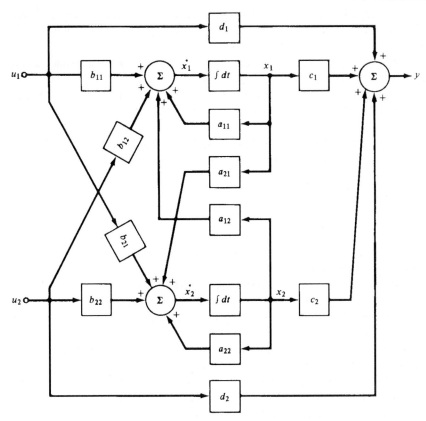

FIGURE 7-1. Simulation diagram of the system of (7-3).

and matrices

$$
\mathbf{A} = \begin{bmatrix} a_{11} & a_{12} & \cdots & a_{1n} \\ a_{21} & a_{22} & \cdots & a_{2n} \\ \vdots & \vdots & & \vdots \\ a_{n1} & a_{n2} & \cdots & a_{nm} \end{bmatrix}, \quad
\mathbf{B} = \begin{bmatrix} b_{11} & b_{12} & \cdots & b_{1m} \\ b_{21} & b_{22} & \cdots & b_{2m} \\ \vdots & \vdots & & \vdots \\ b_{n1} & b_{n2} & \cdots & b_{nm} \end{bmatrix}
$$

$$
\mathbf{C} = \begin{bmatrix} c_{11} & c_{12} & \cdots & c_{1n} \\ c_{21} & c_{22} & \cdots & c_{2n} \\ \vdots & \vdots & & \vdots \\ c_{p1} & c_{p2} & \cdots & c_{pn} \end{bmatrix}, \quad
\mathbf{D} = \begin{bmatrix} d_{11} & d_{12} & \cdots & d_{1m} \\ d_{21} & d_{22} & \cdots & d_{pm} \\ \vdots & \vdots & & \vdots \\ d_{p1} & d_{p2} & \cdots & d_{pm} \end{bmatrix}
$$

(7-5)

Then using the usual rules of matrix algebra[†] it is obvious that the state model of (7-1) and (7-2) may be written compactly as

$$\dot{\mathbf{x}} = \mathbf{A}\mathbf{x} + \mathbf{B}\mathbf{u} \tag{7-6a}$$

$$\mathbf{y} = \mathbf{C}\mathbf{x} + \mathbf{D}\mathbf{u} \tag{7-6b}$$

[†]For those unfamiliar with matrix algebra, Appendix D provides a summary of the necessary results.

These equations may also be indicated schematically by a block diagram as shown in Figure 7-2. The double lines are used to indicate a multiple-variable signal flow path of the appropriate dimension. The blocks represent matrix multiplication of the appropriate vectors and matrices. The block containing the integrator in fact contains n integrators with appropriate connections specified by the **A** and **B** matrices. Of course, to obtain a useful simulation model we must know all the elements of all the matrices, but at least this block diagram indicates the essential structure of the system and the signal flow paths.

A similar statement can be made regarding the state equations themselves. Before we can obtain numerical solutions in any particular case we must know all elements of the four matrices **A, B, C**, and **D**. However, as in most problems, it is generally most convenient to do the majority of the work in terms of arbitrary matrices. Particular numerical data may then be substituted as a last step. Furthermore, the compact matrix notation provides a convenient theoretical tool on those occasions when one is interested only in general properties of the system, without regard to particular numerical data. We exploit this notation in the following section to obtain a solution to the state equations.

7-4 Time-Domain Solution of the State Equations

For the purpose of developing the solution of the state equations, let us consider (7-6) repeated here for convenience:

$$\dot{\mathbf{x}} = \mathbf{A}\mathbf{x} + \mathbf{B}\mathbf{u} \tag{7-7a}$$

$$\mathbf{y} = \mathbf{C}\mathbf{x} + \mathbf{D}\mathbf{u} \tag{7-7b}$$

together with the known initial condition

$$\mathbf{x}(t_0) = \mathbf{x}_0 \tag{7-8}$$

We assume that the inputs are specified for all $t \geq t_0$. It is evident that if the state vector can be found, the output vector **y** can be determined easily from equation (7-7b). The problem then becomes that of solving equation (7-7a) subject to the initial condition of (7-8).

An intuitively pleasing approach to the solution of the state equations is to simply build up the solution as time progresses. Suppose that we divide the time interval $t \geq t_0$ into an infinite number of subintervals, each of duration Δt. If Δt is small, then

$$\dot{\mathbf{x}}(t_1) \simeq \frac{\mathbf{x}(t_1 + \Delta t) - \mathbf{x}(t_1)}{\Delta t} \triangleq \frac{\Delta \mathbf{x}}{\Delta t} \tag{7-9}$$

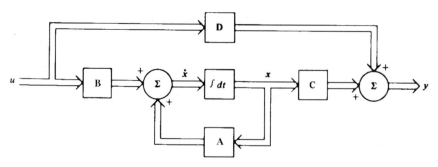

FIGURE 7-2.   Block diagram of the state model of (7-6).

where $\Delta\mathbf{x}$ is the change in $\mathbf{x}$ in the time interval from t_1 to $t_1 + \Delta t$ for any $t_1 = t_0 + k\,\Delta t, k = 0, 1, 2, \ldots$.
Equation (7-9) suggests that, in the time interval t_0 to $t_0 + \Delta t$, $\mathbf{x}$ changes by the amount

$$\Delta\mathbf{x} = \dot{\mathbf{x}}(t_0)\,\Delta t \tag{7-10}$$

so that at $t_0 + \Delta t$,

$$\mathbf{x}(t_0 + \Delta t) \simeq \mathbf{x}(t_0) + \dot{\mathbf{x}}(t_0)\,\Delta t \tag{7-11}$$

Now $\dot{\mathbf{x}}(t_0)$ can be found from (7-7) as

$$\dot{\mathbf{x}}(t_0) = \mathbf{A}\mathbf{x}(t_0) + \mathbf{B}\mathbf{u}(t_0) = \mathbf{A}\mathbf{x}_0 + \mathbf{B}\mathbf{u}(t_0) \tag{7-12}$$

since all quantities on the right-hand side of (7-12) are known. We thus find $\mathbf{x}(t_0 + \Delta t)$. Repeating, we can find $\dot{\mathbf{x}}(t_0 + \Delta t)$ from this knowledge and use it to predict $\mathbf{x}(t_0 + 2\Delta t)$. Continuing the process, we can build up the solution one step at a time for as long as we wish. Of course, the result is only approximate. Its accuracy depends on how small Δt is chosen. As Δt tends toward zero, the approximate solution approaches the actual one. This method of solution, or modifications of it, is useful for digital computation.

Actually, the solution process can be expressed concisely as a recursion relation. Let us write (7-9) as

$$\dot{\mathbf{x}}(t_0 + k\,\Delta t) = \frac{\mathbf{x}(t_0 + (k+1)\,\Delta t) - \mathbf{x}(t_0 + k\,\Delta t)}{\Delta t} \tag{7-13}$$

where k is any nonnegative integer. Upon rewriting the state equations at time $t = t_0 + k\,\Delta t$, we obtain

$$\frac{\mathbf{x}(t_0 + (k+1)\,\Delta t) - \mathbf{x}(t_0 + k\,\Delta t)}{\Delta t} = \mathbf{A}\mathbf{x}(t_0 + k\,\Delta t) + \mathbf{B}\mathbf{u}(t_0 + k\,\Delta t) \tag{7-14}$$

or

$$\mathbf{x}(t_0 + (k+1)\,\Delta t) = \mathbf{x}(t_0 + k\,\Delta t) + \Delta t[\mathbf{A}\mathbf{x}(t_0 + k\,\Delta t) + \mathbf{B}\mathbf{u}(t_0 + k\,\Delta t)] \tag{7-15}$$

Equation (7-15) provides the desired recursion relation. If $\mathbf{x}(t_0 + k\,\Delta t)$ is considered the present value of $\mathbf{x}$, then the next value may be found. Letting $k = 0, 1, 2, \ldots$ allows the solution to be constructed numerically. This equation is an example of a difference equation. We will study such equations in some detail in the following chapter.

Although the solution technique described above has a great deal of intuitive appeal and is computationally convenient, it suffers the disadvantage that it does not provide a closed-form solution. Frequently, a closed-form solution is desired, so we must determine how to solve the equations to yield this result. Of course, we consider only

$$\dot{\mathbf{x}} = \mathbf{A}\mathbf{x} + \mathbf{B}\mathbf{u}, \qquad \mathbf{x}(t_0) = \mathbf{x}_0 \tag{7-16}$$

realizing that once $\mathbf{x}$ is known, $\mathbf{y}$ can be easily found. Using the approach familiar to us from differential equations, we assume that the solution has two parts,

$$\mathbf{x} = \mathbf{x}_h + \mathbf{x}_p \tag{7-17}$$

where $\mathbf{x}_h$ is the solution to the homogeneous equation and $\mathbf{x}_p$ is a particular solution. The homogeneous solution must satisfy the initial condition.

The homogeneous equation is given by, assuming $\mathbf{x}_p(t_0) = \mathbf{0}$,

$$\dot{\mathbf{x}}_h = \mathbf{A}\mathbf{x}_h, \qquad \mathbf{x}_h(t_0) = \mathbf{x}_0 \tag{7-18}$$

By analogy to the scalar equation

$$\dot{x} = ax, \qquad x(t_0) = x_0 \tag{7-19}$$

which has solution

$$x(t) = e^{a(t-t_0)}x_0 \tag{7-20}$$

we might guess the solution of (7-18) to be

$$\mathbf{x}_h(t) = e^{\mathbf{A}(t-t_0)}\mathbf{x}_0 \tag{7-21}$$

Immediately a question arises as to the meaning of the matrix exponential. We shall have more to say about it later, but for now let us assume that it can be differentiated in the same fashion as the scalar exponential, that it is an $n \times n$ matrix, and that it satisfies

$$e^{\mathbf{A}\cdot 0} = \mathbf{I} \tag{7-22}$$

where $\mathbf{I}$ is an $n \times n$ identity matrix. Then

$$\dot{\mathbf{x}}_h(t) = \mathbf{A}e^{\mathbf{A}(t-t_0)}\mathbf{x}_0 = \mathbf{A}\mathbf{x}_h \tag{7-23}$$

and the assumed solution satisfies the homogeneous equation. Furthermore, in view of (7-22),

$$\mathbf{x}_h(t_0) = e^{\mathbf{A}\cdot 0}\mathbf{x}_0 = \mathbf{x}_0 \tag{7-24}$$

Hence (7-21) gives the homogeneous solution.

Before determining the particular solution we digress to note some properties of the matrix exponential $e^{\mathbf{A}t}$. For our purposes we may define $e^{\mathbf{A}t}$ in terms of its infinite series expansion as follows:

$$e^{\mathbf{A}t} = \mathbf{I} + \mathbf{A}t + \frac{\mathbf{A}^2t^2}{2!} + \frac{\mathbf{A}^3t^3}{3!} + \cdots \tag{7-25}$$

Clearly, $e^{\mathbf{A}t}$ is an $n \times n$ matrix. The assumptions we have made regarding $e^{\mathbf{A}t}$ may be easily verified directly from (7-25). The inverse of $e^{\mathbf{A}t}$ may also be found as

$$(e^{\mathbf{A}t})^{-1} = e^{-\mathbf{A}t} \tag{7-26}$$

Now let us complete the solution of the state equations by determining the particular solution. We adopt the method of variation of parameters and assume a particular solution of the form

$$\mathbf{x}_p(t) = e^{\mathbf{A}(t-t_0)}\mathbf{z}(t) \tag{7-27}$$

where $\mathbf{z}(t)$ is a vector function to be found. Differentiate (7-27) and require that $\mathbf{x}_p$ satisfy the original equation:

$$\dot{\mathbf{x}}_p(t) = \mathbf{A}e^{\mathbf{A}(t-t_0)}\mathbf{z}(t) + e^{\mathbf{A}(t-t_0)}\dot{\mathbf{z}}(t) = \mathbf{A}\mathbf{x}_p(t) + \mathbf{B}\mathbf{u}(t)$$
$$= \mathbf{A}e^{\mathbf{A}(t-t_0)}\mathbf{z}(t) + \mathbf{B}\mathbf{u}(t) \tag{7-28}$$

Then

$$e^{\mathbf{A}(t-t_0)}\dot{\mathbf{z}}(t) = \mathbf{B}\mathbf{u}(t) \tag{7-29}$$

or

$$\dot{\mathbf{z}}(t) = e^{\mathbf{A}(t_0-t)}\mathbf{B}\mathbf{u}(t) \tag{7-30}$$

Equation (7-30) can be solved by direct integration to yield

$$\mathbf{z}(t) - \mathbf{z}(t_0) = \int_{t_0}^{t} e^{\mathbf{A}(t_0-\lambda)}\mathbf{B}\mathbf{u}(\lambda)\, d\lambda \tag{7-31}$$

Now the particular solution must be such that

$$\mathbf{x}_p(t_0) = \mathbf{0} \tag{7-32}$$

since the homogeneous solution already satisfies the initial condition. It can be shown that $e^{A(t-t_0)}$ is nonsingular; therefore, from (7-27) we must have $\mathbf{z}(t_0) = \mathbf{0}$. Then

$$\mathbf{z}(t) = \int_{t_0}^{t} e^{A(t_0-\lambda)}\mathbf{Bu}(\lambda)\, d\lambda \tag{7-33}$$

Using this result in (7-27) then yields the particular solution as

$$\mathbf{x}_p(t) = e^{A(t-t_0)}\int_{t_0}^{t} e^{A(t_0-\lambda)}\mathbf{Bu}(\lambda)\, d\lambda \tag{7-34}$$

or

$$\mathbf{x}_p(t) = \int_{t_0}^{t} e^{A(t-\lambda)}\mathbf{Bu}(\lambda)\, d\lambda \tag{7-35}$$

The complete solution is obtained by adding the homogeneous solution and the particular solution to yield

$$\mathbf{x}(t) = e^{A(t-t_0)}\mathbf{x}_0 + \int_{t_0}^{t} e^{A(t-\lambda)}\mathbf{Bu}(\lambda)\, d\lambda \tag{7-36}$$

The matrix exponential arises so often that it is usually given a special name and symbol. It is called the *state transition matrix* and denoted by $\boldsymbol{\Phi}(t)$. Using this notation we may write the solution as

$$\mathbf{x}(t) = \boldsymbol{\Phi}(t - t_0)\mathbf{x}_0 + \int_{t_0}^{t} \boldsymbol{\Phi}(t - \lambda)\mathbf{Bu}(\lambda)\, d\lambda \tag{7-37}$$

The reason for calling $\boldsymbol{\Phi}(t)$ the state transition matrix is clear. Once the input is specified, the state transition matrix uniquely describes the manner in which the system state changes from its value at $t = t_0$ to its value at any time $t > t_0$. After we examine an alternative method of solving the state equations in the following section, we shall describe some techniques of finding the state transition matrix. Using (7-37), we may write the complete system response as

$$\mathbf{y}(t) = \mathbf{C}\boldsymbol{\Phi}(t - t_0)\mathbf{x}_0 + \int_{t_0}^{t} \mathbf{C}\boldsymbol{\Phi}(t - \lambda)\mathbf{Bu}(\lambda)\, d\lambda + \mathbf{Du}(t) \tag{7-38}$$

7-5 Frequency-Domain Solution of the State Equations

An alternative method of solving the state equation makes use of the Laplace transform technique discussed in Chapter 6. This method is quite attractive from an analytical point of view. Again we are concerned primarily with

$$\dot{\mathbf{x}} = \mathbf{Ax} + \mathbf{Bu}, \qquad \mathbf{x}(t_0) = \mathbf{x}_0 \tag{7-39}$$

For convenience, we shall assume that $t_0 = 0$. Laplace transforming (7-39) yields

$$s\mathbf{X}(s) - \mathbf{x}_0 = \mathbf{AX}(s) + \mathbf{BU}(s) \tag{7-40}$$

Rearranging yields

$$(s\mathbf{I} - \mathbf{A})\mathbf{X}(s) = \mathbf{x}_0 + \mathbf{BU}(s) \tag{7-41}$$

or

$$\mathbf{X}(s) = (s\mathbf{I} - \mathbf{A})^{-1}\mathbf{x}_0 + (s\mathbf{I} - \mathbf{A})^{-1}\mathbf{BU}(s) \tag{7-42}$$

When we take the inverse transform we must obtain precisely the $\mathbf{x}(t)$ given by our previous analysis, with $t_0 = 0$. From the first term of (7-42) it is evident that

$$\mathcal{L}^{-1}[(s\mathbf{I} - \mathbf{A})^{-1}] = \boldsymbol{\Phi}(t) \tag{7-43}$$

or

$$\boldsymbol{\Phi}(s) = (s\mathbf{I} - \mathbf{A})^{-1} \tag{7-44}$$

where $\boldsymbol{\Phi}(s)$ is the Laplace transform of the state transition matrix.

The second term of (7-42) must yield the integral term of (7-37), which is consistent with the convolution property of a product of Laplace transforms,

$$\mathcal{L}^{-1}[(s\mathbf{I} - \mathbf{A})^{-1}\mathbf{B}\mathbf{U}(s)] = \mathcal{L}^{-1}[\boldsymbol{\Phi}(s)\mathbf{B}\mathbf{u}(s)]$$

$$= \boldsymbol{\Phi}(t) * \mathbf{B}\mathbf{u}(t)$$

$$= \int_0^t \boldsymbol{\Phi}(t - \lambda)\mathbf{B}\mathbf{u}(\lambda)\, d\lambda \tag{7-45}$$

Therefore, the time-domain solution is exactly as we have written in the preceding section. In the frequency domain, the response of the system is given by

$$\mathbf{Y}(s) = \mathbf{C}(s\mathbf{I} - \mathbf{A})^{-1}\mathbf{x}_0 + \mathbf{C}(s\mathbf{I} - \mathbf{A})^{-1}\mathbf{B}\mathbf{U}(s) + \mathbf{D}\mathbf{U}(s)$$

$$= \mathbf{C}(s\mathbf{I} - \mathbf{A})^{-1}\mathbf{x}_0 + [\mathbf{C}(s\mathbf{I} - \mathbf{A})^{-1}\mathbf{B} + \mathbf{D}]\mathbf{U}(s) \tag{7-46}$$

Equation (7-46) is simply the Laplace transform of (7-38).

For single-input, single-output systems we defined the transfer function as the ratio of the Laplace transform of the output to the Laplace transform of the input, under the requirement of zero initial conditions. We can extend this concept to the multiple-variable case. For zero initial conditions (7-46) becomes

$$\mathbf{Y}(s) = [\mathbf{C}(s\mathbf{I} - \mathbf{A})^{-1}\mathbf{B} + \mathbf{D}]\mathbf{U}(s) \tag{7-47}$$

It is evident that the quantity in brackets plays the role of a transfer function. Accordingly, we define a transfer function matrix as

$$\mathbf{H}(s) \triangleq \mathbf{C}(s\mathbf{I} - \mathbf{A})^{-1}\mathbf{B} + \mathbf{D} = \mathbf{C}\boldsymbol{\Phi}(s)\mathbf{B} + \mathbf{D} \tag{7-48}$$

Hence we may write

$$\mathbf{Y}(s) = \mathbf{H}(s)\mathbf{U}(s) \tag{7-49}$$

Since there are m inputs and p outputs, $\mathbf{H}$ is a $p \times m$ matrix having elements $H_{ij}(s)$, $i = 1, 2, \ldots, p; j = 1, 2, \ldots, m$. The element $H_{ij}(s)$ is the transfer function between the jth input u_j and the ith output y_i.

We may also define an impulse response matrix $\mathbf{H}(t)$ as

$$\mathbf{H}(t) \triangleq \mathcal{L}^{-1}[\mathbf{H}(s)] \tag{7-50}$$

Hence $\mathbf{H}(t)$ may be written as

$$\mathbf{H}(t) = \mathbf{C}\boldsymbol{\Phi}(t)\mathbf{B} + \mathbf{D}\boldsymbol{\delta}(t)$$

$$= \mathbf{C}e^{\mathbf{A}t}\mathbf{B} + \mathbf{D}\boldsymbol{\delta}(t) \tag{7-51}$$

where $\boldsymbol{\delta}(t)$ is an m-vector of unit impulses. $\mathbf{H}(t)$ has the same dimension as $\mathbf{H}(s)$ and its elements $h_{ij}(t)$ satisfy

$$h_{ij}(t) = \mathcal{L}^{-1}[H_{ij}(s)] \tag{7-52}$$

where $h_{ij}(t)$ is equal to $y_i(t)$ when a unit impulse is applied at the jth input.

The formal solution of the state equations is now complete. We have an expression that gives the system output for any input and any set of initial conditions. The matrix multiplications and the integral of (7-38), although perhaps tedious, are straightforward. The integral may be evaluated either directly or with the aid of the Laplace transform form of (7-45), and employing partial-fraction expansions. If we use a time-domain approach, it is evident that we must know the state transition matrix $\mathbf{\Phi}(t) = e^{\mathbf{A}t}$. We now turn to this problem.

7-6 Finding the State Transition Matrix

Two techniques for determining the state transition matrix of a given system have already been mentioned, although not specifically identified as such. These make use of the series definition of $e^{\mathbf{A}t}$ and of the Laplace transform method mentioned in Section 7-5. Recall that we defined the state transition matrix in terms of its infinite series as

$$\mathbf{\Phi}(t) = e^{\mathbf{A}t} = \mathbf{I} + \mathbf{A}t + \frac{\mathbf{A}^2 t^2}{2!} + \frac{\mathbf{A}^3 t^3}{3!} + \cdots \tag{7-53}$$

For a given $\mathbf{A}$, (7-53) may be evaluated to any desired degree of accuracy at any specified time t by performing the indicated matrix multiplications and summation. A simple example will illustrate the procedure.

EXAMPLE 7-1 _____

In this example, we illustrate finding the state transition matrix using series expansion. Consider, for example,

$$\mathbf{A} = \begin{bmatrix} -1 & 0 \\ 0 & -2 \end{bmatrix} \tag{7-54}$$

It is immediately found that

$$\mathbf{A}^2 = \begin{bmatrix} 1 & 0 \\ 0 & 4 \end{bmatrix}, \quad \mathbf{A}^3 = \begin{bmatrix} -1 & 0 \\ 0 & -8 \end{bmatrix}, \quad \cdots, \quad \mathbf{A}^n = \begin{bmatrix} (-1)^n & 0 \\ 0 & (-2)^n \end{bmatrix} \tag{7-55}$$

Therefore,

$$e^{\mathbf{A}t} = \begin{bmatrix} 1 & 0 \\ 0 & 1 \end{bmatrix} + \begin{bmatrix} -1 & 0 \\ 0 & -2 \end{bmatrix} t + \begin{bmatrix} 1 & 0 \\ 0 & 4 \end{bmatrix} \frac{t^2}{2} + \begin{bmatrix} -1 & 0 \\ 0 & -8 \end{bmatrix} \frac{t^3}{6} + \cdots$$

$$= \begin{bmatrix} 1 - t + \dfrac{t^2}{2} - \dfrac{t^3}{6} + \cdots & 0 \\ 0 & 1 - 2t + \dfrac{4t^2}{2} - \dfrac{8t^3}{6} + \cdots \end{bmatrix} \tag{7-56}$$

We immediately recognize the series representations of e^{-t} and e^{-2t}, so

$$e^{\mathbf{A}t} = \begin{bmatrix} e^{-t} & 0 \\ 0 & e^{-2t} \end{bmatrix} \tag{7-57}$$

This approach may be satisfactory for numerical computations in which the computer can carry out the calculations to whatever degree of accuracy is required. However, except in extremely simple cases, it is unsatisfactory for analytical work. The problem revolves around recognizing the closed-form expression of the infinite series. The following example should prove convincing.

EXAMPLE 7-2

In this example the series expansion technique is applied to a more complex problem. Find the state transition matrix for

$$\mathbf{A} = \begin{bmatrix} 0 & 1 \\ -6 & -5 \end{bmatrix} \tag{7-58}$$

We have

$$\mathbf{A}^2 = \begin{bmatrix} -6 & -5 \\ 30 & 19 \end{bmatrix}, \quad \mathbf{A}^3 = \begin{bmatrix} 30 & 19 \\ -114 & -65 \end{bmatrix}, \quad \mathbf{A}^4 = \begin{bmatrix} -114 & -65 \\ 390 & 211 \end{bmatrix} \tag{7-59}$$

and so on. Using this result in (7-53) yields

$$\mathbf{e}^{\mathbf{A}t} = \begin{bmatrix} 1 - \dfrac{6t^2}{2!} + \dfrac{30t^3}{3!} - \dfrac{114t^4}{4!} + \cdots & t - \dfrac{5t^2}{2!} + \dfrac{19t^3}{3!} - \dfrac{65t^4}{4!} + \cdots \\ -6t + \dfrac{30t^2}{2!} - \dfrac{114t^3}{3!} + \dfrac{390t^4}{4!} + \cdots & 1 - 5t + \dfrac{19t^2}{2!} - \dfrac{65t^3}{3!} + \dfrac{211t^4}{4!} + \cdots \end{bmatrix} \tag{7-60}$$

This expression for $\mathbf{e}^{\mathbf{A}t}$ can be shown to be the first five terms of

$$\mathbf{e}^{\mathbf{A}t} = \begin{bmatrix} 3e^{-2t} - 2e^{-3t} & e^{-2t} - e^{-3t} \\ -6e^{-2t} + 6e^{-3t} & -2e^{-2t} + 3e^{-3t} \end{bmatrix} \tag{7-61}$$

but it is highly unlikely that we would realize it unless we already knew the answer.

The second method of determining the state transition matrix makes use of the fact that

$$\mathbf{\Phi}(s) = (s\mathbf{I} - \mathbf{A})^{-1} \tag{7-62}$$

We simply form $(s\mathbf{I} - \mathbf{A})$, take the matrix inverse, and then take the inverse Laplace transform to find $\mathbf{\Phi}(t) = \mathbf{e}^{\mathbf{A}t}$. Partial-fraction-expansion methods are usually helpful in taking the inverse transform. An example will illustrate this method.

EXAMPLE 7-3

The previous example is now worked again to illustrate the Laplace transform for finding the state transition matrix. As before

$$\mathbf{A} = \begin{bmatrix} 0 & 1 \\ -6 & -5 \end{bmatrix} \tag{7-63}$$

We have

$$(s\mathbf{I} - \mathbf{A}) = \begin{bmatrix} s & -1 \\ 6 & s+5 \end{bmatrix} \tag{7-64}$$

and

$$(s\mathbf{I} - \mathbf{A})^{-1} = \begin{bmatrix} \dfrac{s+5}{(s+2)(s+3)} & \dfrac{1}{(s+2)(s+3)} \\[3mm] \dfrac{-6}{(s+2)(s+3)} & \dfrac{s}{(s+2)(s+3)} \end{bmatrix} \tag{7-65}$$

Performing a partial-fraction expansion of each term and inverse transforming yields

$$\mathbf{e}^{\mathbf{A}t} = \mathscr{L}^{-1}[(s\mathbf{I} - \mathbf{A})^{-1}] = \begin{bmatrix} 3e^{-2t} - 2e^{-3t} & e^{-2t} - e^{-3t} \\ -6e^{-2t} + 6e^{-3t} & -2e^{-2t} + 3e^{-3t} \end{bmatrix} u(t) \tag{7-66}$$

as promised in Example 7-2.

The Laplace transform method is quite convenient for analytical work since it yields answers in closed form. However, it is not very useful for machine computation. There exist other techniques for finding $\mathbf{e}^{\mathbf{A}t}$ which are often useful. One such technique involves finding a suitable coordinate transformation which takes $\mathbf{A}$ to diagonal or Jordan canonical form. Once this is done the state transition matrix for the new matrix can be written by inspection. Then $\mathbf{e}^{\mathbf{A}t}$ is found by applying the inverse coordinate transformation. A fourth technique makes use of the fact that $\mathbf{e}^{\mathbf{A}t}$ may be expressed *exactly* by a *finite* power series in $\mathbf{A}$. The coefficients of the series expansion are found by solving a set of simultaneous algebraic equations. These results are based on the Cayley-Hamilton theorem (see Appendix D). The coefficients of the finite power series representation become functions of time. Therefore, it is possible to write

$$\mathbf{\Phi}(t) = \mathbf{e}^{\mathbf{A}t} = \alpha_0(t)\mathbf{I} + \alpha_1(t)\mathbf{A} + \alpha_2(t)\mathbf{A}^2 + \cdots + \alpha_{n-1}(t)\mathbf{A}^{n-1} \tag{7-67}$$

When the eigenvalues of $\mathbf{A}$ are distinct it is straightforward to determine the coefficients $\alpha_i(t)$. Suppose that the eigenvalues λ_i, $i = 1, 2, \ldots, n$, are distinct. Then we use (7-67) with $\mathbf{A}$ replaced by each λ_i in turn to obtain a set of n equations:

$$e^{\lambda_1 t} = \alpha_0(t) + \alpha_1(t)\lambda_1 + \alpha_2(t)\lambda_1^2 + \cdots + \alpha_{n-1}(t)\lambda_1^{n-1}$$

$$e^{\lambda_2 t} = \alpha_0(t) + \alpha_1(t)\lambda_2 + \alpha_2(t)\lambda_2^2 + \cdots + \alpha_{n-1}(t)\lambda_2^{n-1} \tag{7-68}$$

$$e^{\lambda_n t} = \alpha_0(t) + \alpha_1(t)\lambda_n + \alpha_2(t)\lambda_n^2 + \cdots + \alpha_{n-1}(t)\lambda_n^{n-1}$$

Equations (7-68) may be solved for the coefficients $\alpha_i(t)$, $i = 0, 1, 2, \ldots, n - 1$. We will use a second-order example to illustrate the procedure.

EXAMPLE 7-4 _____

The purpose of this example is to illustrate the Caley-Hamilton approach for determining the state transition matrix. Assume that

$$\mathbf{A} = \begin{bmatrix} 0 & 1 \\ -6 & -5 \end{bmatrix} \tag{7-69}$$

The characteristic equation is

$$|\lambda\mathbf{I} - \mathbf{A}| = \begin{vmatrix} \lambda & -1 \\ 6 & \lambda + 5 \end{vmatrix} = \lambda(\lambda + 5) + 6 = \lambda^2 + 5\lambda + 6 = 0 \tag{7-70}$$

Clearly, the eigenvalues are

$$\lambda_1 = -2, \qquad \lambda_2 = -3 \tag{7-71}$$

We have

$$\mathbf{e}^{\mathbf{A}t} = \alpha_0(t)\mathbf{I} + \alpha_1(t)\mathbf{A} \tag{7-72}$$

and from (7-68) we get

$$e^{-2t} = \alpha_0(t) - 2\alpha_1(t) \tag{7-73a}$$

$$e^{-3t} = \alpha_0(t) - 3\alpha_1(t) \tag{7-73b}$$

Subtracting, we obtain

$$\alpha_1(t) = e^{-2t} - e^{-3t} \tag{7-74}$$

It follows that

$$\alpha_0(t) = 3e^{-2t} - 2e^{-3t} \tag{7-75}$$

Then

$$\mathbf{e}^{\mathbf{A}t} = (3e^{-2t} - 2e^{-3t})\begin{bmatrix} 1 & 0 \\ 0 & 1 \end{bmatrix} + (e^{-2t} - e^{-3t})\begin{bmatrix} 0 & 1 \\ -6 & -5 \end{bmatrix} \tag{7-76}$$

or

$$\mathbf{e}^{\mathbf{A}t} = \begin{bmatrix} 3e^{-2t} - 2e^{-3t} & e^{-2t} - e^{-3t} \\ -6e^{-2t} + 6e^{-3t} & -2e^{-2t} + 3e^{-3t} \end{bmatrix} \tag{7-77}$$

which agrees with the results of the preceding two examples.

MATLAB *Application*

Determining eigenvalues for large matrices often requires considerable effort. The MATLAB command eig reduces the required effort considerably. Consider the following:

```
EDU» c7ex4
a = [0 1; -6 -5];        % Define the a matrix
eiga = eig(a);           % Compute eigenvalues
eiga                     % Display eigenvalues

eiga =

   -2
   -3
```

The results clearly agree with (7-71).

If some or all of the eigenvalues are repeated, the method must be modified somewhat. This discussion is not meant to be exhaustive and we refer the interested reader to the references at the end of the chapter.

We now possess all the tools necessary to find the complete response of a system described by a state model. Therefore, we close this section with two examples in which we use the Laplace transform method to solve such a set of equations.

EXAMPLE 7-5

In this example we find the complete response for a system defined by the state equations

$$\begin{bmatrix} \dot{x}_1 \\ \dot{x}_2 \end{bmatrix} = \begin{bmatrix} 0 & 1 \\ -2 & -3 \end{bmatrix}\begin{bmatrix} x_1 \\ x_2 \end{bmatrix} + \begin{bmatrix} 0 \\ 1 \end{bmatrix}u \tag{7-78a}$$

$$\begin{bmatrix} y_1 \\ y_2 \end{bmatrix} = \begin{bmatrix} 1 & 0 \\ 1 & 1 \end{bmatrix}\begin{bmatrix} x_1 \\ x_2 \end{bmatrix} \tag{7-78b}$$

where

$$\mathbf{x}(0) = \begin{bmatrix} 1 \\ 1 \end{bmatrix} \tag{7-79}$$

and u is a unit step input. This is a single-input, two-output system. From (7-46) we have the Laplace transform of the response as

$$\mathbf{Y}(s) = \mathbf{C}(s\mathbf{I} - \mathbf{A})^{-1}\mathbf{x}_0 + \mathbf{C}(s\mathbf{I} - \mathbf{A})^{-1}\mathbf{B}\mathbf{U}(s) \tag{7-80}$$

First, we obtain $(s\mathbf{I} - \mathbf{A})^{-1}$ as

$$(s\mathbf{I} - \mathbf{A})^{-1} = \begin{bmatrix} s & -1 \\ 2 & s+3 \end{bmatrix}^{-1} = \frac{1}{s^2 + 3s + 2}\begin{bmatrix} s+3 & 1 \\ -2 & s \end{bmatrix} \tag{7-81}$$

Then

$$\mathbf{Y}(s) = \frac{\begin{bmatrix} 1 & 0 \\ 1 & 1 \end{bmatrix}\begin{bmatrix} s+3 & 1 \\ -2 & s \end{bmatrix}\begin{bmatrix} 1 \\ 1 \end{bmatrix}}{s^2 + 3s + 2} + \frac{\begin{bmatrix} 1 & 0 \\ 1 & 1 \end{bmatrix}\begin{bmatrix} s+3 & 1 \\ -2 & s \end{bmatrix}\begin{bmatrix} 0 \\ 1 \end{bmatrix}\frac{1}{s}}{s^2 + 3s + 2}$$

$$= \frac{\begin{bmatrix} 1 & 0 \\ 1 & 1 \end{bmatrix}\begin{bmatrix} s+4 \\ s-2 \end{bmatrix}}{(s+1)(s+2)} + \frac{\begin{bmatrix} 1 & 0 \\ 1 & 1 \end{bmatrix}\begin{bmatrix} 1 \\ s \end{bmatrix}}{s(s+1)(s+2)}$$

$$= \begin{bmatrix} \dfrac{s+4}{(s+1)(s+2)} \\ \dfrac{2(s+1)}{(s+1)(s+2)} \end{bmatrix} + \begin{bmatrix} \dfrac{1}{s(s+1)(s+2)} \\ \dfrac{s+1}{s(s+1)(s+2)} \end{bmatrix} \tag{7-82}$$

Now, canceling the $s + 1$ factors in the bottom two terms and applying the partial-fraction expansion yields

$$\mathbf{Y}(s) = \begin{bmatrix} \dfrac{3}{s+1} - \dfrac{2}{s+2} \\ \dfrac{2}{s+2} \end{bmatrix} + \begin{bmatrix} \dfrac{1}{2} - \dfrac{1}{s+1} + \dfrac{1}{2} \\ \dfrac{1}{2} - \dfrac{1}{2} \\ \dfrac{1}{2} \\ \dfrac{1}{2} - \dfrac{1}{2} \\ s+2 \end{bmatrix} \tag{7-83}$$

or

$$\mathbf{y}(t) = \begin{bmatrix} (3e^{-t} - 2e^{-2t})u(t) \\ 2e^{-2t}u(t) \end{bmatrix} + \begin{bmatrix} (\frac{1}{2} - e^{-t} + \frac{1}{2}e^{-2t})u(t) \\ (\frac{1}{2} - \frac{1}{2}e^{-2t})u(t) \end{bmatrix} \qquad (7\text{-}84)$$

The first term is the initial-condition response, while the second is the forced response. Combining the terms yields the complete response in its simplest form as

$$\begin{bmatrix} y_1(t) \\ y_2(t) \end{bmatrix} = \begin{bmatrix} (\frac{1}{2} + 2e^{-t} - \frac{3}{2}e^{-2t})u(t) \\ (\frac{1}{2} + \frac{3}{2}e^{-2t})u(t) \end{bmatrix} \qquad (7\text{-}85)$$

MATLAB Application

In this MATLAB application, we solve (approximately) the state equations given by (7-78a) and (7-78b). Since the state equations are for a continuous-time model and the MATLAB solution, of necessity, will be based on a discrete-time model, the discrete-time model must first be developed. This is accomplished through the use of the MATLAB function c2d. Although the entire problem could be solved using the MATLAB routine lsim, we choose instead to show the complete development in the following MATLAB code. The technique used here will be somewhat clearer to the student after Section 7-9, which deals with discrete-time state models, is studied.

```
EDU» c7ex5
a = [0 1; -2 -3];              % Define a matrix
b = [0; 1];                    % Define b matrix
c = [1 0; 1 1];                % Define c matrix
Ts = 0.01;                     % Define sampling period
nf = 501;                      % Define number of samples
x = zeros(2,nf);               % Initialize x matrix
x(:,1) = [1;1];                % Establish initial condition
u = ones(1,nf);                % Define input signal
[ad,bd] = c2d(a,b,Ts);         % Compute discrete system
% The next three lines perform the simulation
for n = 1:nf-1
        x(:,n+1) = ad*x(:,n)+bd*u(n);
end
y = c*x;                       % Compute system output
y1 = y(1,:);                   % Determine y1(t)
y2 = y(2,:);                   % Determine y2(t)
t = (0:nf-1)*Ts;               % Establish time vector
tf = (nf-1)*Ts;                % Establish 'finish' time
subplot(2,1,1)                 % Generate first subplot
plot(t,y1)                     % Plot y1(t)
axis([0 tf 0 1.5]);            % Set axis parameters
xlabel('Time - seconds')       % Generate x-label
ylabel('y1(t)')                % Generate y-label
subplot(2,1,2)                 % Generate second subplot
plot(t,y2)                     % Plot y2(t)
axis([0 tf 0 2]);              % Set axis parameters
xlabel('Time - seconds')       % Generate x-label
ylabel('y2(t)')                % Generate y-label
```

The correctness of the these waveforms can be verified by comparing the MATLAB output with the results given by (7-85). It is instructive to vary the sampling period and observe the results.

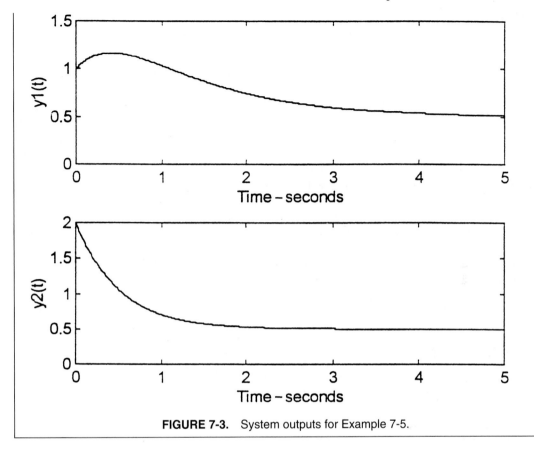

FIGURE 7-3.   System outputs for Example 7-5.

Note that we did not explicitly find e^{At}, but instead carried through all multiplications in terms of the transforms until the final step. Unless e^{At} is explicitly needed, this is usually the easiest procedure, as it leads to somewhat simpler functions which must be inverse transformed.

EXAMPLE 7-6

In this example we consider a system with two inputs and two outputs. Such systems are difficult to handle unless we use state-variable methods. For simplicity, we use a system similar to that in Example 7-5 and let initial conditions be zero. The system equations are assumed to be

$$\begin{bmatrix} \dot{x}_1 \\ \dot{x}_2 \end{bmatrix} = \begin{bmatrix} 0 & 1 \\ -2 & -3 \end{bmatrix} \begin{bmatrix} x_1 \\ x_2 \end{bmatrix} + \begin{bmatrix} 2 & 1 \\ 0 & 1 \end{bmatrix} \begin{bmatrix} u_1 \\ u_2 \end{bmatrix} \qquad (7\text{-}86a)$$

$$\begin{bmatrix} y_1 \\ y_2 \end{bmatrix} = \begin{bmatrix} 1 & 0 \\ 1 & 1 \end{bmatrix} \begin{bmatrix} x_1 \\ x_2 \end{bmatrix} \qquad (7\text{-}86b)$$

We assume that $u_1 = u(t)$, a unit step function, and $u_2 = e^{-3t}u(t)$. For $\mathbf{x}(0) = \mathbf{0}$ we obtain from (7-46),

$$\mathbf{Y}(s) = \mathbf{C}(s\mathbf{I} - \mathbf{A})^{-1}\mathbf{B}\mathbf{U}(s) \qquad (7\text{-}87)$$

Using $(s\mathbf{I} - \mathbf{A})^{-1}$ from Example 7-5, we obtain

$$
\mathbf{Y}(s) = \frac{\begin{bmatrix} 1 & 0 \\ 1 & 1 \end{bmatrix} \begin{bmatrix} s+3 & 1 \\ -2 & s \end{bmatrix} \begin{bmatrix} 2 & 1 \\ 0 & 1 \end{bmatrix} \begin{bmatrix} \dfrac{1}{s} \\ \dfrac{1}{s+3} \end{bmatrix}}{(s+1)(s+2)}
$$

$$
= \frac{\begin{bmatrix} s+3 & 1 \\ s+1 & s+1 \end{bmatrix} \begin{bmatrix} \dfrac{2}{s} + \dfrac{1}{s+3} \\ \dfrac{1}{s+3} \end{bmatrix}}{(s+1)(s+2)} \tag{7-88}
$$

or

$$
\mathbf{Y}(s) = \frac{1}{(s+1)(s+2)} \begin{bmatrix} \dfrac{3s^2 + 16s + 18}{s(s+3)} \\ \dfrac{(s+1)(4s+6)}{s(s+3)} \end{bmatrix} \tag{7-89}
$$

Finally, we obtain

$$
\mathbf{Y}(s) = \begin{bmatrix} \dfrac{3s^2 + 16s + 18}{s(s+1)(s+2)(s+3)} \\ \dfrac{4s+6}{s(s+2)(s+3)} \end{bmatrix} \tag{7-90}
$$

We may take the inverse transform with the aid of the partial-fraction expansion to obtain

$$
\begin{bmatrix} y_1(t) \\ y_2(t) \end{bmatrix} = \begin{bmatrix} (3 - \tfrac{5}{2}e^{-t} - e^{-2t} + \tfrac{1}{2}e^{-3t})u(t) \\ (1 + e^{-2t} - 2e^{-3t})u(t) \end{bmatrix} \tag{7-91}
$$

MATLAB Application

As mentioned in the previous example, a MATLAB simulation of the system requires finding a discrete-time state model for the continuous-time state model given by (7-86a) and (7-86b). The MATLAB simulation follows:

```
EDU» c7ex6
a = [0 1; -2 -3];          % Define a matrix
b = [2 1; 0 1];            % Define b matrix
c = [1 0; 1 1];            % Define c matrix
Ts = 0.01;                 % Define sampling period
nf = 501;                  % Define number of samples
x = zeros(2,nf);           % Initialize x matrix
x(:,1) = [0; 0];           % Establish initial conditions
t = (0:nf-1)*Ts;           % Define sampling point vector
u(1,:) = ones(1,nf);       % Define first input signal
u(2,:) = exp(-3*t);        % Define second input signal
```

```
[ad,bd] = c2d(a,b,Ts);          % Compute discrete equivalent
% The next three lines perform the simulation
for n=1:nf-1
        x(:,n+1) = ad*x(:,n)+bd*u(:,n);
end
y = c*x;                        % Compute system output
y1 = y(1,:);                    % Determine y1(t)
y2 = y(2,:);                    % Determine y2(t)
t = (0:nf-1)*Ts;                % Establish time vector
tf = (nf-1)*Ts;                 % Establish 'finish' time
subplot(2,1,1)                  % Generate first subplot
plot(t,y1)                      % Plot y1(t)
axis([0 tf 0 4]);               % Generate axis parameters
xlabel('Time - seconds)         % Generate x-label
ylabel('y1(t)')                 % Generate second label
subplot(2,1,2)                  % Generate second subplot
plot(t,y2)                      % Plot y2(t)
axis([0 tf 0 1.5]);             % Set axis parameters
xlabel('Time - seconds')        % Generate x-label
ylabel('y2(t)')                 % Generate y-label
```

This yields the following plots:

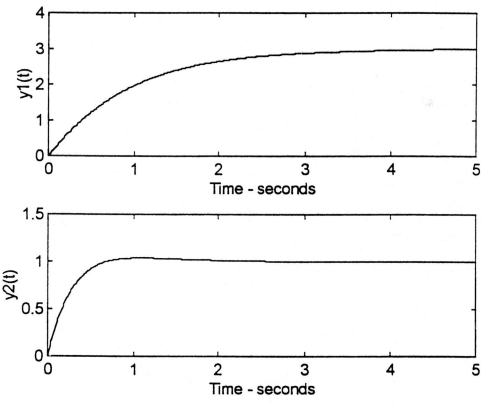

FIGURE 7-4. System Outputs for Example 7-6.

As before, the correctness of the waveforms shown in Figure 7-4 can be verified by comparing them with the results given by (7-91). As mentioned previously, the student should study the effect of changing the sampling period.

Two comments are perhaps in order at this point. First, it would be difficult to solve the equations of this example without the use of state-variable methods. The reader should be convinced of this by drawing a block diagram of the equations and considering how it might be used to obtain y_1 and y_2. Second, although we have only treated a second-order system, the same techniques can be used on systems of any order. The algebra, of course, becomes tedious for higher-order systems.

7-7 State Equations for Electrical Networks

We now consider the problem of obtaining state equations for systems. We consider only the special case of electrical networks here. However, by the use of suitable analogies, the technique can be extended to other classes of lumped-parameter systems (e.g., mechanical systems).

Earlier in this chapter we hinted that a suitable set of state variables for electrical networks were the capacitor voltages and inductor currents. Physically, this choice is attractive because the capacitor voltages and inductor currents specify the stored energy. These are convenient choices from a mathematical point of view as well. We recall the element relations for capacitors and inductors as

$$i_C = C\frac{dv_C}{dt}, \qquad v_L = L\frac{di_L}{dt} \qquad (7\text{-}92)$$

Note that the derivatives of the chosen state variables appear in these expressions. If we can write expressions for the capacitor current and inductor voltage in terms of the state variables and source voltages and/or currents, then we can obtain equations in the form

derivative of state variable = linear combination of state variables and inputs

This is precisely the form of the state equations. The capacitor current and inductor voltage can always be expressed in terms of state variables and source quantities by writing suitable KCL and KVL equations. The output equation can also be obtained in this way. Let us summarize the method in the following algorithm and then consider some examples.

The algorithm is:

1. Choose capacitor voltages and inductor currents as state variables.
2. For each capacitor, write a KCL, expressing the capacitor current $C\,dv/dt$ in terms of state variables, source quantities, and other currents as necessary.
3. For each inductor, write a KVL, expressing the inductor voltage $L\,di/dt$ in terms of state variables, source quantities, and other voltages as necessary.
4. Write other KCL, KVL, and element relation equations as necessary to eliminate the "other" currents and voltages in steps 2 and 3 in terms of state variables and source quantities. The resulting equations, after dividing by C and L as appropriate, are the state equations.
5. Write KCL and KVL equations and use element relations as necessary to express the output(s) in terms of state variables and source quantities. The state model is then complete.

We will use this algorithm to obtain state equations for two electrical networks.

EXAMPLE 7-7

In this example we derive a state model for the network shown in Figure 7-5.

We choose as state variables the capacitor voltage v_C and the inductor current i_L. Writing a KCL for the capacitor yields

$$C\dot{v}_C = i_{R_1} - i_L \tag{7-93}$$

A KVL for the inductor gives

$$L\dot{i}_L = v_C - v_{R_2} \tag{7-94}$$

But, following step 4 of the algorithm, we obtain

$$i_{R_1} = \frac{v_s - v_C}{R_1} \tag{7-95}$$

and

$$v_{R_2} = R_2 i_{R_2} = R_2 i_L \tag{7-96}$$

Hence

$$C\dot{v}_C = \frac{v_s}{R_1} - \frac{v_C}{R_1} - i_L \tag{7-97a}$$

$$L\dot{i}_L = v_C - R_2 i_L \tag{7-97b}$$

or

$$\dot{v}_C = \frac{v_s}{R_1 C} - \frac{v_C}{R_1 C} - \frac{i_L}{C} \tag{7-98a}$$

$$\dot{i}_L = \frac{v_C}{L} - \frac{R_2 i_L}{L} \tag{7-98b}$$

Clearly,

$$v_0 = v_{R_2} = R_2 i_L \tag{7-99}$$

If we wish we may write the equations in matrix form as

$$\begin{bmatrix} \dot{v}_C \\ \dot{i}_L \end{bmatrix} = \begin{bmatrix} -\dfrac{1}{R_1 C} & \dfrac{-1}{C} \\ \dfrac{1}{L} & -\dfrac{R_2}{L} \end{bmatrix} \begin{bmatrix} v_C \\ i_L \end{bmatrix} + \begin{bmatrix} \dfrac{1}{R_1 C} \\ 0 \end{bmatrix} v_s \tag{7-100a}$$

$$v_0 = \begin{bmatrix} 0 & R_2 \end{bmatrix} \begin{bmatrix} v_C \\ i_L \end{bmatrix} \tag{7-100b}$$

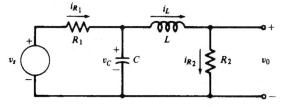

FIGURE 7-5.   Network for Example 7-7.

EXAMPLE 7-8 _____

In this example we consider a more complex case and obtain the state equations for the two-input, one-output circuit shown in Figure 7-6. The inputs are v_{s_1} and v_{s_2}, and the output is i_0.

Again we choose v_C and i_L as state variables. A KCL for the capacitor yields

$$C\dot{v}_C = i_0 - i_{R_3} \tag{7-101}$$

A KVL around the left-hand loop gives

$$L\dot{i}_L = v_{s_1} - v_{R_1} \tag{7-102}$$

Turning our attention to the inductor current equation, we see that we must eliminate v_{R_1}. We have

$$v_{R_1} = R_1 i_{R_1} \tag{7-103}$$

and writing a loop equation around the loop defined by v_{s_1}, R_1, R_2, and C yields

$$R_1 i_{R_1} + R_2 i_0 + v_C = v_{s_1} \tag{7-104}$$

Since

$$i_0 = i_{R_1} - i_L \tag{7-105}$$

we have

$$(R_1 + R_2)i_{R_1} = v_{s_1} - v_C + R_2 i_L \tag{7-106}$$

and

$$i_{R_1} = \frac{v_{s_1} - v_C + R_2 i_L}{R_1 + R_2} \tag{7-107}$$

Then

$$v_{R_1} = \frac{v_{s_1} - v_C + R_2 i_L}{R_1 + R_2} R_1 \tag{7-108}$$

and we have

$$L\dot{i}_L = v_{s_1} - \frac{R_1}{R_1 + R_2} v_{s_1} + \frac{R_1}{R_1 + R_2} v_C - \frac{R_1 R_2}{R_1 + R_2} i_L \tag{7-109}$$

or

$$\dot{i}_L = \frac{R_2}{L(R_1 + R_2)} v_{s_1} + \frac{R_1}{L(R_1 + R_2)} v_C - \frac{R_1 R_2}{L(R_1 + R_2)} i_L \tag{7-110}$$

as one state equation.

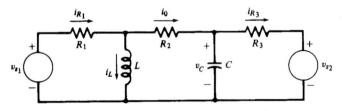

FIGURE 7-6. A two-input, two-output system.

Consider next the capacitor voltage equation

$$C\dot{v}_C = i_0 - i_{R_3} \tag{7-111}$$

Both i_0 and i_{R_3} must be eliminated. Now

$$i_{R_3} = \frac{v_C - v_{s_2}}{R_3} \tag{7-112}$$

so we have

$$C\dot{v}_C = i_0 - \frac{v_C}{R_3} + \frac{v_{s_2}}{R_3} \tag{7-113}$$

We also had

$$i_0 = i_{R_1} - i_L \tag{7-114}$$

Using the result for i_{R_1} yields

$$i_0 = \frac{v_{s_1}}{R_1 + R_2} - \frac{v_C}{R_1 + R_2} + \frac{R_2}{R_1 + R_2} i_L - i_L \tag{7-115}$$

or

$$i_0 = \frac{v_{s_1}}{R_1 + R_2} - \frac{v_C}{R_1 + R_2} + \frac{R_1 i_L}{R_1 + R_2} \tag{7-116}$$

This is the output equation. We may substitute this into the capacitor voltage state equation to yield

$$C\dot{v}_C = \frac{v_{s_1}}{R_1 + R_2} - \frac{v_C}{R_1 + R_2} - \frac{v_C}{R_3} - \frac{R_1 i_L}{R_1 + R_2} + \frac{v_{s_2}}{R_3} \tag{7-117}$$

Then the corresponding state equation is

$$\dot{v}_C = -\left[\frac{1}{CR_3} + \frac{1}{C(R_1 + R_2)}\right]v_C - \frac{R_1}{C(R_1 + R_2)} i_L + \frac{1}{C(R_1 + R_2)} v_{s_1} + \frac{1}{CR_3} v_{s_2} \tag{7-118}$$

We may put all this in matrix form as

$$\begin{bmatrix} \dot{v}_C \\ \dot{i}_L \end{bmatrix} = \begin{bmatrix} -\dfrac{1}{CR_3} - \dfrac{1}{C(R_1 + R_2)} & -\dfrac{R_1}{C(R_1 + R_2)} \\ \dfrac{R_1}{L(R_1 + R_2)} & -\dfrac{R_1 R_2}{L(R_1 + R_2)} \end{bmatrix} \begin{bmatrix} v_C \\ i_L \end{bmatrix}$$

$$+ \begin{bmatrix} \dfrac{1}{C(R_1 + R_2)} & \dfrac{1}{CR_3} \\ \dfrac{R_2}{L(R_1 + R_2)} & 0 \end{bmatrix} \begin{bmatrix} v_{s_1} \\ v_{s_2} \end{bmatrix} \tag{7-119}$$

$$i_0 = \begin{bmatrix} -\dfrac{1}{R_1 + R_2} & -\dfrac{R_1}{R_1 + R_2} \end{bmatrix} \begin{bmatrix} v_C \\ i_L \end{bmatrix} + \begin{bmatrix} \dfrac{1}{R_1 + R_2} & 0 \end{bmatrix} \begin{bmatrix} v_{s_1} \\ v_{s_2} \end{bmatrix} \tag{7-120}$$

While the second of our examples requires considerably more equations to be written in order to eliminate the undesired variables, both of the examples show that the algorithm may be applied to determine the state equations. Of course, more complicated circuits may be handled by the same method. The effort naturally increases.

7-8 State Equations from Transfer Functions

In this section we consider systems described by transfer functions of the form

$$\frac{Y(s)}{U(s)} = \frac{b_m s^m + b_{m-1} s^{m-1} + \cdots + b_1 s + b_0}{s^n + a_{n-1} s^{n-1} + a_{n-2} s^{n-2} + \cdots + a_1 s + a_0} \tag{7-121}$$

where $m < n$ and $Y(s)$ and $U(s)$ are the Laplace transformed system output and input, respectively. We pose the question "How can we generate an equivalent state-variable description?" Note that (7-121) is equivalent to the nth-order differential equation (with zero initial conditions)

$$y^{(n)} + a_{n-1} y^{(n-1)} + a_{n-2} y^{n-2} + \cdots + a_1 \dot{y} + a_0 y = b_m u^{(m)} + b_{m-1} u^{(m-1)} + \cdots + b_1 \dot{u} + b_0 u \tag{7-122}$$

Thus, if we are able to obtain state equations from transfer functions, then we are also able to obtain state-equation representations of nth-order differential equations.

Since the system is single-input, single-output the state model will take the form

$$\dot{\mathbf{x}} = \mathbf{A}\mathbf{x} + \mathbf{b}u \tag{7-123a}$$

$$y = \mathbf{c}\mathbf{x} + \mathbf{d}u \tag{7-123b}$$

The problem is solved when the elements of $\mathbf{A}, \mathbf{b}, \mathbf{c},$ and $\mathbf{d}$ are known. Actually, there are infinitely many possible choices for the elements of the matrices such that $\mathbf{c}[s\mathbf{I} - \mathbf{A}]^{-1}\mathbf{b}+\mathbf{d}$ yields the transfer function in (7-121). Each choice would result from a particular technique of realizing the system described by the transfer function in state-equation form. We are going to examine only one realization technique. It is not necessarily the easiest or the best, but it does lead to state equations that have some nice properties.

Let us first consider a much simplified version of (7-121), namely

$$\frac{Y(s)}{U(s)} = \frac{4}{(s + 1)(s + 2)} \tag{7-124}$$

Using the partial-fraction expansion, we may write (7-124) as

$$Y(s) = \frac{4U(s)}{s + 1} - \frac{4U(s)}{s + 2} \tag{7-125}$$

Now define

$$X_1(s) = \frac{U(s)}{s + 1}, \qquad X_2(s) = \frac{U(s)}{s + 2} \tag{7-126}$$

Rearranging and inverse transforming (7-126) yields

$$\dot{x}_1 = -x_1 + u, \qquad \dot{x}_2 = -2x_2 + u \tag{7-127}$$

These may obviously be written as

$$\begin{bmatrix} \dot{x}_1 \\ \dot{x}_2 \end{bmatrix} = \begin{bmatrix} -1 & 0 \\ 0 & -2 \end{bmatrix} \begin{bmatrix} x_1 \\ x_2 \end{bmatrix} + \begin{bmatrix} 1 \\ 1 \end{bmatrix} u \tag{7-128}$$

Using the definitions (7-126) in (7-125) and inverse transforming yields

$$y = \begin{bmatrix} 4 & -4 \end{bmatrix} \begin{bmatrix} x_1 \\ x_2 \end{bmatrix} \tag{7-129}$$

Equations (7-128) and (7-129) are precisely the state model we have been seeking. Note that the **A** matrix is diagonal. This is a particularly convenient state of affairs (see Problem 7-7). Also, all elements of the vector **b** are unity.

We can readily generalize this example. Returning to (7-121), let us make the further assumption that the numerator and denominator contain no common factors. We will briefly discuss removing this restriction later. Let us also assume temporarily that there are no repeated poles. Then we may factor the denominator and write (7-121) as

$$\frac{Y(s)}{U(s)} = \frac{b_m s^m + b_{m-1} s^{m-1} + \cdots + b_1 s + b_0}{(s - p_1)(s - p_2) \cdots (s - p_n)} \tag{7-130}$$

Using the partial-fraction expansion, we obtain

$$\frac{Y(s)}{U(s)} = \frac{B_1}{s - p_1} + \frac{B_2}{s - p_2} + \cdots + \frac{B_n}{s - p_n} \tag{7-131}$$

Now we define

$$X_i(s) = \frac{U(s)}{s - p_i}, \qquad i = 1, 2, \ldots, n \tag{7-132}$$

and proceeding as in (7-126)–(7-129) we obtain the state model as

$$\begin{bmatrix} \dot{x}_1 \\ \dot{x}_2 \\ \vdots \\ \dot{x}_n \end{bmatrix} = \begin{bmatrix} p_1 & 0 & 0 & \cdots & 0 \\ 0 & p_2 & 0 & \cdots & 0 \\ \vdots & & & & \\ 0 & & & 0 & p_n \end{bmatrix} \begin{bmatrix} x_1 \\ x_2 \\ \vdots \\ x_n \end{bmatrix} + \begin{bmatrix} 1 \\ 1 \\ \vdots \\ 1 \end{bmatrix} u \tag{7-133a}$$

$$y = \begin{bmatrix} B_1 & B_2 & \cdots & B_n \end{bmatrix} \begin{bmatrix} x_1 \\ x_2 \\ \vdots \\ x_n \end{bmatrix} \tag{7-133b}$$

Note that $\mathbf{d} = \mathbf{0}$ due to the fact that $m < n$.

There are at least two drawbacks to the method given, but they are not serious. First, the denominator polynomial must be factored so that the partial-fraction expansion can be carried out. This is certainly not a serious problem if a computer is available. A second problem is that if some of the p_i are complex, then the associated B_i also are complex. Thus the associated states are complex and have no obvious physical significance. The computations associated with the solution of the equations also become more complicated. Although beyond the scope of this text, it is possible to apply a transformation of variables that will convert **A** back to a real matrix, at the same time keeping that portion of the matrix associated with real poles in diagonal form so as to keep the computational difficulty as small as possible.

We now wish to remove the restriction that all the poles be distinct. Again let us consider a simple case and generalize from it. Suppose a transfer function contains a pole p_i of multiplicity two. Then the partial-fraction-expansion terms associated with these poles will be

$$Y_i(s) = \frac{A U(s)}{(s - p_i)^2} + \frac{B U(s)}{s - p_i} \tag{7-134}$$

where $Y_i(s)$ here represents only that part of the output associated with these poles. Now define

$$X_{i_1}(s) = \frac{U(s)}{(s - p_i)^2}, \qquad X_{i_2}(s) = \frac{U(s)}{s - p_i} \tag{7-135}$$

Clearly

$$X_{i_1}(s) = \frac{X_{i_2}(s)}{s - p_i} \tag{7-136}$$

Inverse transforming the second equation of (7-135) and (7-136) yields

$$\dot{x}_{i_1} = p_i x_{i_1} + x_{i_2}$$
$$\dot{x}_{i_2} = p_i x_{i_2} + u \tag{7-137}$$

Equation (7-134) leads to

$$y_i = A x_{i_1} + B x_{i_2} \tag{7-138}$$

Equations (7-137) and (7-138) can be rewritten in matrix form as

$$\begin{bmatrix} \dot{x}_{i_1} \\ \dot{x}_{i_2} \end{bmatrix} = \begin{bmatrix} p_i & 1 \\ 0 & p_i \end{bmatrix} \begin{bmatrix} x_{i_1} \\ x_{i_2} \end{bmatrix} + \begin{bmatrix} 0 \\ 1 \end{bmatrix} u \tag{7-139a}$$

$$y_i = \begin{bmatrix} A & B \end{bmatrix} \begin{bmatrix} x_{i_1} \\ x_{i_2} \end{bmatrix} \tag{7-139b}$$

and we can readily see what changes have been introduced by the repeated roots. The **A** matrix is now in the form of a Jordan block. The Jordan block has the pole on the diagonal, ones on the superdiagonal, and zeros elsewhere. In this case, the Jordan block is 2×2, the multiplicity of the pole. If the pole had multiplicity k, then the Jordan block would be $k \times k$. For reference, third- and fourth-order Jordan blocks are, respectively,

$$\begin{bmatrix} p_i & 1 & 0 \\ 0 & p_i & 1 \\ 0 & 0 & p_i \end{bmatrix} \quad \text{and} \quad \begin{bmatrix} p_i & 1 & 0 & 0 \\ 0 & p_i & 1 & 0 \\ 0 & 0 & p_i & 1 \\ 0 & 0 & 0 & p_i \end{bmatrix} \tag{7-140}$$

The **b** vector in (7-139a) has a one in its last position and zeros elsewhere. This is a general property of the **b** vector associated with repeated poles. The **c** vector is unchanged in that it contains the constants in the partial-fraction expansion. Note that the discussion here pertains only to repeated poles. When the transfer function contains both distinct and repeated poles those parts of **A**, **b**, **c** relating to the distinct poles are unchanged; that is, that portion of **A** remains diagonal. Those parts of **A**, **b**, **c** related to the repeated poles change as indicated here. The following example should clarify matters.

EXAMPLE 7-9 _____

The purpose of this example is to illustrate the technique just discussed for developing a state model of a system from a transfer function. Consider the system defined by the transfer function

$$\frac{Y(s)}{U(s)} = \frac{s^2 + 3s + 9}{s^5 + 8s^4 + 24s^3 + 34s^2 + 23s + 6} \tag{7-141}$$

The denominator contains simple roots at $s = -2$ and $s = -3$ and a root of multiplicity three at $s = -1$. Accordingly, the function may be written as

$$\frac{Y(s)}{U(s)} = \frac{B_1}{(s+1)^3} + \frac{B_2}{(s+1)^2} + \frac{B_3}{s+1} + \frac{B_4}{s+2} + \frac{B_5}{s+3} \tag{7-142}$$

Using techniques of Section 5-4, we may find the constants and write

$$\frac{Y(s)}{U(s)} = \frac{3.5}{(s+1)^3} - \frac{4.75}{(s+1)^2} + \frac{5.875}{s+1} - \frac{7}{s+2} + \frac{1.125}{s+3} \tag{7-143}$$

Then, using the definitions

$$X_1(s) = \frac{U(s)}{(s+1)^3}, \qquad X_2(s) = \frac{U(s)}{(s+1)^2}, \qquad X_3(s) = \frac{U(s)}{s+1}$$

$$X_4(s) = \frac{U(s)}{s+2}, \qquad X_5(s) = \frac{U(s)}{s+3} \tag{7-144}$$

and inverse transforming we obtain

$$\begin{bmatrix} \dot{x}_1 \\ \dot{x}_2 \\ \dot{x}_3 \\ \dot{x}_4 \\ \dot{x}_5 \end{bmatrix} = \begin{bmatrix} -1 & 1 & 0 & 0 & 0 \\ 0 & -1 & 1 & 0 & 0 \\ 0 & 0 & -1 & 0 & 0 \\ 0 & 0 & 0 & -2 & 0 \\ 0 & 0 & 0 & 0 & -3 \end{bmatrix} \begin{bmatrix} x_1 \\ x_2 \\ x_3 \\ x_4 \\ x_5 \end{bmatrix} + \begin{bmatrix} 0 \\ 0 \\ 1 \\ 1 \\ 1 \end{bmatrix} u \tag{7-145}$$

with the output equation

$$y = \begin{bmatrix} 3.5 & -4.75 & 5.875 & -7 & 1.125 \end{bmatrix} \begin{bmatrix} x_1 \\ x_2 \\ x_3 \\ x_4 \\ x_5 \end{bmatrix} \tag{7-146}$$

Note that the matrix **A** and the vectors **b** and **c** do have the structure promised.

MATLAB Application

As mentioned in the text, the state model corresponding to a given transfer function is not unique. We, therefore, will work backwards and show that the state model just derived is equivalent to the transfer function given by (7-141). The MATLAB code follows.

```
EDU» c7ex9
% The following four statements define the state model.
A = [-1 1 0 0 0; 0 -1 1 0 0; 0 0 -1 0 0; 0 0 0 -2 0; 0 0 0 0 -3];
B = [0 0 1 1 1]';
C = [3.5 -4.75 5.875 -7 1.125];
D = 0;
[num,den] = ss2tf(A,B,C,D);        % Determine the transfer function
num                                 % Display numerator
```

```
num =

           0     0.0000      0.0000     1.0000     3.0000     9.0000

den                                         % Display denominator

den =

       1     8     24     34     23     6
```

We see that the numerator and denominator polynomials agree with the transfer function model as expressed by (7-141).

To show that the state model is indeed not unique, it is interesting to derive a state model based on the transfer function model just computed. The MATLAB code is

```
EDU» [A,B,C,D] = tf2ss(num,den)

A =

     -8    -24    -34    -23     -6
      1      0      0      0      0
      0      1      0      0      0
      0      0      1      0      0
      0      0      0      1      0

B =

      1
      0
      0
      0
      0

C =

     0.0000      0.0000     1.0000     3.0000     9.0000

D =

      0
```

This state model is clearly different from the state model expressed by (7-145) and (7-146) but is still equivalent to the transfer function model given by (7-141).

Recall that when we began this discussion regarding obtaining state models from transfer functions we required that the numerator and denominator of (7-121) not have any common factors. Let us see what happens if this assumption is not satisfied. We consider the case of only one such factor, $s - p_i$. For the pole at $s = p_i$, we expect a term in the partial fraction expansion of the form

$$\frac{B_i}{s - p_i} \tag{7-147}$$

where

$$B_i = \lim_{s \to p_i} (s - p_i) \frac{Y(s)}{U(s)} \qquad (7\text{-}148)$$

Since the numerator of $Y(s)/U(s)$ also contains $(s - p_i)$ as a factor, the result is $B_i = 0$. Since the term, therefore, does not appear in the partial-fraction expansion, it seems appropriate to ignore the corresponding state variable and obtain an $(n - 1)$-dimension state model. This would be equivalent to canceling the common factors in $Y(s)/U(s)$ and then obtaining a state model of the reduced system. Actually, however, it is more appropriate to retain the state, still defining

$$X_i(s) = \frac{U(s)}{s - p_i} \qquad (7\text{-}149)$$

which maintains the original system order. The vector **c** will contain a zero corresponding to the state x_i, indicating that the state is not observed in the output. It is also possible to obtain a state model in which the state x_i is observed in the output but is not affected by the system input u. In the first case the system is said to be *controllable* but not *observable,* while in the second case it is *observable* but not *controllable.* When pole-zero cancellation exists, the system cannot be both controllable and observable. A clear understanding of the concepts of controllability and observability is quite important in the study of automatic control systems, but is well beyond the scope of this text. We refer the interested readers to one of the excellent texts listed at the end of this chapter.

*7-9 State Equations for Discrete-Time Systems

Many systems are inherently discrete-time in nature. This is usually because essential data regarding system variables are available only at certain discrete instants of time called the sampling instants. For example, data regarding economic activity are collected only at monthly or quarterly intervals. Such systems must be described by difference equations rather than differential equations.

In state-space form, the equations that describe a discrete-time system become a set of first-order difference equations which constitute a recursion relation. This recursion relation allows computation of the state at sampling instant $(k + 1)T$ from the state at sampling instant kT and the input at time kT, where k is an integer. Of course, an output equation must also be specified, so the discrete-time state model becomes

$$\mathbf{x}_{k+1} = \mathbf{F}\mathbf{x}_k + \mathbf{G}\mathbf{u}_k \qquad (7\text{-}150a)$$

$$\mathbf{y}_k = \mathbf{H}\mathbf{x}_k + \mathbf{J}\mathbf{u}_k \qquad (7\text{-}150b)$$

where we have suppressed the explicit dependence on T for simplicity of notation. For the single-input, single-output system $\mathbf{u}_k$ and $\mathbf{y}_k$ are scalars and **G** and **H** become vectors **g** and **h,** respectively. In most cases **J** is null.

Our first concern is to find a solution to (7-150a) and see if any quantity analogous to the state transition matrix of Section 7-4 exists. Let us assume the initial state is $\mathbf{x}_0$. Then we can compute **x** directly from (7-150a) as

$$\mathbf{x}_1 = \mathbf{F}\mathbf{x}_0 + \mathbf{G}\mathbf{u}_0 \qquad (7\text{-}151)$$

Now we may use (7-151) in (7-150a) to compute $\mathbf{x}_2$, etc.

The results are

$$\mathbf{x}_2 = \mathbf{F}\mathbf{x}_1 + \mathbf{G}\mathbf{u}_1 = \mathbf{F}^2\mathbf{x}_0 + \mathbf{F}\mathbf{G}\mathbf{u}_0 + \mathbf{G}\mathbf{u}_1$$

$$\mathbf{x}_3 = \mathbf{F}\mathbf{x}_2 + \mathbf{G}\mathbf{u}_2 = \mathbf{F}^3\mathbf{x}_0 + \mathbf{F}^2\mathbf{G}\mathbf{u}_0 + \mathbf{F}\mathbf{G}\mathbf{u}_1 + \mathbf{G}\mathbf{u}_2$$

$$\text{(7-152)}$$

$$\mathbf{x}_n = \mathbf{F}^n\mathbf{x}_0 + \mathbf{F}^{n-1}\mathbf{G}\mathbf{u}_0 + \mathbf{F}^{n-2}\mathbf{G}\mathbf{u}_1 + \cdots + \mathbf{F}\mathbf{G}\mathbf{u}_{n-2} + \mathbf{G}\mathbf{u}_{n-1}$$

This last equation is the solution we seek, since it gives the state at time nT in terms of the initial state $\mathbf{x}_0$ and the input up to (but not including) time nT. We may write the result more compactly as

$$\mathbf{x}_n = \mathbf{F}^n\mathbf{x}_0 + \sum_{j=0}^{n-1} \mathbf{F}^{n-j-1}\mathbf{G}\mathbf{u}_j \qquad \text{(7-153)}$$

If we compare (7-153) with (7-150a) or (7-150b), the solution obtained for the continuous-time state equations, the similarities are obvious. Clearly the quantity $\mathbf{F}^n$ is the state transition matrix, denoted $\mathbf{\Phi}(n)$. In the absence of an input the quantity indicates how the initial state is changed by the system to the state at time nT. The term $\mathbf{F}^n\mathbf{x}_0$ is the initial condition response. The second term in (7-153) represents the forced response. It is analogous to the second term (a convolution integral) in (7-36) or (7-37) and is called a *convolution summation.* Once the solution is obtained, (7-150b) may be used directly to compute the system output.

In all honesty, we must point out that (7-153) is not very well suited for hand calculations. For such purposes, methods based on the z-transform, to be discussed in the next chapter, are much more desirable. However, (7-153) can be very convenient for machine computation when the input sequence is specified.

An important application of the discrete-time state model is that of digital computer simulation of a continuous-time system. For this purpose it is necessary to obtain a discrete-time model because the computer only has values available at certain instants of time. This problem is also important in the digital control of continuous-time processes. The problem may be stated concisely as follows. Given a continuous-time system

$$\dot{\mathbf{x}} = \mathbf{A}\mathbf{x} + \mathbf{B}\mathbf{u} \qquad \mathbf{y} = \mathbf{C}\mathbf{x} + \mathbf{D}\mathbf{u} \qquad \text{(7-154)}$$

determine the matrices $\mathbf{F}, \mathbf{G}, \mathbf{H}, \mathbf{J}$ in the state model (7-150), where sample times are T seconds apart.

In solving this problem we note that since the computer cannot obtain values of $\mathbf{u}(t)$ between sample instants it seems sensible to assume that $\mathbf{u}$ does not change between the sample instants; that is, we assume

$$\mathbf{u}(t) = \mathbf{u}_k \qquad \text{for } kT \le t < (k + 1)T \qquad \text{(7-155)}$$

This assumption will allow us to obtain a model for the first equation of (7-154). The second of these equations is easy. Since it is valid for all t it is valid at the sampling instants kT. Thus the second equation of (7-154) becomes

$$\mathbf{y}(kT) = \mathbf{C}\mathbf{x}(kT) + \mathbf{D}\mathbf{u}(kT) \qquad \text{(7-156)}$$

Using the obvious shorthand notation $\mathbf{y}(kT) = \mathbf{y}_k$, etc., and comparing (7-156) with (7-150b), we see that

$$\mathbf{H} = \mathbf{C}, \qquad \mathbf{J} = \mathbf{D} \qquad \text{(7-157)}$$

Now let us return to the first equation of (7-154). In order to obtain the state model (7-150a) we treat time kT as the initial time t_0. Thus the initial state $\mathbf{x}_0$ is just $\mathbf{x}_k$. Then using (7-36) we may write the solution as

$$\mathbf{x}(t) = e^{\mathbf{A}(t-kT)}\mathbf{x}_k + \int^t e^{\mathbf{A}(t-\lambda)}\mathbf{B}\mathbf{u}(\lambda) \, d\lambda \tag{7-158}$$

where $t \geq kT$. Our interest, however, is in the particular time $t = (k + 1)T$, and in view of (7-155) we may replace $\mathbf{u}(\lambda)$ by $\mathbf{u}_k$. Thus we obtain

$$\mathbf{x}[(k + 1)T] = e^{\mathbf{A}[(k+1)T-kT]}\mathbf{x}_k + \int_{kT}^{(k+1)T} e^{\mathbf{A}[(k+1)T-\lambda]}\mathbf{B}\mathbf{u}_k \, d\lambda \tag{7-159}$$

or

$$\mathbf{x}_{k+1} = e^{\mathbf{A}T}\mathbf{x}_k + \int_{kT}^{(k+1)T} e^{\mathbf{A}[(k+1)T-\lambda]}\mathbf{B} \, d\lambda \, \mathbf{u}_k \tag{7-160}$$

since $\mathbf{u}_k$ is a constant. Now, upon comparing (7-160) and (7-150a), we note that

$$\mathbf{F} = e^{\mathbf{A}T} \tag{7-161}$$

and

$$\mathbf{G} = \int_{kT}^{(k+1)T} e^{\mathbf{A}[(k+1)T-\lambda]}\mathbf{B} \, d\lambda \tag{7-162}$$

In those cases in which $\mathbf{A}$ is nonsingular we can simplify (7-162) to

$$\mathbf{G} = (e^{\mathbf{A}T} - \mathbf{I})\mathbf{A}^{-1}\mathbf{B} \tag{7-163}$$

where we have used the fact that

$$\int e^{\mathbf{P}t} \, dt = e^{\mathbf{P}t}\,\mathbf{P}^{-1} \tag{7-164}$$

if $\mathbf{P}^{-1}$ exists. Equations (7-157) and (7-162) or (7-164) specify the matrices of the discrete-time model in terms of the matrices of the original, continuous-time system. Hence the problem is solved.

EXAMPLE 7-10

The purpose of this example is to illustrate the use of the preceeding material for obtaining a discrete-time state model for a continuous-time system. Assume that a two-input, one-output, system is defined by the following equations and that a discrete-time model is to be found for a sampling period of 0.1s:

$$\begin{bmatrix} \dot{x}_1 \\ \dot{x}_2 \end{bmatrix} = \begin{bmatrix} 0 & 1 \\ -6 & -5 \end{bmatrix}\begin{bmatrix} x_1 \\ x_2 \end{bmatrix} + \begin{bmatrix} 1 & 1 \\ 1 & 1 \end{bmatrix}\begin{bmatrix} u_1 \\ u_2 \end{bmatrix} \tag{7-165a}$$

$$y = \begin{bmatrix} 1 & 2 \end{bmatrix}\begin{bmatrix} x_1 \\ x_2 \end{bmatrix} \tag{7-165b}$$

For this system we may compute

$$\mathbf{A}^{-1} = \begin{bmatrix} -5/6 & -1/6 \\ 1 & 0 \end{bmatrix} = \begin{bmatrix} -0.833 & -0.167 \\ 1 & 0 \end{bmatrix} \tag{7-166}$$

so (7-163) may be utilized. We must compute e^{AT}, however. Actually this has been done using three different methods in Section 7-6, so we use the result (see Examples 7-2, 7-3, and 7-4 for e^{At})

$$\mathbf{F} = e^{AT} = \begin{bmatrix} 3e^{-0.2} - 2e^{-0.3} & e^{-0.2} - e^{-0.3} \\ -6e^{-0.2} + 6e^{-0.3} & -2e^{-0.2} + 3e^{-0.3} \end{bmatrix} = \begin{bmatrix} 0.975 & 0.078 \\ -0.467 & 0.585 \end{bmatrix} \quad (7\text{-}167)$$

Then from (7-163) and (7-165)–(7-167) we have

$$\mathbf{G} = \begin{bmatrix} -0.025 & 0.078 \\ -0.467 & -0.415 \end{bmatrix} \begin{bmatrix} -0.833 & -0.167 \\ 1 & 0 \end{bmatrix} \begin{bmatrix} 1 & 1 \\ 1 & 1 \end{bmatrix}$$

$$= \begin{bmatrix} -0.025 & 0.078 \\ -0.467 & -0.415 \end{bmatrix} \begin{bmatrix} -1 & -1 \\ 1 & 1 \end{bmatrix} = \begin{bmatrix} 0.103 & 0.103 \\ 0.052 & 0.052 \end{bmatrix} \quad (7\text{-}168)$$

The desired discrete-time state model is thus

$$\begin{bmatrix} x_{1_{k+1}} \\ x_{2_{k+1}} \end{bmatrix} = \begin{bmatrix} 0.975 & 0.078 \\ -0.467 & 0.585 \end{bmatrix} \begin{bmatrix} x_{1_k} \\ x_{2_k} \end{bmatrix} + \begin{bmatrix} 0.103 & 0.103 \\ 0.052 & 0.052 \end{bmatrix} \begin{bmatrix} u_{1_k} \\ u_{2_k} \end{bmatrix} \quad (7\text{-}169a)$$

$$y_k = \begin{bmatrix} 1 & 2 \end{bmatrix} \begin{bmatrix} x_{1_k} \\ x_{2_k} \end{bmatrix} \quad (7\text{-}169b)$$

MATLAB Application

The MATLAB function `c2d` preforms the operations described by (7-161) and (7-162) and, therefore, derives a discrete-time state model from a continuous-time state model. For the problem being considered, the MATLAB code is as follows:

```
EDU» c7ex10
a = [0 1; -6 -5];          % Define a matrix
b = [1 1; 1 1];            % Define b matrix
Ts = 0.1;                  % Define sampling period
[F,G] = c2d(a,b,Ts);       % Compute F matrix and G matrix
F                          % Display F

F =

    0.9746    0.0779
   -0.4675    0.5850

G                          % Display G

G =

    0.1034    0.1034
    0.0525    0.0525
```

These results agree with (7-167) and (7-168). The results of this problem illustrate the discrete-time model development used in the MATLAB applications for Examples 7-5 and 7-6.

Summary

The principal objective of this chapter was to introduce the basic concepts of systems analysis using state-variable techniques. We saw that system analysis can be carried out by using state equations, which are usually expressed in matrix form. These equations can be solved with either time-domain or Laplace transform techniques. We also investigated methods for finding the state equations of electrical networks, and we briefly discussed state equations for discrete-time systems. The specific points explored in this chapter follow.

1. The state of a system at time t_0 is the set of values at time t_0 of an appropriately chosen set of variables that provide the minimum information necessary to specify the condition of the system at time $t = t_0$ and, given the system input, obtain the system response for $t > t_0$. These variables are referred to as state variables. The state vector is a vector whose elements are the state variables.

2. An nth-order linear differential system with m inputs and p outputs can be represented by n first-order differential equations of the form

$$\dot{\mathbf{x}} = \mathbf{Ax} + \mathbf{Bu}$$

   and p output equations of the form

$$\mathbf{y} = \mathbf{Cx} + \mathbf{Du}$$

3. The time-domain solution of state equations allows the complete system response to be written, for $t > t_0$,

$$\mathbf{y}(t) = \mathbf{C\Phi}(t - t_0)\mathbf{x}_0 + \int_{t_0}^{t} \mathbf{C\Phi}(t - \lambda)\mathbf{Bu}(\lambda)\, d\lambda + \mathbf{Du}(t)$$

   where $\mathbf{\Phi}(t)$ is the state transition matrix and $\mathbf{x}_0$ is the vector of state variables at $t = t_0$.

4. State equations can be solved by Laplace transform techniques. This is referred to as the frequency-domain solution of state equations. In general, the Laplace transform of the system output, denoted $\mathbf{Y}(s)$, can be written

$$\mathbf{Y}(s) = \mathbf{C}(s\mathbf{I} - \mathbf{A})^{-1}\mathbf{x}_0 + [\mathbf{C}(s\mathbf{I} - \mathbf{A})^{-1}\mathbf{B} + \mathbf{D}]\mathbf{U}(s)$$

   The inverse Laplace transform yields the output vector $\mathbf{y}(t)$.

5. There are three basic techniques for determining the state transition matrix. The first technique is to use the definition

$$\mathbf{\Phi}(t) = \mathbf{e}^{\mathbf{A}t}$$

   and expand $\mathbf{e}^{\mathbf{A}t}$ in a series expansion. This technique is typically useful in extremely simple cases in which the closed-form representation for the infinite series resulting from the expansion of $\mathbf{e}^{\mathbf{A}t}$ can be recognized. Recognizing the appropriate closed form is often difficult. This difficulty is overcome by determining $\mathbf{\Phi}(t)$ from the relationship

$$\mathbf{\Phi}(s) = (s\mathbf{I} - \mathbf{A})^{-1}$$

   and inverse transforming $\mathbf{\Phi}(s)$ to obtain $\mathbf{\Phi}(t)$. A third technique makes use of the Caley-Hamilton theorem.

6. The state equations for electrical networks are determined by first choosing as state variables capacitor voltages and inductor currents. KCL and KVL equations are then applied to place the equations in the proper form for state equations.

7. State equations for a system can be derived from the system transfer function. This is typically accomplished by using the Laplace transform. The system realization, and consequently the state equations derived from a given transfer function, is not unique.

8. State equations can be derived for a discrete-time system and take the same form as state equations for continuous-time systems.

Further Reading

A number of excellent textbooks dealing with both linear system theory and control systems contain detailed treatments of state-variable methods. Representative books include the following:

R. J. MAYHAN, *Discrete-Time and Continuous-Time Linear Systems*. Reading, MA: Addison-Wesley, 1984.

B. C. KUO, *Automatic Control Systems*, 7th ed. Englewood Cliffs, NJ: Prentice-Hall, 1995.

G. F. FRANKLIN, J. D. POWELL, and A. EMANI-NAEINI, *Feedback Control of Dynamic Systems*. 3rd. ed. Reading, MA: Addison-Wesley, 1984.

J. G. REID, *Linear System Fundamentals: Continuous and Discrete, Classic and Modern*. New York: McGraw-Hill, 1983.

A number of textbooks have recently been developed that make use of MATLAB for the analysis and design of control systems. One book that stresses the application of MATLAB to a wide variety of control systems problems is D. K. Frederick and J. H. Chow, *Feedback Control Problems Using MATLAB*, Boston, MA: PWS, 1995.

Problems

Section 7-3

7-1. A system is defined by the equations

$$\dot{x}_1 = -4x_1 + 3x_2 + 6u$$

$$\dot{x}_2 = -x_1 - 7x_2 - 4u$$

$$y_1 = 5x_1 - 3x_2 + 2u$$

$$y_2 = 2x_1 + x_2 - 2u$$

(a) Write the equations in standard matrix form and identify the **A, B, C,** and **D** matrices.

(b) Draw a simulation diagram.

7-2. Repeat Problem 7-1 for the set of equations

$$\dot{x}_1 = -3x_1 + 2x_2 + x_3 + 2u_1 + u_2$$

$$\dot{x}_2 = 3x_1 - 2x_2 + x_3 + u_1$$

$$\dot{x}_3 = -4x_2 - x_3 + u_1$$

$$y_1 = 5x_1 - 3x_2 + x_3 + u_1$$

$$y_2 = x_1 + x_3 + u_1 + u_2$$

7-3. Construct a simulation diagram for the system defined by

$$\mathbf{A} = \begin{bmatrix} -1 & 0 & 0 \\ 1 & -3 & 0 \\ 0 & 1 & -4 \end{bmatrix}, \quad \mathbf{B} = \begin{bmatrix} 2 & 3 \\ 1 & 1 \\ 0 & 4 \end{bmatrix}, \quad \mathbf{C} = \begin{bmatrix} 1 & 2 & 1 \\ 0 & 1 & 1 \end{bmatrix}, \quad \mathbf{D} = \begin{bmatrix} 1 & 3 \\ 0 & 2 \end{bmatrix}$$

7-4. Write state equations in matrix form which describe the simulation diagram shown in Figure P7-4.

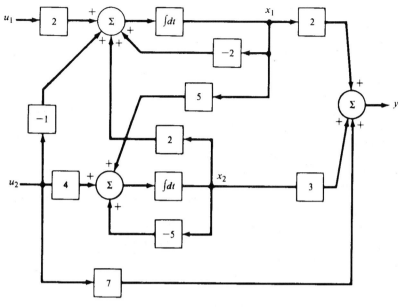

FIGURE P7-4

Sections 7-4-7-6

7-5. Use the series expansion method to compute e^{At} if

$$\mathbf{A} = \begin{bmatrix} -1 & 0 \\ 0 & -4 \end{bmatrix}$$

7-6. Use $\mathbf{\Phi}(s) = (s\mathbf{I} - \mathbf{A})^{-1}$ to compute $\mathbf{\Phi}(t)$ for

$$\mathbf{A} = \begin{bmatrix} -2 & 0 \\ 0 & -5 \end{bmatrix}$$

Check your result using series expansion methods.

7-7. Show that if an $n \times n$ matrix $\mathbf{A}$ is given by

$$\mathbf{A} = \begin{bmatrix} \lambda_1 & 0 & 0 & \cdots & 0 \\ 0 & \lambda_2 & 0 & \cdots & 0 \\ \vdots & & & & \vdots \\ 0 & 0 & 0 & \cdots & \lambda_n \end{bmatrix}$$

then

$$\mathbf{e}^{At} = \begin{bmatrix} e^{\lambda_1 t} & 0 & 0 & \cdots & 0 \\ 0 & e^{\lambda_2 t} & 0 & \cdots & 0 \\ \vdots & & & & \vdots \\ 0 & 0 & 0 & \cdots & e^{\lambda_n t} \end{bmatrix}$$

7-8. For the **A** of Problem 7-5, find $\mathbf{x}(t)$ if $u(t)$ is a unit step function and

$$\mathbf{x}(0) = \begin{bmatrix} 3 \\ 1 \end{bmatrix}, \qquad \mathbf{B} = \begin{bmatrix} 1 \\ 2 \end{bmatrix}$$

Verify your result using MATLAB.

7-9. In (7-15) we gave the recursion relation for computing the approximate solution to the state equations. Let the system be described by

$$\begin{bmatrix} \dot{x}_1 \\ \dot{x}_2 \end{bmatrix} = \begin{bmatrix} -1 & 0 \\ 0 & -0.6 \end{bmatrix} \begin{bmatrix} x_1 \\ x_2 \end{bmatrix}, \qquad \begin{bmatrix} x_1(0) \\ x_2(0) \end{bmatrix} = \begin{bmatrix} 3 \\ 2 \end{bmatrix}$$

(a) With $\Delta t = \frac{1}{2}$ s, use (7-15) to compute $\mathbf{x}(1\ \text{s})$.
(b) Repeat part (a) with $\Delta t = \frac{1}{10}$ s.
(c) Compute the actual value of $\mathbf{x}(1\ \text{s})$ and compare with parts (a) and (b).

7-10. Use the Laplace transform method to find $e^{\mathbf{A}t}$ for the **A** given in Problem 7-5.

7-11. Use the Laplace transform method to find $e^{\mathbf{A}t}$ if

$$\mathbf{A} = \begin{bmatrix} 0 & 1 & 0 \\ 0 & 0 & 1 \\ -6 & -11 & -6 \end{bmatrix}$$

7-12. Use the Cayley-Hamilton approach to find $e^{\mathbf{A}t}$ if

$$\mathbf{A} = \begin{bmatrix} 0 & 1 \\ -4 & -5 \end{bmatrix}$$

7-13. Use the Laplace transform approach to find $y(t)$ for the system given by

$$\begin{bmatrix} \dot{x}_1 \\ \dot{x}_2 \end{bmatrix} = \begin{bmatrix} 0 & 1 \\ -3 & -4 \end{bmatrix} \begin{bmatrix} x_1 \\ x_2 \end{bmatrix} + \begin{bmatrix} 0 \\ 2 \end{bmatrix} u$$

$$y = \begin{bmatrix} 1 & 0 \end{bmatrix} \begin{bmatrix} x_1 \\ x_2 \end{bmatrix}$$

where $\mathbf{x}_0 = \begin{bmatrix} 1 \\ 1 \end{bmatrix}$ and u is a unit step. Verify using MATLAB.

7-14. Find the transfer function matrix for the system

$$\begin{bmatrix} \dot{x}_1 \\ \dot{x}_2 \end{bmatrix} = \begin{bmatrix} 0 & 1 \\ -3 & -4 \end{bmatrix} \begin{bmatrix} x_1 \\ x_2 \end{bmatrix} + \begin{bmatrix} 0 \\ 2 \end{bmatrix} u$$

$$\begin{bmatrix} y_1 \\ y_2 \end{bmatrix} = \begin{bmatrix} 1 & 0 \\ 0 & 1 \end{bmatrix} \begin{bmatrix} x_1 \\ x_2 \end{bmatrix}$$

Section 7-7

7-15. Obtain a state model for the network shown in Figure P7-15.

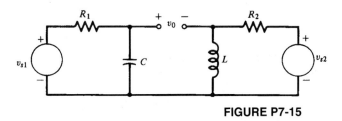

FIGURE P7-15

7-16. Obtain a state model for the network shown in Figure P7-16.

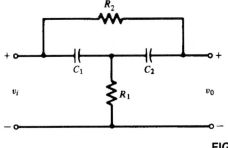

FIGURE P7-16

7-17. Obtain a state model for the network shown in Figure P7-17.

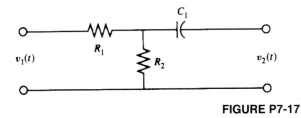

FIGURE P7-17

7-18. Obtain a state model for the network shown in Figure P7-18.

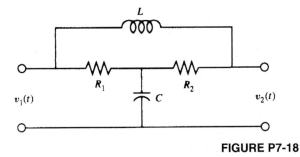

FIGURE P7-18

Section 7-8

7-19. Obtain a state model for the system defined by

$$\frac{Y(s)}{U(s)} = \frac{s+1}{s(s+3)}$$

7-20. Obtain a state model for the systems described by

(a) $\dfrac{Y(s)}{U(s)} = \dfrac{3}{(s+1)(s+3)}$

(b) $\dfrac{Y(s)}{U(s)} = \dfrac{s+1}{(s+1)(s+5)(s+6)}$

7-21. Obtain state models for the systems described by

(a) $\dfrac{Y(s)}{U(s)} = \dfrac{s^2 + 4s + 1}{(s+1)(s+2)}$

(b) $\dfrac{Y(s)}{U(s)} = \dfrac{s^2 + 2s + 2}{s(s+1)(s+3)}$

7-22. Obtain state models for the systems described by

(a) $\dfrac{Y(s)}{U(s)} = \dfrac{s^2 + s + 3}{s(s+1)}$

(b) $\dfrac{Y(s)}{U(s)} = \dfrac{s^2 + 3s + 2}{s(s+1)(s+3)}$

7-23. Obtain a state model for
(a) $\dddot{y} + 2\ddot{y} + 6\dot{y} + 7y = 4u$
(b) $\ddot{y} + 14\dot{y} + 25y = 25u$
(c) $y^{(4)} + 3\dddot{y} + 7\ddot{y} + 10y = u$

7-24. Obtain a state-model for each of the following transfer functions and verify the results using MATLAB.

(a) $\dfrac{Y(s)}{U(s)} = \dfrac{s+2}{(s+1)(s+3)}$

(b) $\dfrac{Y(s)}{U(s)} = \dfrac{s+4}{s(s+2)(s+3)(s+6)}$

(c) $\dfrac{Y(s)}{U(s)} = \dfrac{s^2 + 2s + 7}{s(s+1)(s+3)}$

Section 7-9

7-25. Obtain discrete-time state equations, for the sampling periods specified, for the systems whose state equations are

(a) $\begin{bmatrix} \dot{x}_1 \\ \dot{x}_2 \end{bmatrix} = \begin{bmatrix} -1 & 0 \\ 0 & -2 \end{bmatrix} \begin{bmatrix} x_1 \\ x_2 \end{bmatrix} + \begin{bmatrix} 2 \\ 1 \end{bmatrix} u,$ $T = 0.25$ s

(b) $\begin{bmatrix} \dot{x}_1 \\ \dot{x}_2 \end{bmatrix} = \begin{bmatrix} 0 & 1 \\ -2 & -3 \end{bmatrix} \begin{bmatrix} x_1 \\ x_2 \end{bmatrix} + \begin{bmatrix} 0 \\ 2 \end{bmatrix} u,$ $T = 0.3$ s

7-26. For the system of (7-169) let $u_{1_k} = 0$ and $u_{2_k} = 1$ for $k \geq 0$. For zero initial state, compute the first 10 values of the output y_k by hand. Verify your results using MATLAB.

Computer Exercises

7-1. In this computer exercise we investigate the validity of the results obtained in Examples 7-5 and 7-6 as shown in Figures 7-3 and 7-4, respectively. A straightforward way of validating these results is to make use of the analytical results for $y_1(t)$ and $y_2(t)$ obtained in the examples. These are given by (7-85) for Example 7-5 and by (7-91) for Example 7-6. The MATLAB simulations given in Examples 7-5 and 7-6 must be executed in order to generate sample sequences corresponding to $y_1(t)$ and $y_2(t)$. These are to be compared to the theoretical results for $y_1(t)$ and $y_2(t)$ given by (7-85) and (7-91). Experiment with changing the sampling period T_s in order to gain a feel for the relationship between the sampling period and the accuracy of the simulation. In performing this investigation choose one sampling period that is excessively small and one sampling period that is too large. Comment on the results. How do you go about choosing an appropriate sampling period?

7-2. Much of the MATLAB code used to perform the discrete-time simulations given in Examples 7-5 and 7-6 could have been eliminated by using the MATLAB function `lsim`. Learn to use `lsim` by reworking Examples 7-5 and 7-6 using the `lsim` command. Comment on the results.

7-3. Develop a block diagram for the control system of Example 7-6 in the form illustrated in Figure 7-1. Use the SIMULINK block diagram editor to build the block diagram. Using SIMULINK simulate the system and compare the results with those obtained in Example 7-6. Comment on the use of SIMULINK for problems of this type.

7-4. A single-input single-output system has the transfer function

$$\frac{Y(s)}{U(s)} = \frac{s + 3}{s^5 + 14s^4 + 56s^3 + 85s^2 + 46s + 8}$$

Assume that the system is initially relaxed and that a unit step input is applied. Using state space methods, find and plot the output (or an approximate output that you know to be reasonably accurate). Verify your result using analysis.

CHAPTER 8

Discrete-Time Signals and Systems

8-1 Introduction

In the preceding chapters we have been concerned only with the analysis of continuous-time signals and continuous-time systems. We now turn our attention to discrete-time signals and systems.

A discrete-time signal is a signal defined by specifying the value of the signal only at discrete times, called *sampling instants*. If the sample values are then quantized and encoded, a digital signal results. A digital signal is formed from a continuous-time (analog) signal through the process of analog-to-digital conversion. In the following section we make a slight detour and study the process of analog-to-digital conversion in some detail so that we will fully understand the relationship between discrete-time signals and digital signals. The errors that occur in analog-to-digital conversion must be identified so that all later approximations and assumptions can be appreciated.

In this chapter we consider a discrete-time signal to arise by sampling a continuous-time signal. Quantizing and encoding these resulting sample values yields a digital signal. To start with a continuous-time signal at first appears to be a rather narrow viewpoint, since in many situations digital signals occur naturally and there is no continuous-time signal related to the sample values. Output data from a computer is a possible example. There are many other examples. However, in taking the view that discrete-time signals arise from continuous-time signals we are better able to relate the theoretical concepts in the following two chapters to the previously developed concepts for continuous-time signals and systems.

An important goal of this chapter is to formulate those tools necessary for the analysis of discrete-time signals and systems. The principal tool will be the *z*-transform, which will allow us to specify the parameters of digital signals and formulate the input-output relationships of digital systems in both the time domain and the frequency domain. In Chapter 9 attention will be turned to the more interesting task of the *synthesis* of digital systems. That is, we will design digital systems so that specified tasks will be performed and a given set of specifications will be satisfied.

8-2 Analog-to-Digital Conversion

A block diagram of an analog-to-digital (A/D) converter is shown in Figure 8-1. The first component is a sampler that extracts sample values of the input signal at the sampling times. The output of the sampler is a discrete-time signal but a continuous-amplitude signal, since the sample values assume the same continuous range of values assumed by the input signal $x(t)$. These signals are often referred to as *sampled data signals*. The second component in an A/D converter is a quantizer, which quantizes the continuous range of sample values into a finite number of sample values so that each sample value can

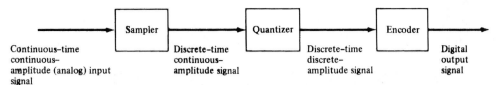

FIGURE 8-1. Analog-to-digital converter.

be represented by a digital word of finite precision or wordlength. The encoder maps each quantized sample value onto a digital word. Each of these processes is now examined in detail.

Sampling

To sample a continuous-time signal $x(t)$ is to represent $x(t)$ at a discrete number of points, $t = nT$, where T is the sampling period, which is the time between samples, and n is an integer that establishes the time position of each sample. This process is illustrated in Figures 8-2a and b, which show both a set of samples of a continuous-time signal and a *sampling switch,* which is our initial model of a sampling device.

In order to extract samples of $x(t)$, the sampling switch closes briefly every T seconds. This yields samples that have a value of $x(t)$ when the switch is closed and a value of zero when the switch is open. For the sampling process to be useful, we must be able to show that it is possible to sample in such a way that the signal $x(t)$ can be reconstructed from the samples. This is most easily accomplished in the frequency domain.

First, the sampled $x(t)$, denoted $x_s(t)$, is written

$$x_s(t) = x(t)p(t) \tag{8-1}$$

where $p(t)$, called the *sampling function,* models the action of the sampling switch as shown in Figure 8-2c. The sampling function $p(t)$ is assumed to be the periodic pulse train of period T illustrated in Figure 8-2d. We will derive the spectrum of $x_s(t)$, and from this spectrum we can choose appropriate values of T.

The first step in the development of the spectrum of $x_s(t)$ is to recognize that since the sampling function $p(t)$ is periodic, it can be represented by a Fourier series. In other words,

$$p(t) = \sum_{n=-\infty}^{\infty} C_n e^{+jn2\pi f_s t} \tag{8-2}$$

where C_n is the nth Fourier coefficient of $p(t)$ and is given by

$$C_n = \frac{1}{T} \int_{-T/2}^{T/2} p(t)e^{-jn2\pi f_s t} \, dt \tag{8-3}$$

In (8-2) and (8-3), f_s is the fundamental frequency of $p(t)$, which is also the sampling frequency, and is given by

$$f_s = \frac{1}{T} \quad \text{hertz} \tag{8-4}$$

Since $x_s(t)$ is the product of $x(t)$ and $p(t)$, we have

$$x_s(t) = \sum_{n=-\infty}^{\infty} C_n x(t)e^{+jn2\pi f_s t} \tag{8-5}$$

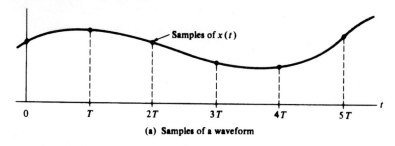

(a) Samples of a waveform

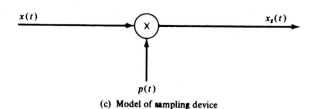

(b) Sampling device

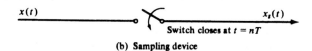

(c) Model of sampling device

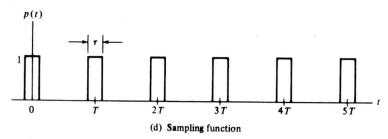

(d) Sampling function

FIGURE 8-2. The sampling operation.

We can now determine the spectrum of $x_s(t)$, which we shall denote by $X_s(f)$, by taking the Fourier transform of (8-5). The Fourier transform of $x_s(t)$ is defined by

$$X_s(f) = \int_{-\infty}^{\infty} x_s(t)e^{-j2\pi ft}\, dt \tag{8-6}$$

which, upon substitution of (8-5) for $x_s(t)$, becomes

$$X_s(f) = \int_{-\infty}^{\infty} \sum_{n=-\infty}^{\infty} C_n x(t)e^{+jn2\pi f_s t}e^{-j2\pi ft}\, dt \tag{8-7}$$

Interchanging the order of integration and summation yields

$$X_s(f) = \sum_{n=-\infty}^{\infty} C_n \int_{-\infty}^{\infty} x(t)e^{-j2\pi(f-nf_s)t}\, dt \tag{8-8}$$

From the definition of the Fourier transform,

$$X(f - nf_s) = \int_{-\infty}^{\infty} x(t)e^{-j2\pi(f-nf_s)t}\, dt \tag{8-9}$$

Thus the Fourier transform of the sampled signal can be written

$$X_s(f) = \sum_{n=-\infty}^{\infty} C_n X(f - nf_s) \tag{8-10}$$

which shows that the spectrum of the sampled continuous-time signal, $x(t)$, is composed of the spectrum of $x(t)$ plus the spectrum of $x(t)$ translated to each harmonic of the sampling frequency. Each of the translated spectra is multiplied by a constant, given by the corresponding term in the Fourier series expansion of $p(t)$. This is illustrated in Figure 8-3 for an assumed $X(f)$.

Also shown in Figure 8-3 is the amplitude response of an assumed reconstruction filter. If the sampled signal is filtered by the reconstruction filter, the output of the filter is, in the frequency domain, $C_0 X(f)$, and the time-domain signal is $C_0 x(t)$.

In Figure 8-3 the assumed $x(t)$ is *bandlimited*. In other words, $X(f)$ is assumed zero for $|f| \geq f_h$. It is clear from Figure 8-3 that if $X(f)$ is to be recoverable from $X_s(f)$, and consequently $x(t)$ from $x_s(t)$, then

$$f_s - f_h \geq f_h \tag{8-11}$$

or

$$f_s \geq 2f_h \quad \text{hertz} \tag{8-12}$$

Thus the minimum sampling frequency is $2f_h$ hertz, where f_h is the highest frequency in $x(t)$.

Thus we have derived the sampling theorem for low-pass *bandlimited* signals.

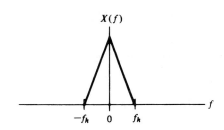

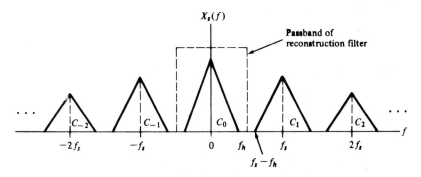

FIGURE 8-3. Spectrum of sampled signal.

Sampling Theorem: *A bandlimited signal x(t), having no frequency components above f_h hertz, is completely specified by samples that are taken at a uniform rate greater than $2f_h$ hertz. In other words, the time between samples is no greater than $1/2f_h$ seconds.*

The frequency $2f_h$ is known as the *Nyquist rate.*

Impulse-Train Sampling Model

The sampling function utilized in the preceding section was general except that it was assumed to be periodic. In practice, the time during which $p(t)$ is nonzero—that is, the pulse width—is small compared to the period T. In digital systems, where the sample is in the form of a number whose magnitude represents the value of the signal $x(t)$ at the sampling *instants,* the pulse width of the sampling function is infinitely small. Because an extremely narrow pulse is a situation *modeled* by an impulse and also because of mathematical simplifications that will result, we shall assume $p(t)$ to be composed of an infinite train of impulse functions of period T. Thus

$$p(t) = \sum_{n=-\infty}^{\infty} \delta(t - nT) \tag{8-13}$$

which is the sampling function illustrated in Figure 8-4. Note that, when using this model, the weight of the impulse carries the sample value.

The values of C_n can be computed from (8-3). This yields

$$C_n = \frac{1}{T} \int_{-T/2}^{T/2} \delta(t)e^{-jn2\pi f_s t}\, dt \tag{8-14}$$

which is

$$C_n = \frac{1}{T} = f_s \tag{8-15}$$

by the sifting property of the delta function. Thus C_n is equal to f_s for all n and the expression for the spectrum of the sampled $x(t)$, as given in general by (8-10), becomes

$$X_s(f) = f_s \sum_{n=-\infty}^{\infty} X(f - nf_s) \tag{8-16}$$

The effect of sampling with an impulse train is illustrated in Figure 8-5. It should be noted that the effect is the same as illustrated in Figure 8-3 except that all translated spectra have the same amplitude.

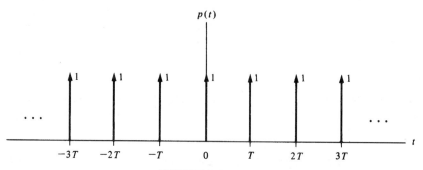

FIGURE 8-4. Impulse sampling function.

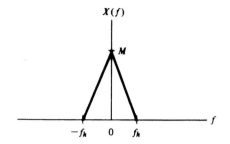

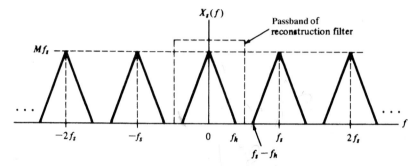

FIGURE 8-5.   Spectrum of impulse-sampled signal.

Once again $X(f)$ can be obtained from $X_s(f)$, and consequently $x(t)$ from $x_s(t)$, by using a low-pass filter for reconstruction of the original continuous-time waveform.

   The effect of sampling at too low a rate can also be seen from Figure 8-5. If a signal is sampled below the Nyquist rate,

$$f_s - f_h \leq f_h \tag{8-17}$$

and adjacent spectra overlap making it impossible to recover $x(t)$ by filtering. This effect is known as *aliasing* and is illustrated in Figure 8-6 for $X(f)$ assumed real.

   In a practical situation it is impossible to sample a signal and reconstruct the original signal from the samples with zero error because no practical signal is strictly bandlimited. However, in all practical signals there is some frequency beyond which the energy is negligible. This frequency is usually taken as the bandwidth. Another source of error arises from the nonexistence of ideal reconstruction filters. This means that some spectral space is required for filtering because of the finite slope of the reconstruction filter beyond the break frequency. All these considerations result in the fact that sampling is usually not performed at the Nyquist rate, but at some significantly higher frequency. Nonbandlimited signals having low-pass spectra are often sampled at approximately 10 times the frequency at which the amplitude spectrum is 3 dB down from its maximum value. The problems associated with reconstruction using practical filters will be examined in detail in the following section.

Data Reconstruction

We saw in the preceding section that the analog signal $x(t)$ can be reconstructed by passing the sampled signal $x_s(t)$ through a low-pass filter. We now investigate this process in more detail. It follows from (8-1) and (8-13) that $x_s(t)$ can be written

$$x_s(t) = x(t) \sum_{k=-\infty}^{\infty} \delta(t - kT) \tag{8-18}$$

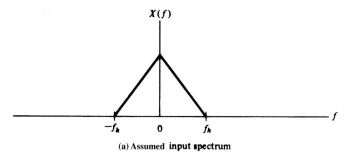

(a) Assumed **input spectrum**

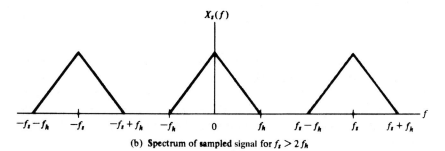

(b) **Spectrum of sampled signal for** $f_s > 2f_h$

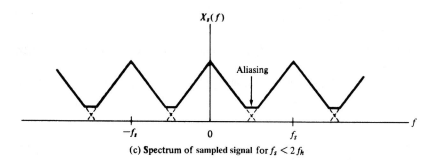

(c) **Spectrum of sampled signal for** $f_s < 2f_h$

FIGURE 8-6.   Illustration of aliasing error for real $X(f)$.

or

$$x_s(t) = \sum_{k=-\infty}^{\infty} x(kT)\delta(t - kT) \tag{8-19}$$

since $\delta(t - kT)$ is zero except at the sampling instants $t = kT$. The reconstruction filter, which is assumed linear and time invariant, has unit impulse response $h(t)$. The reconstruction filter output, $y(t)$, is given by the convolution

$$y(t) = \int_{-\infty}^{\infty} \sum_{k=-\infty}^{\infty} x(kT)\delta(\lambda - kT)h(t - \lambda)\, d\lambda \tag{8-20}$$

or, upon changing the order of summation and integration,

$$y(t) = \sum_{k=-\infty}^{\infty} x(kT) \int_{-\infty}^{\infty} \delta(\lambda - kT)h(t - \lambda)\, d\lambda \tag{8-21}$$

By using the sifting property of the delta function, we can evaluate the integral to obtain

$$y(t) = \sum_{k=-\infty}^{\infty} x(kT)h(t - kT) \tag{8-22}$$

This result is certainly not surprising and could have been obtained by inspection rather than by formal evaluation of the convolution integral. Since the filter input is a weighted sum of impulse functions, it follows that the filter output is a weighted sum of impulse responses. The weights are given by the sample values at the sampling instants. The impulse responses are determined by recognizing that the filter is time invariant so that input $x(kT)\delta(t - kT)$ yields output $x(kT)h(t - kT)$. The output $y(nT)$ is simply the sum of these individual impulse responses. The problem now is to determine the appropriate form of the impulse response $h(t)$ so that $y(t)$ is equal to, or is at least a good approximation to, the original analog signal $x(t)$.

Ideal Reconstruction Filter. Assuming that the signal $x(t)$ is sampled at a frequency exceeding the Nyquist rate, the ideal reconstruction filter is defined in Figure 8-5. As shown, if the sampled signal $x_s(t)$ is passed through an ideal low-pass filter, with bandwidth greater than f_h but less than $f_s - f_h$ and with a passband amplitude response of T, the filter output is $x(t)$. We choose the bandwidth of the reconstruction filter to be $0.5f_s$. This defines the ideal reconstruction filter. The transfer function of this ideal reconstruction filter is therefore

$$H(f) = \begin{cases} T, & |f| < 0.5f_s \\ 0, & \text{otherwise} \end{cases} \tag{8-23}$$

as shown in Figure 8-7a. It follows that the impulse response of the ideal reconstruction filter is given by

$$h(t) = T \int_{-f_s/2}^{f_s/2} e^{j2\pi ft} \, df \tag{8-24}$$

which is

$$h(t) = \frac{T}{j2\pi t} (e^{j\pi f_s t} - e^{-j\pi f_s t}) \tag{8-25}$$

It follows that $h(t)$ can be written

$$h(t) = \frac{\sin \pi f_s t}{\pi f_s t} \tag{8-26}$$

or

$$h(t) = \text{sinc } f_s t \tag{8-27}$$

Substituting (8-27) into (8-22) yields, after noting that $y(t) = x(t)$ for the ideal reconstruction filter,

$$x(t) = \sum_{k=-\infty}^{\infty} x(kT) \text{ sinc } f_s(t - kT) \tag{8-28}$$

A more convenient form for this expression, which is often referred to as an interpolation formula, is

$$x(t) = \sum_{k=-\infty}^{\infty} x(kT) \text{ sinc}\left(\frac{t}{T} - k\right) \tag{8-29}$$

Equation (8-29) shows that the original data signal $x(t)$ can be reconstructed by weighting each sample by a sinc function centered at the sample time and summing. This operation is illustrated in Figure 8-7b.

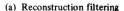

$$\sum_{n=-\infty}^{\infty} x(nT)\delta(t-nT)$$

Ideal reconstruction filter

$$H(f) = \begin{cases} T, & |f| \leq 0.5\,f_s \\ 0, & \text{otherwise} \end{cases}$$

$x(t)$

(a) Reconstruction filtering

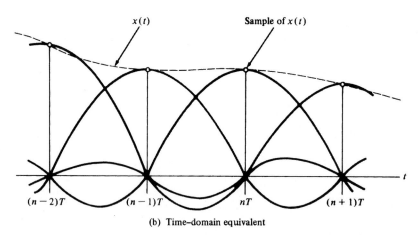

$x(t)$ Sample of $x(t)$

$(n-2)T$ $(n-1)T$ nT $(n+1)T$

(b) Time–domain equivalent

FIGURE 8-7.   Ideal reconstruction.

The reconstruction filter is clearly noncausal, and the unit impulse response is not time limited. These two observations have two serious implications. Since the ideal reconstruction filter is noncausal, it cannot be used for real-time applications. Also, since $h(t)$ is not time limited, an infinite number of impulse responses must be used for interpolating values between samples if exact results are to be obtained. The technique is still conceptually useful, however. Many non-real-time applications arise. For example, suppose you are given a computer disk of sample values and wish to interpolate between samples, which is effectively an increase in the sampling frequency. This is clearly not a real-time application. Also, since $h(t)$ decays as $1/t$, the terms many samples distant from the point being interpolated have little effect on the interpolated value and may therefore be neglected if small errors can be tolerated. If a value is to be interpolated between nT and $nT + T$, and l samples each side of the value to be interpolated are to be used, we have

$$x(t) \approx y(t) = \sum_{k=n-l+1}^{n+l} x(kT)\,\text{sinc}\left(\frac{t}{T} - k\right), \qquad nT < t < nT + T \tag{8-30}$$

Next we consider two simple examples, after which we consider an interpolation technique in which $h(t)$ is still noncausal but of finite extent.

EXAMPLE 8-1 _____

This example illustrates the effect of sampling a periodic signal. Consider the simple signal

$$x(t) = 6 \cos 2\pi(5)t \tag{8-31}$$

sampled at 7 and 14 Hz. Since the highest frequency (the only frequency in this case) is 5 Hz, we will see the effect of sampling a signal at both a frequency less than and greater than twice the highest frequency in $x(t)$. Since

$$X(f) = 3\delta(f - 5) + 3\delta(f + 5) \qquad (8\text{-}32)$$

it follows from (8-16) that the spectrum of the sampled $x(t)$ is given by

$$X_s(f) = 3f_s \sum_{n=-\infty}^{\infty} [\delta(f - 5 - nf_s) + \delta(f + 5 - nf_s)] \qquad (8\text{-}33)$$

The spectrum of $x(t)$ is shown in Figure 8-8a. The spectrum of the sampled signal is shown in Figure 8-8b for a sampling frequency of 7 Hz and in Figure 8-8c for a sampling frequency of 14 Hz. Both spectra extend from $f = -\infty$ to $f = +\infty$, but are illustrated in Figure 8-8 only in the range $|f| \leq 33$ Hz. The assumed reconstruction filter, $H(f)$, is an ideal low-pass filter with a bandwidth of

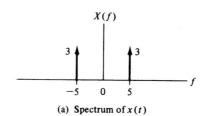

(a) Spectrum of $x(t)$

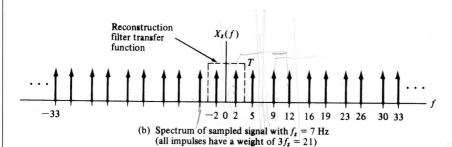

(b) Spectrum of sampled signal with $f_s = 7$ Hz
(all impulses have a weight of $3f_s = 21$)

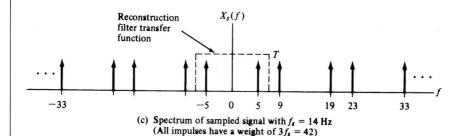

(c) Spectrum of sampled signal with $f_s = 14$ Hz
(All impulses have a weight of $3f_s = 42$)

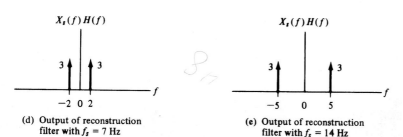

(d) Output of reconstruction filter with $f_s = 7$ Hz

(e) Output of reconstruction filter with $f_s = 14$ Hz

FIGURE 8-8. Sampling a single sinusoid.

0.5f_s and an amplitude response of T. Thus the output of the low-pass filter has the spectrum shown in Figure 8-8d for a sampling frequency of 7 Hz and has the spectrum shown in Figure 8-8e for a sampling frequency of 14 Hz.

Figure 8-8d illustrates the effect of aliasing arising from using a sampling frequency less than twice the highest frequency in $x(t)$. The result is a signal having the proper amplitude but the incorrect frequency. (The student should be convinced that the frequency of the component at the output of the reconstruction filter is always the sampling frequency minus the frequency of the input signal.)

Figure 8-8e illustrates the effect of sampling properly at a frequency greater than twice the highest frequency of $x(t)$. The output of the reconstruction filter is identical to the original signal.

EXAMPLE 8-2

In this example we consider a nonperiodic signal for $x(t)$ so that $X(f)$ has a continuous spectrum. We also assume that $X(f)$ is real.

The spectrum of $x(t)$ is shown in Figure 8-9a. The highest frequency is 5 Hz, so that the minimum acceptable sampling frequency is 10 Hz. Once again we assume sampling frequencies of 7 and 14 Hz.

Equation (8-16) shows that the spectrum of the sampled signal is obtained by multiplying $X(f)$ by f_s and then reproducing $f_s X(f)$ about dc and all harmonics of the sampling frequency. This is shown in Figure 8-9b for a sampling frequency of 7 Hz and in Figure 8-9c for a sampling frequency of 14 Hz. For a sampling frequency of 7 Hz, overlap of the translated spectra (aliasing) occurs and $X_s(f)$ is found by summing spectra in those regions of spectral overlap. For a sampling frequency of 14 Hz, there is no spectral overlap with translated spectra since the sampling frequency exceeds the minimum value of 10 Hz.

As in the previous examples, the reconstruction filter is assumed to be an ideal low-pass filter with an amplitude response of T and a bandwidth of 0.5 f_s. The output spectrum of the reconstruction filter is shown in Figure 8-9d for a sampling frequency of 7 Hz and in Figure 8-9e for a sampling frequency of 14 Hz. The impact of aliasing is clear from a comparison of Figures 8-9d and e.

Linear Interpolation. If the sampling frequency is large compared with the Nyquist rate, i.e., $f_s \gg 2f_h$, linear interpolation can often be used to reconstruct a close approximation of the analog signal $x(t)$ from the sampled signal $x_s(t)$. The unit impulse response of a filter that performs linear interpolation is given by

$$h(t) = \Lambda\left(\frac{t}{T}\right) = \begin{cases} 1 - \dfrac{|t|}{T}, & |t| < T \\ 0, & \text{otherwise} \end{cases} \qquad (8\text{-}34)$$

The situation is shown in Figure 8-10a. The interpolation equation for the linear interpolator is

$$y(t) = \sum_{k=-\infty}^{\infty} x(kT)\Lambda(t - kT) \qquad (8\text{-}35)$$

For t between $t = nT - T$ and $t = nT$ a sketch will show that

$$y(t) = x(nT - T)\left(n - \frac{t}{T}\right) + x(nT)\left(\frac{t}{T}n + 1\right), \qquad (n-1)T \le t \le nT \qquad (8\text{-}36)$$

Note that only two sample values are necessary to interpolate values between samples. Linear interpolation is therefore very fast. While $y(t)$ is only an approximation to the analog signal $x(t)$, it is a very good approximation if the sampling frequency is much greater than the Nyquist rate $2f_h$.

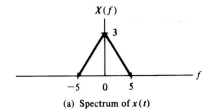

(a) Spectrum of $x(t)$

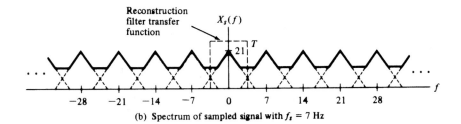

(b) Spectrum of sampled signal with $f_s = 7$ Hz

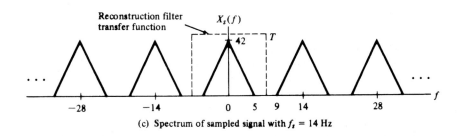

(c) Spectrum of sampled signal with $f_s = 14$ Hz

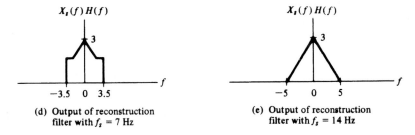

(d) Output of reconstruction filter with $f_s = 7$ Hz

(e) Output of reconstruction filter with $f_s = 14$ Hz

FIGURE 8-9.   Sampling a signal having a continuous spectrum.

The nature of the approximation can also be seen from Figure 8-10b, which illustrates an assumed spectrum of the sampled signal along with the amplitude response of the reconstruction filter, which is

$$H(f) = T \operatorname{sinc}^2 fT \qquad (8\text{-}37)$$

There are two sources of error. The first is that since $H(f)$ is not zero for $|f| > \frac{1}{2}f_h$, interference from translated spectra occurs. This is shown by the shaded areas in Figure 8-10b. The second source of error is that $H(f)$ is not exactly constant for $|f| < f_h$. It is clear that if f_h is small compared with f_s, the reconstruction error can be made negligible.

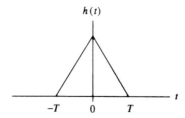

(a) Unit impulse response of linear interpolation filter

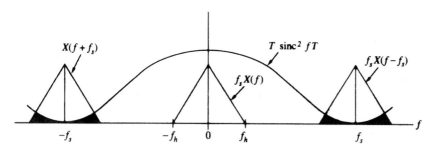

(b) Frequency domain view of linear interpolation

FIGURE 8-10. Linear interpolation.

Reconstruction with a Simple RC Filter. Both of the reconstruction techniques previously considered use noncausal reconstruction filters and are therefore not applicable to real-time applications. For many applications a practical causal reconstruction filter is the first-order RC filter shown in Figure 8-11a. The transfer function of the filter is

$$H(f) = \frac{T}{1 + j(f/f_3)} \tag{8-38}$$

where f_3 is the half-power, or 3-dB, frequency, which is $1/2\pi RC$. The inverse Fourier transform, which is the unit impulse response of the reconstruction filter, is

$$h(t) = 2\pi f_3 T e^{-2\pi f_3 t} u(t) \tag{8-39}$$

Substituting (8-39) into (8-22) yields the interpolation formula

$$y(t) = \sum_{k=-\infty}^{\infty} x_s(kT)(2\pi f_3 T)(e^{-2\pi f_3(t-kT)}u(t - kT) \tag{8-40}$$

The spectrum of the sampled signal and the amplitude response of the reconstruction filter are shown in Figure 8-11b. The error sources for this filter are the same as the error sources for the linear interpolation filter. It can be seen that the translated spectra resulting from the sampling operation are not completely attenuated, and therefore $y(t)$ is only an approximation to the analog signal $x(t)$. This is shown by the shaded area. In addition, $H(f)$ is not exactly constant for $|f| < f_h$. The approximation is very good, however, if f_3 is significantly greater than f_h, the bandwidth of the analog signal, and if $f_s \gg f_3$, so that the translated spectra are greatly attenuated.

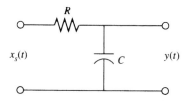

(a) Simple first-order lowpass reconstruction filter

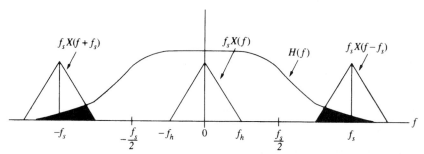

(b) Simple first-order lowpass reconstruction filter

FIGURE 8-11. Reconstruction with a simple first-order filter.

Quantizing and Encoding

The process of quantizing and encoding is illustrated in Figure 8-12. To quantize a sample value is to round it to the nearest of a finite set of permissible values. Encoding is accomplished by representing each of the permissible sample values by a digital word of fixed wordlength. For a binary representation, the number of quantizing levels q and the digital wordlength n are related by

$$q = 2^n \qquad (8\text{-}41)$$

Quantization level number	Encoded output
7	111
6	110
5	101
4	100
3	011
2	010
1	001
0	000

FIGURE 8-12. Quantizing and encoding.

For example, in Figure 8-12 we have eight quantizing levels, each of which is uniquely specified by a three-bit word.

Since the quantized value is the only value retained after sample values are quantized, errors are induced in the quantizing process that cannot be removed by additional processing. A quantitative measure of this error is easily derived. Since we would normally "decode" a quantizing level as the center of that level, the maximum error induced by quantizing a sample is $\pm\frac{1}{2}S$, where S is the width of a quantizing interval. Assume that there are a large number of quantizing intervals, resulting in a small value for S. Then, for *most* quantizing intervals, the signal $x(t)$ will be nearly linear within the quantizing interval, as shown in Figure 8-13a. For the system of interest it is the samples that are quantized and not the continuous-time signal $x(t)$. However, since sampling and reconstruction can be accomplished without error for bandlimited signals, the quantizer is the only error source in our A/D converter model. Thus we can evaluate the error by simply determining the error resulting from quantizing a continuous-time signal. This error is shown in Figure 8-13b, in which $2t_1$ is the time that the continuous-time signal $x(t)$ remains within the quantizing interval. The mean-square error, E, is given by

$$E = \frac{1}{2t_1} \int_{-t_1}^{t_1} \epsilon^2(\alpha)\, d\alpha = \frac{1}{t_1} \int_0^{t_1} \epsilon^2(\alpha)\, d\alpha \tag{8-42}$$

Since the error is given by

$$\epsilon(t) = \frac{S}{2t_1} t \tag{8-43}$$

we have

$$E = \frac{1}{t_1} \int_0^{t_1} \left(\frac{S}{2t_1}\right)^2 \alpha^2\, d\alpha = \frac{S^2}{12} \tag{8-44}$$

which is not a function of t_1. Since E is a mean-square value of a signal, it has dimensions of watts and is therefore interpreted as a noise power.

The quantizing step size S is related to the *dynamic range* of the A/D converter. The dynamic range D is the range of variation of the input signal $x(t)$. This range is defined as

$$D = \max[x(t)] - \min[x(t)] \tag{8-45}$$

Since there are $q = 2^n$ quantizing levels, the quantizing step size is given by

$$S = \frac{D}{2^n} = D2^{-n} \tag{8-46}$$

Substituting (8-46) into (8-44) yields the mean-square error

$$E = \frac{D^2}{12} 2^{-2n} \tag{8-47}$$

It can be seen that the mean-square error decreases exponentially with the A/D converter wordlength, n.

Equation (8-47) is not always a good approximation for the mean-square error, even when the number of quantization levels is large. If a signal stays at a given level for significant time intervals, (8-47) is not valid because it ignores the error contribution at those values. Square waves and constant signals are examples.

A quantitative measure of the quality of a quantized signal is the signal-to-noise ratio (SNR) at the quantizer output. The SNR is defined as the ratio of the signal power to the noise power. The approximation is usually made that the signal power at the quantizer output is identical to the signal power at

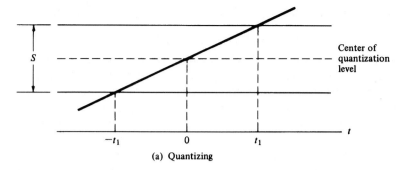

(a) Quantizing

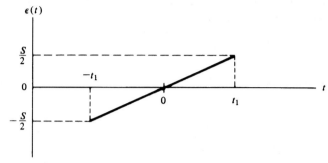

(b) Quantizing error

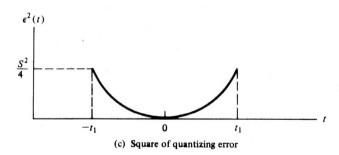

(c) Square of quantizing error

FIGURE 8-13.   Calculation of quantizing error.

the quantizer input. This approximation is very good when the number of quantizing levels or, equivalently, the wordlength, is large. If the signal power is denoted P_s, the SNR is

$$\text{SNR} = \frac{P_s}{E} \tag{8-48}$$

Using (8-47) for E yields

$$\text{SNR} = \frac{P_s}{(D^2/12)2^{-2n}} = 12 P_s D^{-2} 2^{2n} \tag{8-49}$$

The SNR is usually expressed in decibels, defined as $10 \log_{10} (\text{SNR})$, where the logarithm is the base 10 logarithm. Thus

$$(\text{SNR})_{\text{dB}} = 10 \log_{10}(12) + 10 \log_{10} P_s - 20 \log_{10} D + 20n \log_{10}(2) \tag{8-50}$$

or

$$(\text{SNR})_{\text{dB}} = 10.79 + 6.02n + 10 \log_{10} P_s - 20 \log_{10} D \qquad (8\text{-}51)$$

The preceding expression shows that the dynamic range, D, should be matched to the signal. If D is larger than necessary to accommodate the signal, the quantizer output SNR is reduced. An important observation to be drawn from (8-51) is that the SNR increases approximately 6 dB for each bit increase in wordlength. This result is independent of the signal $x(t)$.

EXAMPLE 8-3

In this example we explore the relationship between quantizing noise and wordlength for a sinusiodal signal. Assume that the signal

$$x(t) = A \cos \omega t \qquad (8\text{-}52)$$

is sampled and quantized. The signal power is

$$P_s = \frac{1}{T_p} \int_0^{T_p} (A \cos \omega t)^2 \, dt = \frac{A^2}{2} \qquad (8\text{-}53)$$

where T_p is the period of the waveform. Thus

$$10 \log_{10} P_s = 10 \log_{10} \frac{A^2}{2} = 20 \log_{10} A - 3.01 \qquad (8\text{-}54)$$

The dynamic range for the sinusoidal signal is the peak-to-peak value of the signal. Therefore

$$D = 2A \qquad (8\text{-}55)$$

and

$$20 \log_{10} D = 20 \log_{10} 2A = 6.02 + 20 \log_{10} A \qquad (8\text{-}56)$$

Substituting (8-54) and (8-56) into (8-51) yields

$$(\text{SNR})_{\text{dB}} = 10.79 + 6.02n + 20 \log_{10} A - 3.01 - 6.02 - 20 \log_{10} A \qquad (8\text{-}57)$$

which reduces to

$$(\text{SNR})_{\text{dB}} = 1.76 + 6.02n \qquad (8\text{-}58)$$

Thus the signal-to-noise ratio at the output of the A/D converter increases by approximately 6 dB for each added bit of wordlength. Equation (8-58) is illustrated in Figure 8-14. The importance of using as large a wordlength as possible is obvious. The effect of *not* using the full range of the quantizer is an effective decrease in q and thus in the wordlength, n. Therefore, it is also important that the peak-to-peak value of the signal being processed span the full range (qS) of the quantizer.

Equation (8-58) was derived by assuming a sinusoidal signal. It should be pointed out that a similar result applies to any test signal, since in general, the signal-to-noise ratio increases 6.02 dB for each increment in the wordlength. The bias, 1.76 dB in this case, will change as the waveshape of the test signal is changed (see Problems 8-5 and 8-6).

In the remainder of this book, the assumption will be made that quantizing errors are negligible. Thus no distinction will be made between sampled-data (nonquantized) and digital (quantized) signals. We simply consider the signals to be discrete-time signals.

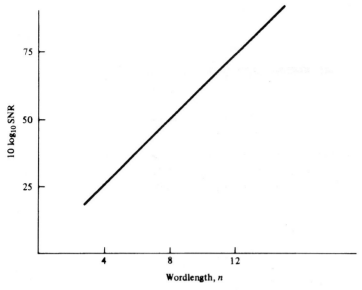

FIGURE 8-14.   Quantizing signal-to-noise ratio assuming a sinusoidal signal.

8-3 The z-Transform

The z-transform is the basic tool for both the analysis and synthesis of discrete-time systems. In the following sections we develop the basic properties of the z-transform and show how to use the z-transform for the characterization and analysis of discrete-time systems. In Chapter 9 we will turn our attention to the more interesting synthesis problem.

Definition of the z-Transform

It has been shown that a suitable model of a sampled signal is

$$x_s(t) = \sum_{n=-\infty}^{\infty} x(t)\delta(t - nT) \tag{8-59}$$

Prior to defining the z-transform, we will make two minor modifications to the preceding expression. First, since $\delta(t - nT)$ is identically zero except at the sampling instants, $t = nT$, $x(t)$ can be replaced by $x(nT)$ if $x(t)$ is continuous at $t = nT$. We will also assume that

$$x(t) \equiv 0, \qquad t < 0 \tag{8-60}$$

which establishes the time reference. Thus (8-59) will be written

$$x_s(t) = \sum_{n=0}^{\infty} x(nT)\delta(t - nT) \tag{8-61}$$

Taking the Laplace transform yields

$$X_s(s) = \int_0^{\infty} \sum_{n=0}^{\infty} x(nT)\delta(t - nT)e^{-st}\, dt \tag{8-62}$$

which is, upon interchanging integration and summation,

$$X_s(s) = \sum_{n=0}^{\infty} x(nT) \int_0^{\infty} \delta(t - nT)e^{-st}\, dt \qquad (8\text{-}63)$$

Using the sifting property of the delta function, this integrates to

$$X_s(s) = \sum_{n=0}^{\infty} x(nT)e^{-snT} \qquad (8\text{-}64)$$

Finally, defining the complex variable z as

$$z = e^{sT} \qquad (8\text{-}65)$$

yields

$$X(z) = \sum_{n=0}^{\infty} x(nT)z^{-n} \qquad (8\text{-}66)$$

In (8-66) $X(z)$ is known as the z-transform of the sequence of samples, $x(nT)$. Note that the subscript s, which has been used to denote a sampled quantity, has been deleted. This can be done without confusion since only sample values are used to compute the z-transform. Therefore, the subscript s is redundant when the argument of the function is z.

Within a sequence of samples, a given sample can be represented by an ordered pair of numbers. One number in the ordered pair is used to represent the value of the sample and the other number in the ordered pair is used to specify the occurrence time of the sample. Each term in the summation of (8-66) has this form. The coefficient, $x(nT)$, denotes the sample value and z^{-n} denotes that the sample occurs n sample periods after the $t = 0$ reference. It is also instructive to recognize that e^{sT} is simply the T-second time shift operator discussed in Chapter 5. In (8-64) the sample value $x(nT)$ is multiplied by e^{-snT}, which places that sample value at $t = nT$. Thus the parameter z is simply shorthand notation for the Laplace time shift operator. As an example, $128.7z^{-37}$ denotes a sample, having value 128.7, which occurs 37 sample periods after the $t = 0$ reference.

Recall from Chapter 5 that the Laplace variable s was given by

$$s = \sigma + j\omega \qquad (8\text{-}67)$$

where σ was a constant used to ensure convergence of the integral defining the Laplace transform and thus the existence of the transform itself. From (8-65),

$$z = e^{\sigma T}e^{j\omega T} \qquad (8\text{-}68)$$

so that the magnitude of z is given by

$$|z| = e^{\sigma T} \qquad (8\text{-}69)$$

Thus, the right-half s-plane, $\sigma > 0$, corresponds to $|z| > 1$, while the left-half s-plane, $\sigma < 0$, corresponds to $|z| < 1$. We see that the left-half s-plane maps into the interior of the unit circle in the z-plane and that the right-half s-plane maps outside the unit circle in the z-plane. This mapping of the Laplace variable s into the z-plane through $z = e^{sT}$ is illustrated in Figure 8-15.

Recall from Chapter 5 that functions of the Laplace variable s having poles with negative real parts (left-half plane poles) decay to zero as $t \to \infty$ in the time domain. In a similar manner, functions of z having poles with magnitudes less than one decay to zero in the time domain as $n \to \infty$. Poles on the $j\omega$ axis in the s-plane correspond to poles on the unit circle in the z plane, and imply time-domain functions that oscillate at a frequency determined by the angle of the pole. A function having a pole at the

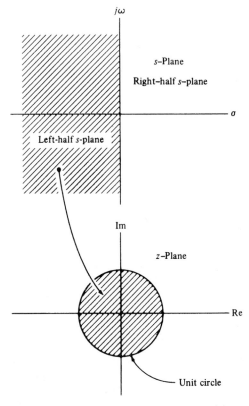

FIGURE 8-15. Mapping defined by $z = e^{sT}$.

origin in the s-plane is a unit step in the time domain and, as we will see in an example, a function having a pole in the z-plane at $z = 1$ corresponds to a sampled unit step in the time domain. This is not surprising since $s = 0$ corresponds to $z = 1$.

As mentioned at the beginning of this chapter, we could define the z-transform, and develop the appropriate theory, without reference to continuous-time functions and the s-plane. We could simply take (8-66) as the definition of the z-transform. As we saw in the previous paragraph, however, relating z to the Laplace variable s allows us to make good use of our previously developed theory of continuous-time signals and systems. We will continue to use this relationship in this chapter and in the two chapters that follow.

The following three examples illustrate the derivation of three very basic z-transform pairs. The results of these three examples will serve as building blocks for deriving other z-transform relationships.

EXAMPLE 8-4 _____

The unit pulse sequence is defined by the sample values

$$x(nT) = \begin{cases} 1, & n = 0 \\ 0, & n \neq 0 \end{cases} \triangleq \delta(n) \tag{8-70}$$

and is illustrated in Figure 8-16a. Substituting $x(nT)$ into the defining equation for the z-transform, (8-66), yields

$$X(z) = 1 + 0z^{-1} + 0z^{-2} + \cdots \tag{8-71}$$

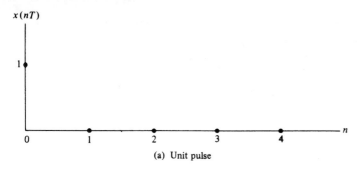

(a) Unit pulse

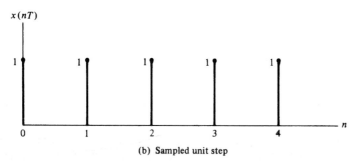

(b) Sampled unit step

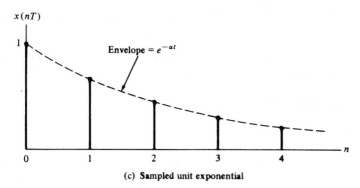

(c) Sampled unit exponential

FIGURE 8-16. Elementary sample sequences.

Thus

$$X(z) = 1 \qquad (8\text{-}72)$$

This is an extremely important result. Note that from (8-70) we can write

$$\delta(n - k) = \begin{cases} 1, & n = k \\ 0, & n \neq k \end{cases} \qquad (8\text{-}73)$$

Thus

$$x(nT) = \sum_{k=-\infty}^{\infty} x(kT)\delta(n - k) \qquad (8\text{-}74)$$

This should remind us of the sifting property of the unit impulse function, $\delta(t)$, as defined in (1-48). We will see that the unit pulse sequence plays a similar role in discrete-time systems as the unit impulse function does in continuous-time systems.

EXAMPLE 8-5 _____

The unit step sample sequence is defined by the sample values

$$x(nT) = 1, \qquad n \geq 0 \tag{8-75}$$

and is illustrated in Figure 8-16b. Substitution into (8-66) yields

$$X(z) = \sum_{n=0}^{\infty} z^{-n} \tag{8-76}$$

We know that for $|x| < 1$

$$\sum_{n=0}^{\infty} x^n = \frac{1}{1-x} \tag{8-77}$$

Thus, with $x = z^{-1}$, we have[†]

$$X(z) = \sum_{n=0}^{\infty} z^{-n} = \frac{1}{1-z^{-1}}, \qquad |z| > 1 \tag{8-78}$$

It should be noted that the unit step sample sequence is defined such that $x(0) = 1$. Therefore, it is not simply a sampled $u(t)$ unless $u(0)$ is defined as 1, since $u(t)$ is not continuous at $t = 0$. The unit step sample sequence is, however, often denoted $u(n)$.

EXAMPLE 8-6 _____

The unit exponential sequence is defined by the sample values

$$x(nT) = e^{-\alpha nT}, \qquad \alpha > 0, \quad n \geq 0 \tag{8-79}$$

which is illustrated in Figure 8-16c and has the z-transform

$$X(z) = \sum_{n=0}^{\infty} e^{-\alpha nT} z^{-n} = \sum_{n=0}^{\infty} (e^{-\alpha T} z^{-1})^n \tag{8-80}$$

Using (8-77) with $x = e^{-\alpha T} z^{-1}$ yields

$$X(z) = \frac{1}{1 - e^{-\alpha T} z^{-1}}, \qquad |z| > e^{-\alpha T} \tag{8-81}$$

With the sampling period T and the parameter α fixed, $e^{-\alpha T}$ is a constant. Letting

$$K = e^{-\alpha T} \tag{8-82}$$

gives the z-transform pair

$$X(z) = \mathscr{Z}[K^n] = \frac{1}{1 - Kz^{-1}}, \qquad |z| > K \tag{8-83}$$

_____

[†]The sum $\sum_{n=0}^{\infty} z^{-n}$ converges absolutely to $1/(1 - z^{-1})$ outside the unit circle ($|z| > 1$). The function $1/(1 - z^{-1})$ is well behaved everywhere in the complex z-plane except at the pole $z = 1$. We take (8-78) as the definition of the z-transform of the sampled unit step. This result could have been derived rigorously by other methods beyond the scope of this text.

This is often the most convenient form to use. Equation (8-83) can be written

$$\frac{1}{1 - Kz^{-1}} = \frac{z}{z - K} \tag{8-84}$$

which shows that $X(z)$ has a zero at $z = 0$ and a pole at $z = K$. Thus only for $|K| < 1$ does K^n go to zero for large n.

A short table of z-transforms is given in Table 8-1. The first three transforms have been derived in the preceding examples. The remaining transform pairs are easily proved (see Problems 8-13 through 8-15).

EXAMPLE 8-7

The purpose of this example is to consider the z-transform of a sequence somewhat more complicated than the simple examples previously considered and to illustrate the use of MATLAB for determining z-transforms. The z-transform of the sequence

$$x(nT) = a^n \cos\left(\frac{n\pi}{2}\right) \tag{8-85}$$

TABLE 8-1
Short Table of z-Transforms

Transform Pair Number	Continuous-time Function $f(t)$ for $t > 0$	Sample Values $f(nT)$ for $n \geq 0$	z-Transform of $f(nT)$
1.	—	$f(nT) = \begin{cases} 1, & n = 0 \\ 0, & n \neq 0 \end{cases} \triangleq \delta(n)$	1
2.	1 (unit step)	1	$\dfrac{1}{1 - z^{-1}}$
3.	e^{-at}	$e^{-anT} = (e^{-aT})^n = K^n$	$\dfrac{1}{1 - e^{-aT}z^{-1}} = \dfrac{1}{1 - Kz^{-1}}$
4.	t	nT	$\dfrac{Tz^{-1}}{(1 - z^{-1})^2}$
5.†	te^{-at}	nTe^{-anT}	$\dfrac{Te^{-aT}z^{-1}}{(1 - e^{-aT}z^{-1})^2}$
6.†	$\sin bt$	$\sin bnT$	$\dfrac{(\sin bT)z^{-1}}{1 - 2(\cos bT)z^{-1} + z^{-2}}$
7.†	$\cos bt$	$\cos bnT$	$\dfrac{1 - (\cos bT)z^{-1}}{1 - 2(\cos bT)z^{-1} + z^{-2}}$
8.†	$e^{-at}\sin bt$	$e^{-anT}\sin bnT$	$\dfrac{e^{-aT}(\sin bT)z^{-1}}{1 - 2e^{-aT}(\cos bT)z^{-1} + e^{-2aT}z^{-2}}$
9.†	$e^{-at}\cos bt$	$e^{-anT}\cos bnT$	$\dfrac{1 - e^{-aT}(\cos bT)z^{-1}}{1 - 2e^{-aT}(\cos bT)z^{-1} + e^{-2aT}z^{-2}}$

†In transforms 5 through 9, e^{-aT} and bT can be replaced by constants, K_1 and K_2, respectively, as was done in transform 3. Convergence of the z-transform requires $|K_1| < 1$.

is defined by

$$X(z) = \sum_{n=0}^{\infty} a^n \cos\left(\frac{n\pi}{2}\right) z^{-n} \tag{8-86}$$

Note that $\cos(n\pi/2) = 0$ for n odd and is ± 1, depending upon the value of n, for n even. It follows that $X(z)$ can be written in the form

$$X(z) = \sum_{k=0}^{\infty} a^{2k}(-1)^k z^{-2k} \tag{8-87}$$

which can be expressed

$$X(z) = \sum_{k=0}^{\infty} (-a^2 z^{-2})^k \tag{8-88}$$

Assuming convergence of the series[†] we obtain

$$X(z) = \frac{1}{1 + a^2 z^{-2}} \tag{8-89}$$

Note this also follows from Table 8-1 [entry 9] by replacing e^{-aT} by a and bT by $\pi/2$.

MATLAB Application

The MATLAB Symbolic Math Toolbox provides routines for performing z-transforms and inverse z-transforms. The following code illustrates the technique:

```
EDU» c8ex7

syms a n z                    % Declare symbolic
xn = a^n*cos(n*pi/2);         % Define x(n)
xz = ztrans(xn,n,z);          % Determine X(z)
xz                            % Display result

xz =

z^2/(a^2+z^2)
```

This result, although written in terms of positive powers of z instead of negative powers of z, agrees with the previously obtained result. MATLAB always expresses results in terms of positive powers of z.

Linearity

The z-transformation is a linear operation. In other words, if A and B are constants,

$$\sum_{n=0}^{\infty} [Ax_1(nT) + Bx_2(nT)] z^{-n} = AX_1(z) + BX_2(z) \tag{8-90}$$

[†]The series in question converges for $|z| > |\alpha|$.

where $X_1(z)$ and $X_2(z)$ are the z-transforms of $x_1(nT)$ and $x_2(nT)$, respectively. This is easily seen by recognizing that the left-hand side of (8-90) can be written

$$\sum_{n=0}^{\infty} [Ax_1(nT) + Bx_2(nT)]\, z^{-n} = A \sum_{n=0}^{\infty} x_1(nT)z^{-n} + B \sum_{n=0}^{\infty} x_2(nT)z^{-n} \qquad (8\text{-}91)$$

By definition, the two sums on the right-hand side of (8-91) are $X_1(z)$ and $X_2(z)$.

Initial Value and Final Value Theorems

The initial value and the final value theorems are useful in that they provide quick insight into the behavior of a time sequence, $x(nT)$, without having to compute the inverse z-transform of $X(z)$. When the inverse z-transform of $X(z)$ is computed, the initial value and final value theorems provide a partial check on the results.

The initial value theorem states that

$$x(0) = \lim_{z \to \infty} X(z) \qquad (8\text{-}92)$$

This result is easily derived. By definition

$$X(z) = \sum_{n=0}^{\infty} x(nT)z^{-n} = x(0) + \sum_{n=1}^{\infty} x(nT)z^{-n} \qquad (8\text{-}93)$$

As $z \to \infty$, the summation on the right vanishes and (8-92) results.

The final value theorem states that

$$x(\infty) = \lim_{z \to 1} (1 - z^{-1})X(z) \qquad (8\text{-}94)$$

The proof of this theorem is more difficult than the proof of the initial value theorem, and we shall not take the time and space to construct the proof here. However, (8-94) can be easily justified. If $X(z)$ has any of its poles outside the unit circle, $X(z)$ corresponds to an unstable function and $x(\infty) = \infty$. If $X(z)$ has all its poles inside the unit circle, $x(nT)$ represents samples of a function that decays with time and $x(\infty) = 0$. If $X(z)$ has *simple* poles on the unit circle but not at $z = 1$, the resulting time response is oscillatory and $x(\infty)$ is not defined, although it is bounded by the limits of the oscillation. By continuing this argument we see that a nonzero steady-state constant value must result from a simple pole in the z-plane at $z = 1$. Thus assume that $X(z)$ has a simple pole at $z = 1$ and decompose $X(z)$ as

$$X(z) = \frac{K}{1 - z^{-1}} + G(z) \qquad (8\text{-}95)$$

where all the poles of $G(z)$ lie inside the unit circle. Clearly, by the preceding argument, K is the steady-state value of $X(z)$. Applying (8-94) yields

$$x(\infty) = \lim_{z \to 1} (1 - z^{-1})X(z) = K + \lim_{z \to 1} (1 - z^{-1})G(z) = K \qquad (8\text{-}96)$$

which is the expected result.

Several interesting proofs of the final value theorem are given in the literature.[†]

[†]As an example of two very different proofs of the final value theorem, the interested student should refer to S. R. Atre, "An Alternate Derivation of Final and Initial Value Theorems in the z Domain," *Proceedings of the IEEE,* Vol. 63, No. 3, March 1975, pp. 537–538, and J. R. Ragazzini and G. F. Franklin, *Sampled-Data Control Systems* (New York: McGraw-Hill, 1958), pp. 61–63.

Inverse z-Transform

The z-transform of a sample sequence is written, by definition,

$$X(z) = x(0) + x(T)z^{-1} + x(2T)z^{-2} + \cdots \tag{8-97}$$

If we can manipulate $X(z)$ into this form, the sample values, $x(nT)$, can be determined by inspection. This is easily accomplished by long division when $X(z)$ is expressed as a ratio of polynomials in z. Before the long division is carried out, it is usually most convenient to arrange both the numerator and the denominator in ascending powers of z^{-1}. A simple example will illustrate the method.

EXAMPLE 8-8

The purpose of this example is to illustrate the determination of the inverse z-transform using long division. Assume that

$$X(z) = \frac{z^2}{(z - 1)(z - 0.2)} \tag{8-98}$$

First $X(z)$ is written in the form

$$X(z) = \frac{z^2}{z^2 - 1.2z + 0.2} \tag{8-99}$$

which is, after multiplying numerator and denominator by z^{-2},

$$X(z) = \frac{1}{1 - 1.2z^{-1} + 0.2z^{-2}} \tag{8-100}$$

The division is illustrated below:

$$
\begin{array}{r}
1 + 1.2z^{-1} + 1.24z^{-2} + 1.248z^{-3} \cdots \\
\hline
1 - 1.2z^{-1} + 0.2z^{-2} \enclose{longdiv}{1 } \\
1 - 1.2z^{-1} + 0.2z^{-2} \\
\hline
1.2z^{-1} - 0.20z^{-2} \\
1.2z^{-1} - 1.44z^{-2} + 0.240z^{-3} \\
\hline
1.24z^{-2} - 0.240z^{-3} \\
1.24z^{-2} - 1.488z^{-3} + 0.248z^{-4} \\
\hline
1.248z^{-3} \quad \cdots \\
\cdots \qquad \cdots
\end{array}
$$

Thus

$$X(z) = 1 + 1.2z^{-1} + 1.24z^{-2} + 1.248z^{-3} + \cdots \tag{8-101}$$

from which

$$x(0) = 1 \tag{8-102a}$$

$$x(T) = 1.2 \tag{8-102b}$$

$$x(2T) = 1.24 \tag{8-102c}$$

$$x(3T) = 1.248 \tag{8-102d}$$

The division can obviously be continued to give as many terms as desired. Although long division is generally the easiest method for inverting a given $X(z)$, the method has the disadvantage of not yielding the general sample value term, $x(nT)$, in closed form.

Note that from the initial value theorem

$$x(0) = \lim_{z \to \infty} X(z) = 1 \qquad (8\text{-}103)$$

which checks with our previous result. The final value theorem yields

$$x(\infty) = \lim_{z \to 1} \frac{z - 1}{z} X(z) = \lim_{z \to 1} \frac{z}{z - 0.2} = 1.25 \qquad (8\text{-}104)$$

MATLAB Application

Long division can be accomplished in MATLAB using the command `dimpulse` in the Signals and Systems Toolbox included in the Student Edition of MATLAB. We will see later that we are simply determining the unit pulse response of a system determined by the arguments of the `dimpulse` function. The following MATLAB code illustrates the application to the example considered here.

```
EDU» c8ex8
num = [1 0 0];                    % Define numerator
den = [1 -1.2 0.2];              % Define denominator
n = 8;                            % Display n terms
x = dimpulse(num,den,n);         % Compute implse response
x                                 % Display result

x =

    1.0000
    1.2000
    1.2400
    1.2480
    1.2496
    1.2499
    1.2500
    1.2500
```

We see that these results are in agreement with the previously computed results.

Although the long vision method of inverse z-transforming a function does not yield the general term in closed form, it is sometimes sufficient. For example, if the time sequence approaches an asymptote in a well-behaved manner, often one only needs to compute the first several values of the time sequence and then compute the asymptote by the final value theorem.

Partial-fraction expansion overcomes the disadvantage of the long-division method in that partial-fraction expansion yields the general term of the time expansion, $x(nT)$. The basic disadvantage of the partial-fraction-expansion method is that the denominator of $X(z)$ must be factored. The technique is exactly the same method as that used to find inverse Laplace transforms by partial-fraction expansion. The idea is to manipulate $X(z)$ into a form that can be inverse z-transformed by using Table 8-1.

Because of the form of the transforms given in Table 8-1, it is usually best to perform a partial fraction expansion of $H(z)/z$. As an alternative z^{-1} can be treated as the variable in the partial fraction expansion. We will, however, expand in terms of z in the following examples.

EXAMPLE 8-9

In this example, the use of partial-fraction expansion for performing inverse z-transform operations is illustrated. We use the same $X(z)$ that was used in the previous example so that

$$X(z) = \frac{1}{1 - 1.2z^{-1} + 0.2z^{-2}} = \frac{z^2}{(z - 1)(z - 0.2)} \tag{8-105}$$

First, we write

$$\frac{X(z)}{z} = \frac{z}{(z - 1)(z - 0.2)} = \frac{K_1}{z - 1} + \frac{K_2}{z - 0.2} \tag{8-106}$$

The constants K_1 and K_2 are given by

$$K_1 = \lim_{z \to 1}(z - 1)\frac{X(z)}{z} = 1.25 \tag{8-107a}$$

and

$$K_2 = \lim_{z \to 0.2}(z - 0.2)\frac{X(z)}{z} = -0.25 \tag{8-107b}$$

Thus

$$X(z) = \frac{1.25z}{z - 1} - \frac{0.25z}{z - 0.2} \tag{8-108}$$

or

$$X(z) = \frac{1.25}{1 - z^{-1}} - \frac{0.25}{1 - 0.2z^{-1}} \tag{8-109}$$

We can find the inverse transform of $X(z)$ by using transform pair 3 in Table 8-1. The first term has $K = 1$, which is a unit step sample sequence, and the second term has $K = 0.2$, which is a sampled exponential. Thus

$$x(nT) = 1.25 - 0.25(0.2)^n, \quad n \geq 0 \tag{8-110}$$

The values of $x(nT)$ are

$$x(0) = 1.25 - 0.25(0.2)^0 = 1 \tag{8-111a}$$

$$x(T) = 1.25 - 0.25(0.2)^1 = 1.2 \tag{8-111b}$$

$$x(2T) = 1.25 - 0.25(0.2)^2 = 1.24 \tag{8-111c}$$

$$x(3T) = 1.25 - 0.25(0.2)^3 = 1.248 \tag{8-111d}$$

It should be noted that these values agree with the values determined by long division in the preceding example. Also, note from (8-110) that

$$x(\infty) = 1.25 \tag{8-111e}$$

MATLAB Application

The MATLAB command residuez allows the residues, poles and the **k** vector[†] to be computed for a given $X(z)$. Consider the following MATLAB code:

```
EDU» c8ex9
b = 1;                          % Define numerator
a = [1 -1.2 0.2];               % Define denominator
[r,p,k] = residuez (b,a);       % Determine residues
r                               % Display residues


r =

    1.2500
   -0.2500


p                               % Display poles

p =

    1.0000
    0.2000


k                               % Display k vector

k =

    []
```

The residues and poles determined through the use of residuez allow (8-109) to be written directly. The inverse transform is then obvious.

It is interesting to continue. The MATLAB Symbolic Math Toolbox contains routines that allow us to take the inverse z-transform of functions of z. The following short MATLAB session illustrates the use of the command invztrans for performing inverse z-transforms. The routines in the Symbolic Math Toolbox for inverse z-transforms require that the polynomial coefficients be expressed as rational numbers rather than as floating-point numbers. A moments thought tells us that this is the only way that pole (and zero) locations can be exact.[‡] Thus 1.2 and 0.2 must be written as 6/5 and 1/5, respectively, or as any rational equivalent to these numbers.

```
EDU» c8ex9b

    syms n z                              % Declare symbolic
    xz = 1/(1-(6/5)*(z^(-1))+(1/5)*z^(-2));   % Define X(z)
```

[†]See the MATLAB manual for an explanation of the **k** vector. In this example, the **k** vector is empty. In the following chapter (Example 9-1) we will use residuez to compute residues for a case in which the **k** vector is not empty. This will illustrate the use of the information returned in the **k** vector.

[‡]Recall that 1.2 represents a number between $1.15+e$ and $1.25-e$, where e is an arbitrarly small positive number, to a precision of two significant figures. Thus 1.2 is not an exact quantity but 6/5 is an exact quantity.

```
xn = iztrans (xz,z,n);              % Compute x(n)
xn                                   % Display x(n)


xn =

5/4-1/4*(1/5)^n
```

This result is identical to (8-110) for non-negative values of n.

EXAMPLE 8-10

The purpose of this example is observe the effect of multiplying the $X(z)$ by z^{-2}. This will lead to the delay operator. Thus we consider

$$Y(z) = \frac{1}{z^2 - 1.2z + 0.2} = \frac{1}{(z - 1)(z - 0.2)} \tag{8-112}$$

The first step in performing the inverse z-transform is to divide both sides by z. This yields

$$\frac{Y(z)}{z} = \frac{1}{z(z - 1)(z - 0.2)} = \frac{K_1}{z} + \frac{K_2}{z - 1} + \frac{K_3}{z - 0.2} \tag{8-113}$$

from which

$$K_1 = \lim_{z \to 0} z \frac{Y(z)}{z} = \frac{1}{0.2} = 5 \tag{8-114a}$$

$$K_2 = \lim_{z \to 1} (z - 1) \frac{Y(z)}{z} = \frac{1}{0.8} = 1.25 \tag{8-114b}$$

and

$$K_3 = \lim_{z \to 0.2} (z - 0.2) \frac{Y(z)}{z} = -6.25 \tag{8-114c}$$

Thus

$$Y(z) = 5 + 1.25 \frac{z}{z - 1} - 6.25 \frac{z}{z - 0.2} \tag{8-115}$$

or

$$Y(z) = 5 + 1.25 \frac{1}{1 - z^{-1}} - 6.25 \frac{1}{1 - 0.2z^{-1}} \tag{8-116}$$

We can now inverse transform using transform pairs 1 and 3 in Table 8-1. This yields

$$y(nT) = 5\delta(n) + 1.25 - 6.25(0.2)^n, \qquad n \geq 0 \tag{8-117}$$

The student should check this result by using long division. Upon doing so, it will be noted that $y(0) = 0$, $y(T) = 0$, and $y(2T) = 1$. This is obvious from an inspection of $Y(z)$, since the leading term in the division process is z^{-2}. Simple observations of this type are useful in that computational errors are often revealed.

MATLAB Application

The MATLAB code for executing the inverse z-transform of $Y(z)$ follows.

```
EDU» c8ex10
syms n z                                    % Declare symbolic
xz1 = 1/(1+(6/5)*(z^(-1))+(1/5)*z^(-2));    % Define X1(z)
xz2 = z^(-2);                               % Define X2(z)
xz = xz1*xz2;                               % Determine X(z)
xn = iztrans(xz,z,n);                       % Compute x(n)
xn                                          % Display x(n)


xn =

5*Delta(n)+5/4-25/4*(1/5)^n
```

which is equivalent to (8-117) for non-negative values of n.

The essential difference between Examples 8-9 and 8-10 is that division by z to form $Y(z)/z$ introduced an additional pole, and therefore an additional term in the partial-fraction expansion, in Example 8-10.

The preceding examples treated partial-fraction expansion of z-domain transfer functions having first-order poles. Partial-fraction expansion of z-domain transfer functions with higher-order poles proceeds in exactly the same manner as partial-fraction expansion of s-domain transfer functions with higher-order poles.

Delay Operator

The delay operation is of fundamental importance in the analysis and synthesis of discrete-time or digital systems, as we shall shortly see. We now show that if the time sequence $[x(nT)]$ is delayed by K sample periods, the effect in the z-domain is to multiply $X(z)$ by z^{-K}.

This is easily proved. Since

$$X(z) = \sum_{n=0}^{\infty} x(nT)z^{-n} \tag{8-118}$$

the z-transform of the sequence $\{x(nT - KT)\}$ is

$$\mathscr{L}[x(nT - KT)] = \sum_{n=0}^{\infty} x(nT - KT)z^{-n} \tag{8-119}$$

Letting $m = n - K$ yields

$$\mathscr{L}\{x(nT - KT)\} = \sum_{m=-K}^{\infty} x(mT)z^{-m-K} \tag{8-120}$$

Since $x(mT)$ is assumed zero for $m < 0$, we have

$$\mathscr{L}\{x(nT - KT)\} = \sum_{m=0}^{\infty} x(mT)z^{-m-K} \tag{8-121}$$

or

$$\mathscr{L}\{x(nT - KT)\} = z^{-K} \sum_{m=0}^{\infty} x(mT)z^{-m} = z^{-K}X(z) \tag{8-122}$$

The delay operation is illustrated in Figure 8-17, where it is clear that delay by K sample periods is equivalent to multiplication by z^{-K}.

EXAMPLE 8-11

We shall now use the delay operator and $x(t)$ from Example 8-9 to work Example 8-10. Note that $Y(z)$ from Example 8-10 can be written

$$Y(z) = \left[\frac{z^2}{z^2 - 1.2z + 0.2} \right] z^{-2} \tag{8-123}$$

However, the term in brackets is $X(z)$ from Example 8-9. Thus

$$Y(z) = X(z)z^{-2} \tag{8-124}$$

Inverse z-transforming yields

$$y(nT) = x[(n - 2)T] \tag{8-125}$$

Using the result of Example 8-9 yields

$$y(nT) = 1.25 - 0.25(0.2)^{n-2}, \qquad n \geq 2 \tag{8-126}$$

The student should verify that this agrees with the result of Example 8-10.

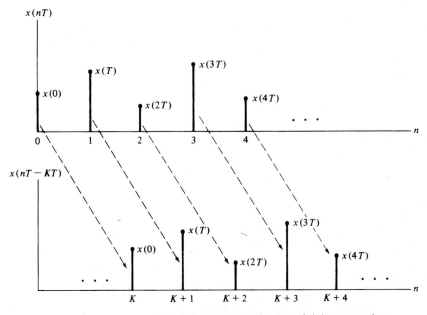

FIGURE 8-17.   Illustration of delay operation.

8-4 Difference Equations and Discrete-Time Systems

Now that we have an understanding of the z-transform, we have the tools necessary to explore discrete-time systems. The development will closely parallel the development of continuous-time systems.

Properties of Systems

In Chapter 1, the properties of continuous-time systems were defined. As in Chapter 1, the dependence of the system output, $y(nT)$, on the system input, $x(nT)$, is symbolically expressed as

$$y(nT) = \mathcal{H}[x(nT)] \tag{8-127}$$

where $\mathcal{H}$ is an *operator* that both identifies the system and specifies the operation to be performed on the input, $x(nT)$, to produce the output $y(nT)$.

Shift-Invariant Systems. A system is *fixed* or *time invariant* if the input–output relationship does not change with time. Discrete-time systems that are fixed are referred to as *shift-invariant* systems. In other words, a discrete-time system is shift invariant if a shift of the input by n_o sample periods results in a shift of the output by the same n_o sample periods. Expressed as an equation, a discrete-time system is shift invariant if and only if

$$\mathcal{H}[x(nT - n_oT)] = y(nT - n_oT) \tag{8-128}$$

for any finite value of n_o.

Causal and Noncausal Systems. A system is *causal* or *nonanticipatory* if the system response to an input does not depend on future values of the input. Parallel to the definition for continuous-time systems, a discrete-time system is causal if and only if the condition

$$x_1(nT) = x_2(nT) \qquad \text{for } n \leq n_o \tag{8-129a}$$

implies the condition

$$\mathcal{H}[x_1(nT)] = \mathcal{H}[x_2(nT)] \qquad \text{for } n \leq n_o \tag{8-129b}$$

for any $x_1(nT)$, $x_2(nT)$ and n_o. In other words, if the difference between two system inputs is zero for $n \leq n_o$, the difference between the respective outputs must be zero for $n \leq n_o$. If this condition is not satisfied, then the system is noncausal.

Linear Systems. Also parallel to continuous-time systems, a discrete-time system is linear if and only if superposition holds. In other words, a discrete-time system is linear if and only if

$$\mathcal{H}[\alpha_1 x_1(nT) + \alpha_2 x_2(nT)] = \mathcal{H}[\alpha_1 x_1(nT)] + \mathcal{H}[\alpha_2 x_2(nT)]$$

$$= \alpha_1 \mathcal{H}[x_1(nT)] + \alpha_2 \mathcal{H}[x_2(nT)]$$

$$= \alpha_1 y_1(nT) + \alpha_2 y_2(nT) \tag{8-130}$$

where

$$y_i(nT) = \mathcal{H}[x_i(nT)] \qquad \text{for } i = 1,2 \tag{8-131}$$

Two observations follow from (8-131). First, the system operation, $\mathcal{H}$, performed on a sum of signals is equivalent to the sum of the operations performed on the individual signals if the system is linear. In addition, we see that multiplication of the system input by a constant results in the multiplication of the output by the same constant for a linear system.

It follows from (8-74) that we can write

$$y(nT) = \mathcal{H}[x(nT)] = \mathcal{H}\left[\sum_{k=-\infty}^{\infty} x(kT)\delta(n-k)\right] \qquad (8\text{-}132)$$

where $\delta(n)$ is the unit pulse. Assuming a linear system, (8-132) can be written

$$y(nT) = \sum_{k=-\infty}^{\infty} \mathcal{H}[x(kT)\delta(n-k)] \qquad (8\text{-}133)$$

For a given value of k, $x(kT)$ is simply a constant and we can therefore write

$$y(nT) = \sum_{k=-\infty}^{\infty} x(kT)\mathcal{H}[\delta(n-k)] \qquad (8\text{-}134)$$

The *unit pulse response* of a system is of fundamental importance, and is defined as

$$h(nT) = \mathcal{H}[\delta(n)] \qquad (8\text{-}135)$$

If the system is shift invariant, we have

$$h(nT - kT) = \mathcal{H}[\delta(n-k)] \qquad (8\text{-}136)$$

and (8-134) becomes

$$y(nT) = \sum_{k=-\infty}^{\infty} x(kT)h(nT - kT) \qquad (8\text{-}137)$$

This is known as the convolution sum and defines the input-output relationship for linear shift-invariant systems. The change of variable $l = n - k$ shows that

$$y(nT) = \sum_{l=-\infty}^{\infty} h(lT)x(nT - lT) \qquad (8\text{-}138)$$

and, as in the case of continuous-time systems, convolution is commutative.

All the systems we consider will be causal, and all signals will be zero for negative index. Thus, in (8-137)

$$x(kT) \equiv 0, \qquad k < 0 \qquad (8\text{-}139a)$$

and

$$h(nT - kT) \equiv 0, \qquad n - k < 0 \quad \text{or} \quad k > n \qquad (8\text{-}139b)$$

For this case the convolution sum can be written

$$y(nT) = \sum_{k=0}^{n} h(kT)x(nT - kT) = \sum_{k=0}^{n} x(kT)h(nT - kT) \qquad (8\text{-}140)$$

EXAMPLE 8-12

In this example we examine the steps involved in discrete convolution for a specific set of sample sequences. We see that the steps involved in discrete convolution are essentially the same as those used for convolution of continuous-time waveforms. The basic difference is that multiplication of waveforms and integration is replaced by multiplication of sample values and summation. The second part of this example explores the use of MATLAB for performing discrete convolution and for plotting the results.

We will use discrete-time convolution to compute the output of a discrete-time processor whose input $x(nT)$ and unit pulse response $h(nT)$ are shown in Figure 8-18a. The terms in the summation

$$y(nT) = \sum_{m=0}^{n} x(mT)h(nT - mT) \tag{8-141a}$$

are shown in Figure 8-18b for $n < 0$. Clearly, $y(nT) \equiv 0$ for $n < 0$ since there is no overlap of the functions $x(mT)$ and $h(nT - mT)$. The table shown in Figure 8-18c is used to evaluate $y(nT)$ for $0 \le n \le 6$. The sample values of $x(mT)$ are shown across the top and the samples of $h(nT - mT)$ are shown in the table for $n = 0, 1, 2, 3, 4, 5, 6$. The samples of $h(nT - mT)$ shift to the right 1 unit for each increment of n. The value of the convolution sum is determined by multiplying and summing the two sets of sample values for each value of n. For example,

$$y(0) = 1 \cdot 3 = 3 \tag{8-141b}$$

$$y(T) = 1 \cdot 2 + 2 \cdot 3 = 8 \tag{8-141c}$$

$$y(2T) = 1 \cdot 1 + 2 \cdot 2 + 2 \cdot 3 = 11 \tag{8-141d}$$

$$y(6T) = 1 \cdot 1 = 1 \tag{8-141e}$$

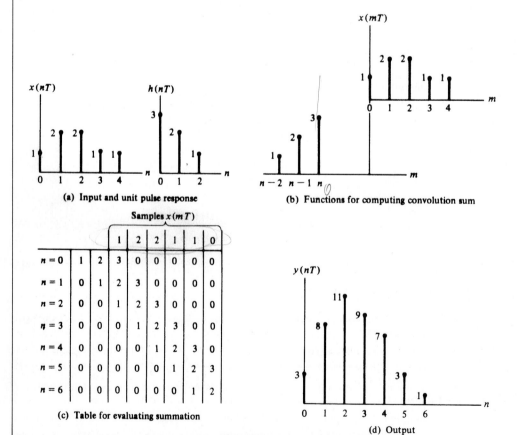

(a) Input and unit pulse response

(b) Functions for computing convolution sum

Samples $x(mT)$

	1	2	2	1	1	0		
$n = 0$	1	2	3	0	0	0	0	0
$n = 1$	0	1	2	3	0	0	0	0
$n = 2$	0	0	1	2	3	0	0	0
$n = 3$	0	0	0	1	2	3	0	0
$n = 4$	0	0	0	0	1	2	3	0
$n = 5$	0	0	0	0	0	1	2	3
$n = 6$	0	0	0	0	0	0	1	2

(c) Table for evaluating summation

(d) Output

FIGURE 8-18. Discrete convolution.

For $n > 6$, $y(nT) \equiv 0$ since there is no overlap of $x(mT)$ and $h(nT - mT)$. The result of the convolution is shown in Figure 8-18d.

MATLAB *Application*

MATLAB contains a command `conv` which performs discrete convolution. We now use MATLAB for convolving the sequences just considered and plot the results. Note that only the first three lines of code are required to perform the convolution. The rest of the code is used to format the plot, perform the plot and to label axes. MATLAB allows us to plot all three functions of interest in the same window. It is helpful to have all three sequences the same length when plotting and note how this is accomplished. Note also, the use of `stem` for plotting, that the sample indices begin at zero and that two zero samples are appended to the end of $y(nT)$.

```
EDU» c8ex12
 x = [1 2 2 1 1];                     % Define .samples of x
 h = [3 2 1];                         % Define samples of h
 y = conv(x,h);                       % Convolve x and h
 % Format for plot.
 nz = 2;                              % Extra zeros on y
 ly = length(y)+nz;                   % Length of y
 xz = [x,zeros(1,ly-length(x))];      % Zero pad x
 hz = [h,zeros(1,ly-length(h))];      % Zero pad h
 yz = [y,zeros(1,nz)];                % Zero pad y
 nn = 0:ly-1;                         % Establish x-axis indices
 % Now plot the results.
 subplot(2,2,1)                       % Top left plot
 stem(nn,xz)                          % Plot x as discrete sequence
 xlabel('Samplex Index - n')         % Label x axis
 ylabel('x(nT)')                      % Label y axis
 subplot(2,2,2)                       % Top right plot
 stem(nn,hz)                          % Plot h as discrete sequence
 xlabel('Sample Index - n')          % Label x axis
 ylabel('h(nT)')                      % Label y axis
 subplot(2,2,3)                       % Bottom left plot
 stem(nn,yz)                          % Plot y as discrete sequence
 xlabel('Sample Index - n')          % Label x axis
 ylabel('y(nT)')                      % Label y axis
```

EXAMPLE 8-13

In the previous example, the sequences that were convolved both had finite extent. In this example we consider sequences that have infinite extent. The graphical convolutional method is clearly not suited to problems of this kind unless we wish to determine only a few samples of the result of the convolution. The method illustrated in this example, which is nothing more than a direct application of (8-140), gives us a closed form result.

For this example assume

$$x(nT) = \left(\frac{1}{2}\right)^n u(n), \qquad h(nT) = \left(\frac{1}{3}\right)^n u(n) \tag{8-142}$$

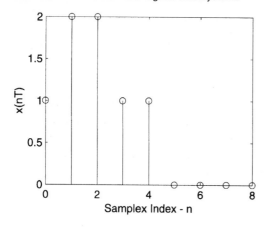

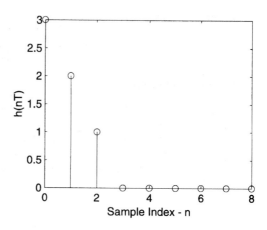

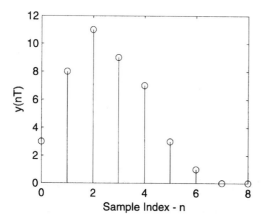

FIGURE 8-19. MATLAB plots for Example 8-12.

For this case the convolution sum becomes

$$y(nT) = \sum_{m=-\infty}^{\infty} \left(\frac{1}{2}\right)^m u(m) \left(\frac{1}{3}\right)^{n-m} u(n-m) \tag{8-143}$$

Since $u(m) = 0$ for $m < 0$ and since $u(n-m) = 0$ for $m > n$, (8-143) becomes

$$y(nT) = \left(\frac{1}{3}\right)^n \sum_{m=0}^{n} \left(\frac{3}{2}\right)^n \tag{8-144}$$

Using the summation formula

$$\sum_{n=0}^{N-1} x^n = \frac{1 - x^N}{1 - x} \tag{8-145}$$

allows $y(nT)$ to be written

$$y(nT) = \left(\frac{1}{3}\right)^n \frac{1 - (\frac{3}{2})^{n+1}}{1 - \frac{3}{2}}, \qquad n \geq 0 \tag{8-146}$$

A simple form for $y(nT)$ is

$$y(nT) = 3\left(\frac{1}{2}\right)^n - 2\left(\frac{1}{3}\right)^n, \qquad n \geq 0 \tag{8-147}$$

MATLAB Application

Even though the MATLAB Symbolic Math Toolbox does not contain a routine for performing symbolic convolution, we can still use MATLAB for solving this example problem by simply recognizing that discrete convolution is equivalent to multiplication in the z-domain. First, we z-transform $x(nT)$ and $h(nT)$ to form $X(z)$ and $H(z)$. Multiplying these yields $Y(z)$ and inverse transforming gives $y(nT)$, which is the desired result. The MATLAB code follows:

```
EDU» c8ex13

syms n z              % Declare symbolic
xn = (1/2)^n;         % Define x(n)
hn = (1/3)^n;         % Define h(n)
xz = ztrans(xn,n,z);  % Z-transform x(n)
hz = ztrans(hn,n,z);  % Z-transform h(n)
yz = xz*hz;           % Multiply
yn = iztrans(yz,z,n); % Inverse transform Y(z)
yn                    % Display y(n)

yn =

3*(1/2)^n-2*(1/3)^n
```

This agrees with (8-147) for non-negative values of n. Note that we have written a routine for symbolic convolution.

Stable Systems. The stability of discrete-time linear systems is defined in the bounded-input, bounded-output (BIBO) sense, as was the case for continuous-time systems in Chapter 2. A linear discrete-time system is BIBO stable if

$$|y(nT)| < \infty, \qquad \text{all } n \tag{8-148}$$

for all bounded inputs. Since we are considering linear shift-invariant systems,

$$y(nT) = \sum_{k=-\infty}^{\infty} x(kT)h(nT - kT) \tag{8-149}$$

Taking the absolute value of both sides of (8-149) yields

$$|y(nT)| = \left| \sum_{k=-\infty}^{\infty} x(kT)h(nT - kT) \right| \tag{8-150}$$

Since the absolute value of a sum is bounded by the sum of the absolute values of the individual terms in the sum, we have

$$|y(nT)| \leq \sum_{k=-\infty}^{\infty} |x(kT)||h(nT - kT)| \tag{8-151}$$

For a bounded input

$$|x(nT)| \leq M < \infty, \qquad \text{all } n \tag{8-152}$$

and (8-151) becomes

$$|y(nT)| \le M \sum_{k=-\infty}^{\infty} |h(nT - kT)|$$

$$= M \sum_{h=-\infty}^{\infty} |h(nT)| \tag{8-153}$$

Thus the system output is bounded if

$$\sum_{n=-\infty}^{\infty} |h(nT)| < \infty \tag{8-154}$$

For causal systems this is equivalent to the requirement that the system poles be inside the unit circle in the z-plane.

EXAMPLE 8-14 _____

In this example we illustrate the use of (8-154) by showing that the system defined by

$$h(nT) = \left[4\left(\frac{1}{3}\right)^n - 3\left(\frac{1}{4}\right)^n \right] u(n) \tag{8-155}$$

is BIBO stable. Since $h(nT) \ge 0$, all n, we have

$$\sum_{n=-\infty}^{\infty} |h(nT)| = \sum_{n=0}^{\infty} \left[4\left(\frac{1}{3}\right)^n - 3\left(\frac{1}{4}\right)^n \right] \tag{8-156}$$

This yields

$$\sum_{n=-\infty}^{\infty} |h(nT)| = \frac{4}{1 - \frac{1}{3}} - \frac{3}{1 - \frac{1}{4}} = 2 \tag{8-157}$$

Since $\sum_{n=-\infty}^{\infty} |h(nT)| < \infty$, the system is BIBO stable.

Difference Equations

Just as the differential equation can be used to model a continuous-time system, the difference equation can be used to model a discrete-time system. Consider, for example, a first-order continuous-time system defined by the differential equation

$$\frac{dy(t)}{dt} + ay(t) = bx(t) \tag{8-158}$$

The system output, $y(t)$, can be expressed in the form

$$y(t) = b \int_{-\infty}^{t} x(\alpha) \, d\alpha - a \int_{-\infty}^{t} y(\alpha) \, d\alpha \tag{8-159}$$

This equation tells us that the present value of the system output, $y(t)$, is a function of all previous values of the system input, $b \int_{-\infty}^{t} x(\alpha) \, d\alpha$, and is also a function of all past values of the system output, $\int_{-\infty}^{t} y(\alpha) \, d\alpha$.

A linear discrete-time system operates in somewhat the same way in that the present system output, $y(nT)$, must be computed using the present input $x(nT)$, past inputs $x(nT - kT)$, and past system outputs

$y(nT - kT)$. The structure of such a processor is illustrated in Figure 8-20. The general difference equation for this processor is

$$y(nT) = L_0x(nT) + L_1x(nT - T) + L_2x(nT - 2T) + \cdots$$
$$+ L_rx(nT - rT) - K_1y(nT - T)$$
$$- K_2y(nT - 2T) - \cdots - K_my(nT - mT) \qquad (8\text{-}160)$$

In other words, the processor illustrated in Figure 8-20 generates an output by weighting the present input, the past r inputs, and the past m outputs.

At the present time, we are concerned with the analysis of discrete-time systems. The analysis problem is usually a problem of determining the system output given the system input and a specification of the system. There are many different ways in which the system may be specified, but the most convenient method is to specify the coefficients of the difference equation.

In Chapter 9 we turn our attention to the synthesis of discrete-time systems, in which the problem is to determine the coefficients of the difference equation to perform some specified task.

The first step in the analysis of a discrete-time processor is to solve the difference equation. This is accomplished by first writing the difference equation in the form

$$y(nT) + K_1y(nT - T) + K_2y(nT - 2T) + \cdots + K_my(nT - mT)$$
$$= L_0x(nT) + L_1x(nT - T) + L_2x(nT - 2T) + \cdots + L_rx(nT - rT) \qquad (8\text{-}161)$$

and z-transforming both sides of the difference equation by using the time-delay operator

$$\mathcal{Z}[x(nT - kT)] = z^{-k}X(z) \qquad (8\text{-}162)$$

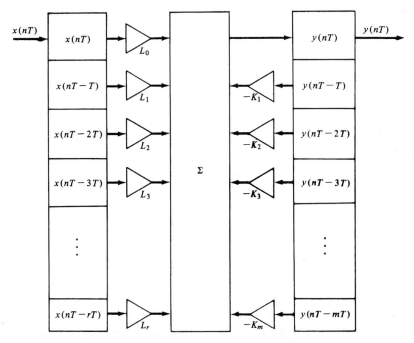

FIGURE 8-20.   Linear digital signal processor.

developed previously. Application of the time-delay operator yields

$$Y(z) + K_1 z^{-1} Y(z) + K_2 z^{-2} Y(z) + \cdots + K_m z^{-m} Y(z)$$
$$= L_0 X(z) + L_1 z^{-1} X(z) + L_2 z^{-2} X(z) + \cdots + L_r z^{-r} X(z) \qquad (8\text{-}163)$$

which is

$$Y(z)(1 + K_1 z^{-1} + K_2 z^{-2} + \cdots + K_m z^{-m}) = X(z)(L_0 + L_1 z^{-1} + L_2 z^{-2} + \cdots + L_r z^{-r}) \qquad (8\text{-}164)$$

The ratio of $Y(z)$ to $X(z)$ is the transfer function of the discrete-time system, which is often referred to as the *pulse transfer function*. For the processor of Figure 8-19 the pulse transfer function is given by

$$\frac{Y(z)}{X(z)} = H(z) = \frac{L_0 + L_1 z^{-1} + L_2 z^{-2} + \cdots + L_r z^{-r}}{1 + K_1 z^{-1} + K_2 z^{-2} + \cdots + K_m z^{-m}} \qquad (8\text{-}165)$$

By the definition of the pulse transfer function, we can write

$$Y(z) = H(z)X(z) \qquad (8\text{-}166)$$

It was shown previously that the z-transform of a unit pulse is unity. Thus if a discrete-time system has a *unit pulse input*, the system output is

$$Y(z) = H(z) \qquad (8\text{-}167)$$

which can be inverse z-transformed to yield

$$y(nT) = h(nT) \qquad (8\text{-}168)$$

The time sequence $\{h(nT)\}$, which is the inverse z-transform of the pulse transfer function $H(z)$, is known as the *unit pulse response*. We will now see that the unit pulse response plays essentially the same role in discrete-time systems that the unit impulse response plays in continuous time systems.

Steady-State Frequency Response of a Linear Discrete-Time System

The steady-state frequency response is an important characterization of a linear fixed discrete-time system, just as it was for a continuous-time system. Recall that the steady-state frequency response of a continuous-time system was determined by placing a sinusoidal signal on the input of the system. Since the system is fixed and linear, the output signal is also sinusoidal at the same frequency as the input signal. Thus, in passing through the system, the input signal is subjected only to an amplitude scaling and a phase shift. The ratio of the output signal amplitude to the input signal amplitude is the amplitude response at the input frequency and the output phase minus the input phase is the phase response at the input frequency.

As a parallel to the previous discussion, the sinusoidal steady-state response of a fixed, linear, discrete-time system is determined by placing the complex sampled sinusoid

$$x(nT) = A e^{j(\omega nT + \theta)} \qquad (8\text{-}169)$$

on the input of a discrete-time system as illustrated in Figure 8-21. Using (8-140), the output of the system can be written

$$y(nT) = A \sum_{m=0}^{n} h(mT) e^{j[\omega(n-m)T + \theta]} \qquad (8\text{-}170)$$

The system output can be placed in the form

$$y(nT) = A e^{j(\omega nT + \theta)} \left[\sum_{m=0}^{n} h(mT) e^{-j\omega mT} \right] \qquad (8\text{-}171)$$

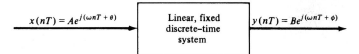

$$x(nT) = Ae^{j(\omega nT + \theta)}$$ → [Linear, fixed discrete-time system] → $$y(nT) = Be^{j(\omega nT + \phi)}$$

FIGURE 8-21. Input-output relationship for a discrete-time system.

For n sufficiently large, the terms $h(nT)$ of the unit pulse response are negligible, and we say that the steady-state response has been reached. The term in brackets is then a complex function depending only on the input frequency, ω. Denoting this function by $G(\omega)$, we can write, for n sufficiently large,

$$y(nT) = G(\omega)Ae^{j(\omega nT+\theta)} \tag{8-172}$$

Since we know that the output $y(nT)$ is a sampled sinusoid, we can write

$$Be^{j(\omega nT+\phi)} = G(\omega)Ae^{j(\omega nT+\theta)} \tag{8-173}$$

in which B is the output amplitude and ϕ is the phase of the output. This yields

$$G(\omega) = \frac{Be^{j(\omega nT+\phi)}}{Ae^{j(\omega nT+\theta)}} = \frac{B}{A}e^{j(\phi-\theta)} \tag{8-174}$$

Thus, at frequency ω,

$$|G(\omega)| = \frac{B}{A} \tag{8-175}$$

and

$$\underline{/G(\omega)} = \phi - \theta \tag{8-176}$$

Since $G(\omega)$ can be represented

$$G(\omega) = \sum_{m=0}^{\infty} h(mT)e^{-j\omega mT} \tag{8-177}$$

and

$$H(z) = \sum_{m=0}^{\infty} h(mT)z^{-m} \tag{8-178}$$

it follows that the sinusoidal steady-state response of a discrete-time system is given by

$$G(\omega) = H(e^{j\omega T}) \tag{8-179}$$

Thus the steady-state response is easily obtained from the pulse transfer function by simply replacing z in the pulse transfer function by $e^{j\omega T}$.

Of fundamental importance is the fact that the sinusoidal steady-state frequency response of a discrete-time system is periodic in the sampling frequency, ω_s. This is easily seen by writing $H(e^{j\omega T})$ with ω replaced by

$$\omega_a = \omega + k\omega_s \tag{8-180}$$

where k is an integer. This yields

$$H[e^{j(\omega+k\omega_s)T}] = H(e^{j\omega T}e^{jk\omega_s T}) \tag{8-181}$$

Since $\omega_s T = 2\pi$, we have

$$H[e^{j(\omega+k\omega_s)T}] = H(e^{j\omega T}e^{jk2\pi}) = H(e^{j\omega T}) \tag{8-182}$$

which shows that $H(e^{j\omega T})$ is periodic with period ω_s.

Since $H(e^{j\omega T})$ is periodic in the sampling frequency, it is often advantageous to normalize the frequency variable with respect to the sampling frequency. Defining the frequency ratio r as

$$r = \frac{\omega}{\omega_s} \tag{8-183}$$

allows ωT to be replaced by

$$\omega T = r\omega_s T = 2\pi r \tag{8-184}$$

so that the sinusoidal steady-state frequency response is expressed as $H(e^{j2\pi r})$.

The concepts developed in this section are illustrated in the following examples.

EXAMPLE 8-15 _____

This example illustrates calculation of the steady-state frequency response. Specifically, we determine the amplitude and phase response of the discrete-time system defined by the difference equation

$$y(nT) = x(nT) + x(nT - 2T) \tag{8-185}$$

It follows from the difference equation that the pulse transfer function of the system is

$$H(z) = 1 + z^{-2} \tag{8-186}$$

Thus the sinusoidal steady-state frequency response is

$$H(e^{j\omega T}) = 1 + e^{-j2\omega T} \tag{8-187}$$

or

$$H(e^{j\omega T}) = (e^{j\omega T} + e^{-j\omega T})e^{-j\omega T} \tag{8-188}$$

which can be written as

$$H(e^{j\omega T}) = (2 \cos \omega T)e^{-j\omega T} \tag{8-189}$$

In terms of the normalized frequency r, (8-189) is

$$H(e^{j2\pi r}) = (2 \cos 2\pi r)e^{-j2\pi r} \tag{8-190}$$

The magnitude and phase responses of the system are illustrated in Figure 8-22 for r between 0 and 0.5. The π radian discontinuity in the phase response at $r = 0.25$ is due to the sign change of $\cos 2\pi r$ at $r = 0.25$.

EXAMPLE 8-16 _____

The purpose of this example is to illustrate the amplitude and phase response of a system that produces an output equal to the input delayed by K sample periods. Such a system is illustrated in Figure 8-23a and has the pulse transfer function

$$H(z) = z^{-K} \tag{8-191}$$

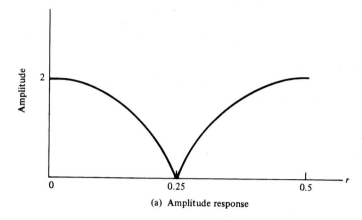

(a) Amplitude response

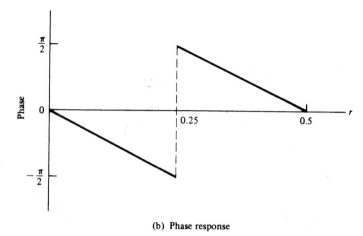

(b) Phase response

FIGURE 8-22. Amplitude and phase responses for Example 8-15.

The sinusoidal steady-state frequency response is, in terms of normalized frequency, given by

$$H(e^{j2\pi r}) = e^{-j2\pi Kr} \qquad (8\text{-}192)$$

so that the time-delay system has the unity amplitude response and the linear phase response illustrated in Figures 8-23b and c, respectively. The student should note that these results are consistent with the Fourier transform result given in Table 4-1 (entry 2), which states that a time shift in the time domain is equivalent to a linear phase shift in the frequency domain.

Frequency Response at $f = 0$ and $f = 0.5f_s$

As can be seen from the previous section, the expression for the frequency response of a discrete-time system of digital filter is often rather complicated. It is easy, however, to determine $H(e^{j2\pi fT})$ at $f = 0$ and $f = 0.5f_s$ if we first recognize that

$$e^{j2\pi fT}\bigg|_{f=0} = e^{j0} = 1 \qquad (8\text{-}193)$$

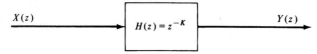

$$X(z) \longrightarrow \boxed{H(z) = z^{-K}} \longrightarrow Y(z)$$

(a) Discrete–time delay system

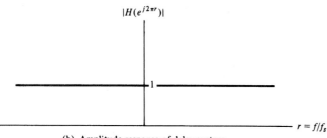

(b) Amplitude response of delay system

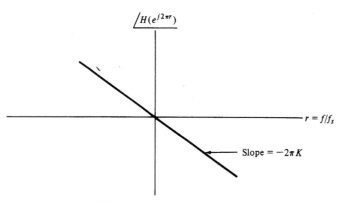

(c) Phase response of delay system

FIGURE 8-23. Discrete-time delay system.

and

$$e^{j2\pi fT}\Big|_{f=0.5f_s} = e^{j\pi f_sT} = e^{j\pi} = -1 \tag{8-194}$$

Thus, the dc response is $H(1)$ and the response at one-half the sampling frequency is $H(-1)$.

These calculations are very simple and are important for two reasons. First, they often offer valuable insight into system performance. In addition they can be used to provide a partial check on frequency-response calculations. If the system transfer function, $H(z)$, is a ratio of polynomials in z with real coefficients, then $H(1)$ and $H(-1)$ are both real.

EXAMPLE 8-17 _____

This example illustrates determination of the frequency response at specific frequency points. Consider the discrete-time system defined by the difference equation

$$y(nT) - 0.5y(nT - T) + 0.38y(nT - 2T) = x(nT) + 0.5x(nT - T) \tag{8-195}$$

The pulse transfer function is

$$H(z) = \frac{1 + 0.5z^{-1}}{1 - 0.5z^{-1} + 0.38z^{-2}} \tag{8-196}$$

The dc response is

$$H(1) = \frac{1 + 0.5}{1 - 0.5 + 0.38} = 1.70 \tag{8-197}$$

and the response at $f = 0.5f_s$ is

$$H(-1) = \frac{1 - 0.5}{1 + 0.5 + 0.38} = 0.27 \tag{8-198}$$

8-5 Example of a Discrete-Time System

As a simple example of a discrete-time system, and as a review of the various techniques that have been studied in this chapter, we will examine the performance of the system illustrated in Figure 8-24. We will determine the difference equation, the pulse transfer function, and the sinusoidal steady-state frequency response (both magnitude and phase) of the system.

It follows from the block diagram of the system that the difference equation defining the operation of the system is

$$y(nT) = kx(nT) + \alpha y(nT - T) \tag{8-199}$$

from which we have

$$y(nT) - \alpha y(nT - T) = kx(nT) \tag{8-200}$$

Since a delay of T is equivalent to multiplication by z^{-1}, $Y(z)$ becomes

$$Y(z) - \alpha z^{-1}Y(z) = kX(z) \tag{8-201}$$

which yields the pulse transfer function

$$H(z) = \frac{Y(z)}{X(z)} = \frac{k}{1 - \alpha z^{-1}} \tag{8-202}$$

As we saw in the preceding section, the sinusoidal steady-state response can be determined from $H(z)$ by setting $z = e^{j\omega T}$. For our example this yields

$$H(e^{j\omega T}) = \frac{k}{1 - \alpha e^{-j\omega T}} \tag{8-203}$$

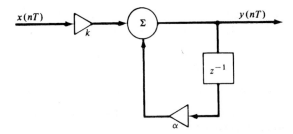

FIGURE 8-24.   Example digital processor.

or

$$H(e^{j\omega T}) = \frac{k}{1 - \alpha \cos \omega T + j\alpha \sin \omega T} \qquad (8\text{-}204)$$

In terms of the normalized frequency ratio, $r = f/f_s$, this becomes

$$H(e^{j2\pi r}) = \frac{k}{1 - \alpha \cos 2\pi r + j\alpha \sin 2\pi r} \qquad (8\text{-}205)$$

To plot the amplitude and phase responses of the system, we place (8-205) in the form

$$H(e^{j2\pi r}) = A(r)e^{j\phi(r)} \qquad (8\text{-}206)$$

in which $A(r)$ and $\phi(r)$ are the amplitude and phase responses, respectively. It follows from (8-206) that the amplitude response $A(r)$ is

$$A(r) = \frac{k}{\sqrt{1 + \alpha^2 - 2\alpha \cos 2\pi r}} \qquad (8\text{-}207)$$

and that the phase response $\phi(r)$ is

$$\phi(r) = -\tan^{-1} \frac{\alpha \sin 2\pi r}{1 - \alpha \cos 2\pi r} \qquad (8\text{-}208)$$

Observation of (8-207) shows that the discrete-time system in Figure 8-24 is a low-pass system since $A(r)$ has a maximum at $r = 0$ and decreases with increasing r for r in the range $0 < r < \frac{1}{2}$. If the filter is to have unity gain at $r = 0$,

$$A(0) = \frac{k}{\sqrt{1 - 2\alpha + \alpha^2}} = \frac{k}{1 - \alpha} = 1 \qquad (8\text{-}209)$$

from which

$$k = 1 - \alpha \qquad (8\text{-}210)$$

It should be noted that any desired dc gain can be achieved by an adjustment of k.

The amplitude and phase responses of the system are illustrated in Figures 8-25 and 8-26, respectively, for r in the range $0 \le r \le \frac{1}{2}$. The amplitude response is even and the phase response is odd since they are the magnitude and phase of a Fourier transform. This, together with the fact that the amplitude and phase responses are periodic, since $H(e^{j2\pi r})$ is periodic, fully defines $A(r)$ and $\phi(r)$ for all values of r. The amplitude response is illustrated with a log frequency scale so that the 6 dB per octave slope characteristic of a first-order system can be observed.

*8-6 Inverse z-Transformation by the Inversion Integral

In Section 5-5 the technique of determining inverse Laplace transforms by the inversion integral was briefly investigated. We now briefly consider the inversion integral method as applied to z-transforms.

The basic inversion integral is easy to derive. By definition, the z-transform of the sample sequence $\{x(nT)\}$ is

$$X(z) = \sum_{m=0}^{\infty} x(mT)z^{-m} \qquad (8\text{-}211)$$

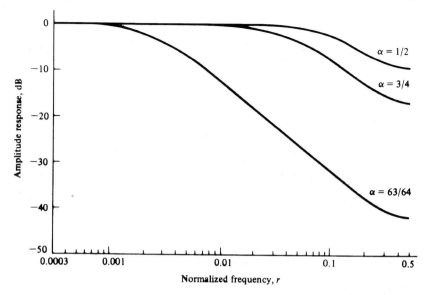

FIGURE 8-25.   Amplitude response of example filter.

Multiplying both sides of (8-211) by z^{n-1} and integrating along a path C in the complex z-plane that lies entirely within the region of convergence of $X(z)$ yields

$$\oint_C X(z)z^{n-1}\, dz = \oint_C \sum_{m=0}^{\infty} x(mT)z^{-m+n-1}\, dz \tag{8-212}$$

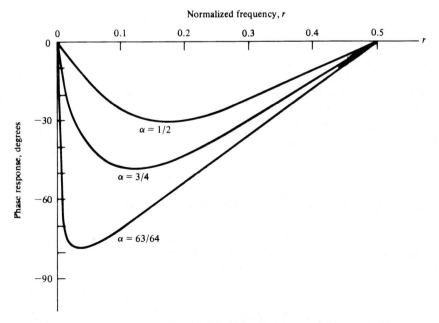

FIGURE 8-26.   Phase response of example filter.

Assuming that

$$\sum_{m=0}^{\infty} |x(mT)| < \infty \tag{8-213}$$

in other words, the sequence $\{x(mT)\}$ is absolutely summable—the order of integration and summation can be interchanged. This yields

$$\oint_C X(z)z^{n-1}\, dz = \sum_{m=0}^{\infty} x(mT)\oint_C z^{-m+n-1}\, dz \tag{8-214}$$

The integral on the right-hand side can be evaluated through the use of the residue theorem presented in Appendix C. For $m = n$, z^{-m+n-1} has a first-order pole at $z = 0$ and the residue is 1. For $m \neq n$, z^{-m+n-1} has no first-order pole, and therefore the residue is zero. Thus

$$\oint_C z^{-m+n-1}\, dz = \begin{cases} 2\pi j, & m = n \\ 0, & m \neq n \end{cases} \tag{8-215}$$

and the only nonzero term in the summation on the right-hand side of (8-209) is the term for which $m = n$. Therefore,

$$\oint_C X(z)z^{n-1}\, dz = x(nT)2\pi j \tag{8-216}$$

which gives

$$x(nT) = \frac{1}{2\pi j}\oint_C X(z)z^{n-1}\, dz \tag{8-217}$$

We can illustrate the inversion integral by means of an example.

EXAMPLE 8-18 _____

This example illustrates the inversion integral for a simple $X(z)$. The inverse z-transform of

$$X(z) = \frac{1}{1 - Kz^{-1}}, \qquad |Kz^{-1}| < 1 \tag{8-218}$$

will be determined by use of the inversion integral. The integral for $x(nT)$ can be written as

$$x(nT) = \frac{1}{2\pi j}\oint_C \frac{z^n}{z - K}\, dz \tag{8-219}$$

For $n \geq 0$, the integrand has only a single pole at $z = K$. The region of convergence is $|Kz^{-1}| < 1$ or $|z| > K$. Assuming that $K \leq 1$, so that the resulting time sequence is bounded, ensures that $X(z)$ converges everywhere outside the unit circle. Thus in (8-219) the path of integration is taken as a circle with $z = 0$ as the center and radius $1 + \epsilon$, in which ϵ is positive. Application of the residue theorem then gives

$$x(nT) = K^n, \qquad n \geq 0 \tag{8-220}$$

Summary

The basic theme of this chapter was the analysis of discrete-time signals and systems. We initiated the chapter with a discussion of analog-to-digital (A/D) converters. We saw that A/D conversion involves the operations of sampling, quantizing, and encoding. If the sampling frequency is sufficiently high, the original signal can be reconstructed from the sequence of samples. We saw that reconstruction can take a variety of forms if small errors can be tolerated.

The basic mathematical tool used in this chapter was the z-transform. A useful table of z-transforms was developed, and techniques for inverting the z-transform were presented. A number of useful z-transform theorems and concepts were then presented. These included the definition of linearity, the initial value and final value theorems, and a discussion of the very important delay operator.

After introducing the z-transform and discrete-time signals, we turned our attention to discrete-time systems. System properties were discussed. These included shift invariance, causality, linearity, and stability. We saw that a linear, shift-invariant system can be represented by a linear difference equation with constant coefficients. The difference equation led immediately to the pulse transfer function and the unit pulse response. The output of a linear, shift-invariant system is given by the convolution of the input sequence with the unit pulse response. We saw that the steady-state frequency response of a discrete-time system is found by evaluating the pulse transfer function on the unit circle. The analysis of a simple discrete-time system served to bring together the basic ideas of discrete-time analysis.

The following items summarize the specific points made in this chapter:

1. Analog-to-digital conversion consists of sampling, quantizing, and encoding.
2. To sample a continuous-time signal $x(t)$ is to represent it at discrete points $t = nT$, where n is an integer and T is referred to as the sampling period. The sampling period is the reciprocal of the sampling frequency, f_s.
3. A continuous-time function is sampled by multiplying $x(t)$ by a periodic waveform, $p(t)$, which is usually taken to be a sequence of pulses. Impulse-train sampling is a model in which

$$p(t) = \sum_{n=-\infty}^{\infty} \delta(t - nT)$$

4. The frequency-domain effect of sampling is to multiply the spectrum (Fourier transform) of $x(t)$, denoted $X(f)$, by f_s, reproduce $X(f)$ at $f = 0$ and all harmonics of the sampling frequency $f = nf_s$. In equation form this gives

$$X_s(f) = f_s \sum_{n=-\infty}^{\infty} X(f - nf_s)$$

where $X_s(f)$ represents the Fourier transform of the sampled signal.

5. A bandlimited signal can be reconstructed without error if the signal is sampled at a sampling frequency $f_s \geq 2f_h$, where f_h is the highest frequency present in the signal being sampled.
6. Reconstruction of a signal $x(t)$ from the sequence of samples is accomplished by passing the samples of $x(t)$ through an ideal low-pass filter with bandwidth W, where W is equal to one-half the sampling frequency. This technique, called ideal reconstruction, is equivalent to weighting each sample by a sinc function and summing the contributions of the individual sinc functions. The ideal reconstruction filter is noncausal, and the unit impulse response has infinite duration. Approximations to the ideal reconstruction filter that overcome these difficulties are possible.
7. The purpose of a quantizer is to allow continuous-amplitude samples to be represented by digital words of finite length. If an n-bit analog-to-digital converter is used, 2^n quantizing levels are available for representing each sample value. The signal-to-noise ratio at the output of an analog-to-digital converter is approximately 6 dB per bit.
8. The z-transform (one-sided) of a discrete-time sequence $x(nT)$ is defined by

$$X(z) = \sum_{n=0}^{\infty} x(nT)z^{-n}$$

and is related to the Laplace transform, s, by

$$z = e^{sT}$$

where T is the sampling period. The z-transform is a linear operator since the z-transform of a sum of discrete-time sequences is equal to the sum of the z-transforms of the individual sequences.

9. The initial value $x(0)$ of a sequence $x(nT)$ that is zero for $n < 0$ is given by

$$x(0) = \lim_{z \to \infty} X(z)$$

The final value, $x(\infty)$, is given by

$$x(\infty) = \lim_{z \to 1} (1 - z^{-1})X(z)$$

10. The inverse z-transform can be computed by using long division or partial-fraction expansion. With partial-fraction expansion it is usually best to expand $H(z)/z$.

11. Delay of a sample sequence $x(nT)$ by k sample periods is equivalent to multiplying $X(z)$ by z^{-k}.

12. A discrete-time system is fixed, or shift invariant, if a shift in the input by n_o samples results in a shift of the output by the same n_o samples.

13. A discrete-time system is causal if the system response to an input does not depend on future values of the input. Stated another way, if the difference between two system inputs is zero for $n \leq n_o$, the difference between the respective outputs will be zero for a causal system for $n \leq n_o$.

14. A system is linear if superposition holds. In other words, a system is linear if

$$\mathcal{H}[\alpha_1 x_1(nT) + \alpha_2 x_2(nT)] = \alpha_1 \mathcal{H}[x_1(nT)] + \alpha_2 \mathcal{H}[x_2(nT)]$$

where α_1 and α_2 are arbitrary constants. If a discrete-time system is both linear and shift variant, the system output, $y(nT)$, is given by the convolution of the system input, $x(nT)$, with the system unit pulse response, $h(nT)$. The convolution is written

$$y(nT) = \sum_{m=-\infty}^{\infty} x(mT)h(nT - mT) = \sum_{m=-\infty}^{\infty} h(mT)x(nT - mT)$$

15. A linear discrete-time system is stable in the bounded-input, bounded-output sense if the unit pulse response $h(nT)$ is absolutely summable.

16. A linear shift-invariant discrete-time system is defined by a difference equation with constant coefficients. Solving this equation for the current output $h(nT)$ allows the current output to be written in terms of the current input, $x(nT)$, past inputs, $x(nT - kT)$ for $k \geq 1$, and past outputs, $y(nT - kT)$ for $k \geq 1$.

17. The pulse transfer function $H(z)$ of a discrete-time system is found by z-transforming the difference equation of the system term by term and solving for the ratio of the z-transform of the system output, $Y(z)$, to the z-transform of the system input, $X(z)$.

18. The steady-state frequency response of a linear shift-invariant system is determined by evaluating the pulse transfer function $H(z)$ at $z = e^{j\omega T}$. The frequency response of a discrete-time system is periodic in the sampling frequency. The frequency response at $f = 0$, the dc response, is $H(1)$. The frequency response at $f = 0.5f_s$, where f_s is the sampling frequency, is $H(-1)$.

19. The inverse z-transform of $X(z)$ is given by the contour integral

$$x(nT) = \frac{1}{2\pi j} \oint_C X(z)z^{n-1}\, dz$$

where the contour C lies in the region of convergence of $X(z)$.

Further Reading

A large number of very readable books have been written that treat the analysis of discrete-time signals and systems. The following are representative:

A. V. OPPENHEIM and R. W. SCHAEFER, *Discrete-Time Signal Processing.* Englewood Cliffs, NJ: Prentice Hall, 1989.

J. G. PAROKIS and D. G. MANOLAKIS, *Digital Signal Processing: Principles, Algorithms and Applications,* 2nd ed. New York: Macmillan, 1992.

R. D. STRUM and D. E. KIRK, *First Principles of Discrete Systems and Digital Signal Processing.* Reading, MA: Addison-Wesley, 1988.

J. A. CADZOW, *Foundations of Digital Signal Processing and Data Analysis.* New York: Macmillan, 1987.

A number of books have been written over the past few years that introduce computer methods for the analysis of linear systems. Many of these make extensive use of MATLAB. The following three books are good examples:

R. D. STRUM and D. E. KIRK, *Contemporary Linear Systems Using MATLAB.* Boston, MA: PWS, 1994.

J. R. BUCK, M. M. DANIEL and A. C. SINGER, *Computer Explorations in Signals and Systems Using MATLAB.* Upper Saddle River, NJ: Prentice Hall, 1997.

C. S. BURRUS, J. H. MCCLELLAN, A. V. OPPENHEIM, T. W. PARKS, R. C. SCHAFER and H. W. SCHUESSLER, *Computer-Based Exercises for Signal Processing Using MATLAB.* Upper Saddle River, NJ: Prentice Hall, 1994.

PROBLEMS

Section 8-2

8-1. The signal

$$x(t) = 4 + 8 \cos 8\pi t$$

is sampled at a rate of 16 samples per second. Plot the amplitude spectrum of the sampled signal showing the weight and the frequency of each component for $|f| < 40$ Hz. Show, by drawing an illustration similar to Figure 8-8, how the signal can be reconstructed from the samples.

8-2. Assume that the signal

$$x(t) = 4 + 8 \cos 8\pi t$$

is sampled at a frequency of 6 samples per second. Plot the amplitude spectrum for $|f| < 17$ Hz. Can the original signal be recovered from the samples? Why or why not?

8-3. The signal

$$x(t) = 3 + 4 \cos 10\pi t + 5 \cos 14\pi t + 2 \cos 20\pi t$$

is sampled at a rate of 30 samples per second. Plot the spectrum of the sampled signal showing all components for $|f| < 80$. Fully explain how $x(t)$ can be reconstructed from the samples.

8-4. The signal having the spectrum shown is sampled. Plot the spectrum of the sampled signal for the three sampling frequencies $f_s = 30$, 40, and 60 hertz. [*Note:* Assume that $X(f)$ is real.] Label all amplitudes and frequencies of interest. Which of the three sampling frequencies are acceptable?

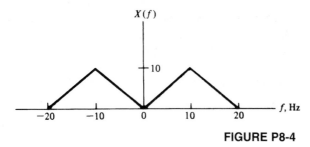

FIGURE P8-4

8-5. An A/D converter has the input signal

$$x(t) = 5 \cos 200\pi t + 4 \cos 600\pi t$$

Determine:
(a) The required dynamic range of the A/D converter.
(b) The output signal-to-noise ratio as a function of the wordlength, n. Express this result in decibels. Plot and compare with the result determined in Example 8-3.

8-6. Repeat Problem 8-5, assuming the signal $x(t)$ shown.

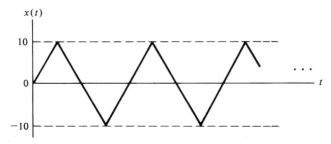

FIGURE P8-6

8-7. An A/D converter has an input signal given by the Fourier series

$$x(t) = \sum_{n=-\infty}^{\infty} X_n e^{jn\omega_0 t}$$

Show that the signal-to-noise ratio at the output of the A/D converter is given by

$$\text{SNR} = \frac{12}{D^2} 2^{2n} \sum_{n=-\infty}^{\infty} |X_n|^2$$

8-8. Assume that the signal $x(t)$ given in Problem 8-1 is digitized with an A/D converter utilizing 8-bit arithmetic. Compute the signal-to-noise ratio at the output of the A/D converter. Compare with the result obtained using 10- and 12-bit arithmetic.

8-9. An RC low-pass filter having transfer function

$$H(f) = \frac{1}{1 + j(f/f_3)}$$

where f_3 is the 3-dB frequency of the filter, is used for reconstruction of a signal from its samples. We can ignore the multiplying constant T since this multiplication can be realized as a multiplier external to the filter. The signal $x(t)$ to be reconstructed is assumed to be a low-pass signal with bandwidth W. We wish to determine an appropriate sampling frequency so that $X(f)$ in (8-16) is passed by the filter with negligible error and that the translated spectra $X(f \pm f_s)$ are sufficiently attenuated.

(a) Let the filter magnitude responses, in dB, at $f = W$ and at $f = f_s - W$, be denoted

$$20 \log_{10} |H(W)| = \delta_1 \quad \text{dB}$$

and

$$20 \log_{10} |H(f_s - W)| = \delta_2 \quad \text{dB}$$

We clearly wish δ_1 to correspond to negligible attenuation relative to the dc gain and δ_2 to a large attenuation. For fixed δ_1, δ_2 and signal bandwidth W, show that the necessary sampling frequency, f_s, is

$$f_s = W \frac{\sqrt{10^{-0.1\delta_1} - 1} + \sqrt{10^{-0.1\delta_2} - 1}}{\sqrt{10^{-0.1\delta_1} - 1}}$$

(b) Use this result to determine the ratio f_s/W for
    (i) $\delta_1 = -3$ dB and $\delta_2 = -30$ dB
    (ii) $\delta_1 = -1$ dB and $\delta_2 = -30$ dB
    (iii) $\delta_1 = -1$ dB and $\delta_2 = -40$ dB
    Comment on these results. How can large values of f_s/W be avoided?

8-10. In this problem we repeat the preceding problem using the second-order Butterworth reconstruction filter defined by

$$H(f) = \frac{(2\pi f_3)^2}{s^2 + \sqrt{2}(2\pi f_3)s + (2\pi f_3)^2}\bigg|_{s=j2\pi f}$$

(a) Show that

$$|H(f)|^2 = \frac{f_3^4}{f_3^4 + f^4}$$

(b) Using the definitions for δ_1 and δ_2 used in Problem 8-9, show that the necessary sampling frequency, f_s, is

$$f_s = W \frac{[10^{-0.1\delta_1} - 1]^{1/4} + [10^{-0.1\delta_2} - 1]^{1/4}}{[10^{-0.1\delta_1} - 1]^{1/4}}$$

(c) Use this result to determine the ratio f_s/W for the same three sets of δ_1 and δ_2 considered in the previous problem. Compare the results with those obtained in Problem 8-9. Comment on this comparison.

8-11. Assume that a reconstruction filter is second order and has Laplace transform transfer function

$$H(s) = \frac{T\omega_3^2}{s^2 + \sqrt{2}s\omega_3 + \omega_3^2}$$

where ω_3 is the half-power frequency in radians per second and T is the sampling period. Determine the interpolation formula corresponding to this filter.

8-12. Assume that a constant signal is sampled so that $x(nT) = a$ for all N. Show that, for $f_s \gg f_3$, the interpolation formula for an RC low-pass reconstruction filter yields an output $y(nT)$ that is equal to the sample values $x(nT)$.

Section 8-3

8-13. Use z-transform pair 3 in Table 8-1 to establish z-transform pairs 4 and 5. (*Hint:* First write

$$\sum_{n=0}^{\infty} e^{-\alpha n T} z^{-n} = \frac{1}{1 - e^{-\alpha T} z^{-1}}$$

and differentiate both sides with respect to α.)

8-14. Use z-transform pair 3 in Table 8-1 to establish z-transform pairs 6 and 7. (*Hint:* Again first write

$$\sum_{n=0}^{\infty} e^{-\alpha n T} z^{-n} = \frac{1}{1 - e^{-\alpha T} z^{-1}}$$

Let $\alpha = jb$ and equate real and imaginary parts.)

8-15. Use z-transform pair 3 in Table 8-1 to establish z-transform pairs 8 and 9. (*Hint:* The procedure is the same as in Problem 8-14. What should you let α be now?)

8-16. Show that the z-transform of $a^n x(nT)$ is $X(z/a)$. Use this result to show that entry 3 in Table 8-1 follows from entry 2.

8-17. Show that the z-transform of $nx(nT)$ is given by $-z\, dX(z)/dz$. Use this result to show that entry 4 in Table 8-1 follows from entry 3.

8-18. The signal below is sampled at 25 samples per second. Determine the z-transform of $x(nT)$ for $0 \le n \le 8$.

$$x(t) = 5 \sin 20\pi t + 2\Pi\left(\frac{t - 0.14}{0.16}\right)$$

8-19. Determine the z-transform for the following sequences of samples:

(a) $x(nT) = \left(\dfrac{1}{5}\right)^n u(n)$

(b) $x(nT) = \left(-\dfrac{1}{5}\right)^n u(n)$

(c) $x(nT) = u(n) + \left(\dfrac{3}{4}\right)^n u(n - 4)$

(d) $x(nT) = 2u(n) - 2u(n - 8)$

8-20. Verify the results of Problem 8-19 using MATLAB.

8-21. Determine the z-transform of the following sequences of samples:

(a) $x(nT) = \left(\dfrac{2}{3}\right)^n u(n - 4)$

(b) $x(nT) = \left(\dfrac{2}{3}\right)^{n-4} u(n - 4)$

8-22. Verify the results of Problem 8-21 using MATLAB.

8-23. Determine $x(5T)$ given that $X(z) = \exp\{az^{-1}\}$ where a is a real constant.

8-24. Determine the z-transform of $x(nT) = a^n \sin\left(\dfrac{\pi}{2}n\right)$ for $n \geq$ where a is a real constant.

8-25. Determine the z-transform and the z-plane poles for the following sequences:
    (a) $x(nT) = (\frac{1}{2})^n[u(n) - u(n - 10)]$
    (b) $x(nT) = (\frac{1}{2})^n[u(n) + u(n - 10)]$

8-26. The energy associated with a sequence of samples, $x(nT)$, is defined to be

$$E = \sum_{n=-\infty}^{\infty} |x(nT)|^2$$

Determine the energy, E, for the following sequences:
    (a) $x(nT) = (\frac{1}{2})^n u(n)$
    (b) $x(nT) = (\frac{1}{2})^n u(n - 6)$
    (c) $x(nT) = (\frac{1}{2})^{n-6} u(n - 6)$
    (d) $x(nT) = [4(\frac{1}{3})^n - 3(\frac{1}{4})^n]u(n)$

8-27. Determine the samples, $x(nT)$, corresponding to each of the following examples of $X(z)$.
    (a) $X(z) = 1 - 0.3z^{-1} + 0.7z^{-2} + 0.8z^{-3} - 0.3z^{-4}$
    (b) $X(z) = (1 - z^{-1})^3$
    (c) $X(z) = (1 - 0.8z^{-1})^3$
    (d) $X(z) = (1 - 0.9z^{-2})^3(1 + 0.9z^{-2})^3$

8-28. Determine the samples, $x(nT)$, corresponding to each $X(z)$ here.
    (a) $X(z) = z^{-2}(1 + 0.3z^{-1} - 0.7z^{-2})(1 + z^{-3})$
    (b) $X(z) = z^{-2}(1 + 0.3z^{-1} - 0.7z^{-2})^2(1 + z^{-3})$

8-29. Use long division to determine $x(0)$, $x(T)$, $x(2T)$, $x(3T)$ and $x(4T)$ for the following functions of z:
    (a) $X(z) = \dfrac{1}{1 - 0.3z^{-1} + 0.3z^{-2} - 0.5z^{-3}}$
    (b) $X(z) = \dfrac{1 - 0.7z^{-1}}{(1 + 0.5z^{-1})^2}$
    (c) $X(z) = \dfrac{z^{-1}(1 + 0.2z^{-1})^2}{(1 - 0.4z^{-2})^2}$

8-30. Use long division to determine $x(0)$, $x(T)$, $x(2T)$, $x(3T)$ and $x(4T)$ for the following function of z:

$$X(z) = \frac{(1 - z^{-1})^4}{1 - 0.3z^{-1} + 0.3z^{-2} - 0.5z^{-3}}$$

8-31. Rework Problems 8-29 and 8-30 using MATLAB.

8-32. Determine the inverse z-transform of the following sequences using partial fraction expansion:
    (a) $X(z) = \dfrac{2}{(1 - z^{-1})(1 + 0.2z^{-1})}$

(b) $X(z) = \dfrac{2}{(1 - z^{-1})(1 + z^{-1})}$

(c) $X(z) = \dfrac{1 + 0.3z^{-1}}{(1 + 0.2z^{-1})(1 - 0.4z^{-1})}$

(d) $X(z) = \dfrac{1 + 0.3z^{-1}}{(1 + 0.2z^{-1})(2 - 0.4z^{-1})}$

8-33. Determine the inverse z-transform of the following sequences using partial fraction expansion:

(a) $X(z) = \dfrac{1}{(1 - z^{-1})(1 + 0.5z^{-1})(1 - 0.2z^{-1})}$

(b) $X(z) = \dfrac{(1 - 0.5z^{-1})}{(1 - z^{-1})(1 + 0.5z^{-1})(1 - 0.2z^{-1})}$

8-34. Work Problem 8-33 using MATLAB.

8-35. Determine the inverse z-transform of the following sequences using partial fraction expansion:

(a) $X(z) = \dfrac{1}{(1 - z^{-1})(1 - 0.5z^{-1})^2}$

(b) $X(z) = \dfrac{1}{(1 - 0.81z^{-1})^2}$

8-36. Determine the inverse z-transform of the following sequence using partial fraction expansion:

$$X(z) = \frac{1}{1 + 0.81z^{-2}}$$

8-37. Work Problems 8-35 and 8-36 using MATLAB.

8-38. Determine the inverse z-transform of the following for all n.

$$H(z) = \frac{z^2 + z}{(z + 0.5)(z - 0.25)}$$

8-39. For the function $H(z)$ given below determine $h(n)$ for all n using partial-fraction expansion. Partially check your result by computing $h(0)$, $h(T)$, $h(2T)$, and $h(\infty)$ by an alternate method.

$$H(z) = \frac{5}{(1 - z^{-1})(1 - \frac{1}{8}z^{-1})(1 + \frac{1}{4}z^{-1})}$$

8-40. Repeat the preceding problem for

$$H(z) = \frac{2}{(1 - z^{-1})^2(1 - \frac{1}{2}z^{-1})}$$

8-41. Determine the inverse z-transform of

$$X(z) = \frac{1}{(1 - z^{-1})(1 + 0.8z^{-1})(1 - 0.5z^{-1})}$$

8-42. Determine the inverse z-transform of

$$X(z) = \frac{z^{-3}}{(1 - z^{-1})(1 + z^{-1})(1 - 0.5z^{-1})(1 - 0.2z^{-1})}$$

8-43. Determine the inverse z-transform of

$$X(z) = \frac{z^{-1}}{(1 - z^{-1})(1 + 0.2z^{-1})^2}$$

8-44. Determine the inverse z-transform of

$$X(z) = \frac{1}{(1 - z^{-1})(1 + 0.2z^{-1})}$$

Section 8-4

8-45. Determine the pulse transfer function, $H(z)$, and the unit pulse response $h(nT)$, for each of the systems defined by the following difference equations.
(a) $y(nT) - y(nT - T) = x(nT)$
(b) $y(nT) - 2y(nT - T) + y(nT - 2T) = x(nT) + 3x(nT - 3T)$
(c) $y(nT) - y(nT - T) + 0.16y(nT - 2T) = x(nT)$
(d) $y(nT) + 0.6y(nT - T) - 0.16y(nT - 2T) = x(nT) + 4x(nT - T)$
(e) $y(nT) - 0.707y(nT - T) + 0.25y(nT - 2T) = x(nT) + x(nT - 2T)$

8-46. Determine the range of values of the parameter K for which the systems defined by the following difference equations are stable.
(a) $y(nT) - Ky(nT - T) + K^2y(nT - 2T) = x(nT)$
(b) $y(nT) - 2Ky(nT - T) + K^2y(nT - 2T) = x(nT)$
(c) $y(nT) - K^2y(nT - 2T) = x(nT - 3T)$

8-47. For each of the systems defined by the difference equations given, classify the system as linear or nonlinear, causal or noncausal, and shift-invariant or non-shift-invariant. The system input is $x(nT)$, and the system output is $y(nT)$.
(a) $y(nT) = ax(nT) + b$, a and b are constants.
(b) $y(nT) = x(nT - n_oT)$, n_o is a constant integer
(c) $y(nT) = x(nT + T) + x(nT - T)$

8-48. The convolution sum (8-132) or (8-133) can be derived by using the z-transform. Start with $Y(z) = H(z)X(z)$ and substitute the defining sums for $H(z)$ and $X(z)$. Perform the indicated multiplication and show that (8-135) results.

8-49. Convolve $h(nT)$ and $x(nT)$ shown in Figure P8-49.

8-50. A digital filter has the unit pulse response and input shown in Figure P8-50. Determine the output.

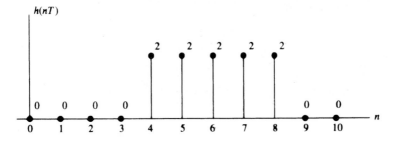

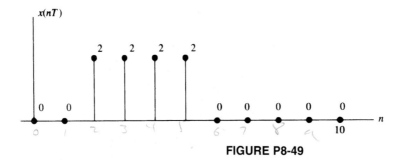

FIGURE P8-49

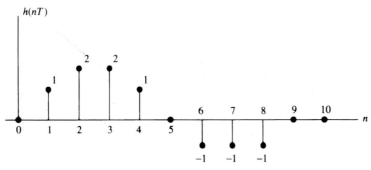

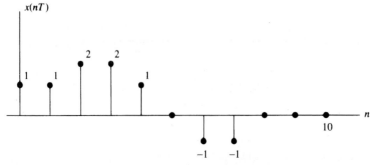

FIGURE P8-50

8-51. Work Problems 8-49 and 8-50 using MATLAB.

8-52. Convolve the two functions shown in Figure P8-52 and plot the result.

8-53. Convolve the two functions shown in Figure P8-53 and plot the result.

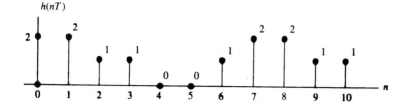

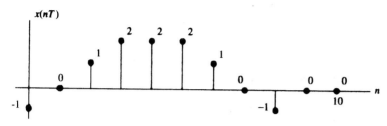

FIGURE P8-52

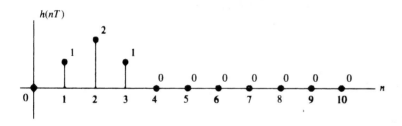

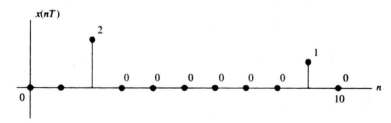

FIGURE P8-53

8-54. Work Problems 8-52 and 8-53 using MATLAB.

8-55. Determine $y(nT)$ as the convolution of $h(nT)$ and $x(nT)$, where

$$x(nT) = (\tfrac{1}{4})^n u(n)$$

$$h(nT) = (\tfrac{1}{3})^n u(n)$$

8-56. Determine $y(nT)$ as the convolution of $h(nT)$ and $x(nT)$, where

$$x(nT) = (\tfrac{1}{4})^{n-3} u(n-3)$$

$$h(nT) = (\tfrac{1}{2})^{n-5}u(n-5)$$

8-57. Determine $y(nT)$ as the convolution of $h(nT)$ and $x(nT)$, where

$$x(nT) = (\tfrac{1}{4})^n u(n-3)$$
$$h(nT) = (\tfrac{1}{3})^n u(n-5)$$

8-58. Determine $y(nT)$ as the convolution of $h(nT)$ and $x(nT)$, where

$$x(nT) = (\tfrac{1}{2})^n u(n)$$
$$h(nT) = (\tfrac{1}{3})^n u(n-1) - (\tfrac{1}{4})^n u(n-3)$$

Check your result using z-transform techniques.

8-59. A linear, shift-invariant, filter has the unit pulse response

$$h(nT) = (\tfrac{1}{3})^n u(n-1) - (\tfrac{1}{4})^n u(n-3)$$

Determine the step response of the filter.

8-60. A linear, shift-invariant, filter has the unit pulse response

$$h(nT) = (\tfrac{1}{4})^n u(n-4)$$

Determine the response to $x(nT) = h(nT)$.

8-61. Determine the amplitude and phase response of the system defined by the difference equation

$$y(nT) = x(nT) - x(nT - 10T)$$

Sketch the amplitude and phase responses as functions of normalized frequency $r = f/f_s$ for r in the range $0 \le r \le \tfrac{1}{2}$.

8-62. Repeat Problem 8-49 for the system defined by the difference equation

$$y(nT) = 2x(nT) + 4x(nT - T) + 2x(nT - 2T)$$

8-63. Repeat Problem 8-49 for the system defined by the difference equation

$$y(nT) + 0.25y(nT - T) = x(nT - T)$$

8-64. A simple discrete-time differentiator is defined by the difference equation

$$y(nT) = \frac{1}{T}[x(nT) - x(nT - T)]$$

Determine the amplitude and phase responses of the discrete-time differentiator. Plot the amplitude and phase responses as functions of normalized frequency $r = f/f_s$. Compare these responses with the amplitude and phase responses of an ideal differentiator. We will consider this problem in more detail in the following chapter.

8-65. A discrete-time filter has the unit pulse response

$$h(nT) = 4[u(n) - u(n-12)]$$

Place the frequency response in the form

$$H(e^{j\omega T}) = K \frac{\sin A(\omega)}{\sin B(\omega)} e^{jC(\omega)}$$

and determine $A(\omega)$, $B(\omega)$, $C(\omega)$, and K. Assuming a sampling frequency of 1000 Hz, determine the amplitude and phase responses of the filter at $f = 0$, 25, and 50 Hz.

8-66. For each of the systems defined below, determine the amplitude and phase responses at $f = 0$ (dc) and $f = \frac{1}{2}f_s$.

(a) $y(nT) = x(nT) + \frac{1}{2}y(nT - T)$

(b) $y(nT) = x(nT) + \frac{1}{2}x(nT - T) + \frac{1}{2}y(nT - T)$

(c) $y(nT) = x(nT) - y(nT - T) + \frac{1}{2}y(nT - 2T)$

(d) $y(nT) = x(nT) - \frac{3}{2}y(nT - T) + \frac{1}{2}y(nT - 2T)$

8-67. Show that the steady-state frequency response at $f = 0.25f_s$ is $H(j1)$, where f_s is the sampling frequency. Use this result to determine the amplitude and phase responses of the system defined by the difference equation

$$y(nT) = x(nT - T) + 0.5y(nT) - 0.2y(nT - 2T)$$

at $f = 400$ Hz, assuming that the sampling frequency is 1600 Hz.

8-68. A discrete-time system is defined by the difference equation

$$y(nT) = x(nT) + 0.2y(nT - T) - 0.5y(nT - 2T)$$

Assuming a sampling frequency of 2000 Hz, determine the amplitude and phase responses of the network at $f = 0$, $f = 500$, and $f = 1000$ Hz.

8-69. A discrete-time system is defined by the difference equation

$$y(nT) = x(nT) + 0.3y(nT - T) - 0.1y(nT - 2T)$$

Assuming that the sampling frequency of 2000 Hz, determine the amplitude and phase responses of the network at $f = 0$, $f = 500$, and $f = 1000$ Hz.

8-70. Repeat Problem 8-69, assuming that the system is defined by

$$y(nT) = \frac{8}{3}x(nT) + \frac{3}{4}y(nT - T) - \frac{1}{8}y(nT - 2T)$$

8-71. A discrete-time system is defined by

$$y(nT) = x(nT) - a_1 y(nT - T) - a_2 y(nT - 2T)$$

Determine a_1 and a_2 such that the amplitude response at $f = 0$ is 1.0 and the amplitude response at $f = 0.5f_s$ is 0.1, where f_s is the sampling frequency.

8-72. A digital filter is defined by the difference equation

$$y(nT) = \frac{3}{4}y(nT - T) + 3x(nT)$$

The input is defined by

$$x(nT) = 5 \cos \frac{\pi}{2} n$$

The steady-state output can be written

$$y_s(nT) = A \cos(\omega nT + \phi)$$

Determine a, ϕ, and the product ωT.

Section 8-5

8-73. Determine the amplitude and phase responses of the digital signal processor illustrated in Figure 8-23 for the case in which $\alpha = -\frac{3}{4}$. Determine k such that the maximum value of the amplitude response is unity. At what normalized frequency does this maximum occur? Plot the resulting amplitude and phase responses.

8-74. Repeat Problem 8-61 for the case in which $\alpha = -\frac{15}{16}$.

Section 8-6

8-75. Use contour integration to find the inverse z-transform of

$$X(z) = \frac{1}{(1 - z^{-1})(1 - 0.2z^{-1})}$$

Compare the result with Example 8-8.

8-76. Use contour integration to find the inverse z-transform of

$$X(z) = \frac{1}{(1 - 0.25z^{-1})(1 - 0.1z^{-1})}$$

Computer Exercises

8-1. In this computer example we consider the problem of reconstructing a waveform from a sequence of samples. The first step is to generate 5,000 samples of a waveform having a known bandwidth. A convenient method of accomplishing this is to generate samples of a series of sinusoids having known frequency. For example let

$$x(nT) = A \sin[2\pi f_1 t + \phi_A] + B \sin[2\pi(2f_1)t + \phi_B] + C \sin[2\pi(3f_1)t + \phi_C]$$

where A, B, C, ϕ_A, ϕ_B and ϕ_C are arbitrary. Let $f_1 T = 0.005$. Such a small value of $f_1 T$ will result in a waveform that is significantly oversampled. We will pretend that this waveform is continuous.

Next we decimate the waveform just generated by forming a vector consisting of every 10th sample of $x(nT)$. These represent samples of the continuous-time waveform. The continuous-time waveform is to be reconstructed from the samples by using (8-22) with an $h(t)$ that is a truncated version of (8-27). The reconstruction error can be evaluated by comparing the reconstructed waveform with the continuous-time waveform from which the samples were extracted.

Evaluate the reconstruction error as a function of the time duration of $h(t)$. It might also be interesting to change the sampling frequency and observe the impact of this change.

8-2. The purpose of this computer exercise is to study the quantizing process and the effect of quantizing errors. As a first step develop a MATLAB M-file that allows you to quantize a signal having a maximum value M and a minimum value of $-M$. In other words the signal has symmetrical maximum and minimum limits with a dynamic range, D, of $2M$. The quantizer input signal is to be quantized into 2^n levels where $n = 2, 4$ or 8.

Using the quantizer just developed, quantize a sinusoidal signal for $n = 2, 4$ and 8. Determine the mean-square value of the quantizing error and compare with the theoretical values.

8-3. In this computer exercise we wish to develop an interactive system that allows us to enter filter parameters and observe important filter characteristics that result from these parameters. We start be generating a unit circle and displaying the z-plane. (Note: For purposes of this exercise it might be useful to represent the z-plane in polar coordinates. The MATLAB routine **polar** might be useful for accomplishing this.)

Develop a routine that allows you to enter the poles, zeros and any appropriate filter constants from the command line prompt. It might be useful to write the program in such a way that you are prompted for the pole and/or zero locations and also be prompted for any necessary constants. Once the poles, zeros and necessary constants have been entered they serve as inputs to a MATLAB routine that generates the amplitude response (in dB), phase response and group delay of the digital filter defined by the entered data. The ambitious student might wish to consider entering the poles and zeros using the mouse.

8-4. Using appropriate functions from the MATLAB Symbolic Math Toolbox, develop an m-file for performing symbolic convolution. The input to the m-file is to be two character strings representing the two functions to be convolved. The output is to be the symbolic convolution of the two functions. Thoroughly test the resulting routine.

8-5. In this computer exercise a digital filter is implemented and used to filter, on a sample-by-sample basis, a vector of sample values. The digital filter to be used is defined by (8-194) with the constraint of (8-205) applied. In other words

$$y(nT) = (1 - \alpha)x(nT) + \alpha y(nT - T)$$

In order to use this filter generate a sampled sinusoid of the form

$$x(nT) = A \sin(2\pi f_i nT)$$

Note that we can represent $f_i T$ by

$$f_i T = \frac{f_i}{f_s} = r_i$$

so that

$$x(nT) = A \sin(2\pi r_i n)$$

We wish to implement the filter on a sample-by-sample basis (not typically desirable in MATLAB). In other words, the expression for $y(nT)$ is embedded in a "for loop" which takes the nth sample of the input from an array and generates the nth sample of the output, $x(nT)$.

Choose a value of α and r_i, execute the program, and generate the output array $y(nT)$. The amplitude of the input sinusoid, A, is arbitrary. After waiting an appropriate period of time for the transient response of the filter to become negligible, plot on the same set of axes one or two periods of the filter input $x(nT)$ and the filter output $y(nT)$. By making appropriate measurements on the filter input $x(nT)$ and the filter output $y(nT)$ determine the amplitude and phase response of the filter at the particular value of r_i previously chosen. By varying r_i plot the filter amplitude response (in dB) and the filter phase response. Do this for several values of α and in each case compare the results with the theoretical results.

8-6. If the contour in (8-219) is chosen as a circle, the integral can be expressed in terms of a real variable of integration. By the simple change of variables $z = r_c e^{j\theta}$ the contour integral (8-129) can be placed in the form

$$x(nT) = \frac{1}{2\pi} \int_0^{2\pi} \frac{r_c^{n+1} e^{j(n+1)\theta}}{r_c e^{j\theta} - K} \, d\theta$$

where r_c is the radius of the contour. After convincing yourself that the preceding equation is correct, write a MATLAB program to evaluate the integral for a given value of K and specific values of n. Do you get the correct result? Experiment with different values of r_c for a fixed value of K. Comment on your results.

CHAPTER 9

Analysis and Design of Digital Filters

9-1 Introduction

Throughout this book we have been concerned with the analysis of continuous-time and discrete-time linear systems. We now turn our attention to a much different type of problem. This is the synthesis problem, wherein one is asked not to analyze a given system but to find the system that satisfies a given set of specifications. Typically, the synthesis problem for discrete-time systems is to determine the pulse transfer function $H(z)$ that satisfies the given specifications. Then, as we will see in the first section of this chapter, a number of different systems (structures or realizations) can be found, all of which have the same pulse transfer function $H(z)$. The system designer then chooses that realization which is most compatible with the requirements.

A number of different techniques that allow us to go from a set of specifications to a pulse transfer function are studied in this chapter. All have their advantages and disadvantages. These will be summarized later. Several synthesis techniques have as their starting point the transfer function of an analog filter or system that satisfies a given set of specifications. Synthesis of the digital filter or system then simply consists of finding the discrete-time equivalent of some given analog system. For those who are not familiar with the terminology and characteristics of analog filters, a brief review is contained in Appendix E.

9-2 Structures of Digital Processors

In this section several different techniques for realizing pulse transfer functions are examined.

Direct-Form Realizations

In Chapter 8 we determined the general form of the pulse transfer function of a fixed discrete-time system. The result was (8-165), which can be written as

$$H(z) = \frac{Y(z)}{X(z)} = \frac{\sum_{i=0}^{r} L_i z^{-i}}{1 + \sum_{j=1}^{m} K_j z^{-j}} \tag{9-1}$$

The difference equation corresponding to (9-1) is

$$y(nT) = \sum_{i=0}^{r} L_i x(nT - iT) - \sum_{j=1}^{m} K_j y(nT - jT) \tag{9-2}$$

and is realized by the structure illustrated in Figure 8-20. If the storage of the present and past inputs and storage of past outputs is represented through the use of the z^{-1} delay operator, Figure 9-1 results.

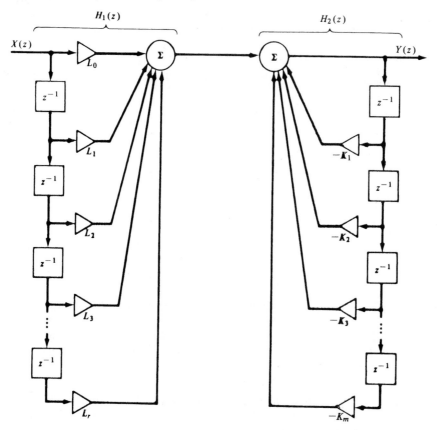

FIGURE 9-1. Direct Form I realization.

The block labeled z^{-1} represents storage of a sample value (a number) for one sample period T. The summing operation illustrated in Figure 8-20 has been broken into two summing junctions, each of which forms one of the summation terms in (9-2). The structure illustrated in Figure 9-1 is known as the Direct Form I realization of the digital signal processor defined by the difference equation (9-2).

Implementation of the processor shown in Figure 9-1 requires $r + m$ unit delays (z^{-1} elements). This requirement can be significantly reduced in many applications by identifying the processor as the cascade of two networks, $H_1(z)$ and $H_2(z)$. The first network, $H_1(z)$, realizes the zeros of the processor and the second network, $H_2(z)$, realizes the poles. Interchanging $H_1(z)$ and $H_2(z)$ results in the structure shown in Figure 9-2. Clearly, the signals at points A_1 and B_1 are equal, as are the signals A_2 and B_2, and so on down the z^{-1} delay elements. Thus the cascaded z^{-1} elements can be combined to yield the realization illustrated in Figure 9-3, which is known as the Direct Form II realization. Figure 9-3 is drawn for the case in which the number of poles, m, exceeds the number of zeros, r.

Cascade and Parallel Realizations

The Direct Form I realization resulted directly from the general discrete-time difference equation, (9-2). A simple manipulation of the block diagram resulted in the more practical Direct Form II realization. The cascade realization results by recognizing that since poles or zeros of the pulse transfer function are either real or occur in complex conjugate pairs, $H(z)$ can be written in the factored form

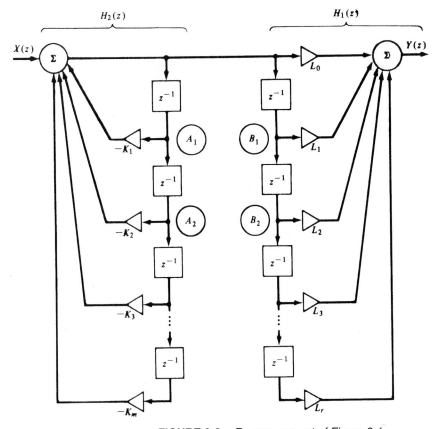

FIGURE 9-2.   Rearrangement of Figure 9-1.

$$H(z) = Kz^{-M} \frac{\prod_{i=1}^{N_1} (1 - a_i z^{-1}) \prod_{j=1}^{N_2} (1 - b_j^* z^{-1})(1 - b_j z^{-1})}{\prod_{k=1}^{D_1} (1 - c_k z^{-1}) \prod_{l=1}^{D_2} (1 - d_l^* z^{-1})(1 - d_l z^{-1})} \qquad (9\text{-}3)$$

In this expression we have N_1 real zeros at values $z = a_i$, N_2 complex conjugate zero pairs at $z = b_j$ and $z = b_j^*$, D_1 real poles at $z = c_k$, and D_2 complex conjugate pole pairs at $z = d_l$ and $z = d_l^*$. The real poles and real zeros are typically realized using the Direct Form II realization. For example, a general term of the form

$$H_1(z) = \frac{1 - a_i z^{-1}}{1 - c_k z^{-1}} \qquad (9\text{-}4)$$

is realized using the structure illustrated in Figure 9-4a. The complex conjugate poles and zeros are realized by pairs. For example, the general term

$$H_2(z) = \frac{(1 - b_j z^{-1})(1 - b_j^* z^{-1})}{(1 - d_l z^{-1})(1 - d_l^* z^{-1})} = \frac{1 - (b_j + b_j^*)z^{-1} + b_j b_j^* z^{-2}}{1 - (d_l + d_l^*)z^{-1} + d_l d_l^* z^{-2}} \qquad (9\text{-}5)$$

is realized by the structure illustrated in Figure 9-4b. Since $A + A^*$ and AA^* are both real numbers, where A is complex, all multipliers in Figure 9-4b are real numbers. The multiplier Kz^{-M} is included in (9-3) to realize any multiplicative constant and also to allow for the fact that $H(z)$ may have a factor of the form z^{-M}, where $M > 0$.

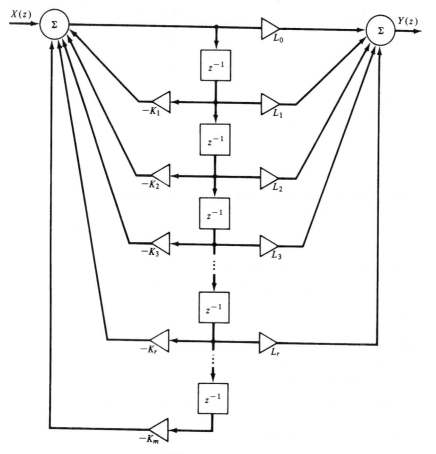

FIGURE 9-3. Direct Form II realization.

The parallel form results by expanding $H(z)$ using partial-fraction expansion. The general form of the partial-fraction expansion for simple poles is

$$H(z) = \sum_{i=0}^{M} A_i z^{-i} + \sum_{k=1}^{D_1} B_k \frac{1}{1 - c_k z^{-1}} + \sum_{l=1}^{D_2} C_l \frac{1 - e_l z^{-1}}{(1 - d_l z^{-1})(1 - d_l^* z^{-1})} \tag{9-6}$$

in which the first summation is included to realize the terms in the partial-fraction expansion that result if $r > m$ in (9-1). The second summation realizes the real poles, and the third summation realizes the complex conjugate pole pairs.

EXAMPLE 9-1 _____

In this example, a cascade and parallel realization of

$$H(z) = \frac{(1 - z^{-1})^3}{(1 - \frac{1}{2}z^{-1})(1 - \frac{1}{8}z^{-1})} \tag{9-7}$$

will be found.

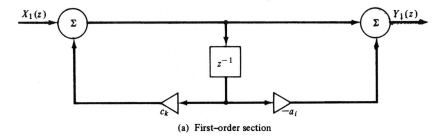

(a) First–order section

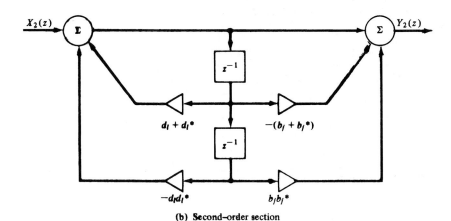

(b) Second–order section

FIGURE 9-4. Basic first- and second-order filters.

The cascade realization is relatively simple. For example, $H(z)$ can be written in the form

$$H(z) = \frac{1 - z^{-1}}{1 - \frac{1}{2}z^{-1}} \frac{1 - z^{-1}}{1 - \frac{1}{8}z^{-1}} (1 - z^{-1})$$

(9-8)

which is realized using the structure illustrated in Figure 9-5a.

The parallel realization is found by first writing $H(z)$ as

$$H(z) = \frac{(z - 1)^3}{z(z - \frac{1}{2})(z - \frac{1}{8})}$$

(9-9)

In order to find the parallel realization, we first determine the partial-fraction expansion of $z^{-1}H(z)$, just as we did to determine inverse z-transforms. This yields

$$\frac{H(z)}{z} = \frac{(z - 1)^3}{z^2(z - \frac{1}{2})(z - \frac{1}{8})} = \frac{A}{z^2} + \frac{B}{z} + \frac{C}{z - \frac{1}{2}} + \frac{D}{z - \frac{1}{8}}$$

(9-10)

The constants A, B, C, and D are

$$A = \lim_{z \to 0} \frac{(z - 1)^3}{(z - \frac{1}{2})(z - \frac{1}{8})} = -16$$

(9-11a)

$$B = \lim_{z \to 0} \frac{d}{dz}\left[\frac{(z - 1)^3}{(z - \frac{1}{2})(z - \frac{1}{8})}\right] = -112$$

(9-11b)

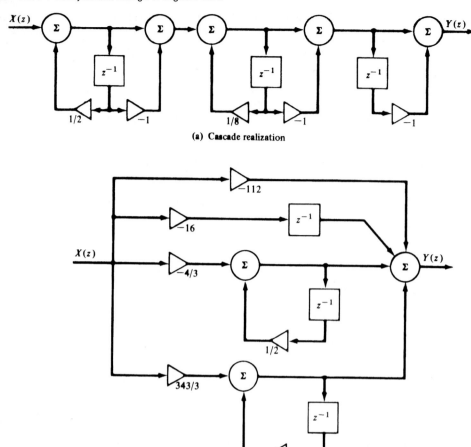

FIGURE 9-5. Cascade and parallel realizations for Example 9-1.

$$C = \lim_{z \to 1/2} \frac{(z - 1)^3}{z^2(z - \frac{1}{8})} = -\frac{4}{3} \qquad (9\text{-}11c)$$

$$D = \lim_{z \to 1/8} \frac{(z - 1)^3}{z^2(z - \frac{1}{2})} = \frac{343}{3} \qquad (9\text{-}11d)$$

Thus

$$H(z) = -112 - 16z^{-1} - \frac{4}{3} \frac{1}{1 - \frac{1}{2}z^{-1}} + \frac{343}{3} \frac{1}{1 - \frac{1}{8}z^{-1}} \qquad (9\text{-}12)$$

which is realized using the structure illustrated in Figure 9-5b. This example problem is chosen to illustrate the role of the first summation in (9-6).

MATLAB *Application*

Since (9-7) gives the numerator and denominator of the transfer function in factored form, it is only the parallel realization that requires any significant computational effort. The MATLAB function **residuez** can determine (9-12) as illustrated by the following MATLAB code. Note the use of convolution for determining the numerator and denominator coefficient vectors. Also note that the k vector provides the first two terms in (9-12), made necessary because of the z^2 factor in the denominator of $H(z)/z$.

```
EDU» c9ex1
n1 = [1 -1];                    % Define numerator factor
num = conv(n1,conv(n1,n1));     % Calculate numerator
d1 = [1 -1/2];                  % Define denominator factor
d2 = [1 -1/8];                  % Define second den. factor
den = conv(d1,d2);             % Calculate denominator
[r,p,k] = residuez(num,den);   % Calculate residues
r                              % Display residues

r =

    -1.3333
   114.3333

p                              % Display poles

p =

    0.5000
    0.1250

k                              % Display k vector

k =

    -112    -16
```

If quantizing errors are neglected, all realizations of the pulse transfer function are equivalent. If quantizing errors are important and must be considered, it is then no longer true that all realizations are equivalent. For example, suppose that the computed value of one of the K_j's is an irrational number. Before the $H(z)$ can be realized, the value of K_j must be truncated or rounded to a value that can be represented by a finite number of bits. This changes the denominator polynomial slightly and as a result *all* poles of the transfer function change by some amount. For high-order systems this quickly becomes a complicated problem. If the original $H(z)$ has poles close to the unit circle, the movement of the poles due to quantization of K_j can actually drive the system unstable.

Consider, however, the corresponding problem if the system is realized using a cascade or parallel combination of first- and second-order sections. In this case, truncating or rounding a coefficient value causes movement of at most two poles. This is a much more tolerable situation and is a basic advantage of cascade or parallel realizations.

EXAMPLE 9-2

Up to this point we have considered only systems with real poles. In this example we consider a system having a complex conjugate pole pair at

$$z = ae^{\pm j\theta} \tag{9-13}$$

as shown in Figure 9-6a. The transfer function for this system is given by

$$H(z) = \frac{z^2}{(z - ae^{+j\theta})(z - ae^{-j\theta})} \tag{9-14}$$

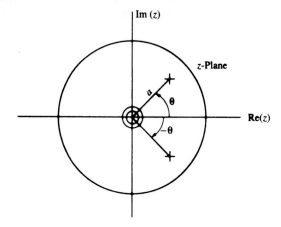

(a) Pole-zero plot for transfer function

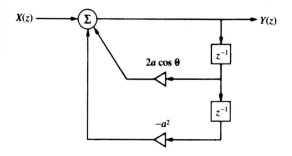

(b) Direct form implementation

FIGURE 9-6. Pole-zero plot and implementation for Example 9-2.

which can be written

$$H(z) = \frac{z^2}{z^2 - 2a(\cos \theta)z + a^2} \tag{9-15}$$

or

$$H(z) = \frac{1}{1 - 2a(\cos \theta)z^{-1} + a^2 z^{-2}} \tag{9-16}$$

The Direct Form realization is shown in Figure 9-6b. Since the numerator is a constant, there is no difference between the Direct Form I and the Direct Form II realizations of this system.

It is interesting to examine the frequency response of this system as the radius vector a and the pole angle θ are varied. In terms of normalized frequency, $r = f/f_s$, the frequency response becomes

$$H(e^{j2\pi r}) = \frac{1}{1 - 2a(\cos \theta)e^{-j2\pi r} + a^2 e^{-j4\pi r}} \tag{9-17}$$

The magnitude response is shown in Figure 9-7a for $\theta = \pi/2$ and various values of a. Note that the maximum gain of the filter increases as the pole approaches the unit circle. The phase responses corresponding to the amplitude responses shown in Figure 9-7a are shown in Figure 9-7b. The

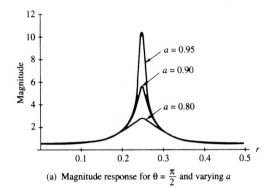

(a) Magnitude response for $\theta = \dfrac{\pi}{2}$ and varying a

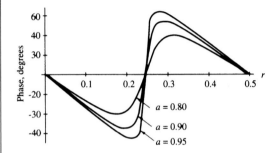

(b) Phase response for $\theta = \dfrac{\pi}{2}$ and varying a

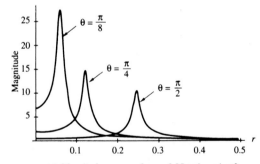

(c) Magnitude response for $a = 0.95$ and varying θ

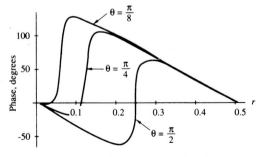

(d) Phase response for $a = 0.95$ and varying θ

FIGURE 9-7. Magnitude and phase response for second-order filter.

magnitude response corresponding to (9-15) is shown in Figure 9-7c for $a = 0.95$ and various values of θ. The corresponding phase responses are shown in Figure 9-7d. Note that the maximum gain of the filter increases as θ, and consequently the resonant frequency of the filter, decreases.

9-3 Discrete-Time Integration

As a specific example of a discrete-time approximation to a continuous-time process, we will now consider the discrete-time integrator. Both rectangular and trapezoidal integration are considered and we will see that while both of these techniques are approximations to continuous-time integration, they have distinctly different characteristics.

Rectangular Integration

A discrete-time integrator is defined by

$$y(nT) = y(nT - T) + \Delta y(nT - T, nT) \tag{9-18}$$

where $\Delta y(nT - T, nT)$ is the value added to the integral in the time increment $(nT - T, nT)$. The algorithm for rectangular integration is illustrated in Figure 9-8a, in which the value of $\Delta y(nT - T, nT)$ is defined to be

$$\Delta y(nT - T, nT) \triangleq Tx(nT - T) \tag{9-19}$$

which results in the difference equation

$$y(nT) = y(nT - T) + Tx(nT - T) \tag{9-20}$$

The pulse transfer function is determined by z-transforming (9-20), which yields

$$Y(z) = z^{-1}Y(z) + z^{-1}TX(z) \tag{9-21}$$

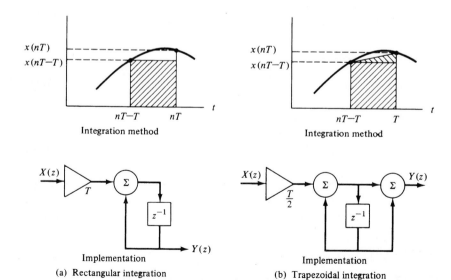

FIGURE 9-8. Rectangular and trapezoidal integration.

Solving for $Y(z)/X(z)$ results in

$$H(z) = \frac{Y(z)}{X(z)} = \frac{z^{-1}T}{1 - z^{-1}} \tag{9-22}$$

The digital processor that accomplishes rectangular integration is also illustrated in Figure 9-8a.

Trapezoidal Integration

The algorithm for trapezoidal integration is illustrated in Figure 9-8b. It follows that $\Delta y(nT - T, nT)$ is

$$\Delta y(nT - T, nT) \triangleq Tx(nT - T) + \tfrac{1}{2}T[x(nT) - x(nT - T)] \tag{9-23}$$

which gives the difference equation

$$y(nT) = y(nT - T) + Tx(nT - T) + \tfrac{1}{2}T[x(nT) - x(nT - T)] \tag{9-24}$$

The z-transform of the difference equation is

$$Y(z) = z^{-1}Y(z) + \frac{T}{2}X(z) + \frac{T}{2}z^{-1}X(z) \tag{9-25}$$

which yields

$$H(z) = \frac{Y(z)}{X(z)} = \frac{T}{2}\frac{1 + z^{-1}}{1 - z^{-1}} \tag{9-26}$$

for the pulse transfer function. The realization of a trapezoidal integrator is illustrated in Figure 9-8b.

EXAMPLE 9-3

As shown in this example the amplitude and phase characteristics of the rectangular and trapezoidal integrators are easily derived. From (9-22) with $z = e^{j\omega T}$, the transfer function of the rectangular integrator, $H_r(e^{j\omega T})$, is

$$H_r(e^{j\omega T}) = \frac{Te^{-j\omega T}}{1 - e^{-j\omega T}} = \frac{Te^{-j\omega T/2}}{e^{+j\omega T/2} - e^{-j\omega T/2}} \tag{9-27}$$

or

$$H_r(e^{j\omega T}) = \frac{Te^{-j\omega T/2}}{2j \sin \omega T/2} \tag{9-28}$$

In terms of the normalized frequency $r = f/f_s$, this becomes

$$H_r(e^{j2\pi r}) = \frac{Te^{-j\pi r}}{2j \sin \pi r} \tag{9-29}$$

In the range $0 \le r \le \tfrac{1}{2}$, $\sin \pi r \ge 0$. Thus the amplitude response $A_r(r)$ of the integrator is given by

$$A_r(r) = \frac{T}{2 \sin \pi r}, \qquad 0 \le r \le \tfrac{1}{2} \tag{9-30a}$$

and the phase, $\phi_r(r)$, is given by

$$\phi_r(r) = -\frac{\pi}{2} - \pi r, \qquad 0 \le r \le \tfrac{1}{2} \tag{9-30b}$$

The sinusoidal steady-state response of the trapezoidal integrator is, from (9-26),

$$H_t(e^{j\omega T}) = \frac{T}{2} \frac{1 + e^{-j\omega T}}{1 - e^{-j\omega T}} \tag{9-31}$$

which can be written

$$H_t(e^{j\omega T}) = \frac{T}{2} \frac{e^{+j\omega T/2} + e^{-j\omega T/2}}{e^{+j\omega T/2} - e^{-j\omega T/2}} \tag{9-32}$$

or

$$H_t(e^{j\omega T}) = \frac{T}{2} \frac{\cos \omega T/2}{j \sin \omega T/2} \tag{9-33}$$

In terms of normalized frequency, (9-33) is

$$H_t(e^{j2\pi r}) = \frac{T}{2} \frac{\cos \pi r}{j \sin \pi r} \tag{9-34}$$

In the range $0 \le r \le \frac{1}{2}$, both $\cos \pi r$ and $\sin \pi r$ are nonnegative. Thus it follows from (9-34) that the amplitude response of a trapezoidal integrator is given by

$$A_t(r) = \frac{T}{2} \frac{\cos \pi r}{\sin \pi r}, \qquad 0 \le r \le \frac{1}{2} \tag{9-35}$$

and the phase response is

$$\phi_t(r) = -\frac{\pi}{2}, \qquad 0 \le r \le \frac{1}{2} \tag{9-36}$$

The magnitude and phase responses of the rectangular and trapezoidal integrators are given in Figure 9-9 for the case where the sampling period T is equal to unity. Also illustrated are the amplitude and phase responses of the ideal continuous integrator, defined by $H(f) = 1/j2\pi f$. Both the rectangular and trapezoidal integrators exhibit amplitude distortion but neither exhibit phase distortion. The rectangular integrator does, however, delay the input signal by $T/2$ seconds as can be determined from the slope of the phase response.

It is interesting to consider the form of (9-28) and (9-33) for input frequencies much less than the sampling frequency (i.e., for $\omega T \ll 1$). For this case

$$\frac{\cos \omega T/2}{\sin \omega T/2} \simeq \frac{1}{\sin \omega T/2} \simeq \frac{2}{\omega T} \tag{9-37}$$

so that (9-27) and (9-33) become

$$H_r(e^{j\omega T}) \simeq \frac{1}{j\omega} e^{-j\omega T/2}, \qquad \omega T \ll 1 \tag{9-38}$$

and

$$H_t(e^{j\omega T}) = \frac{1}{j\omega}, \qquad \omega T \ll 1 \tag{9-39}$$

respectively. Thus, for low input frequencies, the rectangular and the trapezoidal integrator both closely approximate an ideal integrator. The basic difference for low input frequencies is

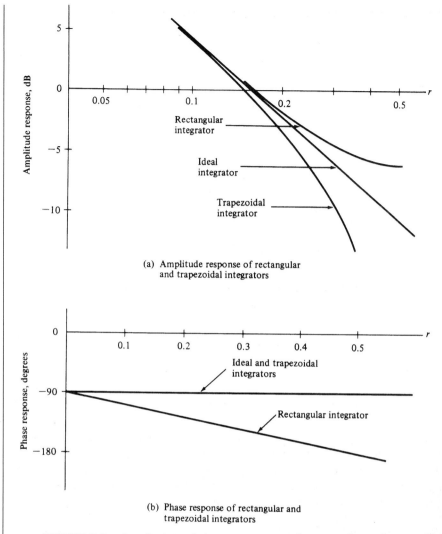

(a) Amplitude response of rectangular
and trapezoidal integrators

(b) Phase response of rectangular and
trapezoidal integrators

FIGURE 9-9. Amplitude and phase responses of rectangular and trapezoidal integrators.

that the rectangular integrator has a $T/2$-second group delay and the trapezoidal integrator has zero group delay.

Example 9-3 illustrates the importance of investigating the performance of a system by studying both the time- and frequency-domain characteristics of the system. A simple comparison of the integration methods illustrated in Figure 9-8 leads us to the conclusion that trapezoidal integration yields a better approximation to the area under a curve. However, a comparison of the frequency responses in Figure 9-9 provides additional insight. We see that the rectangular integrator provides less attenuation of high frequencies than an ideal integrator, whereas the trapezoidal integrator provides more high-frequency attenuation than an ideal integrator.

The trapezoidal integration algorithm will gain additional significance when the bilinear z-transform is studied in the following section.

EXAMPLE 9-4

In this example we determine a digital equivalent (one of many) for the system defined by the differential equation

$$\frac{dy}{dt} = x(t) - \beta y(t) \tag{9-40}$$

The block diagram of the system represented by the differential equation is shown in Figure 9-10a. The only block in the system having memory is the integrator. A simple digital equivalent is derived by substituting a trapezoidal integrator for the analog integrator. Using (9-26) results in the block diagram illustrated in Figure 9-10b.

The transfer function of the system illustrated in Figure 9-10b is easily derived. From the block diagram it follows that

$$Y(z) = \frac{T}{2}\frac{1 + z^{-1}}{1 - z^{-1}}[X(z) - \beta Y(z)] \tag{9-41}$$

which reduces to

$$H(z) = \frac{Y(z)}{X(z)} = \frac{T(1 + z^{-1})}{(\beta T + 2) + (\beta T - 2)z^{-1}} \tag{9-42}$$

We will return to this problem in a later section to derive this same result by a different method.

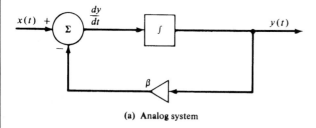

(a) Analog system

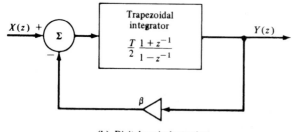

(b) Digital equivalent using trapezoidal integration

FIGURE 9-10. Analog and digital equivalents using trapezoidal integration.

9-4 Infinite Impulse Response (IIR) Filter Design

We now turn our attention to a fundamental problem, the problem of determining the required coefficients of a linear difference equation in order that a specified task be performed. This *is* the synthesis problem. The resulting structures are referred to as *digital filters* and are classified as either *infinite-duration* unit pulse response (IIR) filters or *finite-duration* unit pulse response (FIR) filters, depending on the form of the unit pulse response of the system. We will see that IIR filters are usually implemented using structures having feedback (recursive structures) and FIR filters usually are implemented using structures with no feedback (nonrecursive structures), but this is not a necessary restriction. In this section we examine the synthesis of IIR filters, and in the following section we examine the synthesis of FIR filters.

In designing IIR filters, the usual starting point will be an analog (continuous-time) system transfer function, $H_a(s)$. The problem will be to determine a discrete-time system, in other words, to determine $H(z)$, which in some sense approximates the performance of the analog system. The process of deriving a digital filter from an analog prototype can be performed using either time-domain or frequency-domain techniques and we will see that satisfactory correspondence between the analog and digital systems in the time domain does not imply that satisfactory correspondence exists in the frequency domain.

Synthesis in the Time Domain—Invariant Design

The concept of time-domain invariance is a simple one. A digital filter is equivalent to an analog filter, in the time-domain-invariance sense, if equivalent inputs yield equivalent outputs. The simplest, as well as the most often used, invariant design is the impulse-invariant digital filter. After a quick look at the impulse-invariant filter we will consider the general invariant synthesis technique.

Impulse-Invariant Design. The synthesis technique for an impulse-invariant digital filter is illustrated in Figure 9-11. Assume an impulse function signal source. The output of the analog filter will be the unit impulse response $h_a(t)$. Sampling this impulse response yields the sample values $h_a(nT)$.

We now consider a second signal path in which the analog filter and sampler are replaced by a sampler and a digital filter. The discrete-time equivalent of a unit impulse function is a unit pulse (the weight

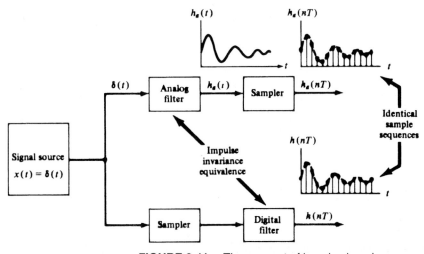

FIGURE 9-11. The concept of impulse invariance.

of the impulse denotes the sample value). Therefore, the input to the digital filter is a unit pulse, and the digital filter output is the unit pulse response of the digital filter. If the parameters of the digital filter are adjusted so that this unit pulse response is identical to the previously specified $h_a(nT)$, the analog filter and sampler are equivalent to the sampler and digital filter. With this condition we say that the digital filter is the *impulse-invariant* equivalent of the analog filter.

To illustrate the impulse-invariant synthesis technique, assume that the transfer function of an analog filter has m distinct real poles such that the partial-fraction expansion of $H_a(s)$ has the form

$$H_a(s) = \sum_{i=1}^{m} \frac{K_i}{s + s_i} \tag{9-43}$$

in which $s = -s_i$ are the pole locations and K_i is the residue of the pole at $-s_i$. Taking the inverse Laplace transform yields

$$h_a(t) = \sum_{i=1}^{m} K_i e^{-s_i t}, \qquad t \geq 0 \tag{9-44}$$

for the unit impulse response of the analog filter. The z-transform of the sampled unit impulse response is

$$H_1(z) = \sum_{n=0}^{\infty} h_a(nT)z^{-n} \tag{9-45}$$

which, upon substitution of (9-44), yields

$$H_1(z) = \sum_{n=0}^{\infty} \sum_{i=1}^{m} K_i (e^{-s_i T} z^{-1})^n \tag{9-46}$$

Interchanging the order of summation allows (9-46) to be written

$$H_1(z) = \sum_{i=1}^{m} K_i \sum_{n=0}^{\infty} (e^{-s_i T} z^{-1})^n \tag{9-47}$$

which gives, after summing on n,

$$H_1(z) = \sum_{i=1}^{m} \frac{K_i}{1 - e^{-s_i T} z^{-1}} \tag{9-48}$$

The filter described in (9-48) has a unit pulse response equivalent to the sampled impulse response of the analog filter from which it was derived. We will later see, however, that the amplitude response of the digital filter will be scaled by f_s due to the sampling operation. Therefore, scaling the amplitude response of the digital filter to approximate the amplitude response of the analog filter requires multiplication of $H_1(z)$ by $T = 1/f_s$. Thus the pulse transfer function of the impulse invariant digital filter equivalent to (9-43) is

$$H(z) = T \sum_{i=1}^{m} \frac{K_i}{1 - e^{-s_i T} z^{-1}} \tag{9-49}$$

It should be noted that this synthesis technique yields the parallel realization illustrated in Figure 9-12.

Now that the impulse-invariant design technique has been illustrated, we examine the general concept of invariant design. Then we shall consider several example problems.

General Time-Invariant Synthesis. The general concept of time-domain invariance is illustrated in Figure 9-13. The input to the analog filter is $x(t)$, and the input to the digital filter is $x(nT)$, the sam-

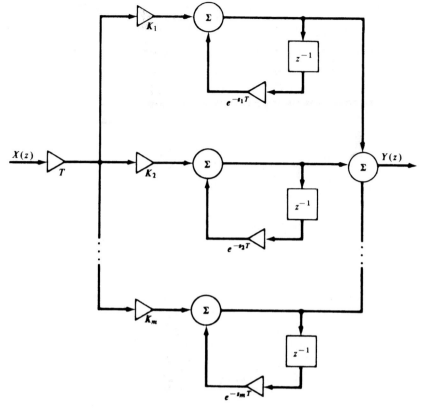

FIGURE 9-12. Impulse-invariant digital filter equivalent to (9-43).

pled version of $x(t)$. With these equivalent inputs applied to the analog and digital filters, the filter co-efficients that determine $H(z)$ are adjusted until the sampled output of the analog filter corresponds to the output of the digital filter.

The required $H(z)$ is determined by writing

$$y(t) = \mathcal{L}^{-1}[H_a(s)X_a(s)] \tag{9-50}$$

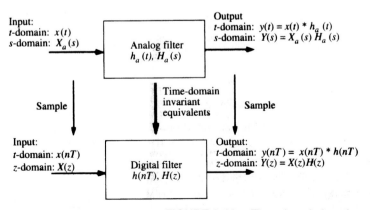

FIGURE 9-13. Time-domain invariance.

where $\mathcal{L}^{-1}$ denotes the inverse Laplace transform. The output samples of the digital filter are defined to be

$$y(nT) = [\mathcal{L}^{-1}[H_a(s)X_a(s)]]|_{t = nT} \tag{9-51}$$

The z-transform of this quantity yields the z-domain output of the digital filter. This gives

$$Y(z) = H(z)X(z) = G\mathcal{Z}\{[\mathcal{L}^{-1}[H_a(s)X_a(s)]]|_{t = nT}\} \tag{9-52}$$

where $\mathcal{Z}$ denotes the z-transform. The constant, G, has been included to give similar frequency responses for both the analog and digital filters. Solving (9-52) for $H(z)$ yields the general synthesis equation

$$H(z) = \frac{G}{X(z)} \mathcal{Z}\{[\mathcal{L}^{-1}[H_a(s)X_a(s)]]|_{t = nT}\} \tag{9-53}$$

The impulse-invariant filter is perhaps the most popular digital filter synthesized using a time-domain invariant synthesis technique. For impulse invariance, (9-53) is used with

$$X(z) = X_a(s) = 1 \tag{9-54}$$

and

$$G = T \tag{9-55}$$

To understand the role of G in this example, recall that the sampling operation, assuming the impulse-train sampling function, multiplies the spectrum of the sampled signal by $C_n = 1/T = f_s$ [see (8-16)]. Multiplication by T is required if the spectrum, which in this case is the transfer function of the filter, is to be scaled so that the analog and digital filters have approximately equivalent amplitude responses. Thus the synthesis equation is

$$H(z) = T\mathcal{Z}\{[\mathcal{L}^{-1}[H_a(s)]]|_{t = nT}\} \tag{9-56}$$

An example illustrates the method.

EXAMPLE 9-5

The purpose of this example is to illustrate the design of impulse invariant digital filters. Particular attention will be paid to scaling the amplitude response and compensating for the effects of aliasing. We begin by considering the analog prototype

$$H_a(s) = \frac{0.5(s + 4)}{(s + 1)(s + 2)} \tag{9-57}$$

By partial-fraction expansion $H_a(s)$ can be written

$$H_a(s) = \frac{1.5}{s + 1} - \frac{1}{s + 2} \tag{9-58}$$

The unit impulse response of the analog filter is, for $t \geq 0$,

$$h_a(t) = 1.5e^{-t} - e^{-2t} \tag{9-59}$$

Thus the unit pulse response of the desired digital filter is, for $n \geq 0$,

$$h_a(nT) = 1.5e^{-nT} - e^{-2nT} \tag{9-60}$$

which upon z-transforming and multiplying by T yields

$$H(z) = \frac{1.5T}{1 - e^{-T}z^{-1}} - \frac{T}{1 - e^{-2T}z^{-1}} \tag{9-61}$$

The parallel realization of the digital filter is illustrated in Figure 9-14.

It is interesting to compare the frequency response of the two filters. For the analog filter, the frequency response is

$$H_a(j\omega) = \frac{0.5(4 + j\omega)}{(1 + j\omega)(2 + j\omega)} \tag{9-62}$$

while, for the digital filter, it is

$$H(e^{j\omega T}) = \frac{1.5T}{1 - e^{-T}e^{-j\omega T}} - \frac{T}{1 - e^{-2T}e^{-j\omega T}} \tag{9-63}$$

It follows that dc responses of the analog and digital filters are given by

$$H_a(0) = 1 \tag{9-64}$$

and

$$H(e^{j0}) = H(1) = \frac{1.5T}{1 - e^{-T}} - \frac{T}{1 - e^{-2T}} \tag{9-65}$$

respectively. Thus the dc responses are different due to aliasing at dc. However, if the sampling frequency is high, T is small, which yields

$$e^{-T} \simeq 1 - T \tag{9-66}$$

and

$$e^{-2T} \simeq 1 - 2T \tag{9-67}$$

which results in

$$H(1) \simeq \frac{1.5T}{1 - (1 - T)} - \frac{T}{1 - (1 - 2T)} = 1 \tag{9-68}$$

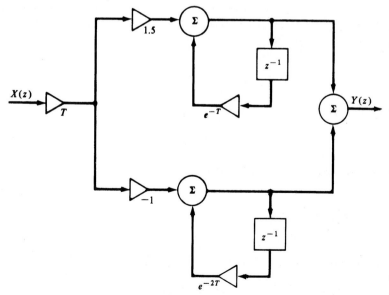

FIGURE 9-14. Impulse-invariant filter for Example 9-5.

We can illustrate this concept using a numerical example. Let the sampling frequency be 20 rad/s so that

$$T = \frac{2\pi}{20} = 0.31416 \quad s \qquad (9\text{-}69)$$

$$e^{-T} = 0.7304 \qquad (9\text{-}70)$$

and

$$e^{-2T} = 0.5335 \qquad (9\text{-}71)$$

With these values, the dc gain of the impulse-invariant digital filter is, from (9-65)

$$H(1) = 1.0745 \qquad (9\text{-}72)$$

which is slightly larger than the gain of the analog filter, as expected, due to the aliasing at dc. For a large sampling frequency, T is small and the approximation $e^{-x} \simeq 1 - x$ becomes valid. For sufficiently large sampling frequency, the aliasing effect becomes negligible and the dc gain is unity as indicated by (9-68).

Figure 9-15 illustrates the amplitude response of both the analog and digital filters. From (9-63) the amplitude response can be written as

$$|H(e^{j2\pi r})| = \frac{\pi}{10} \sqrt{\frac{0.25488 - 0.06983 \cos 2\pi r}{2.74925 - 3.51275 \cos 2\pi r + 0.77932 \cos 4\pi r}} \qquad (9\text{-}73)$$

Also, the frequency response of the analog filter is

$$|H_a(j\omega)| = 0.5 \sqrt{\frac{16 + \omega^2}{4 + 5\omega^2 + \omega^4}} \qquad (9\text{-}74)$$

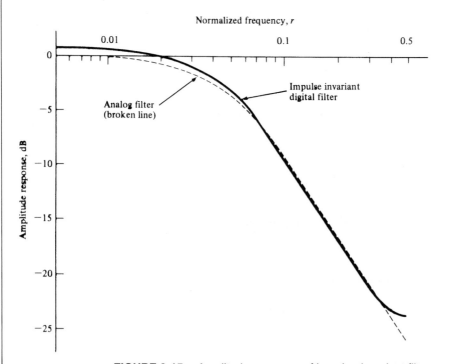

FIGURE 9-15. Amplitude response of impulse-invariant filter.

As expected, the amplitude response of the digital filter is a very good approximation to the analog filter for most values of r. The phase responses are illustrated in Figure 9-16.

In some applications it is desirable that both filters have the same dc gain. Note that this could have been accomplished in this example by multiplying by

$$G = \frac{T}{1.0745} = 0.2924 \tag{9-75}$$

in (9-73) instead of multiplying by

$$G = T = \frac{\pi}{10} = 0.3142 \tag{9-76}$$

MATLAB Application

MATLAB supports the design of impulse invariant digital filters through the command `impinvar` in the Signals and Systems Toolbox. The s-domain transfer function is first defined along with the sampling frequency. The command `impinvar` determines the numerator and denominator of the z-domain transfer function. Since the transfer function is to be expressed in series form as shown in (9-61) partial fraction expansion is next applied. The MATLAB code follows:

```
EDU» c9ex5
num = 0.5*[1 4];              % Define numerator
den = conv([1 1],[1 2]);     % Define denominator
Fs = 20/(2*pi);              % Define samp. freq. in Hz
[bz,az] = impinvar(num,den,Fs);  % Determine imp. invar. filter
[r,p,k] = residuez(bz,az);   % Place in parallel form
r                            % Display residues
```

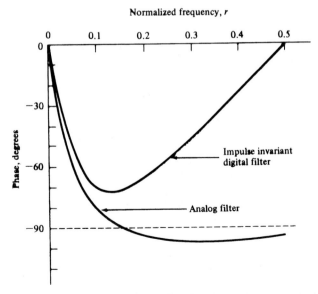

FIGURE 9-16.   Phase response of impulse-invariant filter.

```
r =


      1.5000
    -1.0000


p                                        % Display poles


p =


      0.7304
      0.5335


k                                        % Display k vector


k =


      []
```

Several comments are in order. First note that the pole locations as determined by `impinvar` agree with the preceding analysis as shown by (9-70) and (9-71). The residues, as determined by `impinvar`, are as shown in (9-58). We see that `impinvar` produces a true impulse invariant digital filter in which samples of the analog response match exactly the impulse response of the digital filter. No scaling of the frequency response by T and no compensation for aliasing effects is carried out in `impinvar`. If desired, this can be done as a separate operation.

Another popular digital filter synthesis procedure using a time-domain criterion is the step-invariance synthesis procedure, which is a direct application of (9-53). The step-invariant filter is derived by placing a unit step on the input of an analog filter and a sampled unit step on the input to a digital filter. The pulse transfer function of the digital filter, $H(z)$, is adjusted until the output of the digital filter represents samples of the output of the analog filter. Thus, in (9-53), we let

$$G = 1 \tag{9-77}$$

$$X_a(s) = \frac{1}{s} \tag{9-78}$$

and

$$X(z) = \frac{1}{1 - z^{-1}} \tag{9-79}$$

This yields

$$H(z) = (1 - z^{-1})\mathscr{Z}\left\{ \mathscr{L}^{-1}\left[\frac{1}{s} H_a(s) \right] \Big|_{t=nT} \right\} \tag{9-80}$$

EXAMPLE 9-6 _____

The step-invariant equivalent of

$$H_a(s) = \frac{0.5(s + 4)}{(s + 1)(s + 2)} \tag{9-81}$$

will now be found. Since

$$\frac{1}{s} H_a(s) = \frac{0.5(s + 4)}{s(s + 1)(s + 2)} = \frac{1}{s} - \frac{1.5}{s + 1} + \frac{0.5}{s + 2} \tag{9-82}$$

we have

$$\mathcal{L}^{-1}\left[\frac{1}{s} H_a(s)\right]\Bigg|_{t=nT} = 1 - 1.5e^{-nT} + 0.5e^{-2nT} \tag{9-83}$$

so that

$$H(z) = (1 - z^{-1})\left(\frac{1}{1 - z^{-1}} - \frac{1.5}{1 - e^{-T}z^{-1}} + \frac{0.5}{1 - e^{-2T}z^{-1}}\right) \tag{9-84}$$

or

$$H(z) = 1 - 1.5\frac{1 - z^{-1}}{1 - e^{-T}z^{-1}} + 0.5\frac{1 - z^{-1}}{1 - e^{-2T}z^{-1}} \tag{9-85}$$

To compare this result with the result of Example 9-5, let

$$T = 0.31416 \quad s \tag{9-86}$$

This yields

$$H(z) = \frac{0.17115z^{-1} + 0.04538z^{-2}}{1 - 1.2639z^{-1} + 0.3897z^{-2}} \tag{9-87}$$

Figure 9-17 compares the amplitude responses of the step-invariant and impulse-invariant digital filters.

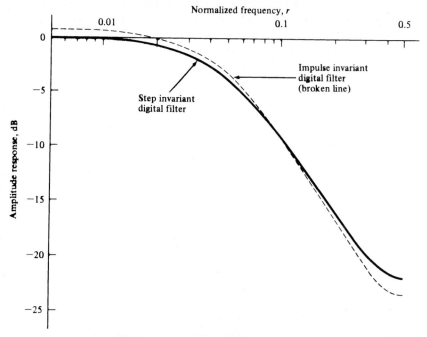

FIGURE 9-17.   Amplitude response of step-invariant filter.

Design in the Frequency Domain—The Bilinear z-Transform

The time-domain invariance method of filter synthesis was simple to apply, but for many applications the effects of aliasing preclude its use. We now examine a synthesis technique that overcomes the aliasing problem and is even easier to apply than the time-domain invariance method since it is not necessary to inverse transform $H_a(s)$. This synthesis method is known as the *bilinear z-transform method.*

To avoid the effects of aliasing, the transfer function of the analog filter must be bandlimited to the range

$$-\tfrac{1}{2}f_s \le f_1 \le \tfrac{1}{2}f_s \tag{9-88}$$

Since typical analog transfer functions do not satisfy this property, they must first be modified using a nonlinear transformation so that they are bandlimited. The technique is to transform the entire complex s-plane into the s_1-plane such that the entire $j\omega$-axis in the s-plane is mapped into the region

$$-\tfrac{1}{2}f_s \le f_1 \le f_s \tag{9-89}$$

in the s_1-plane. Although there are many transformations that will accomplish this, we seek one that is easy to apply and yields a pulse transfer function for the digital filter directly as a ratio of polynomials in z^{-1}.

A transformation that satisfies these requirements is

$$\omega = C \tan\left[\frac{\omega_1}{\tfrac{1}{2}\omega_s} \frac{\pi}{2}\right] = C \tan\frac{\omega_1 T}{2} \tag{9-90}$$

or

$$C = \omega \cot\frac{\omega_1 T}{2} \tag{9-91}$$

This transformation is illustrated in Figure 9-18. It can be seen that $\omega = \infty$ for the analog filter maps into $\omega_1 = \tfrac{1}{2}\omega_s$ for the digital filter. The constant C can be chosen so that the correspondence $\omega = \omega_1$ can be established at any desired frequency. As an example, if $\omega = \omega_1 = \omega_r$, we have, upon solving for C,

$$C = \omega_r \cot\frac{\omega_r T}{2} \tag{9-92}$$

For small ω_r such that

$$\frac{\omega_r T}{2} << 1 \tag{9-93}$$

we can write

$$C = \omega_r\left(\frac{2}{\omega_r T}\right) = \frac{2}{T} \tag{9-94}$$

since $\cot x \simeq 1/x$ for small x. This relationship is indicated by the $C = 2/T$ curve in Figure 9-18 and shows that for this value of C, $\omega = \omega_1$ for frequencies that are small compared to the sampling frequency. On the other hand, if $C = 1/T$ we have

$$\omega = \frac{1}{T} \tan\frac{\omega_1 T}{2} \tag{9-95}$$

and, as shown in Figure 9-18, $\omega = \omega_1$ at a normalized frequency of 0.74, where normalization is with respect to $\tfrac{1}{2}\omega_s$. It should be noted that ω and ω_1 are always equal at $f = 0$. Thus the dc response of the bilinear z-transform digital filter is always equal to the dc response of the analog filter from which it is derived.

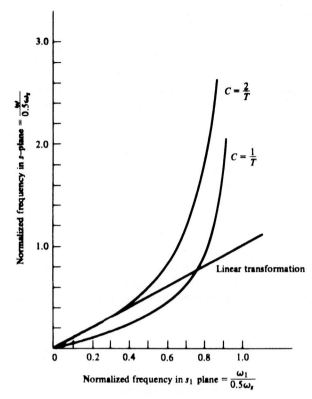

FIGURE 9-18. Bilinear z-transformation frequency mapping.

The relationship between s and s_1 is easily determined from (9-90). First, recall that $\tan x$ can be written as

$$\tan x = \frac{\sin x}{\cos x} = -j\frac{e^{jx} - e^{-jx}}{e^{jx} + e^{-jx}} \qquad (9\text{-}96)$$

Letting $s = j\omega$ and $s_1 = j\omega_1$ in (9-90) then yields

$$-js = -jC\frac{e^{+s_1T/2} - e^{-s_1T/2}}{e^{+s_1T/2} + e^{-s_1T/2}} \qquad (9\text{-}97)$$

which can be written as

$$s = C\frac{e^{s_1T/2} - e^{-s_1T/2}}{e^{s_1T/2} + e^{-s_1T/2}} = C\tanh\frac{s_1T}{2} \qquad (9\text{-}98)$$

The digital filter is determined from $H_a(s_1)$ by letting

$$z = e^{s_1T} \qquad (9\text{-}99)$$

With this substitution, (9-98) becomes

$$s = C\frac{1 - e^{-s_1T}}{1 + e^{-s_1T}} = C\frac{1 - z^{-1}}{1 + z^{-1}} \qquad (9\text{-}100)$$

Thus the digital filter $H(z)$ is determined from the analog filter $H_a(s)$ by simply making the substitution

$$s \Rightarrow C \frac{1 - z^{-1}}{1 + z^{-1}} \tag{9-101}$$

for s in $H_a(s)$. Since $H_a(s)$ is expressed as a ratio of polynomials in s, the resulting $H(z)$ is a ratio of polynomials in z^{-1}. One only needs to determine the appropriate value of the scaling constant C to apply the bilinear z-transform synthesis method.

The effect of transforming the complex s-plane into the s_1-plane is illustrated in Figure 9-19. Note that since $H_a(s)$ is not bandlimited, aliasing occurs when $H_a(s)$ is digitized. However, since $H_a(s_1)$ is bandlimited, no aliasing occurs. It is also important to note that $H_a(s_1)$ is not explicitly determined as part of the synthesis process. The complex s-plane is transformed into the s_1-plane, and $H_a(s_1)$ is digitized to form $H(z)$ all in one step by application of (9-100).

EXAMPLE 9-7

This example illustrates the development of a bilinear z-transform digital filter based on a second-order Butterworth analog prototype. The analog prototype is derived in Example E-3.

A second-order low-pass Butterworth filter, having a 3-dB bandwidth of ω_c radians per second, is given by the system function

$$H_a(s) = \frac{\omega_c^2}{s^2 + \sqrt{2}\omega_c s + \omega_c^2} \tag{9-102}$$

as shown in Appendix E. In this example the bilinear z-transform digital filter equivalent to (9-102) will be determined.

We wish the digital filter also to have a 3-dB bandwidth of ω_c, so that

$$\omega_r = \omega_c \tag{9-103}$$

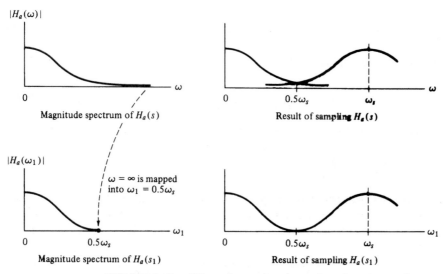

FIGURE 9-19. Effect of mapping the s-plane into the s_1-plane.

yielding

$$C = \omega_c \cot \frac{\omega_c T}{2} = \omega_c \cot \frac{\pi \omega_c}{\omega_s} \tag{9-104}$$

Thus, for s in (9-102) we substitute

$$\omega_c \cot \frac{\pi \omega_c}{\omega_s} \frac{1 - z^{-1}}{1 + z^{-1}} \tag{9-105}$$

This substitution is simplified by defining

$$m = \cot \frac{\pi \omega_c}{\omega_s} = \cot \frac{\pi f_c}{f_s} \tag{9-106}$$

so that

$$s = \omega_c m \frac{1 - z^{-1}}{1 + z^{-1}} \tag{9-107}$$

This yields, in (9-102), the pulse transfer function

$$H(z) = \frac{\omega_c^2}{\left(\omega_c m \dfrac{1 - z^{-1}}{1 + z^{-1}}\right)^2 + \sqrt{2}\omega_c \left(\omega_c m \dfrac{1 - z^{-1}}{1 + z^{-1}}\right) + \omega_c^2} \tag{9-108}$$

It is important to note that ω_c cancels in (9-108). This illustrates the fact that the pulse transfer function is only a function of the ratio of ω_c to ω_s through the parameter m as defined in (9-106). Multiplying the numerator and denominator of (9-108) by $(1 + z^{-1})^2$ and grouping like terms yields

$$H(z) = \frac{1 + 2z^{-1} + z^{-2}}{(m^2 + \sqrt{2}m + 1) + 2(1 - m^2)z^{-1} + (m^2 - \sqrt{2}m + 1)z^{-2}} \tag{9-109}$$

Thus $H(z)$ is determined by the ratio of f_c to f_s. For the case where $f_c = 500$ Hz and $f_s = 2000$ Hz,

$$m = \cot \frac{\pi}{4} = 1 \tag{9-110}$$

and the pulse transfer function becomes

$$H(z) = \frac{1 + 2z^{-1} + z^{-2}}{3.414214 + 0.585786z^{-2}} \tag{9-111}$$

Dividing numerator and denominator by 3.414214 yields the pulse transfer function in standard form:

$$H(z) = \frac{0.292893 + 0.585786z^{-1} + 0.292893z^{-2}}{1 + 0.171573z^{-2}} \tag{9-112}$$

The amplitude response is determined by substituting $e^{j2\pi r}$ for z in (9-112) and taking the magnitude of the resulting expression. This was done and the result is illustrated in Figure 9-20. Also illustrated in Figure 9-20 is the amplitude response of the analog second-order Butterworth filter, which is given by

$$|H_a(f)| = \frac{1}{\sqrt{1 + (f/f_c)^4}} \tag{9-113}$$

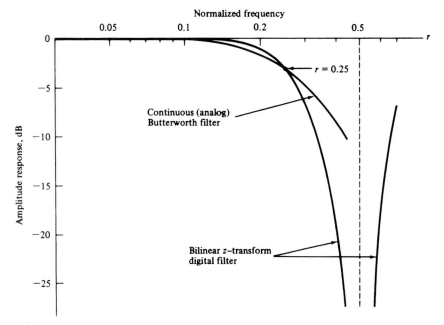

FIGURE 9-20. Amplitude response of bilinear z-transform filter.

Since for the case analyzed

$$f_c = \tfrac{1}{4} f_s \qquad (9\text{-}114)$$

f/f_c in (9-113) reduces to

$$\frac{f}{f_c} = \frac{4f}{f_s} = 4r \qquad (9\text{-}115)$$

Thus, $f = 4rf_c$, and

$$|H_a(4rf_c)| = \frac{1}{\sqrt{1 + 256r^4}} \qquad (9\text{-}116)$$

MATLAB *Application*

The MATLAB Signals and Systems Toolbox provides the capability to design bilinear z-transform digital filters. The transfer function for the analog prototype is first determined. The numerator and denominator polynomials of the analog prototype are then mapped to the numerator and denominator polynomials for the bilinear z-transform digital filter. The example problem is solved using the following code:

```
EDU» c9ex7
Wn = 2*pi*500;              % 3-dB freq. analog prototype
n = 2;                      % Order of analog prototype
Fp = Wn/(2*pi);            % Prewarping frequency in Hertz
Fs = 2000;                  % Sampling frequency in Hertz
[b,a] = butter(n,Wn,'s');   % Define analog prototype
[num,den] = bilinear(b,a,Fs,Fp);  % Determine digital filter
```

```
num                                     % Display numerator

num =

    0.2929      0.5858      0.2929

den                                     % Display denominator

den =

    1.0000      0.0000      0.1716
```

Note that the numerator and denominator polynomials of the resulting digital filter agree with (9-112).

The analog and digital filter responses illustrated in Figure 9-20 compare poorly because the 3-dB break frequency of the digital filter is relatively close to half the sampling frequency. The amplitude responses of bilinear z-transform digital filters derived from a low-pass Butterworth prototype were determined for $n = 1, 2, 3, 4,$ and 5. The results are shown in Figure 9-21 for $f_c/f_s = 0.01$ and in Figure 9-22 for $f_c/f_s = 0.05$. It can be seen that the nonlinear effects of the bilinear z-transformation are negligible for $r < 0.1$.

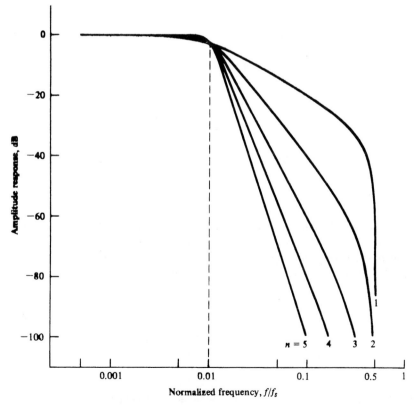

FIGURE 9-21.   Digital Butterworth filter responses with $f_c/f_s = 0.01$.

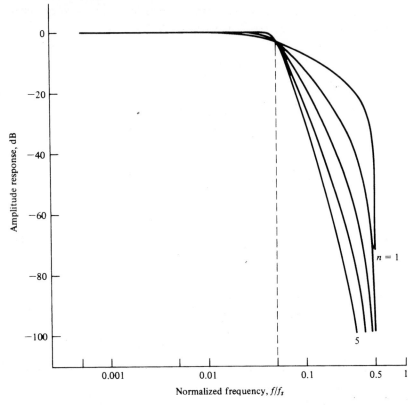

FIGURE 9-22. Digital Butterworth filter responses with $f_c/f_s = 0.05$.

Thus for input frequencies less than $0.1f_s$, the amplitude response of the digital filter closely approximates the amplitude response of the analog Butterworth prototype.

EXAMPLE 9-8

An analog integrator is defined by

$$H_a(s) = \frac{1}{s} \tag{9-117}$$

so that the bilinear z-transform integrator is

$$H(z) = \frac{1}{s}\bigg|_{s = C[(1-z^{-1})/(1+z^{-1})]} = \frac{1}{C}\frac{1 + z^{-1}}{1 - z^{-1}} \tag{9-118}$$

If C is set equal to the limiting value of $2/T$, we have

$$H(z) = \frac{T}{2}\frac{1 + z^{-1}}{1 - z^{-1}} \tag{9-119}$$

Comparison of this expression with (9-26) illustrates the equivalence of trapezoidal integration and the bilinear z-transformation.

As an additional illustration of this equivalence, let us return to Example 9-4, in which we determined the transfer function of the digital system based on the differential equation

$$\frac{dy}{dt} = x(t) - \beta y(t) \tag{9-120}$$

Laplace transforming this equation yields

$$sY(s) = X(s) - \beta Y(s) \tag{9-121}$$

or

$$H_a(s) = \frac{Y(s)}{X(s)} = \frac{1}{s + \beta} \tag{9-122}$$

With s replaced by

$$\frac{2}{T} \frac{1 - z^{-1}}{1 + z^{-1}} \tag{9-123}$$

we have

$$H(z) = \frac{T(1 + z^{-1})}{(\beta T + 2) + (\beta T - 2)z^{-1}} \tag{9-124}$$

which is exactly the same result achieved by replacing the continuous-time integrator by the trapezoidal integrator.

Bilinear z-Transform Bandpass Filters

Bandpass filters can also be developed using the bilinear z-transform. As shown in Appendix E, a bandpass filter can be generated from a low-pass filter by substituting $(s^2 + \omega_c^2)/(s\omega_b)$ for the Laplace variable s in the system function for the low-pass prototype. The parameter ω_c is the geometric center frequency and ω_b is the bandwidth of the analog bandpass filter. The bandpass digital filter is then generated by replacing s by $C(1 - z^{-1})/(1 + z^{-1})$. These two steps can be combined into a single step by going from s to z directly. This yields

$$\frac{s^2 + \omega_c^2}{s\omega_b} \Rightarrow \frac{\left[C\dfrac{1 - z^{-1}}{1 + z^{-1}} \right]^2 + \omega_c^2}{C\left[\dfrac{1 - z^{-1}}{1 + z^{-1}} \right]\omega_b} \tag{9-125}$$

which can be written as

$$\frac{s^2 + \omega_c^2}{s\omega_b} \Rightarrow \frac{C^2(1 - z^{-1})^2 + \omega_c^2(1 + z^{-1})^2}{C\omega_b(1 - z^{-2})} \tag{9-126}$$

Collecting coefficients of like powers of z results in the transformation

$$\frac{s^2 + \omega_c^2}{s\omega_b} \Rightarrow \frac{C^2 + \omega_c^2}{C\omega_b} \frac{1 - 2\left[\dfrac{C^2 - \omega_c^2}{C^2 + \omega_c^2} \right]z^{-1} + z^{-2}}{1 - z^{-2}} \tag{9-127}$$

or

$$\frac{s^2 + \omega_c^2}{s\omega_b} \Rightarrow A \frac{1 - Bz^{-1} + z^{-2}}{1 - z^{-2}} \qquad (9\text{-}128)$$

where

$$A = \frac{C^2 + \omega_c^2}{C\omega_b} \qquad (9\text{-}129)$$

and

$$B = 2\frac{C^2 - \omega_c^2}{C^2 + \omega_c^2} \qquad (9\text{-}130)$$

The problem now is to determine the appropriate values of A and B.

First we consider the value of A. The center frequency ω_c in the s-plane must be replaced by C tan $(\omega_c T/2)$ to map the value of the center frequency into the s_1-plane. Since ω_c is the geometric frequency, as shown in Appendix E,

$$\omega_c^2 = \omega_u \omega_l \qquad (9\text{-}131)$$

where ω_u and ω_l are the upper and lower critical frequencies, respectively. The equivalent expression in the s_1-plane is

$$\omega_c^2 \Rightarrow C^2 \tan \frac{\omega_u T}{2} \tan \frac{\omega_l T}{2} \qquad (9\text{-}132)$$

The bandwidth ω_b in the s-plane becomes

$$\omega_b \Rightarrow C \tan \frac{\omega_u T}{2} - C \tan \frac{\omega_l T}{2} \qquad (9\text{-}133)$$

in the s_1-plane. Thus A becomes

$$A = \frac{C^2 + C^2 \tan \dfrac{\omega_u T}{2} \tan \dfrac{\omega_l T}{2}}{C^2 \left(\tan \dfrac{\omega_u T}{2} - \tan \dfrac{\omega_l T}{2} \right)} \qquad (9\text{-}134)$$

or

$$A = \frac{1 + \tan \dfrac{\omega_u T}{2} \tan \dfrac{\omega_t T}{2}}{\tan \dfrac{\omega_u T}{2} - \tan \dfrac{\omega_t T}{2}} \qquad (9\text{-}135)$$

Using the trigonometric identity

$$\tan (x - y) = \frac{\tan x - \tan y}{1 + \tan x \tan y} \qquad (9\text{-}136)$$

yields the convenient form

$$A = \cot \left[(\omega_u - \omega_l) \frac{T}{2} \right] \qquad (9\text{-}137)$$

We must now determine the value of B as defined by (9-130). From (9-130) and (9-132) we can write

$$B = 2 \frac{C^2 - C^2 \tan \frac{\omega_u T}{2} \tan \frac{\omega_l T}{2}}{C^2 + C^2 \tan \frac{\omega_u T}{2} \tan \frac{\omega_l T}{2}} \qquad (9\text{-}138)$$

or, upon dividing numerator and denominator by C^2,

$$B = 2 \frac{1 - 1 \tan \frac{\omega_u T}{2} \tan \frac{\omega_l T}{2}}{1 + 1 \tan \frac{\omega_u T}{2} \tan \frac{\omega_l T}{2}} \qquad (9\text{-}139)$$

Since

$$\frac{1 - \tan x \tan y}{1 + \tan x \tan y} = \frac{\cos(x + y)}{\cos(x - y)} \qquad (9\text{-}140)$$

we have the simple relationship

$$B = 2 \frac{\cos(\omega_u + \omega_l)\dfrac{T}{2}}{\cos(\omega_u - \omega_l)\dfrac{T}{2}} \qquad (9\text{-}141)$$

We can therefore easily calculate the necessary values of A and B for use in the transformation defined by (9-128).

The expressions for A and B can be simplified by recognizing that

$$\omega_x T = 2\pi r_x \qquad (9\text{-}142)$$

where ω_x is some specific frequency and r_x is that specific frequency normalized to the sampling frequency so that $r_x = f_x/f_s$. Using this relationship gives

$$A = \cot \pi(r_u - r_l) \qquad (9\text{-}143)$$

and

$$B = 2 \frac{\cos \pi(r_u + r_l)}{\cos \pi(r_u - r_l)} \qquad (9\text{-}144)$$

where r_u and r_l are the upper and lower frequencies used to define the filter bandwidth.

We can now illustrate these concepts in the design of a simple bandpass digital filter.

EXAMPLE 9-9 _____

In this example a bilinear z-transform digital filter is designed from the first-order analog prototype

$$H_a(s) = \frac{1}{s + 1} \qquad (9\text{-}145)$$

Using the transformation defined in (9-128) yields the pulse transfer function

$$H(z) = \frac{1}{A\left[\dfrac{1 - Bz^{-1} + z^{-2}}{1 - z^{-2}}\right] + 1} \tag{9-146}$$

which can be written in the form

$$H(z) = \frac{1 - z^{-2}}{(A + 1) - ABz^{-1} + (A - 1)z^{-2}} \tag{9-147}$$

A sampling frequency of 5,000 Hz is assumed, along with upper and lower 3-dB frequencies of 1,000 and 500 Hz, respectively. Using these values gives the normalized frequency values of 0.2 and 0.1 for r_u and r_l. Thus

$$A = \cot \pi(0.2 - 0.1) = 3.0776835 \tag{9-148}$$

and

$$B = 2\,\frac{\cos \pi(0.2 + 0.1)}{\cos \pi(0.2 - 0.1)} = 1.2360680 \tag{9-149}$$

The pulse transfer function then becomes

$$H(z) = \frac{1 - z^{-2}}{4.0776835 - 3.8042261z^{-1} + 2.0776835z^{-2}} \tag{9-150}$$

which, in standard form, can be written

$$H(z) = \frac{0.2452373 - 0.2452373z^{-2}}{1 - 0.9329380z^{-1} + 0.5095254z^{-2}} \tag{9-151}$$

The magnitude and phase responses of this filter are shown in Figure 9-23. The response goes to $-\infty$ at $z = 0.5$ because of the zero of $H_a(s)$ at $s = \infty$. Note that a first-order low-pass prototype gave rise to a second-order bandpass filter.

9-5 Design of Finite-Duration Impulse Response (FIR) Digital Filters

In the previous section design methods for infinite-duration impulse response (IIR) digital filters were studied. We found that, by starting with an analog prototype filter, the design techniques were very easy to apply. We now turn our attention to a distinctly different problem, the problem of the design of finite-duration impulse response filters. We will see that a very different implementation will result. The design technique to be studied in this section differs from those previously considered in that our starting point is not an analog filter but an arbitrary frequency response that represents the desired frequency response of the digital filter being designed. It will therefore be possible to design digital filters that have no analog prototype equivalent.

A number of computer-aided design tools are available for the synthesis of digital filters and a couple of these techniques will be explored in the following section. Computer-aided design techniques for FIR digital filters are, as we will see, especially important. However, even though such methods are widely available and easy to use, the student should understand the basic characteristics of FIR digital filters, and be familiar with fundamental techniques used for the design of FIR digital filters. In this sec-

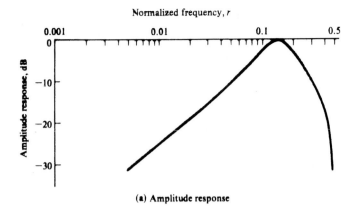

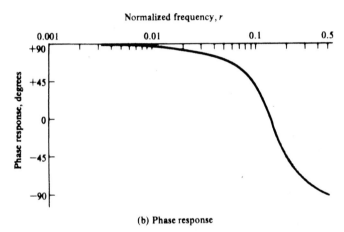

FIGURE 9-23.   Amplitude and phase responses for example bandpass filter.

tion the Fourier series and windowing technique is studied. This technique is derived by first realizing that the transfer function of a digital filter is periodic in the sampling frequency. Thus the transfer function can be expanded in a Fourier series. We will show that the Fourier coefficients resulting from this expansion give the unit pulse response of the digital filter.

Since we have typically used Fourier series techniques to determine an expansion for a periodic time-domain waveform, it may seem strange to expand a function of frequency in a Fourier series. Our discomfort is only a result of conditioning, and Fourier series techniques are valid for any periodic function satisfying the conditions discussed in Chapter 3. Before formally presenting this design technique, we introduce it by a simple example.

EXAMPLE 9-10

Suppose we wish to design an FIR digital filter having the desired transfer function

$$H_d(e^{j2\pi r}) = \tfrac{1}{2}(1 + \cos 2\pi r) \qquad (9\text{-}152)$$

in which r is the normalized frequency defined by $r = f/f_s$. Note that the transfer function is real and positive so that the phase response is zero for all values of r. The amplitude and phase responses are shown in Figures 9-24a and b.

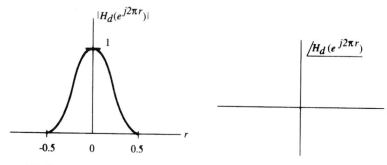

(a) Desired amplitude response

(b) Desired phase response

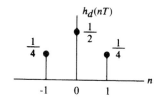

(c) Unit pulse response of noncausal filter

(d) Unit pulse response of causal filter

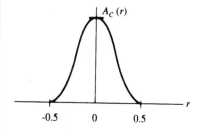

(e) Amplitude response of causal filter

(f) Phase response of causal filter

FIGURE 9-24. Amplitude and phase responses for Example 9-10.

By expanding $\cos 2\pi r$ in (9-152), using Euler's expansion, we obtain

$$H_d(e^{j2\pi r}) = \tfrac{1}{4}e^{j2\pi r} + \tfrac{1}{2} + \tfrac{1}{4}e^{-j2\pi r} \qquad (9\text{-}153)$$

which is the Fourier series expansion of the transfer function of the digital filter. It follows by inspection that the z-domain transfer function corresponding to the desired frequency response is

$$H_d(z) = \tfrac{1}{4}z + \tfrac{1}{2} + \tfrac{1}{4}z^{-1} \qquad (9\text{-}154)$$

Thus the unit pulse response of the example filter is

$$h_d(nT) = \tfrac{1}{4}\delta(nT + T) + \tfrac{1}{2}\delta(nT) + \tfrac{1}{4}\delta(nT - T) \qquad (9\text{-}155)$$

The unit pulse response is shown in Figure 9-24c. It can be seen that the filter is not causal because of the term at $n = -1$.

Observation of the unit pulse response shown in Figure 9-24c reveals that the filter can be made causal by shifting the unit pulse response to the right one sample period. Thus the filter defined by

$$H_c(z) = H_d(z)z^{-1} \tag{9-156}$$

will be causal and have the unit pulse response

$$h_c(nT) = \tfrac{1}{4}\delta(nT) + \tfrac{1}{2}\delta(nT - T) + \tfrac{1}{4}\delta(nT - 2T) \tag{9-157}$$

which is shown in Figure 9-24d. The frequency response of the causal filter is defined by

$$H_c(e^{j2\pi r}) = H_d(e^{j2\pi r})e^{-j2\pi r} \tag{9-158}$$

or

$$H_c(e^{j2\pi r}) = \tfrac{1}{2}(1 + \cos 2\pi r)e^{-j2\pi r} \tag{9-159}$$

The amplitude response is

$$A_c(r) = \tfrac{1}{2}(1 + \cos 2\pi r) \tag{9-160}$$

and the phase response is

$$\phi_c(r) = -2\pi r \tag{9-161}$$

The amplitude and phase responses are shown in Figures 9-24e and f, respectively. Note that the amplitude responses of the noncausal and causal filters are identical to the design specification in (9-152). The phase response of the causal filter deviates from the desired phase response of $\phi(r) = 0$ by a linear function of frequency. This linear phase characteristic, as we have seen, is simply equivalent to a delay of one sample period and causes no signal distortion. This can also be seen by calculating the group delay. Since $r = fT$, the phase response can also be expressed as

$$\phi(f) = -2\pi fT \tag{9-162}$$

The group delay is then

$$T_g(f) = -\frac{1}{2\pi}\frac{d\phi(f)}{df} = T \tag{9-163}$$

indicating a delay of one sample period.

The causal FIR digital filter can be implemented as shown in Figure 9-25. It is important to note that the multiplying factors, which are known as tap weights, since the cascade of z^{-1} elements forms

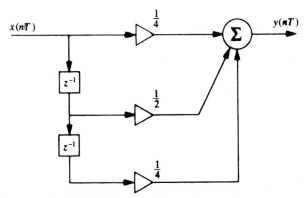

FIGURE 9-25.   Implementation for FIR digital filter studied in Example 9-10.

a tapped delay line, are exactly equal to the Fourier series coefficients of the desired frequency response. Also note from Figure 9-25 that the FIR filter has only feedforward signal paths. Thus, the implementation has no feedback signal paths and therefore has no poles except at $z = 0$. The FIR filter is therefore always stable.

We have seen that an FIR filter can be designed by expanding the desired amplitude response in a Fourier series and then shifting the resulting unit pulse response so that a causal filter results. This example is extremely simple because the desired frequency response can be expressed exactly by a Fourier series having a finite number of terms. In general, an exact representation of the desired frequency response is not possible with a finite number of terms, and the Fourier series must be truncated if a useful filter is to result. Thus, an approximation to the desired frequency response is obtained. Using this example as motivation, we now consider the general case and see the nature of this approximation.

FIR Filters—A General Design Technique

As we observed in the preceding problem, the desired frequency response, which is periodic in the sampling frequency, corresponds to a noncausal digital filter. In general we can write

$$H_d(z)\Big|_{z=e^{j\omega T}} = H_d(e^{j\omega T}) = \sum_{n=-\infty}^{\infty} h_d(nT)e^{-jn\omega T} \tag{9-164}$$

where we have allowed the filter to be noncausal by starting the summation at $n = -\infty$ rather than $n = 0$, as we do for causal filters.

As in our example problem, it is usually convenient to express the frequency variable in terms of the normalized frequency r. Letting $2\pi r = \omega T$ in (9-164) yields

$$H_d(e^{j2\pi r}) = \sum_{n=-\infty}^{\infty} h_d(nT)e^{-j2\pi nr} \tag{9-165}$$

An expression for the Fourier coefficients of the desired frequency response $H_d(e^{j2\pi r})$ can be found by multiplying (9-165) by $e^{j2\pi lr}$ and integrating over one period of the frequency response. This yields

$$\int_{-1/2}^{1/2} H_d(e^{j2\pi r})e^{j2\pi lr}\, dr = \sum_{n=-\infty}^{\infty} h_d(nT) \int_{1/2}^{1/2} e^{j(l-n)2\pi r}\, dr \tag{9-166}$$

Since

$$\int_{-1/2}^{1/2} e^{j(l-n)2\pi r}\, dr = \begin{cases} 1, & l = n \\ 0, & l \neq n \end{cases} \tag{9-167}$$

we have

$$\int_{-1/2}^{1/2} H_d(e^{j2\pi r})e^{j2\pi lr}\, dr = h_d(lT) \tag{9-168}$$

or, upon replacing l by n

$$h_d(nT) = \int_{-1/2}^{1/2} H_d(e^{j2\pi r})e^{j2\pi nr}\, dr \tag{9-169}$$

It follows that $h_d(nT)$ represents the values of the unit pulse response of the noncausal digital filter having the desired frequency response.

At this point we have the weights necessary to implement the FIR digital filter. There are two problems, however. The first is that the digital filter is noncausal, since the unit pulse response, $h_d(nT)$, is nonzero for at least one negative index n. This problem was encountered in Example 9-10, where we saw that its solution is to shift the noncausal unit pulse response so that a causal unit pulse response results. Before this can be done, however, the unit pulse response must be of finite extent. This points to the second problem. Unless the desired frequency response, $H_d(e^{j2\pi r})$, can be expressed exactly with a finite number of terms, which is usually not the case, the Fourier series coefficients must be truncated to a series having a finite number of terms. This truncation introduces errors, and therefore we have a trade-off between accuracy and complexity.

Truncating the Fourier coefficients to a finite number of terms yields the noncausal pulse transfer function

$$H_{nc}(z) = \sum_{n=-M}^{M} h_d(nT)z^{-n} \tag{9-170}$$

Note that within the range $-M \leq n \leq M$ the unit pulse response is still in terms of the Fourier coefficients of the desired frequency response. The series has simply been truncated to $2M + 1$ terms. Another way to view truncation is to recognize that the pulse transfer function can be written

$$H_{nc}(z) = \sum_{n=-M}^{M} h_d(nT)z^{-n} = \sum_{n=-\infty}^{\infty} w_r(n)h_d(nT)z^{-n} \tag{9-171}$$

where $w_r(n)$ is known as the rectangular window function and is defined by

$$w_r(n) = \begin{cases} 1, & |n| \leq M \\ 0, & |n| > M \end{cases} \tag{9-172}$$

The time-domain multiplication by a rectangular window function has an impact in the frequency domain. Since multiplication in the time domain is equivalent to convolution in the frequency domain, the frequency response of the resulting digital filter is the convolution of the desired frequency response with the Fourier transform of the window function. The Fourier transform of $w_r(n)$ is

$$W_r(e^{j2\pi r}) = \sum_{n=-M}^{M} e^{-j2\pi nr} = \frac{\sin \pi(2M + 1)r}{\sin \pi r} \tag{9-173}$$

in which we have written the transform in terms of the normalized frequency variable. We will see the effect of this convolution in the examples.

Causal Filters

A causal filter, denoted $H_c(z)$, can be generated from $H_{nc}(z)$ by multiplying $H_{nc}(z)$ by z^{-M}. Using (9-171) for $H_{nc}(z)$ yields

$$H_c(z) = z^{-M} \sum_{n=-M}^{M} h_d(nT)z^{-n} = \sum_{n=-M}^{M} h_d(nT)z^{-(n+M)} \tag{9-174}$$

For $k = n + M$, (9-174) becomes

$$H_c(z) = \sum_{k=0}^{2M} h_d(kT - MT)z^{-k} \tag{9-175}$$

Defining the filter weights of the causal filter as L_k, where

$$L_k = h_d(kT - MT) \tag{9-176}$$

gives

$$H_c(z) = \sum_{k=0}^{2M} L_k z^{-k} \tag{9-177}$$

The resulting causal FIR digital filter and its unit pulse response are shown in Figures 9-26a and b, respectively. Note that the unit pulse response is identically equal to zero outside of the interval $0 \le k \le 2M$. This, of course, is a result of applying a rectangular window. Thus, the causal FIR digital filter has a time duration of $2M$ sample periods or $2MT$ seconds. The implementation of the causal FIR digital filter is simply the feedforward network, $H_1(z)$, shown in Figure 9-1.

The sinusoidal steady-state frequency response of the causal filter is closely related to the frequency response of the noncausal filter. Since

$$H_c(z) = H_{nc}(z)z^{-M} \tag{9-178}$$

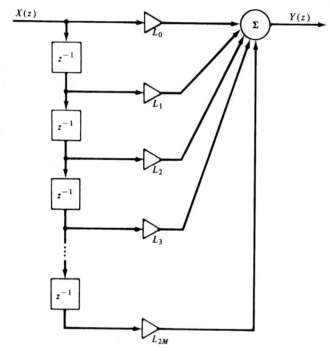

(a) Implementation of FIR filter

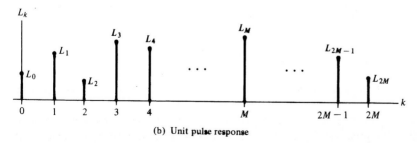

(b) Unit pulse response

FIGURE 9-26. FIR filter and unit pulse response.

it follows that

$$H_c(e^{j2\pi r}) = H_{nc}(e^{j2\pi r})e^{-j2\pi Mr} \qquad (9\text{-}179)$$

Letting

$$H_c(e^{j2\pi r}) = A_c(r)e^{j\phi_c(r)} \qquad (9\text{-}180)$$

and

$$H_{nc}(e^{j2\pi r}) = A_{nc}(r)e^{j\phi_{nc}(r)} \qquad (9\text{-}181)$$

gives

$$A_c(r)e^{j\phi_c(r)} = A_{nc}(r)e^{j\phi_{nc}(r)}e^{-j2\pi Mr} \qquad (9\text{-}182)$$

Thus

$$A_c(r) = A_{nc}(r) \qquad (9\text{-}183)$$

and

$$\phi_c(r) = \phi_{nc}(r) - j2\pi Mr \qquad (9\text{-}184)$$

We see from these results that the causal and noncausal filters have the same amplitude response. The phase of the causal digital filter differs from that of the noncausal filter only by a linear function of frequency. This was to be expected, of course, since multiplication by z^{-M} is a delay of M sample periods, which in turn is equivalent to a linear phase shift.

Most applications that will be encountered fall into one of two classes.

1. *Filtering.* For filtering applications the desired transfer function is usually defined in terms of the amplitude response of the filter. Typically, some portion of the input signal spectrum is to be attenuated, and some portion of the input signal spectrum is to be passed to the filter output with no attenuation. Ideally, this is accomplished with no phase distortion. The ideal transfer function is therefore real, corresponding to an even unit pulse response as shown in Figure 9-27a. Since the transfer function is real, the phase response corresponding to the unit pulse response shown in Figure 9-27a is zero for all frequencies, as shown in Figure 9-27b. Shifting the unit pulse response to the right to obtain a causal filter imparts a linear phase shift as shown in Figure 9-27c.

2. *Filtering plus Quadrature Phase Shift.* These applications include integrators, differentiators, and Hilbert transform devices. The desired frequency response, $H_d(e^{j2\pi r})$, is imaginary. The unit pulse response of the noncausal filter is odd, as shown in Figure 9-28a. The phase response is $\pm\pi/2$ for all frequencies, as shown in Figures 9-28b and c, respectively. Shifting the unit pulse response to form a noncausal filter results in the phase responses shown in Figures 9-28d and e. Figure 9-28d applies if the desired phase is $-\pi/2$ for $f > 0$, as would be the case for an integrator or a Hilbert transformer. Figure 9-28e applies if the desired phase is $+\pi/2$ for $f > 0$, as would be the case for a differentiator.

Design Summary

In summary, the procedure for the design of an FIR digital filter consists of the following steps:

1. A sampling frequency is first selected. This selection is usually made after considering the bandwidth of the signals to be processed by the filter. Once the sampling frequency is selected, the normalized frequency variable, r, is defined.

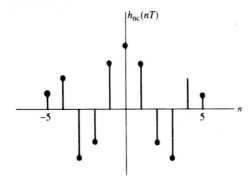

(a) Even unit-pulse response

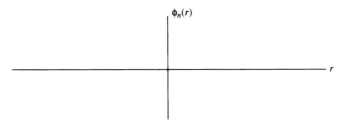

(b) Phase response for even unit-pulse response

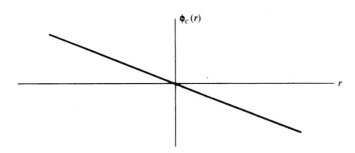

(c) **Phase response for causal filter derived from noncausal**
filter having even unit-pulse response

FIGURE 9-27. Unit pulse response and phase responses for filtering operations.

2. The desired frequency response, $H_d(e^{j2\pi r})$, is then expanded in an exponential Fourier series. It is recognized that the resulting Fourier coefficients, $h_d(nT)$, are the filter weights (the unit pulse response) of a noncausal digital filter that satisfies the design specifications. The unit pulse response corresponding to the desired digital filter is also infinite in extent.

3. The sequence, $h_d(nT)$, is next made finite in extent by multiplying the sequence by an appropriate window function. The number of nonzero terms in the window function is a trade-off between accuracy, allowable delay and complexity. The result of this step is a pulse transfer function $H_{nc}(z)$, that is noncausal but which has a finite number of nonzero terms.

4. The noncausal transfer function resulting from the previous step, $H_{nc}(z)$, is then multiplied by z^{-M}, with M chosen to ensure that a causal digital filter results.

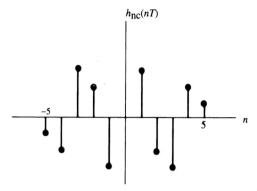

(a)   Odd unit-pulse response

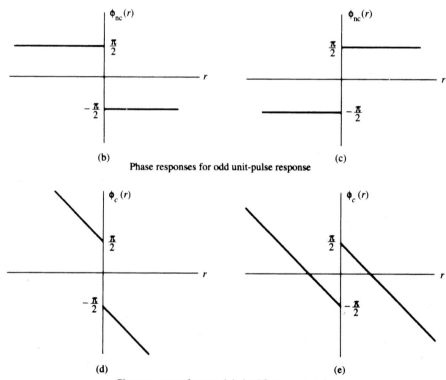

(b)   (c)

Phase responses for odd unit-pulse response

(d)   (e)

Phase responses for causal derived from noncausal
filter having odd unit-pulse responses

FIGURE 9-28.   Unit pulse response and phase responses for systems requiring a quadrature phase shift.

After this procedure is applied, the amplitude and phase responses of the resulting digital filter should be computed to ensure that the original design specifications are satisfied.

EXAMPLE 9-11 _____

In this example we will design a FIR digital differentiator. The ideal analog differentiator is defined by the transfer function

$$H_a(f) = j2\pi f \tag{9-185}$$

Since $r = f/f_s$, so that $f = rf_s$, we have for the desired digital filter response

$$H_d(e^{j2\pi r}) = j2\pi f_s r \tag{9-186}$$

The desired unit pulse response (we drop the subscript) is

$$h(nT) = \int_{-1/2}^{1/2} (j2\pi f_s r) e^{j2\pi mr} \, dt \tag{9-187}$$

Integrating by parts yields

$$h(nT) = j2\pi f_s \left\{ \frac{r}{j2\pi n} e^{j2\pi mr} \Big|_{r=-1/2}^{r=1/2} - \frac{1}{j2\pi n} \int_{-1/2}^{1/2} e^{j2\pi mr} \, dr \right\} \tag{9-188}$$

This becomes

$$h(nT) = \frac{f_s}{n} \left\{ \left(\frac{1}{2} e^{j\pi n} + \frac{1}{2} e^{-j\pi n} \right) - \frac{1}{j2\pi n} (e^{j\pi n} - e^{-j\pi n}) \right\} \tag{9-189}$$

or

$$h(nT) = \frac{f_s}{n} \left\{ \cos \pi n - \frac{\sin \pi n}{\pi n} \right\} \tag{9-190}$$

Recognizing that

$$\cos \pi n = (-1)^n \tag{9-191}$$

and

$$\frac{\sin \pi n}{\pi n} = \delta(n) \tag{9-192}$$

gives

$$h(nT) = \frac{f_s}{n} \{(-1)^n - \delta(n)\} \tag{9-193}$$

In other words,

$$h(nT) = \begin{cases} 0, & n = 0 \\ \dfrac{f_s}{n} (-1)^n, & n \neq 0 \end{cases} \tag{9-194}$$

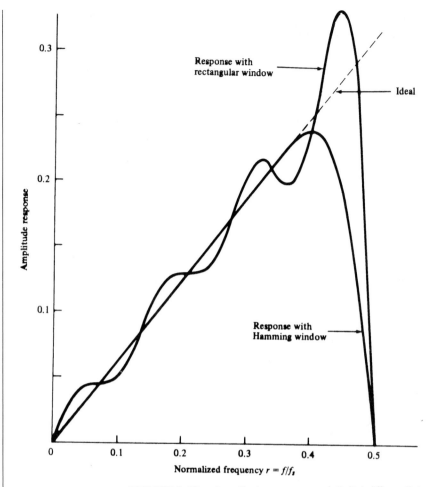

FIGURE 9-29. Amplitude response of digital differentiator.

Figure 9-29 illustrates the performance of the differentiator for the case in which 15 terms or weights are used in the realization. In other words, $h(nT) = 0, |n| \geq 8$. A Hamming window for this case is given by

$$w_h(n) = 0.54 + 0.46 \cos \frac{n\pi}{7}, \qquad |n| \leq 7 \qquad (9\text{-}195)$$

Table 9-1 illustrates the terms of the unit pulse response for the FIR differentiator with rectangular and Hamming windows. The phase response is shown in Figure 9-30. In reading Table 9-1, keep in mind that $M = 7$, and, therefore, the filter weights for the causal digital differentiator are given by

$$L_n = h(nT - 7T) \qquad (9\text{-}196)$$

TABLE 9-1
Filter Weights for FIR Differentiator

n	Unit Pulse Response with Rectangular Window, $h(nT)$	Hamming Window Function, $w_h(nT)$	Unit Pulse Response with Hamming Window, $w_h(nT)h(nT)$
-7	-0.142857	0.08	$+0.011429$
-6	-0.166667	0.125554	-0.020926
-5	$+0.2$	0.253195	$+0.050639$
-4	-0.25	0.437640	-0.109410
-3	$+0.333333$	0.642360	$+0.214120$
-2	-0.5	0.826805	-0.413403
-1	$+1.0$	0.954446	$+0.954446$
0	0	1.0	0
1	-1.0	0.954446	-0.954446
2	$+0.5$	0.826805	$+0.413403$
3	-0.333333	0.642360	-0.214120
4	$+0.25$	0.437640	$+0.109410$
5	-0.2	0.253195	-0.050639
6	$+0.166667$	0.125554	$+0.020926$
7	-0.142857	0.08	-0.011429

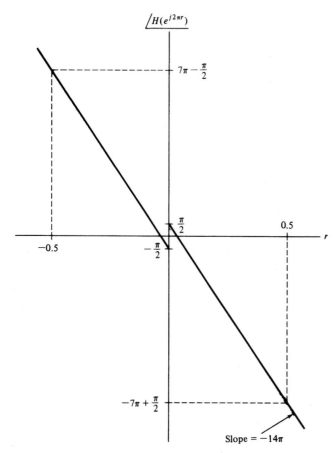

FIGURE 9-30. Phase response for $M = 7$ FIR differentiator.

EXAMPLE 9-12

As another example of a FIR digital filter design, we approximate an ideal low-pass filter using 17 weights. Assuming that the filter bandwidth is $0.15f_s$, we have, in terms of the normalized frequency variable r,

$$H(e^{j2\pi r}) = \begin{cases} 1, & |r| \le 0.15 \\ 0, & 0.15 < |r| < 0.5 \end{cases} \tag{9-197}$$

The filter weights, $h(nT)$, are therefore

$$h(nT) = \int_{-0.15}^{0.15} e^{j2\pi n r}\, dr \tag{9-198}$$

which is, upon performing the integration,

$$h(nT) = \frac{1}{j2\pi n}\left(e^{j0.3\pi} - e^{-j0.3\pi}\right) \tag{9-199}$$

The filter weights can therefore be expressed as

$$h(nT) = \frac{1}{\pi n}\sin 0.3\pi n \tag{9-200}$$

Since we are assuming 17 weights, the Hamming window for this application can be written

$$w_h(n) = 0.54 + 0.46\cos\frac{\pi n}{8}, \qquad |n| \le 8 \tag{9-201}$$

Figure 9-31 illustrates the amplitude response with and without the Hamming window. In drawing Figure 9-31 we have allowed the amplitude response to be negative over several frequency ranges in order to avoid having to draw the phase. The phase will be $\pm 180°$ or $\pm \pi$ rad for those frequency ranges where the amplitude responses are drawn negative. Table 9-2 gives the terms of the unit pulse response with a rectangular window and with a Hamming window. The filter weights of the causal filter are

$$L_n = h(nT - 7T) \tag{9-202}$$

EXAMPLE 9-13

As our last example of the design of a FIR digital filter, consider a 90° phase shifter. This filter is frequently referred to as Hilbert transform filter, since the output of a 90° phase shifter is defined as the Hilbert transform of the input signal. The desired sinusoidal steady-state frequency response is

$$H_d(e^{j2\pi r}) = \begin{cases} -j, & 0 < r < 0.5 \\ +j, & -0.5 < r < 0 \end{cases} \tag{9-203}$$

Note that, as in the case of the digital differentiator, $H_d(e^{j2\pi r})$ is imaginary for all r. It follows that the unit pulse response is

$$h(nT) = \int_{-1/2}^{0} (+j)e^{j2\pi n r}\, dr + \int_{0}^{1/2} (-j)e^{j2\pi n r} \tag{9-204}$$

Performing the integration yields

$$h(nT) = \frac{1}{2\pi n}(1 - e^{-j\pi n}) - \frac{1}{2\pi n}(e^{j\pi n} - 1) \tag{9-205}$$

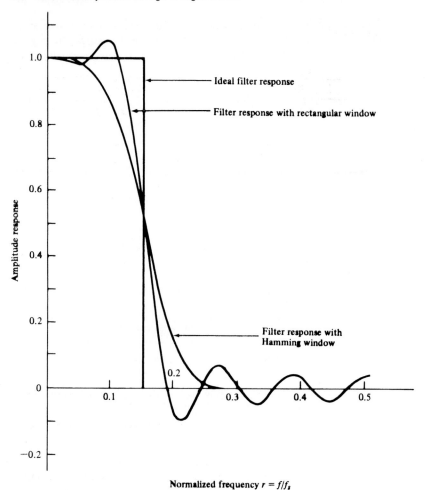

FIGURE 9-31. Amplitude response of digital low-pass filter. (The negative portions are shown negative for convenience.)

or

$$h(nT) = \frac{1}{\pi n} \{1 - \cos \pi n\} \qquad (9\text{-}206)$$

Thus

$$h(nT) = \begin{cases} 0, & n \text{ even} \\ \dfrac{2}{\pi n}, & n \text{ odd} \end{cases} \qquad (9\text{-}207)$$

The amplitude response is shown in Figure 9-32 for $M = 7$ for rectangular and Hamming windows. The Hamming window for this case is given by (9-195). The filter weights are given in Table 9-3.

TABLE 9-2
Filter Weights for FIR Low-Pass Filter ($f_c = 0.15f_s$)

n	Unit Pulse Response with Rectangular Window, $h(nT)$	Hamming Window Function, $w_h(nT)$	Unit Pulse Response with Hamming Window, $w_h(nT)h(nT)$
-8	0.037841	0.08	0.003027
-7	0.014052	0.115015	0.001616
-6	-0.031183	0.214731	-0.006696
-5	-0.063662	0.363966	-0.023171
-4	-0.046774	0.54	-0.025258
-3	0.032788	0.716034	0.023477
-2	0.151365	0.865269	0.130972
-1	0.257518	0.964985	0.248501
0	0.3	1.0	0.3
1	0.257518	0.964985	0.248501
2	0.151365	0.865269	0.130972
3	0.032788	0.716034	0.023477
4	-0.046774	0.54	-0.025258
5	-0.063662	0.363966	-0.023171
6	-0.031183	0.214731	-0.006696
7	0.014052	0.115015	0.001616
8	0.037841	0.08	0.003027

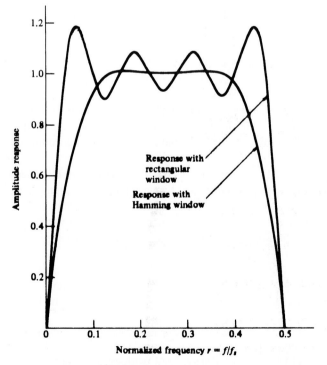

FIGURE 9-32.   Amplitude response of digital 90° phase shifter.

TABLE 9-3
Filter Weights for FIR 90° Phase Shifter

n	Unit Pulse Response with Rectangular Window, $h(nT)$	Hamming Window Function, $w_h(nT)$	Unit Pulse Response with Hamming Window, $w_h(nT)h(nT)$
−7	−0.090946	0.08	−0.007276
−6	0	0.125554	0
−5	−0.127324	0.253195	−0.032238
−4	0	0.437640	0
−3	−0.212207	0.642360	−0.136313
−2	0	0.826805	0
−1	−0.636620	0.954446	−0.607619
0	0	1.0	0
1	0.636620	0.954446	0.607619
2	0	0.826805	0
3	0.212207	0.642360	0.136313
4	0	0.437640	0
5	0.127324	0.253195	0.032238
6	0	0.125554	0
7	0.090946	0.08	0.007276

9-6 Computer-Aided Design of Digital Filters

As we saw in the previous section, digital filters can be designed to satisfy a set of specifications that are not compatible with any known analog prototype. Filters having an arbitrary amplitude response while maintaining a perfectly linear phase response are an example. Simply stated, we can design digital filters for which analog equivalents do not exist. A number of design algorithms have been developed that provide the tools for developing these filters. Since the algorithms are usually realized as computer programs, we refer to their use as a computer-aided design. Design techniques exist for design of both IIR and FIR digital filters and MATLAB provides examples for each.

The MATLAB algorithm used for the design of IIR filters is designated `yulewalk` and is based on a minimum mean-square error fit in the time domain. Thus, the inverse Fourier transform of the desired amplitude response is used. The MATLAB manual provides somewhat more detail on the algorithm and specifies the reference upon which the algorithm is based.[†]

The MATLAB algorithm used for the design of FIR digital filters is based on the Parks-McClellan[‡] algorithm, which for many years has been the most popular algorithm for designing FIR digital filters. The Parks-McClellan algorithm makes use of Chebyshev approximation theory, which yields a filter design in which the maximum error between the desired and actual frequency responses is minimized (referred to as *minimax approximations*). In addition, as we shall see, the approximation error is uniformly distributed over the filter passbands and stopbands.

The MATLAB implementation of the Parks-McClellan algorithm is contained in the routine `remez` in the Signals and Systems Toolbox. The name `remez` results because the Remez exchange algorithm is used in the polynomial interpolation problem that plays a key role in the implementation of the Parks-

[†]B. Friedlander and B. Porat, "The Modified Yule-Walker Method of ARMA Spectral Estimation," *IEEE Transactions on Aerospace and Electronic Systems,* Vol. AES-20, No. 2, March 1984.

[‡]L. R. Rabiner, J. H. McClellan and T. W. Parks, "FIR Filter Design Techniques Using Weighted Chebyshev Approximations," *Proceedings of the IEEE,* Vol. 63, 1975.

McClellan algorithm. Although, in the following example, we consider the design of a dual-band lowpass filter, the Parks-McClellan algorithm can be used to design a wide variety of filters, including those that have transfer functions that are purely imaginary, such as the differentiator and the Hilbert transform filter.

In order to demonstrate the use of MATLAB for designing both IIR and FIR digital filters we consider the desired amplitude response to be the two-level passband characteristic given in Figure 9-33. The phase response is not specified but we desire the phase response to be as linear as possible. The particular amplitude response given in Figure 9-33 is chosen simply because it is more interesting than the simple boxcar (ideal) lowpass filter considered in Example 9-12 and no specific application is in mind. The parameter r, as usual, represents frequency normalized with respect to the sampling frequency. For the examples to follow we choose $r_1 = 0.1$, $r_2 = 0.11$, $r_3 = 0.25$ and $r_4 = 0.26$. (Note that we define r as frequency normalized with respect to the sampling frequency while the vector specifying the critical frequencies in the MATLAB programs specify the critical frequencies in terms of the folding or Nyquist frequency.) We choose $B = 0.4$. The critical frequencies chosen for the examples to follow result in relatively narrow transition bands. This will tend to result in higher errors for a given specified order.

EXAMPLE 9-14

As the first example of computer-aided design, we develop an IIR digital filter having the desired amplitude response given in Figure 9-33. A 10th-order filter is specified. The MATLAB code follows.

```
EDU» c9ex14
f = [0 0.2 0.22 0.5 0.52 1];          % Specify f vector
m = [1 1 0.4 0.4 0 0];                % Specify amplitude response
n = 10;                               % Specify order
[b,a] = yulewalk(n,f,m);             % Compute transfer function
[h,w] = freqz(b,a,512);              % Determine filter response
subplot(2,1,1)                        % Designate first subplot
plot(f/2,m,w/(2*pi),abs(h))          % Plot amplitude response
xlabel('Normalized Frequency - r')   % Label x-axis
ylabel('Amplitude')                   % Label y-axis
phase = angle(h);                     % Calculate phase
phase = unwrap(phase);                % Unwrap phase
phasedeg = phase*180/pi;              % Express phase in degrees
subplot(2,1,2)                        % Designate second subplot
plot(w/(2*pi),phasedeg)              % Plot phase response
xlabel('Normalized Frequency - r')   % Label x-axis
ylabel('Phase - degrees')            % Label y-axis
```

The results of this design are shown in Figure 9-34.

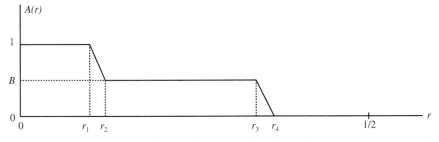

FIGURE 9-33.   Desired amplitude response.

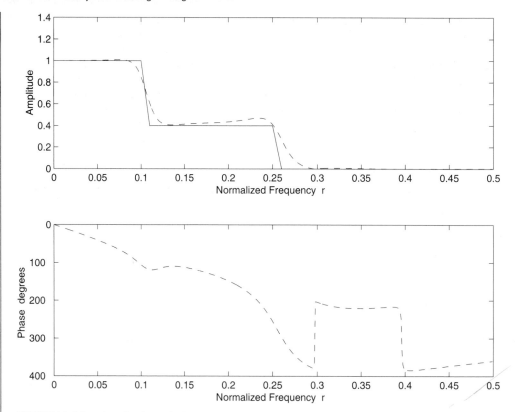

FIGURE 9-34. Amplitude and phase response for IIR filter. (The desired filter characteristic is shown using a solid line and the actual filter is shown using a dashed line.)

We see that the error in the amplitude response, especially in the first of the two passbands, is small. In addition, the phase response is reasonably linear up to a frequency of about $r = 0.29$. Past this point, the amplitude is small so that the phase response is of little or no interest. One should, of course, be aware that the validity of these statements depends upon the application.

EXAMPLE 9-15

In this example we repeat the previous example with the design of a FIR digital filter based on the Parks-McClellan algorithm. The MATLAB code follows.

```
EDU» c9ex15
f = [0 0.2 0.22 0.5 0.52 1];        % Specify f vector
m = [1 1 0.4 0.4 0 0];              % Specify amplitude response
n = 30;                             % Define n
b = remez(n,f,m);                   % Compute impulse response
[h,w] = freqz(b,1,512);            % Compute frequency response
amp = abs(h);                       % Compute amplitude response
subplot(2,1,1)                      % Designate first subplot
plot(f/2,m,w/(2*pi),amp)           % Plot amplitude response
```

```
xlabel('Normalized Frequency - r')        % Label x-axis
ylabel('Amplitude')                       % Label y-axis
phase = angle(h);                         % Calculate phase
phase = unwrap(phase);                    % Unwrap phase
phasedeg = phase*180/pi;                  % Express phase in degrees
subplot(2,1,2)                            % Designate second subplot
plot(w/(2*pi),phasedeg)                   % Plot phase response
xlabel('Normalized Frequency - r')        % Label x-axis
ylabel('Phase - degrees')                 % Label y-axis
```

The results of this design are shown in Figure 9-35.

It is interesting to look at the unit pulse response of the filter just derived. This is easily accomplished by entering the MATLAB code

```
EDU» sb = size(b);
EDU» nn = 0:sb(2)-1;
EDU» stem(nn,b)
EDU» xlabel('Discrete Time Index - n')
EDU» ylabel('Weight')
```

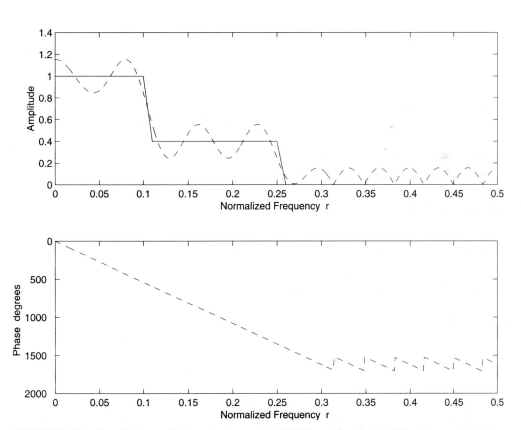

FIGURE 9-35.   Amplitude and phase response for FIR filter. (The desired filter characteristic is shown using a solid line and the actual filter is shown using a dashed line.)

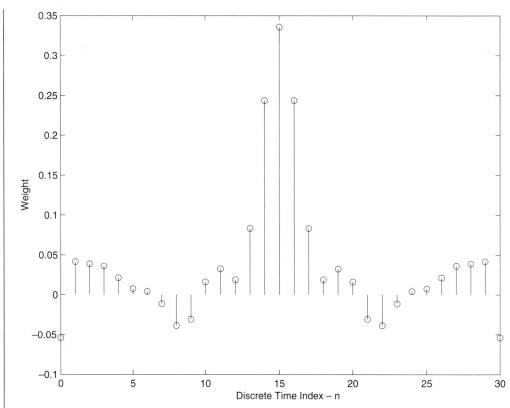

FIGURE 9-36. Unit pulse response of IIR digital filter.

The first two lines of code are necessary for the plot to begin at zero. Since n, the filter order, was set equal to 30, the filter has 31 weights, $b(0)$, $b(1)$, ..., $b(30)$. These are shown in Figure 9-36.

Note that the filter unit pulse response is even about the center weight at $n = 15$. Thus, as we saw in the last section, the non-causal version of the filter, which is determined by placing the center weight at $n = 0$, has a transfer function that is purely real. The linear phase then results from translating the unit pulse response to the right 15 sample periods.

In comparing the results of Examples 9-14 and 9-15, it should be noted that although the order of the FIR filter is three times the order of the IIR, the error in the amplitude response for the IIR filter is much larger than the error in the amplitude response for FIR filter. A basic advantage of IIR filters over FIR filters is that a desired amplitude response characteristic usually can be approximated with a given error constraint using fewer delays in the IIR filter than would be required in the FIR realization. This reduced number of delay elements comes at the cost of non-linear phase in the IIR realization. The phase response, as expected, is perfectly linear in the FIR filter. The sawtooth pattern of the phase response in the stopband is due to the fact that the amplitude response is negative for periodic segments. Thus each of the phase jumps is 180°. Also note that the error in the two passbands and the error in the stopband are equal for the IIR filter. This demonstrates the equiripple characteristic of filters designed using the Parks-McClellan algorithm.

Summary

This chapter basically considered two topics: the implementation of digital signal processors from the pulse transfer function, $H(z)$, and the design of digital signal processors to meet some performance specification. The four main implementations considered were Direct Form I, Direct Form II, cascade, and parallel.

The design or synthesis problem usually involves the development of a digital signal processor that meets some time-domain or frequency-domain specification. The impulse-invariant and step-invariant digital filters are based on a time-domain specification, while the bilinear z-transform digital filter is based on a frequency-domain specification. All of these filters are infinite-duration impulse response (IIR) digital filters.

The finite-duration impulse response (FIR) digital filter is based on a frequency-response specification, and the filter implementation is accomplished by taking the Fourier transform of the desired frequency-response specification.

There are advantages to the use of both IIR and FIR digital filters. The main advantages of IIR filters are as follows:

1. The design techniques for IIR digital filters are very easy to apply. The design is initiated with an analog prototype, and one who is familiar with analog filter theory will usually have a good feel for the performance of a given filter in a given application.
2. Hardware requirements for an IIR digital filter are usually less than the hardware requirements for a comparable FIR filter. However, with modern LSI and VLSI techniques, hardware considerations are becoming less important.

The main advantages of FIR digital filters are the following:

1. FIR filters can be designed that have perfectly linear phase. Therefore, phase distortion is eliminated.
2. Since FIR filters have no feedback, they have no poles and are therefore always stable.
3. The fast Fourier transform (FFT), which is the main topic covered in the next chapter, gives the filter designer a very simple and efficient tool for determining the filter weights.
4. Since no analog prototype is required in the synthesis procedure, digital filters can be designed that have no analog equivalent.

There are also disadvantages that are often important. The main disadvantages of the IIR synthesis techniques treated in this chapter are these:

1. Since the design procedure is initiated with an analog filter function, it is first necessary to determine an analog filter that meets the desired specifications.
2. Phase distortion is frequently a problem.

The main disadvantages of the FIR filter synthesis techniques discussed in this chapter are that:

1. If the digital filter is to have an extremely small bandwidth, a large number of filter weights may be necessary. The result will be a digital filter with a large group delay.
2. The selection of an appropriate window function may be difficult.

It should be clear that the designer of a digital filter often has many options available. Choosing the appropriate technique for an application requires a good understanding of digital filter theory and the requirements of the specific application of interest.

The main points made in this chapter include the following:

1. Two digital filter implementations that follow directly from the difference equation of a digital signal processor or filter are the Direct Form I and the Direct Form II implementations. The Direct Form I implementation is actually a cascade of two networks. The first of these networks realizes the zeros of the pulse transfer function, and the second realizes the poles of the transfer function. The number of unit delay elements, z^{-1} elements, necessary for a Direct Form I implementation is the sum of the number of poles and the number of zeros. If the order of the cascade is reversed, the Direct Form II realization results. The total number of z^{-1} elements necessary for the Direct Form II implementation is the number of poles or the number of zeros, whichever is greater.

2. The cascade realization results from writing the transfer function as a product of poles and zeros. The implementation of the filter is then the cascade of first-order or second-order digital filters. First-order sections are used for the implementation of simple poles and zeros, and second-order sections are used for the implementation of complex conjugate poles and zeros.

3. The parallel realization results from expanding the transfer function of the digital filter by using partial-fraction expansion. Each term in the partial-fraction expansion is realized by using a first- or second-order digital filter. First-order sections are used for the implementation of simple poles, and second-order sections are used for the implementation of complex conjugate poles.

4. The integrator is a basic building block for many systems. We investigated two discrete-time integrators, the rectangular integrator and the trapezoidal integrator. Both of these are approximations to the ideal integrator. The trapezoidal integrator exhibits a constant phase shift of $-\pi/2$ rad for $f > 0$ and thus exhibits zero group delay. The rectangular integrator exhibits a group delay of $1/2T$. The rectangular integrator exhibits less attenuation of very high frequencies than does the ideal integrator, while the trapezoidal integrator exhibits greater attenuation of high frequencies than does the ideal integrator.

5. Infinite-duration impulse response digital filters can be designed by using either a time-domain or a frequency-domain criterion. If a time-domain criterion is used, the design technique is to determine the digital filter having an output that is equivalent to the sampled output of an analog filter for some input signal. If the input signal is a unit impulse, the resulting digital filter is an impulse-invariant digital filter. If the signal is a step function, a step-invariant digital filter results.

6. The impulse-invariant digital filter is designed by first expanding the transfer function, $H_a(s)$, of an analog prototype using partial-fraction expansion. Taking the inverse Laplace transform yields the unit impulse response $h_a(t)$. This is then sampled, z-transformed, and multiplied by T. In equation form

$$H(z) = T\mathcal{Z}\{\mathcal{L}^{-1}\{H_a(s)\}|_{t=nT}\}$$

The result of this design technique is a digital filter having a unit pulse response equivalent to the sampled impulse response of the analog prototype.

7. The step-invariant digital filter is designed by first determining the step response of the analog prototype from which the digital filter is to be derived. This is then sampled, z-transformed, and multiplied by $(1 - z^{-1})$. In equation form

$$H(z) = (1 - z^{-1})\mathcal{Z}\left\{\mathcal{L}^{-1}\left\{\frac{H_a(s)}{s}\right\}\bigg|_{t=nT}\right\}$$

The result of this design technique is a digital filter having a unit step response equivalent to the sampled unit step response of the prototype analog filter.

8. The bilinear z-transform technique is an attempt to design a digital filter having a steady-state frequency response that approximates the steady-state response of the analog filter from which the digital filter is derived. The procedure is to use the substitution

$$s \Rightarrow C \frac{1 - z^{-1}}{1 + z^{-1}}$$

in the transfer function of the analog prototype $H_a(s)$. By proper choice of C, it is possible to match the responses of the analog and digital filters at some reference frequency. The appropriate value of C is

$$C = \omega_r \cot \frac{\omega_r T}{2}$$

in which ω_r is the reference frequency and T is the sampling period. A major attribute of the bilinear z-transform synthesis technique is that a digital filter can be developed with strictly algebraic techniques. If $H_a(s)$ is expressed as a ratio of polynomials in s, the resulting $H(z)$ is a ratio of polynomials in z.

9. Finite-duration impulse response digital filters can be implemented as shift registers with tap weights given by the Fourier series coefficients of the desired frequency response of the digital filter. Since there are, in general, an infinite number of nonzero Fourier coefficients, it is necessary to truncate the Fourier series expansion of the desired frequency response at a finite number of terms. The error introduced by this truncation can be partially controlled by using an appropriate window function. The unit pulse response must then be multiplied by z^{-M}, with M chosen to ensure that a causal digital filter results. This multiplication by z^{-M} introduces a linear phase shift into the frequency-response characteristic and therefore does not contribute to phase distortion.

10. We studied FIR filter design techniques for the implementation to two types of filters. For the first type the desired transfer function of the noncausal filter was real, and for the second type the desired transfer function of the noncausal filter was imaginary.

11. Computer-aided design techniques exist for both IIR and FIR digital filters. FIR digital filters designed using the Parks-McClellan method, which is the most popular of these techniques, result in linear phase filters having equiripple error characteristics. It typically takes more delay elements (higher filter order) to realize an FIR digital filter than it takes to realize an IIR digital filter given a desired filter characteristic and a constraint on the maximum error.

Further Reading

The four textbooks listed in the previous chapter treat the design and analysis of digital filters:

A. V. OPPENHEIM and R. W. SCHAFER, *Discrete-Time Signal Processing.* Englewood Cliffs, NJ: Prentice-Hall, 1989.

J. G. PAROKIS and D. G. MANOLAKIS, *Digital Signal Processing: Principles, Algorithms and Applications,* 2nd ed. New York: Macmillan, 1992.

R. D. STRUM and D. E. KIRK, *First Principles of Discrete Systems and Digital Signal Processing.* Reading, MA: Addison-Wesley, 1988.

J. A. CADZOW, *Foundations of Digital Signal Processing and Data Analysis.* New York: Macmillan, 1987.

Two additional books focus almost exclusively on the digital filter synthesis problem:

R. W. HAMMING, *Digital Filters,* 3rd ed. Englewood Cliffs, NJ: Prentice-Hall, 1989.

E. P. CUNNINGHAM, *Digital Filtering: An Introduction.* Boston: Houghton Mifflin, 1992.

A number of books have recently been published that provide examples of digital filter synthesis using MATLAB. The following book is especially recommended:

V. K. INGLE and J. G. PROAKIS, *Digital Signal Processing Using MATLAB V.4.* Boston, MA: PWS, 1997.

PROBLEMS

Section 9-2

9-1. Determine and sketch the Direct Form I and the Direct Form II realizations for each of the following pulse transfer functions:

(a) $H(z) = \dfrac{1 + 0.3z^{-1} - 0.7z^{-2}}{1 + 0.6z^{-1}}$

(b) $H(z) = \dfrac{1 + 0.3z^{-1}}{1 - 0.6z^{-1} + 0.9z^{-2}}$

(c) $H(z) = \dfrac{(1 + 0.3z^{-1})^2(1 - 0.1z^{-1})}{1 - 0.2z^{-2} + 0.3z^{-4}}$

(d) $H(z) = \dfrac{1 + 0.3z^{-1} - 0.6z^{-2} - 0.7z^{-3}}{(1 + 0.2z^{-1})^3}$

9-2. Determine and sketch the Direct Form I and Direct Form II realizations for each of the following pulse transfer functions:

(a) $H(z) = \dfrac{3 + 0.4z^{-1} + 0.5z^{-2}}{2 - 0.6z^{-1} + 0.8z^{-2}}$

(b) $H(z) = \dfrac{1 - 0.2z^{-2}}{4 - 0.4z^{-1}} + \dfrac{1}{1 - 0.2z^{-2}}$

(c) $H(z) = (1 - z^{-1})^3 + \dfrac{1}{1 - 0.3z^{-2}}$

(d) $H(z) = 2 + 3z^{-1} - 0.4z^{-2}$

9-3. Determine and sketch the Direct Form I and the Direct Form II realizations for each of the following pulse transfer functions:

(a) $H(z) = \dfrac{2}{1 - z^{-1}} + \dfrac{1}{(1 - 0.2z^{-1})^2}$

(b) $H(z) = (1 - z^{-1}) + \dfrac{3}{(1 - 0.3z^{-1})^2}$

(c) $H(z) = \dfrac{1}{1 - 0.6z^{-1} + 0.4z^{-2} + 0.8z^{-3}}$

9-4. Determine a cascade and parallel realization, using only first-order structures for the following pulse transfer function.

$$H(z) = \dfrac{z(z - 1)}{(z - \frac{1}{2})(z - \frac{1}{8})}$$

9-5. Determine a parallel realization for the following pulse transfer function using structures having minimum order.

$$H(z) = \dfrac{1}{(1 - 0.1z^{-1})(1 - 0.9z^{-1} + 0.81z^{-2})}$$

9-6. Determine a parallel realization for

$$H(z) = \dfrac{z^{-3}}{(1 - 0.1z^{-1})(1 - 0.5z^{-1})^2}$$

9-7. Determine a parallel realization for

$$H(z) = \frac{z^{-3}}{(1 - 0.1z^{-1})(1 - 0.4z^{-1})^3}$$

9-8. Determine a parallel realization for

$$H(z) = \frac{z^{-3}}{(1 - z^{-1})(2 - z^{-1})(4 + z^{-1})}$$

9-9. Determine a parallel realization for

$$H(z) = \frac{z^{-2}}{(1 - 0.3z^{-1})^2}$$

9-10. Determine a parallel and cascade realization for

$$H(z) = \frac{z^{-4}}{(1 - z^{-1})(1 - z^{-2})}$$

using structures having minimum order.

9-11. For the system shown determine the pulse transfer function and determine the Direct Form II realization.

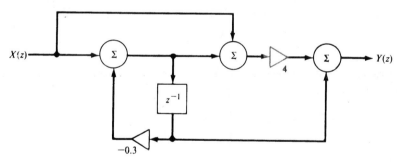

FIGURE P9-11

9-12. Realize the system shown using only one z^{-1} element.

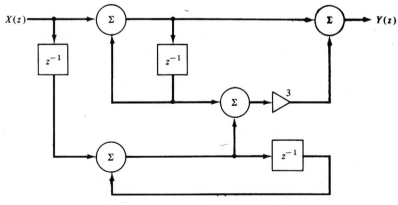

FIGURE P9-12

9-13. Determine the pulse transfer function of the system shown as a ratio of polynomials in z^{-1}. The parameters a, b, c, and d are arbitrary.

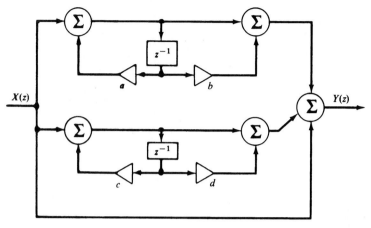

FIGURE P9-13

9-14. Determine the pulse transfer function of the system shown and draw the Direct Form II realization.

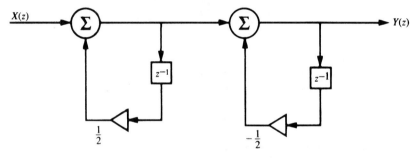

FIGURE P9-14

9-15. Determine the pulse transfer function of the system shown in Figure P9-15 and draw the Direct Form II realization.

9-16. Determine the pulse transfer function of the digital filter shown in Figure P9-16 and draw the Direct Form II realization.

9-17. Determine the pulse transfer function of the system shown in Figure P9-17 and draw the Direct Form II realization.

9-18. Determine the pulse transfer function of the digital filter shown in Figure P9-18 and draw the Direct Form II realization.

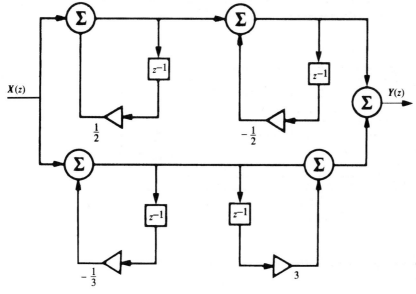

FIGURE P9-15

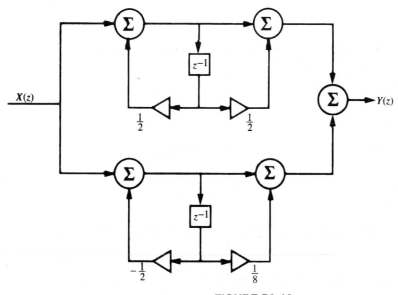

FIGURE P9-16

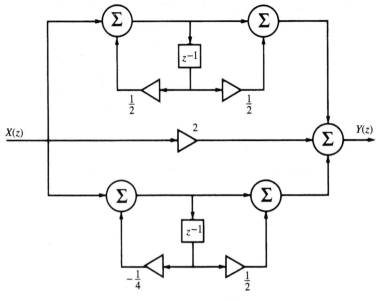

FIGURE P9-17

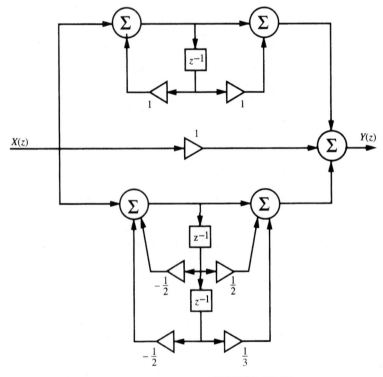

FIGURE P9-18

9-19. Show that the digital filter examined in Example 9-2 has the unit pulse response

$$h(nT) = \frac{a^n}{\sin \theta} \sin[(n + 1)\theta]u(n)$$

9-20. Show that the amplitude response of the digital filter examined in Example 9-2, evaluated at $\omega T = \theta$, is

$$|H(e^{j\theta})| = \frac{1}{1 - a} \sqrt{\frac{1}{1 + a^2 - 2a \cos 2\theta}}$$

Use this result to show that, at $\theta = \pi/2$, the amplitude response is

$$H(e^{j\pi/2}) = \frac{1}{1 - a^2}$$

Section 9-3

9-21. The two-term running average filter is an approximation to an integrator. It is defined as

$$y(nT) = \tfrac{1}{2}[x(nT) + x(nT - T)]$$

Determine the pulse transfer function and plot the amplitude and phase responses as functions of $r = f/f_s$ in the range $0 \leq r \leq 0.5$. In what sense is the two-term running average filter an integrator?

9-22. Repeat Problem 9-21 for the four-term running average filter defined by

$$y(nT) = \tfrac{1}{4}[x(nT) + x(nT - T) + x(nT - 2T) + x(nT - 3T)]$$

Is this a better integrator than the two-term running average filter? Why or why not?

9-23. Determine the pulse transfer function for the N-term running average filter defined by

$$y(nT) = \frac{1}{N} \sum_{k=0}^{N-1} x(nT - kT)$$

Express the amplitude response as a function of the normalized frequency variable $r = f/f_s$.

9-24. Determine and plot the unit pulse response of both a rectangular and trapezoidal integrator. Compare with the unit impulse response of an ideal analog integrator. Comment on your results.

9-25. Assume that an ideal analog integrator has input

$$x(t) = Ae^{-\alpha t}$$

Determine and plot the output. Now assume that samples of $x(t)$ defined by

$$x(nT) = A(e^{-\alpha T})^n$$

are applied to both a rectangular and trapezoidal integrator. Determine and plot the output sequence, $y(nT)$, for both integrators and compare with the sampled output of an ideal integrator. Comment on the results.

9-26. A first approximation to a digital differentiator is a system that generates an output equal to the slope of the line connecting the sample points $x(nT)$ and $x(nT - T)$.

(a) Show that the defining difference equation is

$$y(nT) = \frac{1}{T} \{x(nT) - x(nT - T)\}$$

(b) Determine the amplitude response and phase response of the system. Compare with an ideal differentiator.

Section 9-4

9-27. Determine the pulse transfer function of both an impulse-invariant and step-invariant integrator. Determine and plot both the amplitude and phase responses of each design in the range $0 \le r \le 0.5$. Compare your results with the results of Example 9-3.

9-28. Design a digital integrator by using the bilinear z-transform synthesis technique. Show that the result is trapezoidal integration.

9-29. Design a digital differentiator by using the step-invariant method. Plot the amplitude response and phase response of your design. Compare your results with the differentiator designed in Problem 9-26.

9-30. Determine the pulse transfer function, using impulse-invariant synthesis, corresponding to the following analog transfer functions. The sampling period, T, is to be left arbitrary.

(a) $H_a(s) = \dfrac{8}{(s + 2)(s + 4)}$

(b) $H_a(s) = \dfrac{8}{s(s + 2)(s + 4)}$

(c) $H_a(s) = \dfrac{s + 1}{(s + 0.5)(s + 4)}$

9-31. Assuming a sampling frequency of 40 radians/second, plot the amplitude and phase responses for each of the digital filters found in Problem 9-30. On the plots include the amplitude and phase responses of the analog filters, from which each of the digital filters is derived. You may wish to use MATLAB for calculating and plotting the responses.

9-32. Repeat Problems 9-30 and 9-31 using step-invariant synthesis.

9-33. For the analog transfer function

$$H_a(s) = \frac{20}{(s + 0.5)(s + 2)(s + 20)}$$

determine the impulse-invariant digital filter. Assume a sampling frequency of 15 Hz. Draw the parallel realization of the filter using only first-order structures.

9-34. Plot the amplitude and phase responses the digital filter found in Problem 9-32. On the plots include the amplitude and phase responses of the analog filter from which the digital filter was derived. Once again, you may wish to use MATLAB for calculating and plotting the responses.

9-35. Repeat Problems 9-32 and 9-33 using step-invariant synthesis.

9-36. Determine an impulse-invariant and a step-invariant digital filter corresponding to the analog transfer function

$$H(s) = \frac{10(s + 1)}{(s + 2)(s + 6)}$$

Leave the sampling frequency arbitrary.

9-37. Determine the pulse transfer function, using both impulse-invariant and step-invariant synthesis, for each of the following analog transfer functions. The sampling period, T, is to be left arbitrary. Draw the Direct Form II realizations for each filter.

(a) $H(s) = \dfrac{5}{s + 5}$

(b) $H(s) = \dfrac{5}{s(s + 5)}$

9-38. Determine an impulse-invariant and a step-invariant digital filter corresponding to the following analog transfer function. Assume a sampling frequency of 20 Hz. Draw the Direct Form II implementation for each digital filter.

$$H_a(s) = \frac{100}{(s + 10)^2}$$

9-39. Repeat Problem 9-36, assuming

$$H_a(s) = \frac{100}{(s + 1)(s + 10)^2}$$

9-40. In this problem we wish to design an "exponential-invariant" digital filter corresponding to the analog transfer function

$$H_a(s) = \frac{5}{s + 5}$$

We wish the output of the digital filter to be equivalent to the sampled output of the analog filter with input

$$x(t) = e^{-3t}u(t)$$

The dc ($f = 0$) gains of the analog filter and the digital filter are to be equal. Derive $H(z)$ of the digital filter, assuming that the sampling frequency is 20 Hz.

9-41. Design a digital differentiator using the bilinear z-transform synthesis technique. Plot the amplitude and phase responses of both the digital filter and the ideal analog filter. Tell why this is not an effective design even for low-input frequencies.

9-42. Determine $H(z)$, using the bilinear z-transform synthesis technique, for the analog transfer function

$$H_a(s) = \frac{5}{(s + 1)(s + 5)}$$

Assume a sampling frequency of 12 Hz and express your result as a ratio of polynomials in z^{-1}. Consider the following two cases:

(a) The digital filter is to have a response closely approximating the analog filter for very low frequencies.

(b) The digital filter and the analog filter are to have the same response at a frequency of 2 Hz.

9-43. Work Problem 9-41 using MATLAB.

9-44. A third-order Butterworth filter having a 3-dB break frequency of rad/s is given by

$$H_a(s) = \frac{1}{s^3 + 2s^2 + 2s + 1}$$

Assuming a sampling frequency of 6 Hz, determine the pulse transfer function of the bilinear z-transform digital filter equivalent to the given $H_a(s)$. Both the analog and digital filters are to have the same response at $\omega = 1$ rad/s.

9-45. Work Problem 9-43 using MATLAB.

9-46. Repeat Problem 9-44 assuming that the digital filter is to have a response closely approximating the analog filter response at very low frequencies.

9-47. Consider the analog system shown.

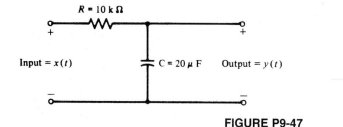

FIGURE P9-47

(a) For a sampling frequency of 15 Hz, determine the pulse transfer function for the impulse-invariant, step-invariant, and bilinear z-transform digital filter corresponding to the given analog system. For the bilinear z-transform filter, match the response of the digital filter to the response of the analog filter at $\omega = 5$ rad/s.

(b) Determine and plot the amplitude response (in decibels) of all three filters found in part (a). Also plot the amplitude response of the analog filter. Compare the responses.

(c) Repeat part (b) for the phase responses.

9-48. Determine $H(z)$, using the bilinear z-transform synthesis technique, corresponding to the analog transfer function

$$H_a(s) = \frac{16}{s^2 + 12s + 16}$$

Both the analog and the digital filters are to have the same magnitude and phase responses at $f = 4$ Hz. Assume a sampling frequency of 50 Hz.

9-49. Using the bilinear z-transform synthesis technique, determine the pulse transfer function corresponding to

$$H_a(s) = \frac{2(s + 4)}{(s + 1)(s + 8)}$$

Both the analog and digital filters are to have the same magnitude and phase responses at $f = 4/\pi$ Hz. Assume a sampling frequency of 14 Hz.

9-50. Work Problem 9-48 using MATLAB.

9-51. Using the bilinear z-transform synthesis technique, determine the pulse transfer function corresponding to

$$H_a(s) = \frac{2(s + 3)}{3(s + 1)(s + 2)}$$

Both the analog filter and the digital filter are to have the same response at $\omega = 3$ rad/s. Assume a sampling frequency of $10/\pi$ Hz. Plot the amplitude response of the resulting filter as a function of r in the range $0 \le r \le 0.5$.

9-52. Work Problem 9-51 using MATLAB.

9-53. A third-order Butterworth filter having a 3-dB break frequency of 1 rad/s is given by

$$H_a(s) = \frac{1}{s^3 + 2s^2 + 2s + 1}$$

Assuming a sampling frequency of 10 Hz, determine the pulse transfer function of the bilinear z-transform digital filter equivalent to the given $H_a(s)$. Both the analog and digital filters are to have the same response at $\omega = 1$ rad/s.

9-54. Work Problem 9-53 using MATLAB.

9-55. Derive $H(z)$, the pulse transfer function for a fourth-order bandpass filter based on the second-order low-pass Butterworth prototype. Give the result in terms of the parameters A and B. Next evaluate A and B assuming a sampling frequency of 5,000 Hz, a lower 3-dB frequency of 800 Hz, and an upper 3-dB frequency of 1200 Hz. Using these values, write $H(z)$ as a ratio of polynomials in z, giving the numerical values of each coefficient.

9-56. A bilinear z-transform digital filter uses $C = 1.5f_s$. Determine the normalized frequency correct to two decimal places, at which $\omega = \omega_1$.

9-57. Show that a notch (band reject) filter can be developed from a low-pass prototype by using the correspondence

$$s \Rightarrow D\frac{1 - z^{-2}}{1 - Ez^{-1} + z^{-2}}$$

Determine the relationship between D and the bandwidth and between E and the center frequency of the notch.

9-58. Use the relationship developed in the previous problem to design a notch filter based on a first-order Butterworth prototype. The sampling frequency is to be 2,000 Hz, and the bandedges of the notch are to be 200 and 300 Hz. Plot the amplitude and phase responses of the resulting filter.

Section 9-5

9-59. The triangular window function can be defined by

$$w_t(n) = 1 - \frac{|n|}{M+1}, \qquad -M \le n \le M$$

Apply the triangular window to the ideal low-pass filter developed in Example 9-12. Plot the resulting amplitude response and compare with the responses obtained with the rectangular and Hamming windows illustrated in Figure 9-31. (Note: Window functions can be defined in various ways. A more general definition will be discussed in the next chapter. The results are summarized in Table 10-4.)

9-60. Derive Equation (9-173).

9-61. An FIR digital filter is to be designed that has the rather strange desired frequency response

$$H_d(e^{j2\pi r}) = \begin{cases} 6, & 0 \le r < \frac{1}{2} \\ 3, & -\frac{1}{2} < r < 0 \end{cases}$$

Determine the unit pulse response $h_d(nT)$. Assuming that a rectangular window is used with $M = 8$ (17 total weights), plot the amplitude and phase responses as functions of r for the causal filter.

9-62. Determine and draw the amplitude and phase responses of the FIR digital filter shown.

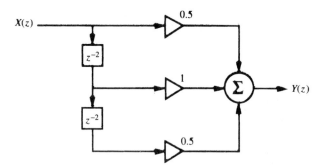

FIGURE P9-62

9-63. Repeat Problem 9-55 for the filter shown.

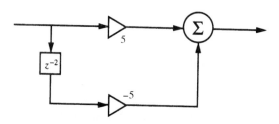

FIGURE P9-63

9-64. Write an equation giving the filter weights, L_k, for the FIR filter defined by the desired amplitude response shown. Assume that a Hamming window is to be used, and let $M = 12$.

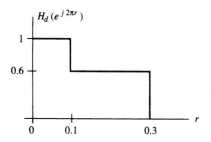

FIGURE P9-64

9-65. Plot the amplitude and phase responses of the filter in Problem 9-64.

9-66. Determine the filter weights, L_k, for the FIR filter defined by the desired amplitude response shown. Assume that a Hamming window is to be used, and let $M = 12$.

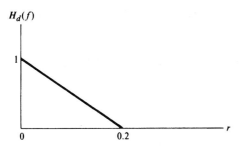

FIGURE P9-66

9-67. Plot the amplitude and phase response of the filter determined in Problem 9-66.

Section 9-6

9-68. Find an IIR and FIR digital filter, using computer-aided design techniques, that realize the ideal lowpass filter of Example 9-12. See how well you can do using the same number of weights as used in Example 9-12.

9-69. Design FIR and IIR digital filters using the specifications given for Problem 9-66.

Computer Exercises

9-1. Develop a MATLAB M-file that generates a step invariant digital filter from a s-domain transfer function. The input to the M-file is to be vectors containing the numerator and denominator coefficients of the continuous-time filter. The MATLAB program generates the numerator and denominator coefficients of the resulting digital filter. Test the resulting program by using the program to develop a step invariant digital filter using (9-81) as the analog prototype and comparing the result with (9-87).

9-2. The purpose of this computer exercise is to develop a program for designing a bilinear z-transform bandpass filter. Using the technique illustrated by (9-128) with A and B defined by (9-143) and (9-144), respectively, develop a MATLAB M-file that computes the coefficients of the numerator and denominator polynomials of the pulse transfer function for a bandpass digital filter. Use the resulting computer program to verify the results obtained in Example 9-9.

9-3. In this computer exercise a fourth-order bilinear z-transform digital filter is derived based on a lowpass Chebychev I analog prototype. The cutoff of the Chebyshev analog filter is 100 Hz and the passband ripple is 1 dB. The digital filter is designed for a sampling frequency of 2,000 Hz. Both the analog filter and the digital filter are to have the same amplitude and phase response at 100 Hz.
(a) Determine the pulse transfer function, $H(z)$, of the digital filter.
(b) On the same set of axes plot the amplitude response in dB for both the analog and digital filters.
(c) Repeat (b) for the phase response.
(d) Plot the group and the phase delay for both the analog and digital filters.
(e) Plot the pole and zero locations for both the analog and the digital filters.

9-4. Repeat the preceding computer exercise using a fourth-order elliptic prototype. Assume that the analog elliptic filter has 1 dB ripple in the passband and that the stopband attenuation is 25 dB down from the peak amplitude response in the passband.

9-5. In Section 9-6 it was noted that it is usually possible to satisfy a given amplitude response specification with an IIR filter having fewer delay elements than would be required for an FIR digital filter. In this computer exercise we quantify this statement. The first step is to develop a MATLAB M-file that determines the squared error between a filter specification (the desired characteristic) and the actual digital filter designed using computer-aided techniques. Using this M-file determine the mean-square error between the desired filter amplitude response and the amplitude response from the IIR filter designed in Exercise 9-14. Repeat for the FIR digital filter developed in Exercise 9-15. Experimentally determine the number of delay elements necessary to approximate the filter specification given in Figure 9-33 using an FIR filter and having a mean-square error less than or equal to that achieved with the IIR filter. Comment on the results of this exercise.

9-6. Develop both FIR and IIR digital filters based on the following amplitude response specification. Experiment with the number of delay elements used in the implementation. Comment on the results.

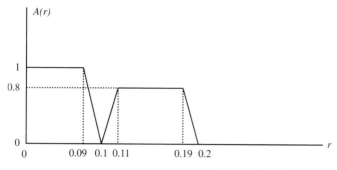

FIGURE P9-6

9-7. Design an IIR Hilbert transform filter using the `remez` routine introduced in Section 9-6. Evaluate the result by applying the using the test signal

$$x(nT) = 3 \cos\left(2\pi t + \frac{\pi}{4}\right) + 5 \sin\left(4\pi t + \frac{\pi}{6}\right)$$

to the resulting FIR filter. Since we can easily compute the exact Hilbert transform of $x(nT)$, the error resulting from computing the Hilbert transform by using the filter is easily determined.

9-8. Repeat the preceding computer exercise for a discrete-time differentiator. Use the same test signal.

10

The Discrete Fourier Transform and Fast Fourier Transform Algorithms

10-1 Introduction

The discrete Fourier transform (DFT) was derived in Section 4-13 as a doubly sampled version of the continuous-time Fourier transform (i.e., sampled in both time and frequency). The DFT transform pair is repeated here for convenience:

$$X_k = \sum_{n=0}^{N-1} x_n e^{-j2\pi kn/N}, \qquad k = 1, 2, \ldots, N \tag{10-1}$$

for the direct transform, and

$$x_n = \frac{1}{N} \sum_{k=0}^{N-1} X_k e^{j2\pi kn/N}, \qquad n = 1, 2, \ldots, N \tag{10-2}$$

for the inverse transform.

To derive the DFT from the continuous-time Fourier transform, the time-sampling interval was taken as $T/N = \Delta t$, where N is the total number of samples taken in a T-second interval (time window), and the frequency-sampling interval was taken as $1/T$. Thus, both relationships imply periodicity—the first in frequency with period $f_s = N/T = 1/\Delta t$, and the second in time with period $T = N\Delta t = N/f_s$. In terms of the sample number, the first is periodic in k with period N, and the second is periodic in n with period N.

In this chapter, we look further at the differences between the continuous-time Fourier transform and the DFT, derive two fast algorithms for computing the DFT (these are referred to as fast Fourier transform (FFT) algorithms), and illustrate several applications of the DFT and FFT.

Since MATLAB includes an FFT algorithm, we are not required to go through the computation of the FFT longhand as in several examples in this chapter. We do it merely to illustrate the DFT and FFT.

10-2 The DFT Compared with the Exponential Fourier Series; Error Sources in the DFT

To emphasize the difference between the Fourier transform and the DFT, we note that the Fourier transform is used to represent a continuous-time energy signal that lasts for $-\infty < t < \infty$, and the resulting spectrum generally covers $-\infty < f < \infty$. The DFT represents a finite number of sample values in the finite observation interval $0 \le nT/N < T$, and the resulting line spectrum is limited to $0 \le k/T < N/T$. Of course, as noted previously, (10-1) and (10-2) repeat each N points due to the periodicity of $e^{\pm j2\pi kn/N}$.

We now illustrate by example the effect of sampling and finite observation interval by examining in some detail the approximation of a Fourier transform by a DFT. Specifically, the results from Fourier transform theory that we wish to make use of are the theorems repeated below for convenience.

1. Transform of the ideal sampling waveform (4-42):

$$y_s(t) \triangleq \sum_{m=-\infty}^{\infty} \delta(t - mT_s) \leftrightarrow f_s \sum_{n=-\infty}^{\infty} \delta(f - nf_s), \qquad f_s = \frac{1}{T_s} \tag{10-3}$$

2. Transform of a rectangular pulse (Example 4-6 and the time-delay theorem):

$$\Pi\left(\frac{t - t_0}{T}\right) \leftrightarrow T \operatorname{sinc}(Tf) \exp(-j2\pi t_0 f) \tag{10-4}$$

3. The convolution theorem of Fourier transforms:

$$x_1(t) * x_2(t) \leftrightarrow X_1(f)X_2(f) \tag{10-5}$$

4. The multiplication theorem of Fourier transforms:

$$x_1(t)x_2(t) \leftrightarrow X_1(f) * X_2(f) \tag{10-6}$$

EXAMPLE 10-1

As a specific example, consider the two-sided exponential signal

$$x(t) = \exp\left(\frac{-|t|}{\tau}\right) \tag{10-7}$$

with Fourier transform

$$X(f) = \frac{2\tau}{1 + (2\pi f\tau)^2} \tag{10-8}$$

These are sketched in Figure 10-1a.

Since we are considering discrete-time signals, we multiply $x(t)$ by the ideal sampling waveform, $y_s(t)$, to produce the sampled exponential signal

$$x_s(t) = y_s(t) \exp\left(\frac{-|t|}{\tau}\right) \tag{10-9}$$

By the multiplication theorem, the Fourier transform of $x_s(t)$ is the convolution of $\mathscr{F}[y_s(t)] = f_s \sum_{n=-\infty}^{\infty} \delta(f - nf_s)$ and $X(f)$. The result of this convolution is

$$X_s(f) = 2\tau f_s \sum_{n=-\infty}^{\infty} \{1 + [2\pi\tau(f - nf_s)]^2\}^{-1} \tag{10-10}$$

The sampled signal $x_s(t)$ and its spectrum $X_s(f)$ are shown in Figure 10-1b for the case where $f_s = 1$ Hz.

In calculating a DFT only a T-second section of the sampled signal is observed. In effect, $x_s(t)$ is multiplied by a window function $\Pi(t/T)$. In the frequency domain, this corresponds to convolution of $X_s(f)$ with the Fourier transform of the window function, which is $T \operatorname{sinc} Tf$. Denoting the Fourier transform of the windowed and sampled signal by $X_{sw}(f)$, we may write it as

$$X_{sw}(f) = 2\tau f_s \sum_{n=-\infty}^{\infty} \{1 + [2\pi\tau(f - nf_s)]^2\}^{-1} * T \operatorname{sinc} Tf \tag{10-11}$$

The sampled, windowed signal and its spectrum are shown in Figure 10-1c.

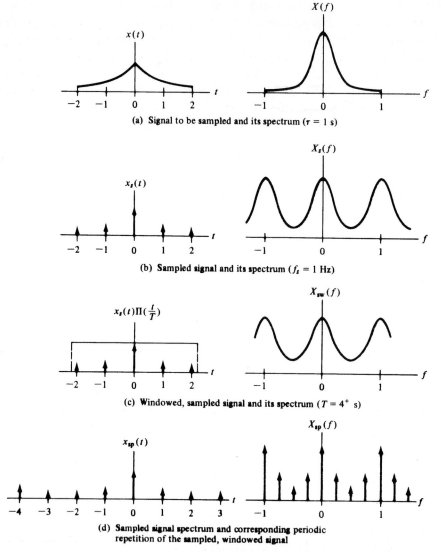

FIGURE 10-1. Signals and spectra illustrating the derivation of the DFT from the Fourier transform.

Finally, the result of a DFT operation effectively samples the spectrum $X_{sw}(f)$ at a discrete set of frequencies separated by the reciprocal of the observation interval (window duration), $1/T$, resulting in the sampled spectrum $X_{sp}(f)$. This corresponds to convolution in the time domain with a sequence of δ-functions since

$$T \sum_{m=-\infty}^{\infty} \delta(t - mT) \leftrightarrow \sum_{n=-\infty}^{\infty} \delta(f - nT^{-1}) \tag{10-12}$$

This produces a periodic sampled sequence, $x_{sp}(t)$, in the time domain. The resultant sampled signal and its spectrum are shown in Figure 10-1d.

MATLAB Application

The following MATLAB program illustrates the errors incurred through the sampling operations in the time and frequency domains when using the FFT. The sampled signal is a double-sided exponential, which we sample at 1, 2, and 4 times per time unit. The time window width is 4 (-2 to 2) for the first example computed as illustrated in Figure 10-2, and 8 (-4 to 4) for the second example computed as illustrated in Figure 10-3. The FFT-computed spectrum is compared with the analytically computed Fourier transform magnitude, which is plotted in the right-hand set of figures.

```
%     c10ex1
%
clg
T=input('Enter the time window width');
Tss=input('Enter the largest time sampling interval');
for k=1:3
k_even=2*k;
k_odd=2*k-1;
Ts=Tss/(2^(k-1));
fs=1/Ts;
tn=-T/2:Ts:T/2;
```

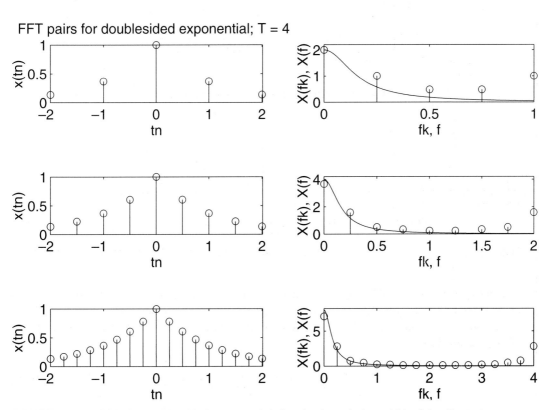

FFT pairs for doublesided exponential; T = 4

FIGURE 10-2. FFT of a double-sided exponential signal using window width of 4 to illustrate sampling errors. Solid lines in right-hand plots show the spectra of the analog signals.

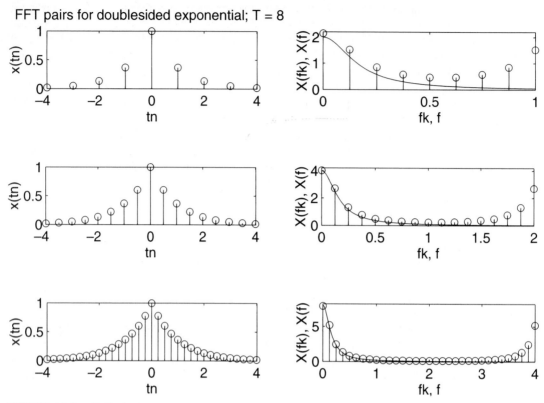

FIGURE 10-3. FFT of a double-sided exponential signal using window width of 8 to illustrate sampling errors. Solid lines in right-hand plots show the spectra of the analog signals.

```
L_t=length(tn);
del_f=1/T;
f_max=(L_t-1)*del_f;
fn=0:del_f:f_max;
L_fn=length(fn);
f=0:.01:f_max;
L_f=length(f);
X=fs*2./(1+(2*pi*f).^2);
L_X=length(X);
xs=exp(-abs(tn));
L_xs=length(xs);
Xs=fft(xs);
L_Xs=length(Xs);
subplot(4,2,k_odd),stem(tn,xs),xlabel('tn'),ylabel('x(tn)'), . . .
axis([-T/2 T/2+.01 0 1]), . . .
if k==1
      title(['FFT pairs for double-sided exponential; T = ', num2str(T)])
end
subplot(4,2,k_even),stem(fn,abs(Xs)),xlabel('fk,f'), . . .
      ylabel('X(fk),X(f)'),axis([0 f_max 0 2^k]), hold
subplot(4,2,k_even),plot(f,X)
end
```

Over the years, a descriptive terminology has come into common use to denote the effects illustrated in Figure 10-1. Because of sampling, the spectrum of the analog signal is repeated about integer multiples of the sampling frequency to produce the spectrum of the sampled signal. The resultant spectrum of the sampled signal therefore consists of a superposition of the analog signal spectrum and its translates in the sampling frequency. The overlapping of the analog signal spectrum with its translates, which results in the spectrum shown in Figure 10-1b, is called *aliasing.*

Windowing the samples of the signal in the time domain corresponds to convolution of the sampled-signal spectrum with the Fourier transform of the window that produces the spectrum of Figure 10-1c. Thus the value of the spectrum for the windowed, sampled signal at a given frequency really is the result of integrating over all spectral components of the windowed spectrum after weighting by the Fourier transform of the rectangular window. Thus a large spectral component in the periodic spectrum shown in Figure 10-1b, say at $f = 1$ Hz, can have a large influence on the spectrum at some other frequency, say $f = 0.5$ Hz, of the windowed signal. This effect is referred to as the *leakage effect,* which is descriptive of how spectral energy "leaks" from one frequency to another as a result of the windowing.

The final effect to be discussed is referred to as the *picket-fence effect.* It simply refers to the fact that the spectrum of the sampled, windowed signal is viewed at a discrete set of frequencies. In effect, the spectrum is viewed through a picket fence. Figure 10-1d illustrates a rather fortunate state of affairs in that the peaks and valleys of the spectrum $X_{sw}(f)$ are viewed through the picket fence of impulses. Had there been a large spectral component or a spectral null at, say, $f = 0.375$ Hz, its presence would not be apparent from the spectral samples shown in Figure 10-1d.

With all these effects being present any time an analog signal is sampled and the DFT of a finite sequence of samples taken, it is indeed surprising that the DFT of the samples comes anywhere close to approximating the spectrum of the original signal. Various measures can be taken to minimize the detrimental effects of aliasing, leakage, and the picket-fence effect. These cures are outlined in Table 10-1.

Note that the various effects derived in Example 10-1 are summarized in Table 10-1. With the time-sampling interval of 1 second in Figure 10-2, considerable aliasing results as noted at the center of the frequency spectrum plot. Even for a time sampling interval of 1/4 second, error is evident at zero frequency. This is due primarily to the leakage effect because from the time-window width of 4 seconds. In Figure 10-3, we still have considerable aliasing for the time-sampling interval of 1 second. The effect of leakage is considerably less than that for the results of Figure 10-2, as noted from the bottom spectral plot of Figure 10-3. This is because the time-window width is twice that of Figure 10-2.

TABLE 10-1
Minimizing DFT Error Effects

Problem	Possible Cure
1. Excessive errors due to aliasing	a. Increase the sampling rate b. Prefilter the signal to minimize high-frequency spectral content
2. Spectral distortion due to leakage	a. Increase the window width by increasing the number of DFT points b. Use window functions that have Fourier transforms with low sidelobes (discussed in Section 10-8) c. If large periodic components are present in the signal, eliminate these before windowing by filtering
3. Picket-fence effect results in important spectral components being missed	Increase the number of DFT points while holding the sampling rate fixed. This places "pickets" closer together

10-3 Examples Illustrating the Computation of the DFT

Before developing efficient algorithms for computing the DFT sum given in (10-1), we consider several examples in which explicit mathematical expressions for the DFT are developed. Usually this is not possible, and the DFT of a sequence must be evaluated numerically. For large sums, this can be exceedingly time consuming, even for a computer. This is the reason that fast Fourier transform (FFT) algorithms were developed, an idea first attributed to J. W. Tukey.[†]

We now write the DFT sum as

$$X(k) = \sum_{n=0}^{N-1} x(n) W_N^{nk}, \qquad k = 0, 1, \ldots, N-1 \tag{10-13}$$

where

$$W_N = \exp[-j2\pi/N] \tag{10-14}$$

For a discrete-time sequence $\{x(n)\}$ of length N, the sum (10-13) gives a discrete-frequency sequence $\{X(k)\}$ of length N.

EXAMPLE 10-2

Find the $N = 8$ DFT of the signal

$$x(n) = 4 + 3\sin(\pi n/2)$$

$$= 4 + 3\sin[2\pi(2n/8)], \qquad 0 \le n \le 7 \tag{10-15}$$

Note that if extended beyond the interval $0 \le n \le 7$, this is a periodic discrete-time signal that goes through two cycles for each 8 values of n.

Solution: We first write $x(n)$ as

$$x(n) = 4 + 3\,\frac{e^{j\pi n/2} - e^{-j\pi n/2}}{2j}$$

$$= 4 - j\frac{3}{2}e^{j\pi n/2} + j\frac{3}{2}e^{-j\pi n/2} \tag{10-16}$$

and note that the sum for $X(k)$ can be written as the sum of three terms:

$$X(k) = \sum_{n=0}^{7} x(n)\, e^{-j(2\pi/8)kn}, \qquad 0 \le k \le 7$$

$$= 4\sum_{n=0}^{7} e^{-j\pi kn/4} - j\frac{3}{2}\sum_{n=0}^{7} e^{j\pi n/2}e^{-j\pi kn/4} + j\frac{3}{2}\sum_{n=0}^{7} e^{-j\pi n/2}e^{-j\pi kn/4} \tag{10-17}$$

To evaluate these sums, consider

$$S_N(l, k) = \sum_{n=0}^{N-1} e^{j2\pi ln/N} W_N^{nk}$$

$$= \sum_{n=0}^{N-1} e^{j2\pi(l-k)n/N}$$

[†]For an interesting article that goes into some of the history of the FFT along with several alternative derivations, see "A Guided Tour of the Fast Fourier Transform" by G. D. Bergland in the *IEEE Spectrum,* Vol. 6, July 1969, pp 41–52.

$$= \sum_{n=0}^{N-1} [e^{j2\pi(l-k)/N}]^n \tag{10-18}$$

Using the result for the sum of a geometric series given by (4-136), this sum can be evaluated for $k \neq l$ as

$$S_N(l, k) = \frac{1 - e^{j2\pi(l-k)}}{1 - e^{j2\pi(l-k)/N}} \equiv 0, \qquad k \neq l \tag{10-19}$$

which follows because $e^{j2\pi(l-k)} = 1$ when $l - k$ is any integer. If $l = k$ the denominator is also 0 and this must be handled as a special case. If $l - k = 0$ (i.e., $k = l$), then the sum for $S_N(l, k)$ becomes

$$S_N(l, \ell) = \sum_{n=0}^{N-1} 1 = N \tag{10-20}$$

Thus, $S_N(l, k)$ can be written compactly as

$$S_N(l, k) = N\delta_{kl} \tag{10-21}$$

where $\delta_{kl} = 1$ for $k = \ell$ and zero otherwise and is the Kronecker delta defined previously. Using this result in (10-17) gives

$$X(k) = 4(8\delta_{k,0}) - j\frac{3}{2}(8\delta_{k,2}) + j\frac{3}{2}(8\delta_{k,-2})$$

$$= 32\delta_{k,0} - j12\delta_{k,2} + j12\delta_{k,-2} \tag{10-22}$$

Note that the last term is nonzero for $k = -2$, which is not in the range $0 \leq k \leq 7$. However, $\{X(k)\}$ is periodic with period 8 (this can be checked by replacing k with $k \pm 8$ in (10-17)), and $\delta_{k,-2}$ can be replaced by $\delta_{k,6}$. This is shown in Figure 10-4 in which two periods of $\{X(k)\}$ are depicted. The period for $0 \leq k \leq 7$ is shown by the solid lines and the period for $-8 \leq k \leq -1$ is given by dashed lines. Noting this periodicity and keeping in mind that the DFT is defined for $0 \leq k \leq N - 1$, we can see that the result for the DFT of the sequence of this example is

$$X(0) = 32 \qquad X(1) = 0 \qquad X(2) = -j12 \qquad X(3) = 0$$

$$X(4) = 0 \qquad X(5) = 0 \qquad X(6) = j12 \qquad X(7) = 0$$

We recall from Chapters 3 and 4 that the amplitude spectrum of a real signal should be even and the phase spectrum should be odd. Although the values listed above for $X(k)$ do not, at first glance, appear to obey these requirements, we see that these symmetry properties are indeed obeyed by the sequence $\{X(k)\}$ if it is periodically extended.

EXAMPLE 10-3

In this example the two-point DFT algorithm, which has two input time-domain samples, $x(0)$ and $x(1)$, and two output frequency-domain samples, $X(0)$ and $X(1)$, is derived. This is illustrated as a block diagram in Figure 10-3a. For this case $N = 2$ and the frequency-domain samples are given by

$$X(k) = \sum_{n=0}^{1} x(n)W_2^{nk}, \qquad k = 0, 1 \tag{10-23}$$

Performing the summation for $k = 0$ and $k = 1$, we obtain

$$X(0) = x(0) + x(1) \tag{10-24a}$$

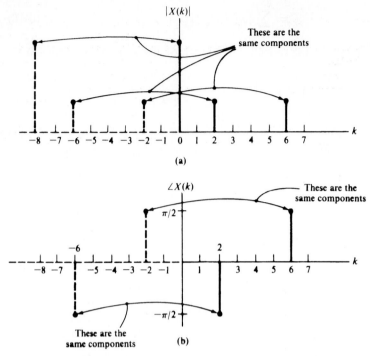

FIGURE 10-4.   Spectra pertaining to Example 10-2: (a) amplitude; (b) phase.

and

$$X(1) = x(0) - x(1) \tag{10-24b}$$

where (10-24b) results because

$$W_2^1 = e^{-j2\pi/2} = e^{-\pi} = -1 \tag{10-25}$$

Equations (10-24a) and (10-24b) are represented by the signal flow graph in Figure 10-5b. The arrows represent both direction of the signal flow and a multiplying factor (gain). If a multiplying

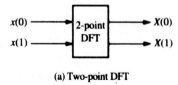

(a) Two-point DFT

$x(0)$ $X0) = x(0) + x(1)$

$x(1)$ $X(1) = x(0) - x(1)$
       -1

(b) Signal flow graph for two-point DFT

FIGURE 10-5.   Two-point DFT.

factor is not shown, it is understood to be $+1$. The calculation in Figure 10-5b is known as a butter-fly calculation because of the shape of the diagram. In the following sections we will see that the butterfly is the basic building block of the FFT algorithm.

Before going on to the next example, we examine W_N^k for three specific values of k. For $k = N/2$ we have

$$W_N^{N/2} = \exp\left[-j\frac{2\pi}{N}\frac{N}{2}\right] = e^{-j\pi} = -1 \qquad (10\text{-}26)$$

Two other special cases of interest are

$$W_N^{N/4} = \exp\left[-j\frac{2\pi}{N}\frac{N}{4}\right] = e^{-j\pi/2} = -j \qquad (10\text{-}27)$$

and

$$W_N^{3N/4} = \exp\left[-j\frac{2\pi}{N}\frac{3N}{4}\right] = e^{-j3\pi/2} = e^{j\pi/2} = j \qquad (10\text{-}28)$$

These values for W_N^k will now be used in deriving the four-point DFT algorithm.

EXAMPLE 10-4

Generalize the derivation of Example 10-3 to a four-point DFT, which is indicated pictorially by the block diagram of Figure 10-6a.

Solution: For a four-point DFT, $N = 4$ in (10-1), which yields

$$X(k) = \sum_{n=0}^{3} x(n)W_4^{nk}, \qquad k = 0, 1, 2, 3 \qquad (10\text{-}29)$$

This equation can be expanded with the aid of (10-26)–(10-28) as the four equations

$$X(0) = x(0) + x(1) + x(2) + x(3) \qquad (10\text{-}30\text{a})$$

$$X(1) = x(0) - jx(1) - x(2) + jx(3) \qquad (10\text{-}30\text{b})$$

$$X(2) = x(0) - x(1) + x(2) - x(3) \qquad (10\text{-}30\text{c})$$

$$X(3) = x(0) + jx(1) - x(2) - jx(3) \qquad (10\text{-}30\text{d})$$

Rearranging these equations gives

$$X(0) = [x(0) + x(2)] + [x(1) + x(3)] \qquad (10\text{-}31\text{a})$$

$$X(1) = [x(0) - x(2)] - j[x(1) - x(3)] \qquad (10\text{-}31\text{b})$$

$$X(2) = [x(0) + x(2)] - [x(1) + x(3)] \qquad (10\text{-}31\text{c})$$

$$X(3) = [x(0) - x(2)] + j[x(1) - x(3)] \qquad (10\text{-}31\text{d})$$

Careful study of the terms in brackets shows that, with proper ordering of the input samples, they can be generated by using the two-point DFT algorithm developed in Example 10-3. This is shown in Figure 10-6b as the left set of two butterflies. The results of this first series of computations, called a *recursion,* can then be put into a second pair of butterflies to produce the final output

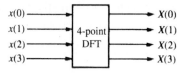

(a) Four-point DFT

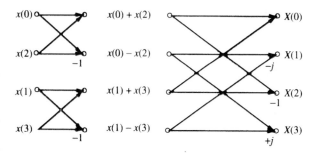

(b) **Signal flow graph for four-point DFT**

FIGURE 10-6.   Four-point DFT.

frequency-domain samples. This set of calculations constitutes the second recursion. The systematic process represented by Figure 10-6b is one form of a fast Fourier transform for four input data points.

Before considering the generalization of the FFT algorithm development of Example 10-4, we consider one additional example in order to illustrate the computations involved in the FFT algorithm just developed. In considering this example, we will also illustrate an interesting application.

EXAMPLE 10-5 _____

As a simple example to illustrate the computation of a four-point DFT consider the complex signal

$$z(n) = x(n) + jy(n) \tag{10-32}$$

where $x(n)$ and $y(n)$ are given by

$$x(n) = \cos \frac{n\pi}{2}, \qquad n = 0, 1, 2, 3 \tag{10-33a}$$

and

$$y(n) = \sin \frac{n\pi}{2}, \qquad n = 0, 1, 2, 3 \tag{10-33b}$$

Thus, $z(n)$ becomes

$$z(n) = \cos \frac{n\pi}{2} + j \sin \frac{n\pi}{2} \tag{10-34}$$

These two sample sequences are shown in Figure 10-7a, and taken together they completely specify $z(n)$. In this case, the complex signal is equivalent to $\exp(jn\pi/2)$. One might form such a complex

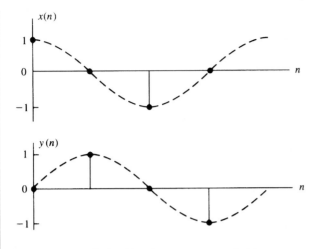

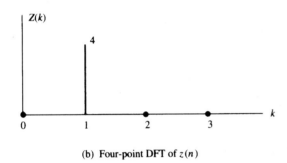

(a) **Real and imaginary components of** $z(n)$

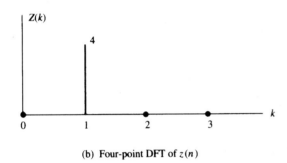

(b) **Four-point DFT of** $z(n)$

FIGURE 10-7.   Four-point DFT of a discrete-time analytic signal.

sample sequence, for example, when processing data from the output of a receiver such as a radar system. The sequence $\{x(n)\}$ might represent samples from the antenna signal multiplied by a cosine reference, whereupon it is low-pass filtered (this operation is referred to as mixing). The sequence $\{y(n)\}$ might represent samples from the antenna signal multiplied by a sine function and low-pass filtered. From the definitions of $x(n)$ and $y(n)$, it follows that

$$z(0) = 1 \qquad z(1) = j \qquad z(2) = -1 \qquad z(3) = -j \qquad (10\text{-}35)$$

Substitution of these values into (10-31) (or using the signal flow graph of Figure 10-6b) yields the four-point transform:

$$Z(0) = [1 - 1] + [j - j] = 0$$

$$Z(1) = [1 + 1] - j[j + j] = 4$$

$$Z(2) = [1 - 1] - [j - j] = 0$$

$$Z(3) = [1 + 1] + j[j + j] = 0 \qquad (10\text{-}36)$$

A plot of this spectrum is given in Figure 10-7b. Note that it consists of a single line at $k = 1$ in keeping with the fact that the signal is a rotating phasor signal (recall the discussion in Chapter 1 in regard to spectra of continuous-time rotating phasor signals). Another way of looking at this signal is that its real and imaginary parts constitute a Hilbert transform pair, and $z(n)$ is therefore an analytic signal. As discussed in Section 4-12, analytic signals have spectra that are identically zero for either negative or positive frequencies, depending on the sign of the complex part of the signal (which is the Hilbert transform of the real part of the signal). This signal is therefore an example of a discrete-time analytic signal.

The examples in this section have illustrated the calculation of the DFT. In the following section, we provide a general derivation of two FFT algorithms. Some of our derivation has already been accomplished, however. In Example 10-3, we derived the butterfly calculation, which was a building block for the four-point DFT computation obtained in Example 10-4. Example 10-4, it will be seen, is a special case ($N = 4$) of a decimation-in-time FFT algorithm.

10-4 Mathematical Derivation of the FFT

In this section we provide a mathematical derivation of two FFT algorithms known as the decimation-in-time and decimation-in-frequency FFT algorithms. These are two different algorithms that perform the same operation, namely, that of finding the DFT of a sequence. To find the inverse DFT, these same algorithms can be used as discussed below (4-142) by conjugating the frequency samples, performing the DFT computation, conjugating the output samples from the DFT operation, and dividing by N.

Decimation-in-Time FFT Algorithm

Consider the DFT sum of (10-13) with the summation carried out separately over the even- and odd-indexed terms in the sum. Letting $n = 2r$ for the even-indexed terms and $n = 2r + 1$ for the odd-indexed terms, we may write the DFT of $x(n)$ as

$$X(k) = \sum_{r=0}^{N/2-1} x(2r)W_N^{2rk} + \sum_{r=0}^{N/2-1} x(2r+1)W_N^{(2r+1)k}$$

$$= \sum_{r=0}^{N/2-1} x(2r)W_N^{2rk} + W_N^k \sum_{r=0}^{N/2-1} x(2r+1)W_N^{2rk} \tag{10-37}$$

Now

$$W_N^{2rk} = e^{-j(2\pi/N)(2rk)} = e^{-j(2\pi/(N/2))(rk)} = W_{N/2}^{rk} \tag{10-38}$$

which allows (10-37) to be written as

$$X(k) = \sum_{r=0}^{N/2-1} x(2r)W_{N/2}^{rk} + W_N^k \sum_{r=0}^{N/2-1} x(2r+1)W_{N/2}^{rk}$$

$$= G(k) + W_N^k H(k), \qquad k = 0, 1, \ldots, N-1 \tag{10-39}$$

where

$$G(k) = \sum_{r=0}^{N/2-1} x(2r)W_{N/2}^{rk} \tag{10-40}$$

and

$$H(k) = \sum_{r=0}^{N/2-1} x(2r+1)W_{N/2}^{rk} \tag{10-41}$$

are $(N/2)$-point DFTs of the even- and odd-indexed points of the sequence $x(n)$, respectively. Although the frequency index k ranges over N values from 0 to $N - 1$, only $N/2$ values of $G(k)$ and $H(k)$ need to be computed since $G(k)$ and $H(k)$ are periodic in k with period $N/2$. Assuming that an algorithm is available for computing an $(N/2)$-point DFT, we may combine the $(N/2)$-point DFTs, $G(k)$ and $H(k)$, in accordance with (10-39) to provide the N-point DFT of the time sequence $x(n)$. In Figure 10-8a, the lines with arrows indicate the quantities to be added, with the power of W_N alongside the arrow indicating

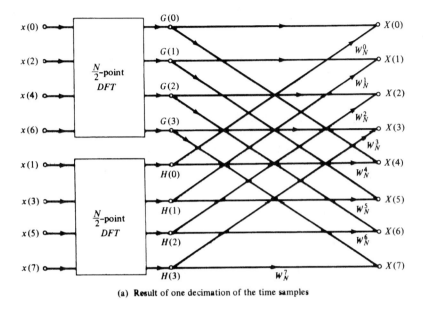

(a) Result of one decimation of the time samples

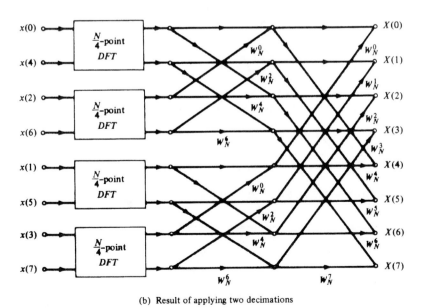

(b) Result of applying two decimations

FIGURE 10-8. Flow graphs showing the decimation-in-time decomposition of an N-point DFT computation ($N = 8$).

the multiplication of $H(k)$ in (10-39) by W_N^k. Arrows with nothing beside them are understood to be multiplications by unity (i.e., the $G(k)$'s are transmitted as they appear at the output of the DFT boxes). This is illustrated in Figure 10-8a. Furthermore, we can break up each of the $(N/2)$-point DFTs shown in Figure 10-8a into two $(N/4)$-point DFTs, which results in the flow graph of Figure 10-8b. Note that in combining the sums for $G(k)$ and $H(k)$, the relation $W_{N/2}^r = W_N^{2r}$ is used.

The decimation procedure can be continued until a series of $N/2$ two-point DFTs results for the first stage of the N-point DFT computation. The flow graph that results for $N = 8$ is precisely that of Figure 10-9. Since this flow graph was arrived at by separating the input samples into successively smaller sets, the resulting algorithm is referred to as a *decimation-in-time* (DIT) algorithm.

We note that the bit-reversed ordering of the input samples results because at each stage of decimation the sequence of input samples is separated into even- and odd-indexed samples. Each even-indexed sample for the first decimation has a least significant bit (LSB) of zero, while each odd-indexed point has an LSB of one. On the second decimation it is the second-most LSB that determines whether a sequence number is included in the even- or odd-indexed DFT, and so on until the last decimation takes place, whereupon it is the most significant bits of the sequence-member indices that determine what DFT sum the sequence members are associated with. What has just been described, therefore, is an ordering of the input samples in bit-reversed order.

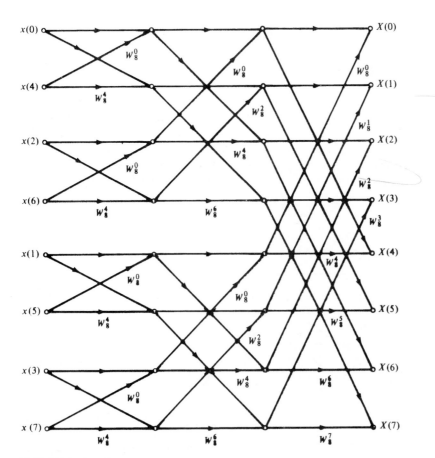

FIGURE 10-9. Complete flow graph for an FFT developed by applying decimation in time ($N = 8$). (Arrows not labeled have multipliers of unity.)

Decimation-in-Frequency FFT Algorithm

To derive another type of algorithm for finding the DFT we now consider the DFT sum (10-13) with the sum carried out over the first half and the last half of the input samples separately. This results in the sums

$$X(k) = \sum_{n=0}^{N/2-1} x(n)W_N^{kn} + \sum_{n=N/2}^{N-1} x(n)W_N^{kn}$$

$$= \sum_{n=0}^{N/2-1} x(n)W_N^{kn} + W_N^{(N/2)k}\sum_{m=0}^{N/2-1} x\left(m + \frac{N}{2}\right)W_N^{mk} \tag{10-42}$$

where the substitution $m = n - N/2$ has been made to obtain the second sum of the second equation above. We may use the fact that $W_N^{(N/2)k} = (-1)^k$ to combine the two sums in (10-42) and obtain

$$X(k) = \sum_{n=0}^{N/2-1}\left[x(n) + (-1)^k x\left(n + \frac{N}{2}\right)\right]W_N^{nk} \tag{10-43}$$

We now consider k even and odd separately. Letting $k = 2r$ and $k = 2r + 1$, respectively, to indicate the even- and odd-indexed output points, we have

$$X(2r) = \sum_{n=0}^{N/2-1}\left[x(n) + x\left(n + \frac{N}{2}\right)\right]W_N^{2rn} \tag{10-44}$$

and

$$X(2r + 1) = \sum_{n=0}^{N/2-1}\left[x(n) - x\left(n + \frac{N}{2}\right)\right]W_N^n W_N^{2rn} \qquad r = 0, 1, \ldots, N/2 - 1 \tag{10-45}$$

Since $W_N^{2rn} = W_{N/2}^n$, as noted previously, we recognize (10-44) and (10-45) as $(N/2)$-point DFTs. The flow graph for obtaining $X(k)$ in this fashion is shown in Figure 10-10. This process can be used to replace each $(N/2)$-point DFT by two $(N/4)$-point DFTs, and so on until only two points are left in each DFT. The resulting flow graph, shown in Figure 10-11, is precisely the reverse of Figure 10-9, with all arrows reversed and input and output interchanged. Since the output, or frequency, points were subdivided to obtain this algorithm, it is referred to as a *decimation-in-frequency* (DIF) FFT algorithm.

The following terminology is sometimes used with respect to FFT algorithms. Each basic computation in Figures 10-9 and 10-11 is referred to as a *butterfly* because of its criss-cross appearance. For the DIT FFT algorithm, a butterfly computation is of the form

$$X_{m+1}(p) = X_m(p) + W_N^s X_m(q)$$
$$X_{m+1}(q) = X_m(p) - W_N^s X_m(q) \tag{10-46}$$

and involves a pair of nodes, denoted as p and q, called *dual nodes*. For the DIF FFT algorithm, a butterfly computation is of the form

$$X_{m+1}(p) = X_m(p) + X_m(q)$$
$$X_{m+1}(p) = [X_m(p) - X_m(q)]W_N^s \tag{10-47}$$

In either case, a butterfly computation requires one complex multiplication and two complex additions (a subtraction is really an addition after negation of the second term in the summation).

Carrying out all the computations in going from one column of nodes to the next in an FFT flowgraph is referred to as completion of a *recursion*. For the particular FFT algorithms shown in Figures 10-9 and 10-11, once all the values for a particular recursion are computed, the old values that were used to obtain them are never required again. Such FFT algorithms are called *in-place algorithms*. A result of the in-place property of the DIT and DIF algorithms illustrated in Figures 10-9 and 10-11 is the

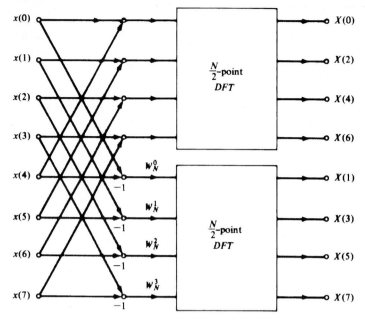

FIGURE 10-10. Flow graph after a single decimation.

scrambled (bit-reversed) input required for the DIT algorithm and the scrambled output required for the DIF algorithm. With a little imagination it should be apparent to the student that if the input nodes of Figure 10-9 or the output nodes of Figure 10-11 are rearranged to avoid the scrambling, then it is impossible to have an in-place algorithm.

In computing the FFT, additional time savings can be obtained by precomputing W_N^m and storing the required values. This, of course, requires additional memory. Whether precomputed and stored or computed as needed, use can be made of the symmetry properties of W_N^m, as illustrated in Table 10-2 for $N = 8$, to save time. There it is noted that $W_N^{m+N/2} = -W_N^m$, $0 \leq m \leq N/2 - 1$.

By examining (10-1), we see that $(N - 1)$ complex multiplications and $(N - 1)$ complex additions are required to compute each output point for the DFT (the first term in the sum involves $\exp(j0) = 1$ and therefore does not require a multiplication). Thus, to compute N output points, we require $N(N - 1)$ complex multiplications and the same number of complex additions. Now each complex multiplication is of the form

$$(A + jB)(C + jD) = (AC - BD) + j(BC + AD) \tag{10-48}$$

and therefore requires four real multiplications and two real additions. Hence computation of all the output points for an N-point DFT requires $4N(N - 1)$ real multiplications and $4N(N - 1)$ real additions.

To compare the number of computations required for the FFT with the number required for the DFT, we note that an N-point FFT, where N is a power of 2, requires $\log_2 N$ recursions which are each composed of $N/2$ butterflies. Regardless of whether the DIT or DIF algorithm is used, each butterfly involves one complex multiplication and two complex additions, or four real multiplications and six real additions. Thus each FFT requires $2N \log_2 N$ real multiplications and $3N \log_2 N$ real additions. Actually, the number of multiplications is less than this because several of the multiplications by W^r are really multiplications by ± 1 or $\pm j$, which are not multiplications at all.

We can get a rough comparison of the speed advantage of an FFT over a DFT by computing the number of multiplications for each since these are usually more time consuming than additions. The comparison is given in Table 10-3.

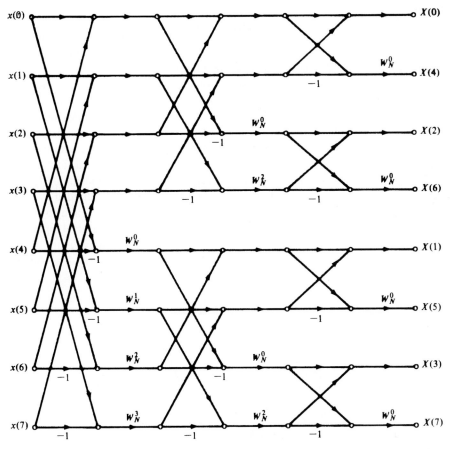

FIGURE 10-11. Flow graph pertaining to decimation-in-frequency FFT ($N = 8$).

TABLE 10-2
Symmetry Properties
of W_N^m for $N = 8$

$m =$	W_8^m	
0	1	
1	$\dfrac{1-j}{\sqrt{2}}$	
2	$-j$	
3	$-\dfrac{1+j}{\sqrt{2}}$	
4	-1	$= -W^0$
5	$-\dfrac{1-j}{\sqrt{2}}$	$= -W^1$
6	j	$= -W^2$
7	$\dfrac{1+j}{\sqrt{2}}$	$= -W^3$

TABLE 10-3
Comparison of the Number of Real Multiplications for the DFT and FFT

Number of Points	DFT	FFT	Speed Factor
2	8	4	2
4	48	16	3
8	224	48	5
16	960	128	8
32	3,968	320	12
64	16,128	768	21
128	65,024	1,792	36
256	261,120	4,096	64
512	1,046,528	9,216	114
1,024	4,190,208	20,480	205
2,048	16,769,024	45,056	372
4,096	67,092,480	98,304	683

EXAMPLE 10-6 _____

To illustrate the use of the algorithms shown by the flow graphs of Figures 10-9 and 10-11, we consider the DFT of the sequence $\{x_n\} = \{1, 0, 0, 0, 0, 0, 0, 0\}$ by means of the DIT and DIF algorithms. Figure 10-12a illustrates the computation of the DFT using the DIT algorithm, with the numbers above the circles indicating the results at each recursion. Note that all 1's are obtained, which is in keeping with the fact that the spectrum of a unit pulse is a constant. Figure 10-12b illustrates the computation of the inverse DFT using the DIT algorithm, with the numbers above the circles again indicating the results at each recursion. Since the DFT of a unit pulse turned out to be real, there is no need to conjugate the input samples. Also, since all values are unity, there is no need to exercise care in putting the input samples into the algorithm in bit-reversed order. Note that we get a single output sample which is nonzero, namely, the zeroth sample, which has value 8. Recalling that the DFT, when used to obtain the inverse DFT, requires the output samples to be conjugated and divided by N, we see that this result is correct. There is no need to conjugate the output since it is real. Scaling by $N = 8$ gives $x_0 = 1$ and all other values zero, which is the same sample sequence as the original input.

The use of the DIF algorithm to compute the DFT of the unit pulse is illustrated by Figure 10-13a. Again, the numbers above the circles indicate the results at each recursion. The same result as obtained

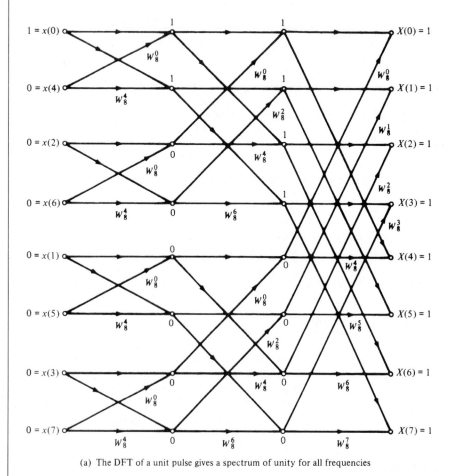

(a) The DFT of a unit pulse gives a spectrum of unity for all frequencies

FIGURE 10-12. Computation of the DFT of a unit pulse using the DIT algorithm.

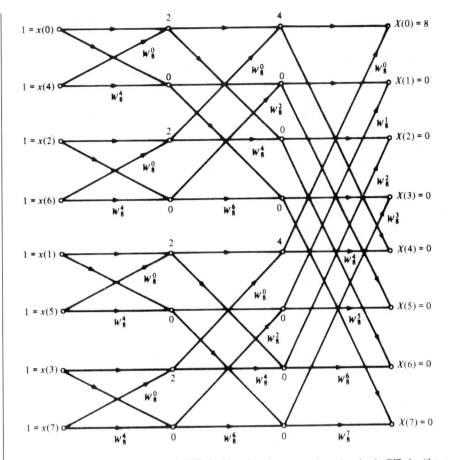

(b) The inverse DFT computation using the DIF algorithm gives the same result as given by the DIT algorithm.

FIGURE 10-12.   Continued.

with the DIT algorithm is obtained by again using the DIF algorithm. The computation of the inverse DFT using the DIF algorithm is shown graphically in Figure 10-13b; the same result is obtained as with the DIT algorithm.

10-5 Properties of the DFT

As in the case of the Fourier transform, several useful properties can be proved for the DFT. The proofs of these properties may be carried out in a similar fashion to the proofs of the analogous properties of the Fourier transform by using the defining relationships for the DFT given by (10-1) and (10-2). Proofs of the properties stated below are left to the problems. The following notation will be employed in stating the properties of the DFT, where all sequences are assumed periodic with period N:

Discrete-time sequences are denoted as $x(n)$ and $y(n)$.

Their DFTs are denoted as $X(k)$ and $Y(k)$.

N is the length of the sequence or size of the DFT.

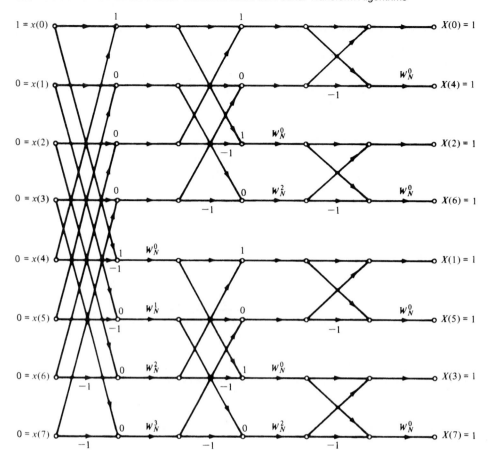

(a) The DFT using the DIF algorithm gives the same result as with the DIT algorithm

FIGURE 10-13. Computation of the DFT of a unit pulse using the DIF algorithm.

A and B are arbitrary constants.

The subscript e denotes a sequence that is even about the point $(N - 1)/2$ for N even and $N/2$ for N odd.

The subscript o denotes a sequence that is odd.

Any real sequence can be expressed in terms of its even and odd parts according to

$$x(k) = x_e(n) + x_o(n)$$
$$= \tfrac{1}{2}[x(n) + x(N - n)] + \tfrac{1}{2}[x(n) - x(N - n)] \qquad (10\text{-}49)$$

The subscript r denotes a real sequence (or the real part of a complex sequence).

The subscript i similarly denotes the imaginary part of a complex sequence.

A double-headed arrow ($\leftrightarrow$) denotes a discrete-Fourier transform pair: for example, $x(n) \leftrightarrow X(k)$. Sequences are assumed periodically repeated if necessary.

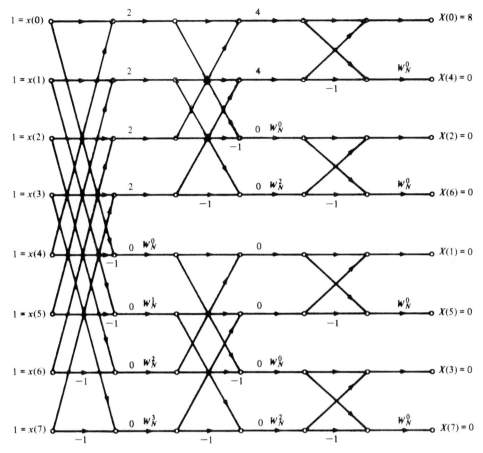

(b) The inverse DFT computation gives back the original signal multiplied by N

FIGURE 10-13.   Continued.

With this notation, the following properties of the DFT may be stated.

1. Linearity:

$$Ax(n) + By(n) \leftrightarrow AX(k) + BY(k) \tag{10-50}$$

2. Time shift:

$$x(n - m) \leftrightarrow X(k) \exp(-j2\pi km/N) = X(k)W_N^{km} \tag{10-51}$$

3. Frequency shift:

$$x(n) \exp(j2\pi nm/N) \leftrightarrow X(k - m) \tag{10-52}$$

4. Duality:

$$N^{-1}X(n) \leftrightarrow x(-k) \tag{10-53}$$

5. Circular convolution:†

$$\sum_{m=0}^{N-1} x(m)y(n-m) \triangleq x(n) \, \bigcirc\!\!\!\!N \, y(n) \leftrightarrow X(k)Y(k) \tag{10-54}$$

6. Multiplication:

$$x(n)y(n) \leftrightarrow N^{-1} \sum_{m=0}^{N-1} X(m)Y(k-m) \triangleq N^{-1}X(k) \, \bigcirc\!\!\!\!N \, Y(k) \tag{10-55}$$

7. Parseval's theorem:

$$\sum_{n=0}^{N-1} |x(n)|^2 = N^{-1} \sum_{k=0}^{N-1} |X(k)|^2 \tag{10-56}$$

8. Transforms of even, real functions:

$$x_{er}(n) \leftrightarrow X_{er}(k) \tag{10-57}$$

(The DFT of an even, real sequence is even and real.)

9. Transforms of odd, real functions:

$$x_{or}(n) \leftrightarrow jX_{oi}(k) \tag{10-58}$$

(The DFT of an odd, real sequence is odd and imaginary.)

10. Assume that $x(n)$ and $y(n)$ are the real and imaginary parts, respectively, of a complex sequence $z(n)$; that is,

$$z(n) = x(n) + jy(n) \tag{10-59}$$

Then

$$z(n) \leftrightarrow Z(k) = X(k) + jY(k) \tag{10-60}$$

To see how to proceed in separating out the DFTs of $x(n)$ and $y(n)$ using property 10, we write both in terms of their even and odd parts as

$$z(n) = x(n) + jy(n) = [x_e(n) + x_o(n)] + j[y_e(n) + y_o(n)] \tag{10-61}$$

where the subscripts "e" and "o" denote "even" and "odd", respectively. Recalling properties 8 and 9, we may write the DFT of the real and imaginary parts of $z(n)$ as

$$x_e(n) + x_o(n) \overset{\text{DFT}}{\longleftrightarrow} X_{er}(k) + jX_{oi}(k) = X(k)$$

$$\text{and} \quad y_e(n) + y_o(n) \overset{\text{DFT}}{\longleftrightarrow} Y_{er}(k) + jY_{oi}(k) = Y(k) \tag{10-62}$$

where $X_{er}(k)$ is a even, real function, which is the DFT of $x_e(n)$ (property 8), $X_{oi}(k)$ is an odd, real function, which, when multiplied by j, is the DFT of $x_o(n)$ (property 9), and similarly for $Y_{er}(k)$ and $Y_{oi}(k)$. Thus, we may write the DFT of $z(n)$ as

$$Z(k) = X_{er}(k) + jX_{oi}(k) + j[Y_{er}(k) + jY_{oi}(k)]$$

$$= [X_{er}(k) - Y_{oi}(k)] + j[X_{oi}(k) + Y_{er}(k)]$$

$$= \text{Re}[Z(k)] + j\text{Im}[Z(k)] \tag{10-63}$$

---

†The circular convolution operation can be viewed as placing the N samples of $\{x(n)\}$ on the circumference of a cylinder and the N samples of $\{y(n)\}$ in reverse order on another cylinder, sliding one cylinder inside the other, while multiplying coincident samples, and adding the products. It is illustrated by Example 10-9 and discussed further in the next section.

That is,

$$Re[Z(k)] = X_{er}(k) - Y_{oi}(k)$$

$$Im[Z(k)] = X_{oi}(k) + Y_{er}(k) \tag{10-64}$$

Since $X_{er}(k)$ is even and $Y_{oi}(k)$ is odd, we may separate these two sequences from each other in the first equation of (10-64) as follows:

$$X_{er}(k) = \frac{1}{2}[Re[Z(k)] + Re[Z(N - k)]]$$

$$Y_{oi}(k) = -\frac{1}{2}[Re[Z(k)] - [Re[Z(N - k)]] \tag{10-65}$$

Similarly, since $X_{oi}(k)$ is odd and $Y_{er}(k)$ is even, we may separate them from each other in the second equation as

$$X_{oi}(k) = \frac{1}{2}[Im[Z(k)] - Im[Z(N - k)]]$$

$$Y_{er}(k) = \frac{1}{2}[Im[Z(k)] + Im[Z(N - k)]] \tag{10-66}$$

This allows the reconstruction of the separate DFTs of $x(n)$ and $y(n)$ according to (10-62).

It is important to think of the properties listed above as applying to periodic sequences. To illustrate these properties, we now consider several examples. In particular, they will illustrate properties 1, 2, 5, and 10.

EXAMPLE 10-7

Reconsider Example 10-6 which used the signal

$$z(n) = x(n) + jy(n) = \exp(jn\pi/2)$$

$$= \cos(n\pi/2) + j\sin(n\pi/2), \quad n = 0, 1, 2, 3 \tag{10-67}$$

as an illustration of a four-point DFT. In this example, we use superposition to compute the DFTs of $\cos(n\pi/2)$ and $\sin(n\pi/2)$ separately and then combine them in accordance with (10-49) to get the DFT of $\exp(n\pi/2)$. We first evaluate $\cos(n\pi/2)$ for $n = 0, 1, 2, 3$ to get $x(0) = 1$, $x(1) = 0$, $x(2) = -1$, and $x(3) = 0$ and substitute these into (10-31) to get

$$X(0) = (1 - 1) + (0 + 0) = 0$$

$$X(1) = (1 + 1) - j(0 + 0) = 2$$

$$X(2) = (1 - 1) - (0 + 0) = 0$$

$$X(3) = (1 + 1) + j(0 - 0) = 2 \tag{10-68a}$$

We next evaluate $\sin(n\pi/2)$ for $n = 0, 1, 2, 3$ to get $y(0) = 0$, $y(1) = 1$, $y(2) = 0$, and $y(3) = -1$ and substitute these into (10-31) to get

$$Y(0) = (0 + 0) + (1 - 1) = 0$$

$$Y(1) = (0 + 0) - j(1 + 1) = -j2$$

$$Y(2) = (0 + 0) - (1 - 1) = 0$$

$$Y(3) = (0 - 0) + j(1 + 1) = j2 \tag{10-68b}$$

Finally, substituting into (10-50) with $A = 1$ and $B = j$, we obtain the same result as given by (10-36):

$$Z(k) = \begin{cases} 0, & k = 0 \\ 2 + j(-j2) = 4, & k = 1 \\ 0, & k = 2, \\ 2 + j(j2) = 0, & k = 3 \end{cases} \tag{10-69}$$

We now reverse the procedure using (10-63) to see if we can recover $\{X(k)\}$ and $\{Y(k)\}$ from $\{Z(k)\}$. Using the definitions of even and odd parts of a sequence, we have from (10-65):

$$X_{er}(k) = \frac{1}{2}\{\text{Re}[Z(k)] + \text{Re}[Z(4 - k)]\} = \begin{cases} 0, & k = 0 \\ 2, & k = 1 \\ 0, & k = 2 \\ 2, & k = 3 \end{cases} \tag{10-70a}$$

Similarly, (10-66) gives

$$Y_{oi}(k) = -\frac{1}{2}\{\text{Re}[Z(k)] - \text{Re}[Z(4 - k)]\} = \begin{cases} 0, & k = 0 \\ -2, & k = 1 \\ 0, & k = 2 \\ 2, & k = 3 \end{cases} \tag{10-70b}$$

Also, $\{X_{oi}(k)\} = \{Y_{er}(k)\} = 0$ since the imaginary part of $\{Z(k)\} = 0$. Upon substituting into the right-hand side of (10-62), we get $\{X(k)\}$ and $\{Y(k)\}$ as given by (10-68a) and (10-68b), respectively.

EXAMPLE 10-8 _____

Consider the DFT of

$$x(n) = \delta(n) \tag{10-71}$$

which is

$$X(k) = \sum_{n=0}^{N-1} \delta(n)W_N^{nk} = 1, \qquad k = 0, 1, \ldots, N - 1 \tag{10-72}$$

Using the time-shift property we find that

$$\text{DFT}[x(n - n_0)] = \exp\left(\frac{-j2\pi kn_0}{N}\right) = W_N^{kn_0} \tag{10-73}$$

Writing out the DFT sum for $x(n - n_0) = \delta(n - n_0)$, we have

$$\text{DFT}[\delta(n - n_0)] = \sum_{n=0}^{N-1} \delta(n - n_0)W_N^{nk} = W_N^{kn_0} \tag{10-74}$$

which is the same as obtained using the time-shift property. Note that the DFT is periodic with period N.

EXAMPLE 10-9 _____

Consider the circular convolution of the two sequences

$$x_1(n) = x_2(n) = 1, \qquad 0 \le n \le N - 1 \tag{10-75}$$

which is denoted $x_{3C}(n)$. The DFT of each of these sequences is

$$X_1(k) = X_2(k) = \sum_{n=0}^{N-1} W_N^{nk} \tag{10-76}$$

$$= \begin{cases} N, & k = 0 \\ 0, & k = 1, 2, \dots, N - 1 \end{cases} \tag{10-77}$$

where use has been made of (10-19) and (10-20). The *circular convolution* of $x_1(n)$ and $x_2(n)$ is the inverse DFT of

$$X_3(k) = X_1(k)X_2(k) = \begin{cases} N^2, & k = 0 \\ 0, & k = 1, 2, \dots, N - 1 \end{cases} \tag{10-78}$$

which is

$$x_{3C}(n) = \frac{1}{N} \sum_{k=0}^{N-1} N^2 \delta(k) W_N^{-nk} = N, \qquad n = 0, 1, \dots, N - 1 \tag{10-79}$$

This is illustrated in Figure 10-12a for $N = 4$.

The *linear convolution*, denoted $x_{3L}(n)$, of the two sequences $x_1(n)$ and $x_2(n)$ is the triangular sequence given by

$$x_{3L}(n) = \begin{cases} n + 1, & 0 \le n \le N \\ 2N - n - 1, & N \le n \le 2N \\ 0, & \text{otherwise} \end{cases} \tag{10-80}$$

as illustrated in Figure 10-14b. The student can readily verify this result by considering $\{x_1(n)\}$ and $\{x_2(n)\}$ as infinite sequences which are zero for $k < 0$ and $n > N$.

The reason for the difference in the circular and linear convolution results is that in the former case aliasing takes place as a result of the sequences $x_1(n)$ and $x_2(n)$ being treated as periodic sequences because of the DFT operation. To avoid this aliasing, $x_1(n)$ and $x_2(n)$ each can be augmented with N points that are zero to obtain the sequences

$$x_1(n) = x_2(n) = \begin{cases} 1, & 0 \le k \le N - 1 \\ 0, & N \le k \le 2N - 1 \end{cases} \tag{10-81}$$

The resulting circular convolution is then a triangular sequence, as illustrated in Figure 10-14c. Any fewer than $N - 1$ zeros would result in time-domain aliasing (note that $N - 1$ will do it, with the Nth point being 0).

10-6 Applications of the FFT

Several applications of the FFT are discussed in this section. These include filtering, spectrum analysis, convolution, power spectrum estimation, system identification, and deconvolution.

Filtering

The next example illustrates the application of the FFT to filtering.

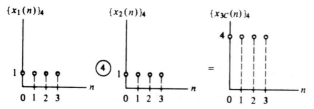

(a) The circular convolution of two four-point sequences whose members are unity is itself a four-point sequence whose members are equal because of aliasing

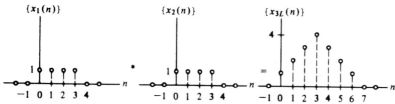

(b) The linear convolution of two sequences consisting of four equal-value samples preceded and followed by zeros is a triangular envelope sequence

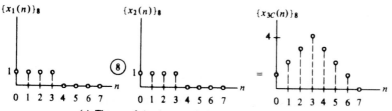

(c) The use of circular convolution to calculate the linear convolution of two sequences requires zero padding with the same number of zeros as nonzero sequence members

FIGURE 10-14. Comparison of circular and linear convolution using the sequences of Example 10-5. In the figures, $\{\cdot\}_m$ indicates a sequence of length m.

EXAMPLE 10-10 _____

To illustrate use of the FFT for filtering, consider the DFT of the 8-point sequence

$$x(n) = \cos\frac{\pi}{4}n + \cos\frac{\pi}{2}n, \qquad 0 \le n \le 7 \tag{10-82}$$

which is sketched in Figure 10-15.

The DFT of this sequence can be obtained using the technique of Example 10-2. Consider the computation of the DFT of the sequence

$$x(n) = \cos(2\pi ln/N), \qquad 0 \le n \le N - 1 \tag{10-83}$$

The DFT sum for this sequence is

$$X(k) = \sum_{n=0}^{N-1} \cos\left(\frac{2\pi ln}{N}\right) e^{-j2\pi nk/N}$$

$$= \frac{1}{2}\left[\sum_{n=0}^{N-1} e^{j2\pi ln/N} e^{-j2\pi kn/N} + \sum_{n=0}^{N-1} e^{-j2\pi ln/N} e^{-j2\pi ln/N}\right] \tag{10-84}$$

$$= \frac{1}{2} [S_N(l,k) + S_N(-l,k)] \tag{10-85}$$

where the terminology of (10-21) has been used.

Using the results of (10-21), we find the 8-point DFT of (10-83) to be

$$X(k) = 4[\delta_{k,1} + \delta_{k,-1} + \delta_{k,2} + \delta_{k,-2}] \tag{10-86}$$

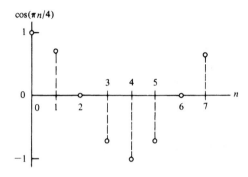

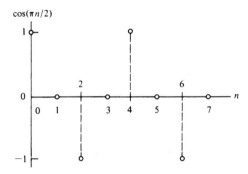

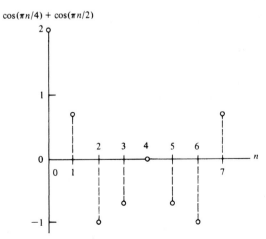

FIGURE 10-15.   Signal sequence to illustrate filtering by means of the FFT.

Next, assume that all frequency components of the spectrum above $k = 1$ are to be eliminated. Note that this entails setting $X(k) = 0$ for $2 \leq k \leq 6$ and $k = 0$. The result for the filtered spectrum is

$$X_f(k) = 4[\delta_{k,1} + \delta_{k,-1}] \tag{10-87}$$

Again using the results of Example 10-2, it is clear that the signal corresponding to the filtered spectrum is

$$x_f(n) = \cos(\pi n/4) \tag{10-88}$$

This example illustrates the implementation of filtering by using the FFT to digitally carry out the operations expressed by (4-67). In implementing digital filters by this method, care must be exercised to invoke the proper symmetry conditions on the transfer function, $H(k)$, as expressed by (4-70) and (4-71). In the discrete-time signal case, however, the center of symmetry is the point $k = N/2$.

Another problem that arises is leakage, which can be combatted by proper window selection. This is discussed in more detail in Section 10-7.

A problem related to window selection is aliasing in the time domain. When an ideal filter, such as in Example 10-10, is implemented in the frequency domain, a $(\sin x)/x$ impulse response as expressed by (4-92) is implied in the time domain. In such cases, the impulse response does not go to zero, which results in the tails of the $(\sin x)/x$ impulse response folding back into the N-point region included in the inverse sum (10-2). The result is aliasing in the time domain caused by undersampling the filter transfer function in the frequency domain. A cure for this problem is to choose a filter with an impulse response that dies out or can be truncated so that at least half of its terms are essentially zero, as illustrated by Example 10-9.

Spectrum Analyzers

In recent years, digital processor technology has advanced to the point where oscilloscopes and spectrum analyzers have been implemented using digital processor technology. The former device utilizes sampling and quantization to display a signal on a monitor. Because the samples can be stored in a digital memory, many operations are possible on the signal that are not possible in analog implementations. The digital spectrum analyzer also operates on a signal that has been sampled and quantized, except it is now the spectrum that is displayed on the monitor rather than the time-domain waveform itself. Thus, the DFT of the signal is taken to display on the monitor as a function of frequency. Again, as in the case of the digital oscilloscope, many operations are possible because of the storage capability of digital data than are possible with an analog spectrum analyzer.

Convolution

When convolutions are implemented by means of the FFT, both the signals being convolved are treated as periodic. This was illustrated by Example 10-9. The result is a circular, or cyclical, convolution that is entirely different from the noncyclical convolution desired in order to avoid time-domain aliasing.

The problem is easily avoided in computing the cyclical convolution of two N-point sequences, say $\{g(n)\}$ and $\{h(n)\}$, by defining $\{\hat{g}(n)\}$ and $\{\hat{h}(n)\}$ as

$$\hat{g}(n) = \begin{cases} g(n), & 0 \leq n \leq N - 1 \\ 0, & N \leq n \leq 2N - 1 \end{cases} \tag{10-89}$$

and

$$\hat{h}(n) = \begin{cases} h(n), & 0 \leq n \leq N - 1 \\ 0, & N \leq n \leq 2N - 1 \end{cases} \tag{10-90}$$

respectively.

If one of the sequences to be convolved is much longer than the other, transforming the entire sequence corresponding to both functions can be avoided by convolving the $N = 2M$-point zero-padded version of the shorter one, with M nonzero points and M points zero, with overlapping $2M$-point segments of the longer one. This is illustrated in Figure 10-16. The shorter sequence, designated $\{\hat{h}(n)\}$, can be thought of as the impulse response of a filter that is acting on the longer sequence, designated $\{g(n)\}$. The present output of the filter is simply a weighted sum of the last M samples it has seen. In the example of Figure 10-16, this means that the $2M$-term sequence $\{\hat{h}(n)\}$ convolved with one of the $2M$-term segments of $g(n)$ would result in M lags for which the correct M-term history is not available, and M lags for which the correct M-term history is available. Thus the first M lags computed are incorrect and the last M lags are correct. Only the rightmost M points are kept, and the next convolution of $\{\hat{h}(n)\}$ with a segment of $\{g(n)\}$ is carried out with M new points and M old points. In each convolution, only the rightmost M points are saved. When all of these sets of M points are pieced together, the result is the desired convolution of $\{g(n)\}$ and $\{h(n)\}$. This procedure avoids having to pad the long sequence with a string of zeros, and the length of the required FFT is determined by the length of the short sequence rather than the long one.[†] Because of the procedure used, this technique is called the *overlap-save* method of convolution. Another commonly used method is the *overlap-add* method, for which the student is referred to the references.

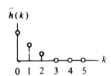

(a) Zero-padded impulse response for the desired convolution operation.

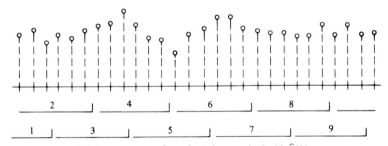

(b) Long sequence of samples to be convolved with $\hat{h}(k)$. Numbers correspond to successive segments of samples

(c) Periodic extension of time-reversed impulse response as implied by the DFT used to implement the circular convolution

FIGURE 10-16.   Illustration of the overlap-save method of convolving a short sequence with a long sequence.

[†]Actually, if an N-point FFT is used and the short sequence is of length M, zero padding with $M - 1$ zeros is sufficient with $N - (M - 1) \geq M$ "good" points obtained for each lag.

EXAMPLE 10-11

Plotted in Figure 10-17 are the various steps in doing a overlap-save convolution of an exponentially decaying signal with a sinusoid which was computed with a MATLAB implementation of the overlap-save method. The MATLAB code is not given here, but rather is left as a computer exercise for the student (Computer Exercise 10-2). The overlap-save convolution shown here uses a 16-point FFT and five overlapping segments of the long signal. The input signal segment and the resulting convolution are shown for the first two steps (note that the first segment is zero-padded for one-half of the window to implement a sinusoid that is zero for negative independent-variable values). After

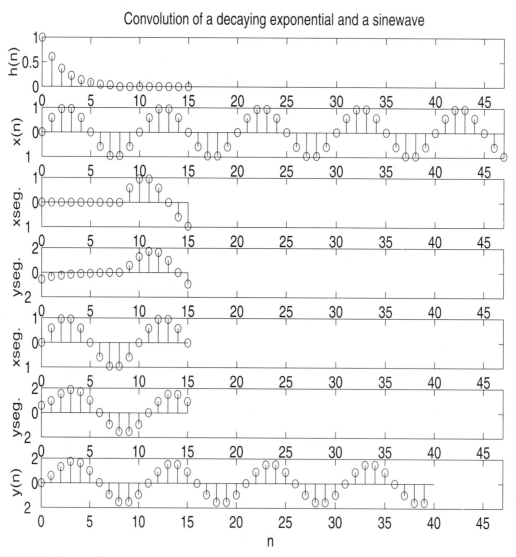

FIGURE 10-17. The convolution of a short decaying exponential signal with a sinewave using the overlap-save method. From top to bottom: short signal; long sinusoid; first segment of long signal; resulting convolution of first segment; second segment of long signal; resulting convolution of second segment; convolution of entire long signal with short signal.

an initial startup transient, the convolution settles down to a steadystate sinusoid as illustrated in the bottom figure.

Energy Spectral Density and Autocorrelation Functions

If we consider a sequence $\{x(n)\}$ of length N whose DFT is $\{X(k)\}$, its energy is defined as

$$E = \sum_{n=0}^{N-1} |x(n)|^2 \tag{10-91}$$

In the frequency domain, the energy is

$$E = \sum_{k=0}^{N-1} \frac{|X(k)|^2}{N} \tag{10-92}$$

where Parseval's theorem, property 7 of the DFT, has been used. Equation (10-92) can be interpreted as the sum of an energy spectral density function

$$S_x(k) = \frac{1}{N} X(k)X^*(k) \tag{10-93}$$

over all the frequency components to give total energy. Using appropriate theorems, it can be shown that

$$S_x(k) = \text{DFT}[R_x(m)] \tag{10-94}$$

where $R_x(m)$, defined as [$x(n)$ is assumed to be zero padded]

$$R_x(m) = \frac{1}{N} \sum_{n=0}^{N-|m|-1} x(n)x^*(n + |m|) \tag{10-95}$$

is called the *autocorrelation function*. Equation (10-94) is known as the *Wiener-Khintchine theorem* for sample-data signals.

If $\{x(n)\}$ is a random sequence, care must be exercised in computing power spectra using the DFT or FFT due to the convergence properties of (10-93). It can be shown that, whereas the mean of a spectral estimate made using (10-93) equals the true spectrum in the limit of an infinitely long sequence $\{x(n)\}$, the variance or mean-square deviation from the mean does not get smaller as the sequence length is extended. One way to improve the convergence of the estimate is to employ window functions. We will not discuss this technique here, but rather will demonstrate a second way.

This second technique breaks $\{x(n)\}$ into a number L of shorter subsegments, computes the power spectrum of each subsegment according to (10-93), and then takes as the spectral estimate the average of the power spectra of these subsegments. The more subsegments used, the smoother the spectral estimate, but the less the frequency resolution for a fixed length of $\{x(n)\}$, since each subsegment is of length $1/L$ of the length of the original sequence and the length of the FFT used on each subsegment is correspondingly shorter than would be used on the entire sequence. This technique for estimating spectra is called *Welch's method.*

EXAMPLE 10-12

MATLAB has a power spectrum estimation function that uses the Welch method. We demonstrate that function with the MATLAB program given below and the estimates of a unit-amplitude cosine wave by itself and in noise shown in Figures 10-18 and 10-19, respectively.

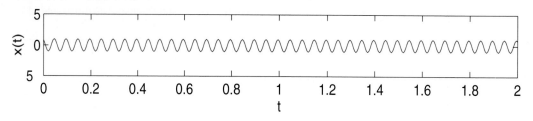

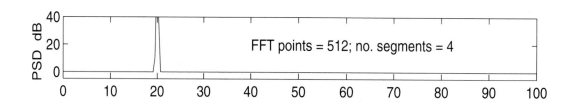

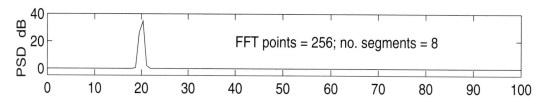

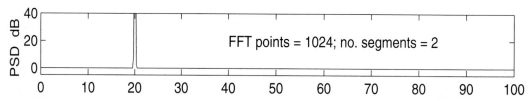

FIGURE 10-18. Power spectral density of a unit-amplitude cosine waveform using Welch's method.

```
EDU» c10ex12
%    This program utilizes the psd function to plot the power spectral
%    density of a cosine signal in Gaussian noise for various
%    combinations of FFT size and number of segments averaged.
%
I_noise=input('Enter 1 for Gaussian noise of variance 5 added to
signal');
N_tot=2048;
N_seg_0=8;
k=1:N_tot;
f_0=20;
del_t=0.005;
f_samp=1/del_t;
t=0:del_t:(N_tot-1)*del_t;
max_t=max(t);
x=cos(2*pi*f_0*k*del_t);
n=sqrt(5)*randn(size(x));
```

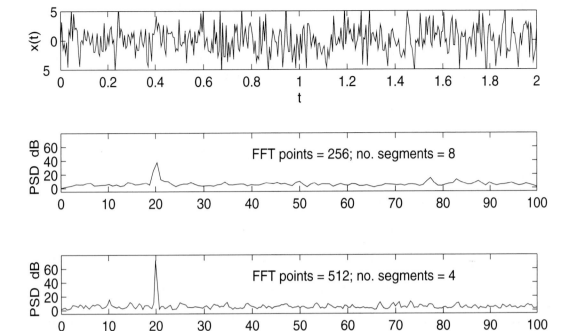

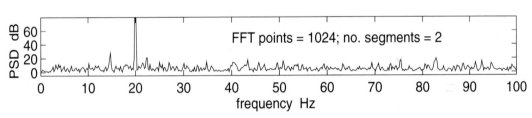

FIGURE 10-19. Power spectral density of a unit-amplitude cosine waveform in Gaussian noise of variance 5 using Welch's method.

```
if I_noise==1
        x=x+n;
else
        x=x;
end
subplot(4,1,1),plot(t,x),axis([0 2-5 5]),xlabel('t'),ylabel('x(t)')
for l=1:3
N_seg=N_seg_0/2^(l-1);
NFFT=N_tot/N_seg;
[P,F]=PSD(x,NFFT,f_samp);
subplot(4,1,l+1),plot(F,P),axis([0 100 -5 80]),ylabel('PSD-dB'), . . .
        text(40, 50,['FFT points = ',num2str(NFFT),'; no. segments = ',
        num2str(N_seg)])
        if l==3
                xlabel('frequency-Hz')
        end
end
```

From Figure 10-18 we observe that the resolution (as shown by the narrowness of the peak at the frequency of the cosine waveform) of frequency components within the spectrum improves with an increased number of FFT points, which should come as no surprise. From this one might conclude that the longest possible FFT length should be used. However, Figure 10-19 verifies the earlier assertion made that averaging the power spectra of more segments produces a smoother estimate than for fewer segment used. With what data we have in Figure 10-19, it appears that averaging the 512-point power spectrum estimates of 4 segments gives the best compromise between resolution and smoothness in this case.

System Identification[†]

We recall that the output of a linear system is given by the convolution of the input with the impulse response. Alternatively, as discussed in Chapter 4, we may obtain the Fourier transform of the output, $Y(f)$, by multiplying the transfer function, $H(f)$, by the Fourier transform of the input, $X(f)$:

$$Y(f) = H(f)X(f) \tag{10-96}$$

This frequency-domain multiplication of $H(f)$ and $X(f)$ can be approximated by employing the overlap-save method illustrated in Figure 10-16.

What if, instead of desiring the output, we wish to find the transfer function of the system? This problem is referred to as *system identification.* To see how we might accomplish this, consider the following series of operations. First, we multiply both sides of (10-96) by $X^*(f)$ to obtain

$$Y(f)X^*(f) = H(f)|X(f)|^2 \tag{10-97}$$

Dividing by the nonnegative, real function $|X(f)|^2$, we obtain

$$H(f) = \frac{Y(f)X^*(f)}{|X(f)|^2} \tag{10-98}$$

which avoids division by a complex function. Comparing this with (10-93), we see a similarity; indeed, the quantity $(1/N)Y(f)X^*(f)$ is defined as the cross-spectrum of output with input and its inverse Fourier transform is the cross-correlation function of output with input. Thus, to "identify" the system, that is, determine $H(f)$, we perform the operations implied by (10-98). To obtain the impulse response [the inverse Fourier transform of (10-98)] we would, in effect, be cross-correlating the output with the normalized input $\mathscr{F}^{-1}[X^*(f)/|X(f)|^2]$. If $|X(f)|^2 = 1$, this corresponds to the operation

$$h(t) = \int_{-\infty}^{\infty} y(\lambda)x(t + \lambda) \, d\lambda \tag{10-99}$$

which is known as the *time-average cross-correlation function* of $x(t)$ and $y(t)$.

What are the pitfalls of this operation when using the DFT? First, we must take care not to divide by zero. If $|X(f)|^2 = 0$ for any f, we must replace these values by a small quantity (perhaps the smallest number that the computer can represent). Second, if $x(t)$ and $y(t)$ are functions that are too long for the FFT operation, the segmentating procedure illustrated by Figure 10-16 must be used.

[†]All independent variables used here are represented as continuous-frequency and time variables for simplicity. Section 10-8 discusses appropriately choosing sampling rate and FFT size.

Signal Restoration or Deconvolution

Given the output of a fixed, linear system and its transfer function, we may determine the Fourier transform of its input by going through a derivation identical to the one which resulted in (10-98) except that $X(f)$ is replaced by $H(f)$. The result is

$$X(f) = \frac{Y(f)H^*(f)}{|H(f)|^2} \tag{10-100}$$

In the time domain, this operation consists of cross-correlation of the output with a normalized impulse response of the filter given by

$$h_n(t) = \mathcal{F}^{-1}\left\{\frac{H^*(f)}{|H(f)|^2}\right\} \tag{10-101}$$

To approximate this operation with the DFT, we may have to use the segmenting operation illustrated by Figure 10-16 since the output would, in general, be arbitrarily long.

10-7 Windows and Their Properties

The discussion given in relation to Figure 10-1 showed that sampling the spectrum of a windowed, time-domain signal implied a periodic extention of the signal. Unless the signal is periodic with an integer number of periods within the window or unless it smoothly approaches zero at each end of the interval, the resulting discontinuities generate additional spectral components. This is illustrated by Problem 10-3, part (b), wherein a noninteger frequency L for the rotating phasor generates spectral components at all DFT output frequencies. This is referred to as *spectral leakage*. The reason can be traced directly to the discontinuities implied at either end of the sampled signal. In order to minimize spectral leakage, the data samples can be multiplied by nonrectangular windows which approach zero smoothly at the beginning and end of the signal. Several window functions are available. A few common ones and their properties are listed in Table 10-4. Their effects on a cosine signal with a noninteger number of cycles within the DFT interval are shown in Figure 10-20. In general, Table 10-4 and Figure 10-20 show that it is not possible to have a narrow mainlobe and low sidelobes for the DFT of a window simultaneously. However, some windows provide a better compromise than others in trading off between these two parameters. For example, Table 10-4 shows that a Kaiser-Bessel window with $\alpha = 2.5$ has a highest sidelobe level of -57 dB and mainlobe 3-dB bandwidth of 1.57 frequency samples. A Hamming window, on the other hand, has a highest sidelobe of -43 dB and a 3-dB mainlobe width of only 1.3 frequency samples.[†]Thus if one wishes to resolve closely spaced frequency components while minimizing leakage from one component to another, the Hamming window would be a logical choice. Alternatively, if suppression of leakage is of primary concern, a Kaiser-Bessel window would be a good choice.

EXAMPLE 10-13 ————————————————————————————————

Several more window functions than listed in Table 10-4 exist. The following MATLAB program contains 10 total window functions. A selected one and its FFT can be displayed. This is done for the Hamming and Kaiser-Bessel window functions in Figures 10-21 and 10-22, respectively. Note that

---

[†]A good discussion of windows and their properties appears in the paper by: F. J. Harris, "On the Use of Windows for Harmonic Analysis with the Discrete Fourier Transform," *Proc. IEEE,* Vol. 66, 1978, pp. 51–83.

TABLE 10-4
Various Windows and Their Characteristics

Window	Highest Sidelobe Level (dB)[a]	3-dB Width Frequency (bins)[b]	Coherent Gain/N: $[\Sigma\, w(n)]^2/\Sigma\, w^2(n)$
1. Rectangular			
$w(n) = 1, n = 0, \ldots, N - 1$	-13	0.89	1.0
2. Triangular			
$w(n) = 2n/N$			
$W(N - n - 1) = w(n)$			
$n = 0, 1, 2, \ldots, N/2$	-27	1.28	0.75
3. Hanning			
$w(n) = \dfrac{1}{2}\left[1 - \cos\dfrac{2n\pi}{N}\right]$			
$n = 0, 1, 2, \ldots, N - 1$	-32	1.44	0.67
4. Hamming			
$w(n) = 0.54 - 0.46 \cos\dfrac{2\pi n}{N}$			
$n = 0, 1, 2, \ldots, N - 1$	-43	1.30	0.74
5. Kaiser-Bessel	$\alpha = 2;$		
$w(n) = I_0(\pi\alpha\beta)/I_0(\pi\alpha)$	-46	1.43	0.67
where	$\alpha = 2.5;$		
	-57	1.57	0.61
$\beta = \sqrt{1 - \left(\dfrac{2n + 1}{N} - 1\right)^2}$	$\alpha = 3.0;$		
	-69	1.71	0.56
$n = 0, 1, 2, \ldots, N - 1$	$\alpha = 3.5$		
	-82	1.83	0.52
$I_0(x) = $ modified Bessel function; order zero			

Source: F. J. Harris, "On the Use of Windows for Harmonic Analysis with the Discrete Fourier Transform," *Proc. IEEE*, Vol. 66, January 1978, pp. 51–83.

[a]To obtain decibels for voltage (current) samples, take 20 times the logarithm to the base 10.

[b]A bin here is taken as one unit of the output independent variable. It is also the equivalent frequency width normalized by the sampling frequency.

low sidelobe level comes at the expense of mainlobe width, but the tradeoff is not a specifiable mathematical function. Also, a large FFT length is used to make the sidelobe structure evident. The student may wish to experiment with the tradeoff between mainlobe width and sidelobe level using this program.

```
%       Study of weighting (window) functions
%
N = input('Enter window size-points');
NFFT = input('Enter number of FFT points');
w = zeros(size(N));
K = menu('Weighting Function Options', 'rectangular',...
         'triangular',...
         'sine lobe',...
         'Hanning',...
```

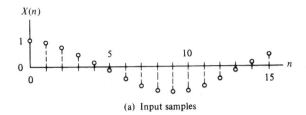

(a) Input samples

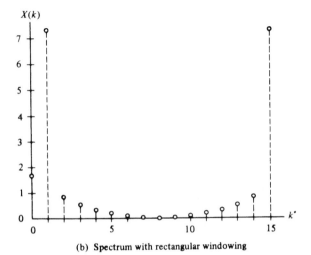

(b) Spectrum with rectangular windowing

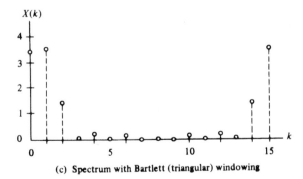

(c) Spectrum with Bartlett (triangular) windowing

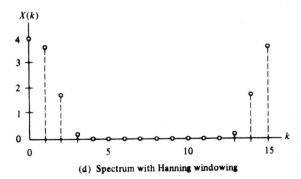

(d) Spectrum with Hanning windowing

FIGURE 10-20. Effects of various windows on the spectrum of a sampled cosine waveform.

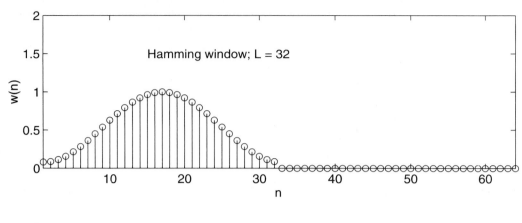

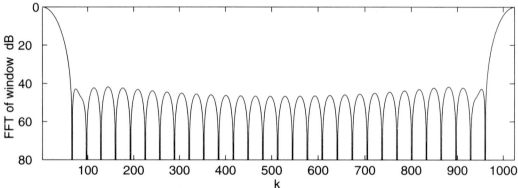

FIGURE 10-21. The Hamming window and its FFT.

```
        'sine cubed', . . .
        'sine to the fourth', . . .
        'Hamming', . . .
        'Blackman', . . .
        'Kaiser-Bessel', . . .
        'Gaussian');
zz = zeros(size(1:NFFT-N));
if K == 1
        w(1:N)=ones(1:N);
elseif K == 2
        for n = 1:N
                if n <= N/2+1
                        w(n) = 2*(n-1)/N;
                else
                        w(n) = 2*(N-n+1)/N;
                end
        end
elseif K  ==  3
        n = 1:N;
        w = sin(pi*(n-1)/N);
elseif K == 4
        n = 1:N;
```

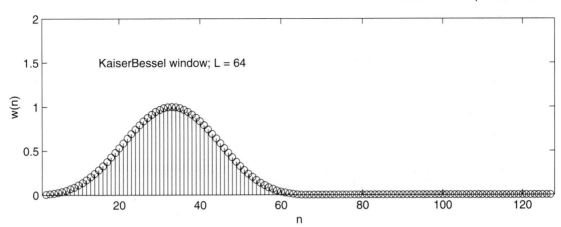

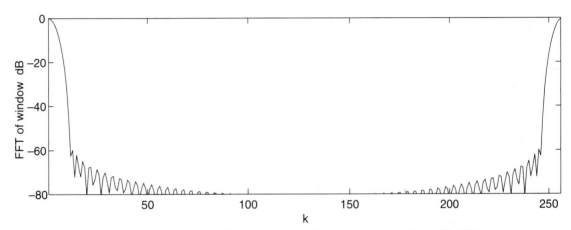

FIGURE 10-22. The Kaiser-Bessel window with parameter $\alpha = 2.5$ and its FFT.

```
       w = (sin(pi*(n-1)/N)).^2;
elseif K == 5
       n = 1:N;
       w = (sin(pi*(n-1)/N)).^3;
elseif K == 6
       n = 1:N;
       w = (sin(pi*(n-1)/N)).^4;
elseif K == 7
       n = 1:N;
       w = 0.54-0.46*cos(2*pi*(n-1)/N);
elseif K == 8
       n = 1:N;
       w = 0.42-0.50*cos(2*pi*(n-1)/N)+0.08*cos(4*pi*(n-1)/N);
elseif K == 9
       alpha = input('Enter value for parameter alpha');
       arg = zeros(size(N));
       for n = 1:N
              arg(n) = pi*alpha*sqrt(1-(2*(n-N/2-1)/N)^2);
```

```
        end
        w = besselI(0,arg)/besselI(0,pi*alpha);
    elseif K == 10
        alpha = input('Enter value for parameter alpha');
        n = 1:N;
        w = exp(-.5*(2*alpha*(n-N/2-1)/N).^2);
    end
    w = [w zz];
    W = fft(w);
    subplot(2,1,1),stem(w),axis([1 2*N 0 2]),xlabel('n'),ylabel('W(n)'), ...
        if K == 1
        text(15, 1.5,['Rect. window; L=',num2str(N)])
        elseif K == 2
        text(15, 1.5,['Triang. window; L=',num2str(N)])
        elseif K == 3
        text(15, 1.5,['Sine window; L=',num2str(N)])
        elseif K == 4
        text(15, 1.5,['Hanning window; L=',num2str(N)])
        elseif K == 5
        text(15, 1.5,['Sine^3 window; L=',num2str(N)])
        elseif K == 6
        text(15, 1.5,['Sine^4 window; L=',num2str(N)])
        elseif K == 7
        text(15, 1.5,['Hamming window; L=',num2str(N)])
        elseif K == 8
        text(15, 1.5,['Blackman window; L=',num2str(N)])
        elseif K == 9
        text(15, 1.5,['Kaiser-Bessel window; L=',num2str(N)])
        elseif K == 10
        text(15, 1.5,['Gaussian window; L=',num2str(N)])
        end
    subplot(2,1,2),plot(20*log10((abs(W)+.00000001)/max(abs(W)))), ...
        axis([1 NFFT -80 0]),xlabel('k'),ylabel('FFT of window-dB')
```

10-8 Selection of Parameters for Signal Processing with the DFT

In processing a signal by means of the DFT or FFT, the sampling theorem requires a sampling rate of $f_s \geq 2W$ samples per second, where W is the bandwidth of the signal (assumed low-pass). Assuming a window width of T seconds and an N-point DFT, the sampling rate is $f_s = N/T$, which must satisfy

$$f_s = \frac{N}{T} \geq 2W \quad \text{Hz} \tag{10-102}$$

The spacing between frequency samples is

$$\Delta f = \frac{f_s}{N} = T^{-1} \quad \text{Hz} \tag{10-103}$$

which is the resolution in frequency.

Combining (10-102) and (10-103), we obtain

$$\Delta f = \frac{1}{T} \geq \frac{2W}{N} \quad \text{Hz} \tag{10-104}$$

Given the signal bandwidth, a desired resolution or frequency spacing dictates a required FFT size. To make T/T_s equal to $N = 2^n$, zeros can be added at the end of the data. This is called zero padding. Although adding to the record length by zero padding increases the FFT size and thereby results in a smaller Δf, one must be cautious to have sufficient record length to support this resolution.

EXAMPLE 10-14

Consider a signal 1 s in duration whose bandwidth is 10 kHz. The spectrum is desired with a frequency resolution of 100 Hz or less. Is this possible? What should N be?

Solution: From (10-102) we have

$$f_s \geq 2W = 20 \quad \text{kHz} \tag{10-105}$$

From (10-103) we have

$$\Delta f = 100 \text{ Hz} = \frac{20,000}{N} \tag{10-106}$$

or

$$N = 200 \tag{10-107}$$

Choosing the next largest power of 2, we let

$$N = 256 \tag{10-108}$$

With $T = N/f_s = 256/20,000 = 1.28 \times 10^{-2}$, we see that the 1-s signal duration is more than sufficient to provide the desired resolution.

EXAMPLE 10-15

The spectrum of a transient signal is to be determined by FFT analysis techniques. Its duration is 2 ms, and it is determined that a sampling rate of 5 kHz should be adequate to avoid significant aliasing. A spectral resolution of 100 Hz is desired. Is this possible?

Solution: The resolution desired, from (10-104), dictates that

$$T = \frac{1}{\Delta f} = 0.01 \text{ s} = 10 \quad \text{ms} \tag{10-109}$$

It is not possible to achieve the desired resolution since the transient signal is only 2 ms in duration. Unless more data are available, a resolution of

$$\Delta f = \frac{1}{T} = 500 \quad \text{Hz} \tag{10-110}$$

must be accepted. This implies

$$N_{\text{samples}} = \frac{T}{T_s} = 10 \tag{10-111}$$

and a 16-point FFT will suffice.

*10-9 The Chirp-z-Transform Algorithm

Several efficient algorithms for computing the DFT of a sequence of finite length, N, have been discussed in the preceding pages. In order to achieve this efficiency, N must be a power of 2 or 4. The chirp-z-transform provides a means for computing the DFT of a time series of length N for a sequence of M frequencies where $M \leq N$. Furthermore, the chirp-z-transform is conveniently implemented as hardware by employing charge-coupled delay lines.

To derive the chirp-z-transform, consider the z-transform of the sequence $x(n)$, $n = 0, 1, 2, \ldots,$ $N - 1$, which for a discrete set of values of $z = z_k$, $k = 0, 1, 2, \ldots, M - 1$, can be expressed as

$$X(z_k) = \sum_{n=0}^{N-1} x(n) z_k^{-n}, \qquad 0 \leq k \leq M - 1 \tag{10-112}$$

Let z_k be a point on a spiral centered at the origin in the z-plane. This can be accomplished by writing

$$z_k = AW^{-k}, \qquad 0 \leq k \leq M - 1 \tag{10-113}$$

where

$$A = A_0 e^{j\theta_0} \tag{10-114}$$

is the point at which the spiral starts and

$$W = W_0 e^{-j\phi_0} \tag{10-115}$$

determines the rate of spiral and the angular separation of the points z_k. Figure 10-23 shows typical cases for $W_0 < 1$, for which the sequence of points $\{z_k\}$ spirals toward the origin, and for $W_0 > 1$, for which the sequence of points $\{z_k\}$ spirals away from the origin. If $|W_0| = 1$, the z_k-sequence lies on a circle of radius A_0.

Substitution of (10-113) into (10-112) yields

$$X(z_k) = \sum_{n=0}^{N-1} x(n) A^{-n} W^{nk}, \qquad 0 \leq k \leq M - 1 \tag{10-116}$$

The next step in the derivation is to make the observation that

$$nk = \tfrac{1}{2}[n^2 + k^2 - (k - n)^2] \tag{10-117}$$

Thus (10-116) can be written as

$$X(z_k) = W^{k^2/2} \sum_{n=0}^{N-1} [x(n) A^{-n} W^{n^2/2}] W^{-(k-n)^2/2}, \qquad 0 \leq k \leq M - 1 \tag{10-118}$$

With the aid of the definitions

$$h(n) \triangleq W^{-n^2/2} \tag{10-119}$$

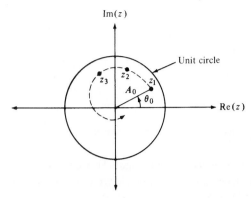

(a) **Spiral toward the origin**

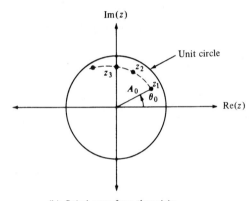

(b) **Spiral away from the origin**

FIGURE 10-23. z-Plane spirals for development of the chirp-z-transform.

and

$$g(n) \triangleq x(n)A^{-n}W^{n^2/2} \tag{10-120}$$

one may put (10-118) into the form

$$X(z_k) = \frac{1}{h(k)} \sum_{n=0}^{N-1} g(n)h(k-n), \qquad 0 \le k \le M-1 \tag{10-121}$$

The summation is recognized as a convolution of the sequences $\{g(n)\}$ and $\{h(n)\}$. Thus (10-121) can be viewed as the series of operations represented by the block diagram of Figure 10-22. If A and W_0 are unity, the sequence $h(n)$ is given by

$$h(n) = (e^{-j\phi_0})^{-n^2/2}$$

$$= e^{jn^2\phi_0/2} \tag{10-122}$$

which can be thought of as a complex exponential sequence with linearly increasing frequency. Such signals are known as chirp signals in radar system applications—hence the name *chirp-z-transform* for (10-121).

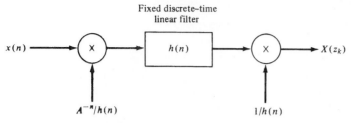

FIGURE 10-24. Block diagram representation of the chirp-z-transform.

All operations implied in Figure 10-24 could, of course, be carried out digitally. However, the convolution operation for the chirp-z-transform can be implemented by means of charge transfer devices (CTDs), and CTD chirp-z transformers are available commercially.

Basically, a CTD can be viewed as a shift register for analog charge samples. As illustrated in Figure 10-25a each stage of capacitance can weight the charge stored on it differently through employment of split electrodes and differential sensing of the charge. With the application of each clock pulse, the charge stored on each capacitor is transferred to the next capacitor in the chain and a new charge sample is stored on the first capacitor in the chain. The impulse response $h(n)$ of the convolution operation is implemented by varying the capacitance values of each stage of the CTD in accordance with the desired weighting for the convolution. Since $h(n)$ is complex, the convolution operation must be implemented to compute the real and imaginary parts of $h(n) * [(A^{-n}/h(n))x(n)]$.

The input and output chirping operations are implemented by means of multiplying analog-to-digital converters, with real and imaginary parts of the chirp coefficients, $h(n)$, being stored in read-only memories.

A schematic representation of a CTD-implemented chirp-z transformer is shown in Figure 10-25b. All complex operations are implemented as equivalent real operations. It is left for the student to show that the block diagram of Figure 10-25b implements the operations of the block diagram of Figure 10-24.

The major advantage of a CTD-implemented chirp-z-transformer over the equivalent digital implementation appears to be cost. The inherent error sources associated with the CTD implementation are charge-transfer inefficiency and the inaccuracy associated with storing the chirping coefficients in the read-only memory as binary numbers. The former error source is due to the fact that the charge-transfer operation in the CTD is not 100% efficient.

EXAMPLE 10-16

The chirp-z-transform may be used to obtain a finer resolution over a fraction of the sampling frequency extent than for the FFT. With the FFT, we get the entire frequency extent from 0 to the sampling frequency f_s, the number of points N is the size of the FFT, and the frequency resolution is f_s/N. With the chirp-z-transform, the starting frequency, resolution, and extent of our frequency analysis for a signal are adjustable. This is illustrated in this example by the following MATLAB implementation of a chirp-z-transform. Some typical outputs are compared with the corresponding outputs of an FFT in Figures 10-26 and 10-27 for the spectral analysis of a pulse signal.

```
%       Implementation of the chirp-z transform of a pulse signal and
%       comparison with the corresponding results for the FFT
%
clf
theta_0 = input('Enter starting point, theta_0, on unit circle in
    radians ');
```

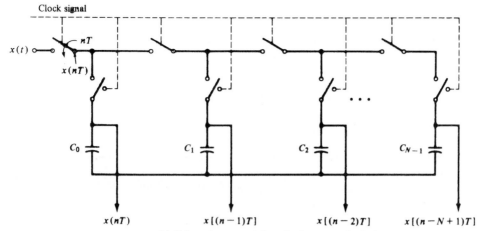

(a) **Schematic representation of a charge transfer device wherein each charge storage element can be weighted independently of the others**

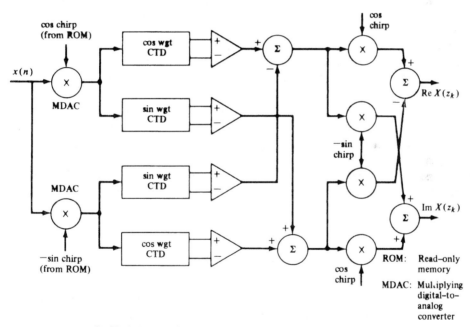

(b) **Block diagram of a chirp–z transformer employing charge transfer devices**

FIGURE 10-25. Illustration of the use of charge transfer devices to implement a chirp-*z* transformer.

```
phi_0 = input('Enter frequency resolution, phi_0, around unit circle in
    radians ');
N = input('Enter total number of points for pulse signal ');
N_pulse = input('Enter number of nonzero points for pulse signal ');
M = input('Enter number of frequency cells for chirp-z transform ');
W = exp(-j*phi_0);
A = exp(j*theta_0);
n = 0:N-1;
```

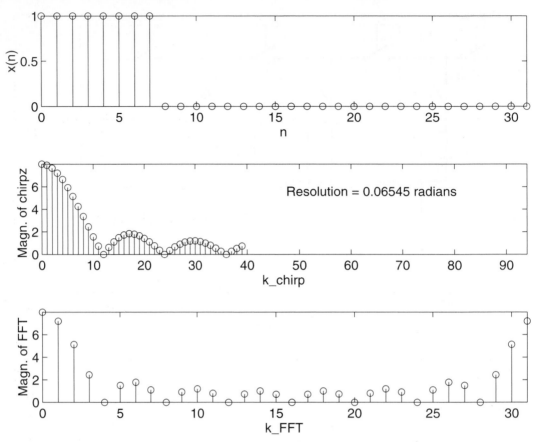

FIGURE 10-26. Chirp-*z* transform (middle) of a 8-point square pulse (top) compared with the pulse's FFT (bottom). The resolution of the chirp-*z* transform is three times that of the FFT with 40 points computed for the chirp-*z* versus 32 for the FFT. Abscissa scales have been adjusted to show the same absolute frequency extent.

```
m = 0:M-1;
x = zeros(size(1:N));
for nn = 1:N_pulse
        x(nn) = 1;
end
k = 1:M;
T = A.*W.^(-(k-1));
Y = ones(size(T));      % Form matrix of (AW^(-1))^(k*n); k = columns &
                            n = rows
for nn = 2:N
        Y = [Y;T.^(nn-1)];
end
X_chirp = x*Y;          % Use matrix multiply to compute chirp-z sum
X_FFT = fft(x);         % Compute FFT for comparison
%          Shift chirp-z and set up axes in plot to make direct comparison
%          between chirp-z and FFT results.
NN = 2*pi/phi_0-1;
```

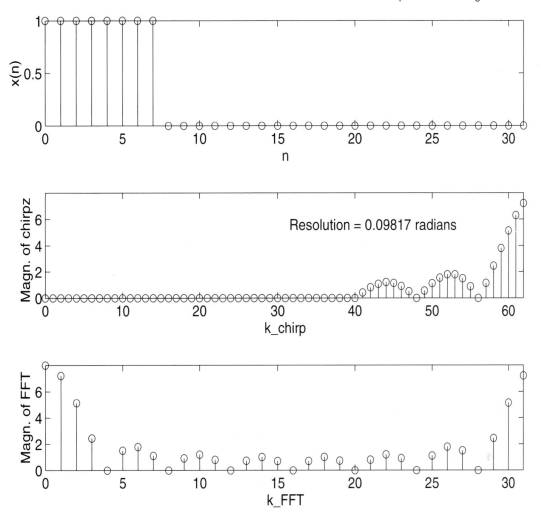

FIGURE 10-27.   Chirp-*z*-transform (middle) of a 8-point square pulse (top) compared with the pulse's FFT (bottom). The resolution of the chirp-*z*-transform is twice that of the FFT with the resolution range taking up the last part of the transform. The initial 0s have been manually placed there to put the transforms in the correct relationship to each other. Abscissa scales have been adjusted to show the same absolute frequency extent.

```
zz = zeros(size(1:theta_0*(NN+1)/(2*pi)));
X_chirp = [zz X_chirp];
L = length(X_chirp);
l = 0:L-1;
subplot(3,1,1),stem(n,x),axis([0 N-1 0 1]),xlabel('n'),ylabel('x(n)')
subplot(3,1,2),stem(l,abs(X_chirp)),axis([0 NN-1 0 N_pulse]),
   xlabel('k_chirp'), . . .
        ylabel('Magn. of chirp-z'), . . .
        text(NN/2,.7*N_pulse,['Resolution = ',num2str(phi_0),' radians'])
subplot(3,1,3),stem(n,abs(X_FFT)),axis([0 N-1 0 N_pulse]),
   xlabel('k_FFT'), . . .
        ylabel('Magn. of FFT')
```

Summary

In this chapter we have considered the discrete Fourier transform (DFT) as a tool for computing the spectra of discrete-time signals. Under suitable restrictions, the DFT closely approximates the spectrum of a continuous-time signal at a discrete set of frequencies. Fast algorithms to compute the DFT, referred to as FFT algorithms, were derived. Several applications of the FFT were discussed. The following are the main points made in this chapter.

1. The DFT of an N-point sequence $\{x_n\}$ is defined as

$$X_k = \sum_{n=0}^{N-1} x_n e^{-j2\pi kn/N}, \qquad k = 0, 1, \ldots, N - 1 \tag{10-1}$$

and the inverse DFT of the sequence $\{X_k\}$ is given by

$$x_n = \frac{1}{N} \sum_{k=0}^{N-1} X_k e^{j2\pi nk/N}, \qquad n = 0, 1, \ldots, N - 1 \tag{10-2}$$

These are often written in terms of $W_N = e^{-j2\pi/N}$.

2. When DFTs are used to process continuous-time signals by sampling, several potential error sources may be important. These are aliasing, spectral leakage, and picket-fence effect. The first one results from the fact that signals that are not strictly bandlimited may be sampled, and the sampling results in overlap between the signal spectrum and its translates about the sampling frequency. Spectral leakage comes about because the signal may be infinite in extent, and multiplication by a window function is necessary to limit the signal in time. This amounts to convolution with the Fourier transform of the window in the frequency domain, which gives rise to a spectral component at one frequency "leaking" or adding to the spectral components at another frequency location. The picket-fence effect is a reflection of the fact that the DFT gives spectral estimates only at the discrete frequences $f_k = k/T$ hertz, where T is the observation window length. Hence, spectral components between these discrete frequencies cannot be observed by means of the DFT.

3. An inverse DFT can be computed using a direct DFT algorithm by conjugating the frequency samples, taking the DFT of these conjugated samples, conjugating the output of the DFT operation, and dividing by N.

4. There are many algorithms for efficiently computing the DFT. These are referred to collectively as the fast Fourier transform (FFT). The decimation-in-time and decimation-in-frequency FFT algorithms were derived in this chapter. Both algorithms produce a DFT with approximately $2N \log_2 N$ real multiplies, whereas direct computation by means of the DFT sum requires about $4N(N - 1)$ real multiplies. For large N this gives tremendous computational savings in performing the transform by an FFT algorithm over a direct evaluation of the DFT.

5. Several properties of the DFT were given in Section 10-5. These include linearity, time shift, frequency shift, duality, circular convolution, multiplication, Parseval's theorem, and various symmetry properties.

6. Applications of the DFT include filtering, spectrum analyzers, convolution (with applications to filtering and system identification), and processing of random signals through computation of such statistics as autocorrelation functions and power spectra.

7. The use of window functions to minimize spectral leakage requires windows having transforms with narrow mainlobes and low sidelobes. Useful window functions having transforms with sidelobes lower than those of rectangular windows include triangular, Hanning, Hamming, and Kaiser-Bessel windows.

8. In applying the DFT to signal analysis, the sampling theorem and the desired resolution impose a constraint between total signal duration, T, signal bandwidth, W, and the number of DFT points, N. This relationship is

$$\frac{1}{T} \geq \frac{2W}{N}$$

An additional requirement for radix-2 FFTs is that $N = 2^n$, where n is an integer.

9. The chirp-z-transform is of interest because it can be used to obtain a set of spectral estimates in a frequency band other than from 0 to the sampling frequency. In fact the number of output points does not have to be equal to the number of input points. A method to compute the chirp-z-transform using charge-coupled device technology was presented. This is becoming less important because of the tremendous advances being made each year in digital signal processor technology; 1990s technology provides a programmable processor that can compute a 1,024-point FFT in 120 μs.

Further Reading

In addition to the references in previous chapters, the following are useful from the standpoint of the DFT and FFT and their applications:

A. OPPENHEIM and R. SCHAFFER, *Discrete-Time Signal Processing.* Englewood Cliffs, NJ: Prentice-Hall, 1989.

O. ERSOY, *Fourier-Related Transforms, Fast Algorithms and Applications,* Upper Saddle River, NJ, Prentice Hall, 1997.

E. O. BRIGHAM, *The Fast Fourier Transform and Its Applications.* Englewood Cliffs, NJ: Prentice-Hall, 1988.

C. S. BURRUS and T. W. PARKS, *DFT/FFT Convolution Algorithms: Theory and Implementation.* New York: Wiley, 1985.

J. S. WALKER, *Fast Fourier Transorms,* 2nd ed., Boca Raton, FL, CRC Press, 1996.

The following reference summarizes FFT algorithms, both for performing the DFT and implementing signal processing functions. It also includes a summary of hardware for performing the FFT and signal processing functions, and includes design examples:

W. W. SMITH and J. M. SMITH, *Handbook of Real-Time Fast Fourier Transforms,* Piscataway, NJ: *IEEE Press,* 1995

PROBLEMS

Section 10-1

10-1. Obtain the DFT of

$$x_n = 1, \qquad 0 \leq n \leq N - 1$$

by using (4-139) to sum the series (10-1). Why is this result reasonable?

10-2. (a) Obtain the DFT of

$$x_n = \begin{cases} 1, & 0 \leq n \leq K < N - 1 \\ 0, & K < n \leq N - 1 \end{cases}$$

by using the sum for a geometric series (4-136), to sum (10-1). Plot the periodically extended time series which corresponds to the inverse DFT of the resulting DFT (see Figure 10-1).

(b) Plot $|X(k)|$ versus k for $N = 8$ and $K = 2, 3$. Explain your plots in terms of the spectral analysis ideas developed in Chapter 4. What departures, if any, are there from the ideas of Chapter 4? (See Figure 10-1.)

Section 10-2

10-3. Let $\tilde{x}_n = A \exp[j(2\pi Ln/N + \theta_0)]$ in (10-1), where $0 \le L \le N - 1$ and A and θ_0 are constant amplitude and phase, respectively.

(a) Use the sum for Σ_N given by (4-139) to sum the series and show that

$$\tilde{X}(k) = \begin{cases} A \dfrac{\sin \pi(L - k)}{\sin[\pi(L - k)/N]} \exp\left[j\left(\theta_0 + \pi(L - k)\left(1 - \dfrac{1}{N}\right)\right)\right], & L \ne k \\ NA \exp(j\theta_0), & L = k \end{cases}$$

(b) Plot $|X(k)|$ versus k for $A = 1$, $N = 8$, and (1) $L = 1$; (2) $L = 1.1$. *Answers:* (1) For $L = 1$, $\{|X(k)|\} = \{0, 8, 0, 0, 0, 0, 0, 0\}$; (2) for $L = 1.1$, $\{|X(k)|\} = \{0.74, 7.87, 0.89, 0.46, 0.34, 0.31, 0.33, 0.42\}$.

(*Note:* Part (b) illustrates the *leakage* phenomenon of the DFT.)

10-4. Using the result of Problem 10-3(a), obtain the DFT of

$$x_n = A \cos(2\pi Ln/N + \theta_0)$$

by writing it as $x_n = \frac{1}{2}\tilde{x}_n + \frac{1}{2}\tilde{x}_n^*$ and noting that the DFT is a linear transform.

10-5. Reproduce Figure 10-1 for the following signals. Plot magnitudes of transforms. Assume convenient values for the parameters involved.

(a) $x(t) = Ae^{-\alpha t}u(t)$
(b) $x(t) = A\Pi(t/\tau)$
(c) $x(t) = A \cos \omega_0 t$

Section 10-3

10-6. (a) Use the technique of Example 10-2 to compute the DFT of

$$x(n) = e^{-n/10} \cos \pi n, \qquad 0 \le n \le N - 1$$

for general N.

(b) Plot the amplitude spectra for the cases $N = 8$ and $N = 16$. Use a scale that is twice as course for $N = 8$ as for $N = 16$.

(c) Suppose you are now told that the discrete-time signal above was obtained by sampling the signal

$$x(t) = e^{-2t} \cos 20\pi t, \qquad t \ge 0$$

From this you should be able to deduce the sampling rate and fix the frequency scale on your plots of part (b). Compare the spectra obtained in part (b) with the amplitude spectrum obtained by Fourier transforming the continuous-time signal $x(t)$.

10-7. Use the technique of Example 10-2 to obtain the DFT of $\cos^2(\pi n/4)$ for $N = 8$ and 16. Plot both the signal and its DFT.

Section 10-4

10-8. (a) Using the flow graph of Figure 10-9, manually compute the FFT of the sequence

$$x(0) = 1 \qquad x(4) = 0$$
$$x(1) = 1 \qquad x(5) = 0$$

$$x(2) = 0 \qquad x(6) = 0$$

$$x(3) = 0 \qquad x(7) = 1$$

(b) Taking into account the periodicity implied by the DFT, plot at least two periods of the signal for which the FFT was found in part (a).

(c) If the signal of part (a) is an approximation of a single continuous-time periodic pulse, discuss the errors in the spectrum of the DFT spectrum as compared with the spectrum of the continuous-time pulse.

10-9. (a) Use the flow graph of Figure 10-11 and repeat Problem 10-8a.

(b) Using the same flow graph as in part (a) and the observation (4-142) about using an FFT algorithm to find an inverse FFT, check your results above by obtaining an inverse FFT found in part (a) and comparing it with the sequence given in Problem 10-8.

10-10. (a) Obtain the eight-point FFT of the following pulse signal using the flow diagram of Figure 10-9.

$$x(0) = x(1) = x(2) = x(3) = 1$$

$$x(4) = 0$$

$$x(5) = x(6) = x(7) = 1$$

Discuss the departure of this FFT from the result for the Fourier transform of the ideal pulse signal by comparing plots of their amplitude spectra.

(b) Obtain the inverse FFT of the result in part (a) by applying (4-142) and scaling by $N = 8$.

10-11. Obtain an algorithm for determining the powers of W shown in Figure 10-9. Generalize to arbitrary N.

10-12. (a) Use the decimation-in-time approach to develop an FFT algorithm based on a four-point basic DFT computation for $N = 16$. Do this by grouping the input samples into four sets of four samples each.

(b) Repeat part (a) using decimation in frequency.

10-13. Write the butterfly computation for the DIT algorithm, given by (10-46), entirely in terms of real computations. For this purpose, let $RX_m(p) = \mathrm{Re}[X_m(p)]$, $IX_m(p) = \mathrm{Im}[X_m(p)]$, etc. Also note that $W_N^s = \cos(2\pi s/N) - j\sin(2\pi s/N)$. Draw a block diagram of the hardware that would be used to compute the butterfly. Use the minimum possible number of multipliers and adders.

10-14. Repeat Problem 10-13 for the DIF butterfly computation expressed by (10-47). Which butterfly requires the least number of real multiplications and additions?

Section 10-5

10-15. Prove the linearity property of the DFT sum.

10-16. (a) Prove the time-shift property of the DFT by using the fact that $x(n - m)$ is a periodic sequence.

(b) Prove the frequency-shift property by using the fact that $X(k - m)$ is a periodic sequence.

10-17. Prove the convolution property of the DFT by substituting for $y(n - m)$ in the convolution sum

$$x(n) \, \text{Ⓝ} \, y(n) = \sum_{m=0}^{N-1} x(m)y(n - m)$$

in terms of the inverse DFT of $Y(k)$, which represents the DFT of $y(n)$. Use the implied periodicity of $Y(k)$.

10-18. Show that

$$\sum_{n=0}^{N-1} |x(n)|^2 = \frac{1}{N} \sum_{k=0}^{N-1} |X(k)|^2$$

by writing the left-hand sum as $\sum_{n=0}^{N-1} x(n)x^*(n)$ and substituting for $x^*(n)$ in terms of the inverse DFT of $X(k)$.

10-19. Prove the multiplication property of the DFT by substituting for $Y(k - m)$ in terms of the DFT of $y(n)$.

10-20. Discuss how property 10 in Section 10-5 could be used to obtain the DFT of a real sequence $2N$ points long by using an N-point DFT.

10-21. For each of the sequence pairs shown, obtain the circular convolution by padding with enough zeros so that the ordinary linear convolution corresponds to the first period of the result.

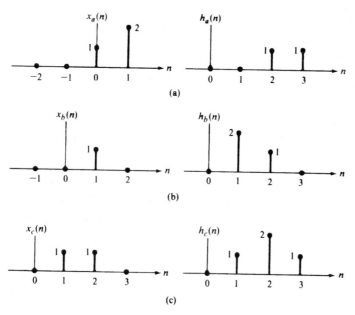

FIGURE P10-21.

10-22. Quadrature sampling. To have a lower sampling rate for bandpass signals, quadrature sampling is often used as illustrated in Fig. P10-22. After the sample/quantizing operation, a complex FFT is taken with the real part of each complex sample corresponding to the cosine (upper) channel and the imaginary part corresponding to the sine (lower) channel.

(a) The FFTs of the cosine and sine channels are desired separately at the output of the FFT. Which property in Section 10-5 applies? Tell how to implement it.

(b) The magnitude of the output spectrum is desired. What are the operations that should be carried out at the FFT output?

(c) The phase of the output spectrum is desired. What operations should be performed at the FFT output?

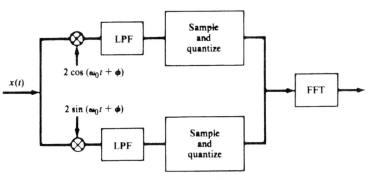

FIGURE P10-22

Sections 10-6

10-23. (a) Referring to Example 10-7, find the four point DFT of

$$z(n) = x(n) + jy(n) = \sin(n\pi/2) + j\cos(n\pi/2)$$

using superposition.

(b) Taking your result for the DFT of $\{z(n)\}$, separate out the transforms of $\{x(n)\}$ and $\{y(n)\}$ and show that the proper results are obtained.

10-24. Given the DFT samples

$$Z(0) = 1 - j, \qquad Z(1) = 2 + j, \qquad Z(2) = 2 - j, \qquad Z(3) = 1 + j$$

The four point signal whose DFT was taken to obtain these DFT values consists of real and imaginary parts, $\{x(n)\}$ and $\{y(n)\}$. Find the FFTs of these real and imaginary parts.

10-25. Consider the convolution of a 50-point impulse response with a 1,000-point random sequence. Choose an appropriate DFT size which is a power of 2 and state how many segments must be chosen for the 1,000-point sequence. How many points are valid for each circular convolution operation when implemented by means of a DFT of the size you have chosen?

10-26. (This problem is meant to illustrate the steps in obtaining an energy spectral estimate as explained in Section 10-6 and does not contain numbers representative of a practical situation.) Consider the eight-point sequence $\{x(n)\}$ defined as

$$x(n) = \begin{cases} 1, & n = 0, 1 \\ 0, & n = 2, 3, 4, 5, 6, 7 \end{cases}$$

If $N = 8$, use (10-95) to compute the autocorrelation function $R_x(n)$. Use an eight-point FFT to compute $S_x(k)$ from (10-93). Next, compute $S_x(k)$ from (10-94) and compare with the result obtained from (10-93).

Section 10-7

10-27. Obtain the DFT of a Hanning window and show that it has the properties quoted in Table 10-4.

10-28. Obtain the DFT of a Hamming window and show that it has the properties quoted in Table 10-4.

Section 10-8

10-29. Compute an appropriate sampling rate and DFT size N in order to analyze a signal spectrally with no significant frequency content above 10 kHz and with a resolution of 100 Hz. Make N a power of 2. Assume an arbitrarily long signal. What is the minimum observation interval length?

Section 10-9

10-30. Sketch the locus of points in the z-plane for the following parameter values for (10-113), (10-114), (10-115):
   (a) $M = 8$, $W_0 = 2$, $\phi_0 = \pi/16$ rad, $A_0 = 2$, and $\theta_0 = \pi/4$ rad.
   (b) $M = 8$, $W_0 = \frac{1}{2}$, $\phi_0 = \pi/16$ rad, $A_0 = 2$, and $\theta_0 = \pi/4$ rad.

10-31. Show that the block diagram of Figure 10-25b provides an implementation of the operations for the chirp-z-transform operation illustrated in Figure 10-24.

10-32. Compare the number of multiplications required for an FFT implementation of the chirp-z-transform with $W_0 = 1$ for various values of N and M with the FFT assuming that the *frequency resolution* for both is the same.

Computer Exercises

10-1. Write a MATLAB program to perform circular convolution using the FFT. Convolve two square pulses, choosing appropriate zero-padding to avoid aliasing, and verify that the proper result for linear convolution occurs. Compare your results with the `conv` function of MATLAB.

10-2. Write a program to carry out convolution of a long sequence with a short sequence employing the overlap-save method illustrated in Figure 10-16. As a test, verify the results shown in Figure 10-17.

10-3. Experiment with the capability of the `psd` function of MATLAB. Use `help psd` to see a description of the various options available. In particular, experiment with the overlap capability to maximize simultaneously the resolution and noise suppression characteristics. Use a sinusoid in Gaussian noise, as in Example 10-12, for your tests.

10-4. Experiment with the chirp z-transform program of Example 10-16. Choose different starting points for the spectral analysis, θ_0, different resolutions, ϕ_0, and different numbers of spectral points. Use a rectangular pulse signal.

10-5. Use the window analysis program of Example 10-13 to extend Table 10-4 to all the windows included in the program.

Comments and Hints on Using MATLAB

A-1 Introduction

The purpose of this appendix is not to give an extensive description of the MATLAB language, but rather to give the student some helpful hints that may make writing and running of MATLAB programs easier. For a relatively inexpensive way to get into using the MATLAB language, if it is not available in an open computer laboratory, the Version 5 User's Guide for *The Student Edition of MATLAB* is recommended. If SIMULINK programs are to be run, the Version 2 User's Guide for *The Student Edition of SIMULINK is recommended*. In addition to the User's Manual, each package includes a scaled-down version of MATLAB or SIMULINK as the case may be. Versions of both programs are available to run under either Microsoft Windows or on a Mac.

A-2 Window and File Management

The discussion in this appendix assumes that you have *The Student Version of MATLAB* running under Microsoft Windows 95.

There are two ways to carry out computations with MATLAB. One is to type the commands directly into the Command window, which is the first window that comes up when the MATLAB is activated by double clicking its icon with the left mouse button or activating its menu listing. For moderately extensive calculations, this becomes unwieldy after a few commands, so an alternative is desired. This alternative is to write a MATLAB *script* file or M-file by using an editor (the one supplied with MATLAB is called Notepad). Such a program is then stored in a directory as "name.m" (just the letters are typed, not the quotes) and then run by typing "name" at the » prompt in the Command window. Note that the ".m" is not typed after the Command window prompt. When debugging a program, it is sometimes useful to type "echo on" as your first command in the Command window before running the program. When the program is run, as described here, each statement will be displayed on the screen as it is executed. When the program gets to a bug, it will stop executing, with the last statement executed displayed on the screen and a diagnostic for the cause of the error displayed next. To stop the echoing of each statement, type "echo off " at the command prompt.

In using MATLAB it is recommended that you create a directory on your hard disk (or use a floppy disk, although this is slower) for your programs and execute the command "cd c:\mystuff" at the MATLAB prompt to direct MATLAB to that directory (given the name "mystuff", the quotation marks are not used, but are employed here to set names used on the computer off from the rest of the text) when retrieving, saving, or running programs. This will save considerable mouse clicking. If you have several programs dealing with different subjects, or even chapters of the book in this course, you will probably benefit from setting up subdirectories for them. Another piece of advice: *Save often and always create a backup of a program that runs before doing extensive modification to it.*

In writing and running programs, three or more windows may be open at one time. These will include the Command window, one or more Notepad windows, and one or more plot windows. Especially in the case of writing programs and testing them, the student will want to have both the Command and Notepad windows open at the same time and use the left mouse button to click on the desired one and bring it to the foreground. This will save considerable time in going back and forth to modifying your program in Notepad and running your program from the Command window.

A-3 Getting Help

If you aren't sure about how MATLAB will treat a certain command or series of commands, it is recommended that you try them in the command window by themselves before putting them into a program. For example, suppose you don't remember whether log( ) or log10( ) is the notation for finding the base-10 logarithm of a number. This can be easily tested in the Command window as follows:

```
» log10(10)
ans =
   1
```

Since the answer returned is 1 for the $\log_{10}(10)$, we conclude that we have guessed correctly at the proper MATLAB notation, and you can now put it in your program assured that you will not have a wrong answer at this point.

The Command window may get very messy at times, and you will have difficulty locating a desired quantity among all the clutter. Clicking the "Edit" drop down menu at the top of the Command window and then clicking the "Clear Session" button under it will clear the screen, but not clear its memory of previously typed and executed commands. These can be displayed after the command prompt and executed again if desired by repeatedly pushing the up arrow ↑ on your keyboard. Of course, it may be easier just to type the command all over again.

MATLAB has an extensive help capability. If one knows the name of a desired operation or function, help is obtained on it simply by typing "help name" where "name" indicates the name of the operation or function (again, only the words should be typed, not the quotation marks).

Just typing "help" at the prompt brings different categories of help topics, which is handy if you can't think of what something might be under. This is illustrated here:

```
» help

HELP topics:

matlab\general      - General purpose commands.
matlab\ops          - Operators and special characters.
matlab\lang         - Programming language constructs.
matlab\elmat        - Elementary matrices and matrix manipulation.
matlab\elfun        - Elementary math functions.
matlab\specfun      - Specialized math functions.
matlab\matfun       - Matrix functions - numerical linear algebra.
matlab\datafun      - Data analysis and Fourier transforms.
matlab\polyfun      - Interpolation and polynomials.
matlab\funfun       - Function functions and ODE solvers.
matlab\sparfun      - Sparse matrices.
matlab\graph2d      - Two dimensional graphs.
```

```
matlab\graph3d       - Three dimensional graphs.
matlab\specgraph     - Specialized graphs.
matlab\graphics      - Handle Graphics.
matlab\uitools       - Graphical user interface tools.
matlab\strfun        - Character strings.
matlab\iofun         - File input/output.
matlab\timefun       - Time and dates.
matlab\datatypes     - Data types and structures.
matlab\dde           - Dynamic data exchange (DDE).
matlab\demos         - Examples and demonstrations.
toolbox\symbolic     - Symbolic Math Toolbox.
fuzzy\fuzzy          - Fuzzy Logic Toolbox.
fuzzy\fuzdemos       - Fuzzy Logic Toolbox Demos.
toolbox\stats        - Statistics Toolbox.
images\images        - Image Processing Toolbox.
images\imdemos       - Image Processing Toolbox --- demos and sample images
nnet\nnet            - Neural Network Toolbox.
nnet\nndemos         - Neural Network Demonstrations and Applications.
toolbox\signal       - Signal Processing Toolbox.
toolbox\optim        - Optimization Toolbox.
toolbox\robust       - Robust Control Toolbox.
toolbox\ident        - System Identification Toolbox.
toolbox\control      - Control System Toolbox.
control\obsolete     - (No table of contents file)
stateflow\stateflow  - Stateflow
stateflow\sfdemos    - (No table of contents file)
simulink\simulink    - Simulink
simulink\blocks      - Simulink block library.
simulink\simdemos    - Simulink demonstrations and samples.
simulink\dee         - Differential Equation Editor
toolbox\tour         - An interface to Matlab demos, installed Toolboxes
                       demos, and information
toolbox\local        - Preferences.

  For more help on directory/topic, type "help topic".
```

Typing "help" with any one of these topics will bring up more detailed help in that category. For example:

```
» help elfun

Elementary math functions.

Trigonometric.
  sin        - Sine.
  sinh       - Hyperbolic sine.
  asin       - Inverse sine.
  asinh      - Inverse hyperbolic sine.
  cos        - Cosine.
  cosh       - Hyperbolic cosine.
  acos       - Inverse cosine.
  acosh      - Inverse hyperbolic cosine.
  tan        - Tangent.
  tanh       - Hyperbolic tangent.
  atan       - Inverse tangent.
```

```
atan2      - Four quadrant inverse tangent.
atanh      - Inverse hyperbolic tangent.
sec        - Secant.
sech       - Hyperbolic secant.
asec       - Inverse secant.
asech      - Inverse hyperbolic secant.
csc        - Cosecant.
csch       - Hyperbolic cosecant.
acsc       - Inverse cosecant.
acsch      - Inverse hyperbolic cosecant.
cot        - Cotangent.
coth       - Hyperbolic cotangent.
acot       - Inverse cotangent.
acoth      - Inverse hyperbolic cotangent.

Exponential.
exp        - Exponential.
log        - Natural logarithm.
log10      - Common (base 10) logarithm.
log2       - Base 2 logarithm and dissect floating point number.
pow2       - Base 2 power and scale floating point number.
sqrt       - Square root.
nextpow2   - Next higher power of 2.

Complex.
abs        - Absolute value.
angle      - Phase angle.
conj       - Complex conjugate.
imag       - Complex imaginary part.
real       - Complex real part.
unwrap     - Unwrap phase angle.
isreal     - True for real array.
cplxpair   - Sort numbers into complex conjugate pairs.

Rounding and remainder.
fix        - Round towards zero.
floor      - Round towards minus infinity.
ceil       - Round towards plus infinity.
round      - Round towards nearest integer.
mod        - Modulus (signed remainder after division).
rem        - Remainder after division.
sign       - Signum.
```

Then typing "help" with anyone of these functions will bring up help on it:

```
» help fix

FIX    Round towards zero.
       FIX(X) rounds the elements of X to the nearest integers
       towards zero.

       See also FLOOR, ROUND, CEIL.
```

Typing "help *operation*" where "*operation*" indicates any operation such as + will provide a list of all such operations and their characteristics.

Another handy way of getting help is to use the "lookfor" command. This will provide a list containing the word following "lookfor." For example,

```
» lookfor sort
CPLXPAIR Sort numbers into complex conjugate pairs.
SORT    Sort in ascending order.
SORTROWS Sort rows in ascending order.
EIGFUN Function to return sorted eigenvalues (used in GOALDEMO).
V2SORT Sorts two vectors and then removes missing elements.
DSORT  Sort complex discrete eigenvalues in descending order.
ESORT  Sort complex continuous eigenvalues in descending order.
PLOTRLOC Sorts points along root locus
SORTRLOC Sorts points along root locus
VSORT Matches two vectors.  Used in RLOCUS.
dlg_sort_uicontrols.m: %DLG_SORT_UICONTROLS( FIG )
fnd_objprop.m: %[OUT1, OUT2] = FND_OBJPROP( IN1 ) Function to perform
  initial sort and build object
objprop.m: %OBJPROPN  Function to perform initial sort and build object
```

Finally, help is available by the menu button "help" at the top right of the Command window. Under this drop down menu is included a table of contents (under which is a search function) and an index of all MATLAB terms alphabetically.

A-4 Programming Hints

Since MATLAB is an interpretive language, it behooves a programmer to spend some time thinking about ways to avoid execution of extra commands. This is particularly true of loops. There are often ways to avoid the use of loops in building up arrays or doing general computation. This procedure is illustrated by the programs of Chapter 3 where partial sums of Fourier series were computed by using the matrix multiply feature of MATLAB (see the MATLAB Application for Example 3-6, for example). This is also true of building arrays. Loops may often be avoided in such cases by using the append, flip, and indexing capabilities of MATLAB.

Another smart programming technique is to make effective use of functions for executing a series of commands over several times in a program. This is illustrated in the text by the special functions stp_fn, pls_fn, etc.

References

The Student Edition of MATLAB, Version 5 User's Guide, Upper Saddle River, NJ: Prentice Hall, 1998.
The Student Edition of SIMULINK, Version 2 User's Guide, Upper Saddle River, NJ: Prentice Hall, 1998.

B

Systematic Procedures for Writing Governing Equations for Lumped Systems

B-1 Introduction

In the analysis of simple systems composed of two or three components, it is usually possible to write down the governing equations by inspection. In more complicated cases, however, we require systematic procedures for obtaining the system equations. For lumped systems, this procedure consists of the following basic steps:

1. Identify the important system elements or components and determine their individual *describing equations* (sometimes referred to as the current-voltage relationships in the case of electrical systems); examples are Ohm's law, $v = iR$, for a resistor, or Coulomb's law, $v = q/C$, for a capacitor.
2. Write down the *combining equations*, which relate the variables of the individual elements to each other; examples are Kirchhoff's voltage and current laws, and application of d'Alembert's principle for mechanical systems.
3. Eliminate all variables that are not of interest by means of the combining equations. The variables of interest will include, but may not be limited to, the input and output variables. Quite often, other auxiliary variables will be involved.
4. Examine the model closely to determine if it is an adequate representation of the physical system. If stray capacitance, lead inductance, or nonlinear effects are of importance, these may be added to the model. Quite often, the first solution for the desired quantities will be carried out with the simplest model that is deemed adequate and refinements will be made in subsequent solutions.
5. Solve the governing equations of the system using a systematic method of solution.

We now discuss further the procedures involved in each of these steps. In order to be able to solve the resulting system equations analytically, it is almost a necessity (except in a few specific cases) that the describing equations be linear. Furthermore, many physical components are well approximated by linear relationships. Therefore, we limit our attention to system components that are described by linear relationships. The resulting system equations then satisfy the superposition principle.

B-2 Lumped, Linear Components and Their Describing Equations

The student is already familiar with the describing equations for lumped, linear circuit elements. These are the basic relations or laws relating the terminal currents and voltages for resistance, inductance, and capacitance. Each of these equations constitutes a model of the physical device by mathematically relating two variables that can be measured at the device terminals; in particular, the voltage *across* the device and the current *through* it. To generalize such describing equations to other physical components, such as mechanical or hydraulic components, it is useful to think of the terminal variables as either *across variables* or *through variables*. An *across* variable is one that is measured by placing an appropriate meter *across* the terminals of the device; a *through* variable is measured by placing an appropriate meter in series with the device so that the measured quantity is transmitted *through* the meter. Thus voltage is an across variable and current is a through variable. For a mechanical device such as a spring, the appropriate variables are force and displacement (or, possibly, the velocity) of one end with respect to the other. To measure the force transmitted by the spring to another object, we would need to place a force meter in series with the spring so that force is a *through* variable. Measurement of the displacement of one end relative to the other, on the other hand, would require that a ruler be placed across the spring. Thus displacement (and relative velocity, since it is the derivative of the displacement) is an *across* variable.

In writing down the describing equations for a device, some convention regarding sign must be adopted. Each device is represented by a schematic diagram symbol with reference directions shown for the through and across variables. Consider, as an example, a resistance of value R, the schematic representation of which, together with reference directions for through (current) and across (voltage) variables, is shown in Figure B-1. The describing equation is Ohm's law, given by

$$v = iR \tag{B-1}$$

The convention we will adopt regarding signs for the through and across variables is the following. To measure the through variable (current), we must insert an instantaneous reading meter in series with the device. If the actual through variable flow is in the direction of the reference arrow on the schematic representation of the device, the meter will read a positive value for the through variable. For the across variable (voltage), the convention is that the meter is connected so that the product of instantaneous across- and through-variable meter readings is *positive* when power is *delivered* to the device.

The describing equations for several types of components are summarized in Table B-1. The terminology and conventions regarding signs is that described above. Symbols that are used in schematic diagrams showing interconnection of various devices are also shown.

In addition to the passive elements defined in Table B-1, ideal sources are also useful models. These include both independent and dependent sources. For electrical systems, the two types of independent sources are constant-voltage and constant-current sources; for translational mechanical systems, the two types of independent sources are constant-velocity and constant-force sources, with analogous sources defined for rotational mechanical systems. By this terminology, of course, we do not mean that a source delivers a time-independent quantity, be it electrical or mechanical, but rather that the delivered quantity does not depend on the load placed on the source. For example, an ideal velocity source given by

$$\dot{\Delta}_s(t) = A \cos \omega_0 t \tag{B-2}$$

maintains the velocity $A \cos \omega_0 t$ regardless of the forces applied.

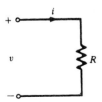

FIGURE B-1.   Schematic representation of a resistor, together with reference directions for the associated through and across variables.

TABLE B-1
Electrical, Translational Mechanical, and Rotational Mechanical Elements

Component; Symbol (Units)	Schematic Symbol	Across Variable; Units	Through Variable; Units	Describing Equations
Resistance, R (ohms)		Voltage (volts)	Current (amperes)	$v = iR$ $i = \dfrac{1}{R} v$
Inductance, L (henries)		Voltage (volts)	Current (amperes)	$v = L \dfrac{di}{dt}$ $i = \dfrac{1}{L} \displaystyle\int_{-\infty}^{t} v(\lambda)\, d\lambda$
Capacitance, C (farads)		Voltage (volts)	Current (amperes)	$v = \dfrac{1}{C} \displaystyle\int_{-\infty}^{t} i(\lambda)\, d\lambda$ $i = C \dfrac{dv}{dt}$
Translational damper, B (kg/s)		Velocity (m/s)	Force (newtons)	$\dot{\Delta} = \dfrac{1}{B} f$ $f = B\,\dot{\Delta}$
Translational spring, K (kg/s²)		Velocity (m/s)	Force (newtons)	$\dot{\Delta} = \dfrac{1}{K} \dfrac{df}{dt}$ $f = K\Delta$ $= K \displaystyle\int_{-\infty}^{t} \dot{\Delta}(\lambda)\, d\lambda$
Mass (inertance), M (kg)		Velocity (m/s) Note: The mass element is always grounded.	Force (newtons)	$\dot{\Delta} = \dfrac{1}{M} \displaystyle\int_{-\infty}^{t} f(\lambda)\, d\lambda$ $f = M \dfrac{d\dot{\Delta}}{dt}$
Rotational damper, B (kg − m²/s)		Angular velocity (rad/s)	Torque (N − m)	$\dot{\theta} = \dfrac{i}{B} T$ $T = B\,\dot{\theta}$
Rotational spring, K (N − m)		Angular velocity (rad/s)	Torque (N − m)	$\dot{\theta} = \dfrac{1}{K} \dfrac{dT}{dt}$ $T = K \displaystyle\int_{-\infty}^{t} \dot{\theta}(\lambda)\, d\lambda$

TABLE B-1 (*Continued*)

Component; Symbol (Units)	Schematic Symbol	Across Variable; Units	Through Variable; Units	Describing Equations
Rotational inertance, J ($kg - m^2$) (moment of inertia)		Angular velocity (rad/s) Note: Rotational inertances are always grounded.	Torque ($N - m$)	$$\dot{\theta} = \frac{1}{J} \int_{-\infty}^{t} T(\lambda)\, d\lambda$$ $$T = J \frac{d\dot{\theta}}{dt}$$

A controlled source produces a terminal quantity whose value is proportional to some other system variable but is independent of the load placed on the source. For example, an ideal angular velocity source that is controlled by the current $i_1(t)$ can be represented by the equation

$$\dot{\theta}_s(t) = KF[i_1(t)] \tag{B-3}$$

independent of the load placed on the source, where $F(\cdot)$ is a specified function and K is a constant of proportionality.

EXAMPLE B-1

To illustrate some of the ideas introduced here and demonstrate, at least in a particular case, that the analogies given in Table B-1 are valid, consider the simple mechanical system shown in Figure B-2. The mass has force F_s acting on it and is assumed to move on rollers without friction. Its displacement from equilibrium is Δ as shown. From d'Alembert's principle we have

$$M \frac{d\dot{\Delta}}{dt} + B\dot{\Delta} + K\Delta = F_s \tag{B-4}$$

where B is the coefficient of viscous friction describing the damper, as shown in Table B-1. We wish to show that by suitable association of elements and variables, this equation may be represented by an electrical network. First we rewrite the equation as

$$M \frac{d\dot{\Delta}}{dt} + B\dot{\Delta} + K \int \dot{\Delta}\, dt = F_s \tag{B-5}$$

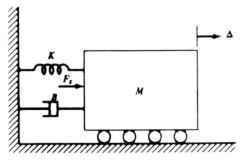

FIGURE B-2. Schematic diagram of a simple mechanical system.

We have chosen to write the equation in terms of velocity, $\dot{\Delta}$, since, as we shall see, it is analogous to voltage. Now consider the parallel RLC network shown in Figure B-3. From KCL and the element relations we obtain the differential equation for the voltage v as

$$C\frac{dv}{dt} + \frac{v}{R} + \frac{1}{L}\int v \, dt = I_s \tag{B-6}$$

Examination of (B-5) and (B-6) shows that the equations are exactly the same if we make the following associations:

$$M \longleftrightarrow C$$

$$B \longleftrightarrow \frac{1}{R}$$

$$K \longleftrightarrow \frac{1}{L} \tag{B-7}$$

$$\dot{\Delta} \longleftrightarrow v$$

$$F_s \longleftrightarrow I_s$$

Since the mechanical and electrical systems are described by the same differential equation, we say that the electrical network and the mechanical system are *analogous* to each other; in particular:

1. Mass M is analogous to capacitance C.
2. Coefficient of friction B is analogous to conductance $G = B = 1/R$.
3. Spring constant K is analogous to the inverse of inductance $1/L = K$.
4. Across-variable velocity is analogous to across-variable voltage.
5. Through-variable force is analogous to through-variable current.

Analogies also exist for other types of systems, as suggested by Table B-1. Clearly, such analogies can aid electrical engineers in understanding and analyzing other kinds of systems. The power of the analogous-network concept lies in the fact that the analysis of electrical networks is particularly well established. Therefore, complicated networks can be handled with relative ease. If the concept of analogous networks can be extended to more complex systems, we will indeed have a powerful tool for systems analysis. To do this we need systematic procedures for obtaining the analogous network or, perhaps more conveniently, the system graph. These topics are examined in more detail in the following section.

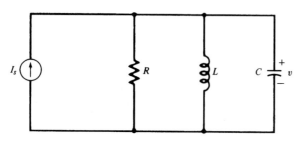

FIGURE B-3. Schematic diagram of a parallel RLC network.

B-3 System Graphs

The concept of a system graph, although not absolutely essential for the writing of equilibrium equations for many lumped systems, nevertheless provides a systematic procedure that one may resort to in cases where these equations are not readily written down by inspection. The graph of a system of two-terminal components is obtained by simply representing each component by a directed line segment. The direction, indicated by an arrow, shows the positive reference sense for the across and through variables of the component. Each line segment is terminated by two nodes. An example is given in Figure B-4, which shows the schematic representation of an electrical system and its graph. The system graph of an electrical network bears a resemblance to the shape of the schematic diagram of the original circuit, a characteristic that is not true of mechanical or other systems.

The describing equations[†] for the components of the system of Figure B-4 are:

$$v_1(t) = v_s(t) \qquad \text{(a)}$$
$$i_2(t) = 0, \quad \text{switch open} \qquad \text{(b)}$$
$$v_2(t) = 0, \quad \text{switch closed} \qquad \text{(c)}$$
$$v_3(t) = R_1 i_3(t) \qquad \text{(d)} \qquad\qquad \text{(B-8)}$$
$$v_4(t) = R_2 i_4(t) \qquad \text{(e)}$$
$$v_5(t) = L \frac{d i_5}{dt} \qquad \text{(f)}$$

and

$$v_6(t) = \frac{1}{C} \int_{-\infty}^{t} i_6(\lambda)\, d\lambda \qquad \text{(g)}$$

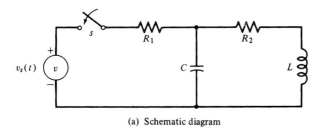

(a) Schematic diagram

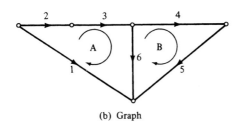

(b) Graph

FIGURE B-4. Example that illustrates the concept of a system graph of an electrical circuit. (In (b), numbers for branches are arbitrarily assigned.)

[†]Subscripts refer to the numbering of the graph branches.

Note that the component describing equations may be linear or nonlinear; there is no restriction from the standpoint of constructing the system graph.

As a second example, consider the translational mechanical system shown schematically in Figure B-5, which represents the suspension system on an automobile. The mass M_1 represents the mass of the automobile, which is coupled to the axle and wheels by a spring and damper system. The smaller mass, M_2, of the axle and wheels is coupled to the ground by the tires, which are represented by a second spring and damper system. The automobile frame is driven by a force generator $f_s(t)$. The system graph is shown in Figure B-5b, where it is noted that the component graph for the masses has nodes located by the mass and "earth." The terminal equations for the system components are, from Table B-1, given by

$$f_1(t) = f_s(t) \qquad \text{(a)}$$

$$f_2(t) = B_1 \dot{\Delta}_2(t) \qquad \text{(b)}$$

$$f_3(t) = K_1 \Delta_3(t) \qquad \text{(c)}$$

$$f_4(t) = M_1 \frac{d\dot{\Delta}_4(t)}{dt} \qquad \text{(d)} \qquad\qquad \text{(B-9)}$$

$$f_5(t) = B_2 \dot{\Delta}_5(t) \qquad \text{(e)}$$

$$f_6(t) = K_2 \Delta_6(t) \qquad \text{(f)}$$

and

$$f_7(t) = M_2 \frac{d\dot{\Delta}_7(t)}{dt} \qquad \text{(g)}$$

Node and Circuit Postulates

We next need a procedure for combining component equations to obtain the system equations. A basis for this is provided by graph theory, a few basic definitions for which are given below.

1. Each line element of a system graph is called a *branch* (also referred to as an edge in some textbooks). An indicated direction on a branch makes it an *oriented branch*.

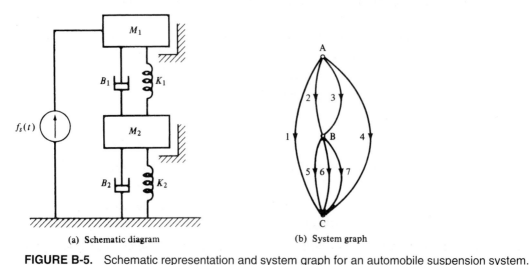

(a) Schematic diagram   (b) System graph

FIGURE B-5. Schematic representation and system graph for an automobile suspension system.

2. A *node* is the end point of a branch.

3. An *oriented graph* is a set of oriented branches, no two of which have a point in common which is not a node.

4. A *subgraph* is any subset of the branches of a graph.

5. A *circuit* is a subgraph such that two and only two branches are incident on each node of the subgraph.

6. The *complement* $\overline{B}$ of a subgraph G is the subgraph remaining in G when the branches of $\overline{B}$ are removed.

7. A connected subgraph of a graph is called a *separate part* if it contains no nodes in common with its complement.

With these definitions stated, we can now give two fundamental properties that relate across and through variables of the elements of a system graph.

1. *The Node Postulate.* Let the system graph of a physical system be composed of n oriented branches. Let $u_j(t)$ be the through variable of the jth element. Then at the kth node of the graph

$$\sum_{j=1}^{n} a_j u_j(t) = 0 \qquad \text{(B-10)}$$

where

$$a_j = \begin{cases} 0 & \text{if the } j\text{th branch is not incident at the } k\text{th node} \\ 1 & \text{if the } j\text{th branch is oriented away from the } k\text{th node} \\ -1 & \text{if the } j\text{th branch is oriented toward the } k\text{th node} \end{cases}$$

We recognize Postulate 1 simply as corresponding to a general statement of Kirchhoff's current law (KCL) for electrical systems. The physical principle on which it is based is the conservation of charge. For translational mechanical systems, Postulate 1 states that the force components along a coordinate direction of each system element connected at a common point must sum algebraically to zero. For example, in Figure B-5, the damper B_1, spring K_1, mass M_1, and force generator $f_s(t)$ all must exert a net force of zero at their common juncture. We recognize this as d'Alembert's principle.

To state the second fundamental property of a system graph, we need to define the concept of *circuit orientation,* which is assigned by an oriented arrow paralleling the circuit. Examples of oriented arrows are shown in Figure B-4b labeled A and B.

2. *The Circuit Postulate.* Let the linear graph of a physical system contain n oriented branches and let $v_j(t)$ represent the across variable of the jth branch; then for the kth circuit

$$\sum_{j=1}^{n} b_j v_j(t) = 0 \qquad \text{(B-11)}$$

where

$$b_j = \begin{cases} 0 & \text{if the } j\text{th branch is not included in the } k\text{th circuit} \\ 1 & \text{if the orientation of the } j\text{th branch is the same as the orientation chosen for the } k\text{th circuit} \\ -1 & \text{if the orientation of the } j\text{th branch is opposite to that of the } k\text{th circuit} \end{cases}$$

Postulate 2 corresponds to Kirchhoff's voltage law (KVL) for electrical systems. The physical principle on which it is based is conservation of energy. For translational mechanical systems, Postulate 2 states that the oriented sum of each component of the vectorial velocity around a circuit of the system graph must sum to zero.

As an example of writing node equations (Postulate 1), we consider the mechanical system and system graph of Figure B-5. Considering node A, and letting the corresponding through variables be $f_1(t)$, $f_2(t), \ldots$, we have

$$f_1(t) + f_2(t) + f_3(t) + f_4(t) = 0 \qquad \text{(node A)} \qquad \text{(B-12a)}$$

For node B, we have

$$f_5(t) + f_6(t) + f_7(t) - f_2(t) - f_3(t) = 0 \qquad \text{(node B)} \qquad \text{(B-12b)}$$

and for node C, we have

$$-f_1(t) - f_5(t) - f_6(t) - f_7(t) - f_4(t) = 0 \qquad \text{(node C)} \qquad \text{(B-12c)}$$

We note that the three equations are not independent; for example, adding the third to the second gives the negative of the first. Therefore, we choose the first two equations. We may eliminate the through variables in each equation in favor of the across variables by using the describing equations. Note that by doing this we do not decrease the number of unknowns to be equal to the number of equations. However, suppose that we define the velocities of nodes A, B, and C relative to a common reference. Since we have eliminated node C as a Node Postulate equation, we choose its velocity as our reference and let it be $v_C(t) = 0$. Then

$$\dot{\Delta}_2(t) = v_A(t) - v_B(t) = \dot{\Delta}_3(t) \qquad \text{(a)}$$
$$\dot{\Delta}_4(t) = v_A(t) \qquad \text{(b)} \qquad \text{(B-13)}$$
$$\dot{\Delta}_5(t) = v_B(t) = \dot{\Delta}_6(t) = \dot{\Delta}_7(t) \qquad \text{(c)}$$

In terms of $v_A(t)$ and $v_B(t)$, the terminal equations become

$$f_1(t) = -f_s(t) \text{ (unknown)} \qquad \text{(a)}$$
$$f_2(t) = B_1(v_A - v_B) \qquad \text{(b)}$$
$$f_3(t) = K_1 \int_{-\infty}^{t} (v_A - v_B) \, d\lambda \qquad \text{(c)}$$
$$f_4(t) = M_1 \frac{dv_A}{dt} \qquad \text{(d)} \qquad \text{(B-14)}$$
$$f_5(t) = B_2 v_B \qquad \text{(e)}$$
$$f_6(t) = K_2 \int_{-\infty}^{t} v_B \, d\lambda \qquad \text{(f)}$$
$$f_7(t) = M_2 \frac{dv_B}{dt} \qquad \text{(g)}$$

Substituting these into the node A and node B equations, we obtain

$$-f_s + B_1(v_A - v_B) + K_1 \int_{-\infty}^{t} (v_A - v_B) \, d\lambda + M_1 \frac{dv_A}{dt} = 0 \qquad \text{(a)}$$
$$\qquad \qquad \qquad \qquad \qquad \qquad \qquad \qquad \qquad \qquad \qquad \qquad \qquad \text{(B-15)}$$
$$B_2 v_B + K_2 \int_{-\infty}^{t} v_B \, d\lambda + M_2 \frac{dv_B}{dt} - B_1(v_A - v_B) - K_1 \int_{-\infty}^{t} (v_A - v_B) \, d\lambda = 0 \qquad \text{(b)}$$

Rearranging and differentiating, we have

$$M_1 \frac{d^2 v_A}{dt^2} + B_1 \frac{dv_A}{dt} + K_1 v_A - B_1 \frac{dv_B}{dt} - K_1 v_B = \frac{df_s}{dt} \qquad \text{(a)}$$

$$-B_1 \frac{dv_A}{dt} - K_1 v_A + M_2 \frac{d^2 v_B}{dt^2} + (B_1 + B_2) \frac{dv_B}{dt} + (K_1 + K_2) v_B = 0 \qquad \text{(b)}$$

(B-16)

If we wish, we may eliminate $v_A(t)$ or $v_B(t)$ to obtain a differential equation involving one unknown variable alone. The elimination is not trivial, however. A much easier method for elimination and solution is discussed in Chapter 6, which deals with applications of the Laplace transform.

As an example of writing circuit equations (Postulate 2), we consider the circuit and system graph of Figure B-4. The circuit equation for A is

$$-v_1 + v_2 + v_3 + v_6 = 0 \qquad \text{(B-17a)}$$

and for B the circuit equation is

$$v_4 + v_5 - v_6 = 0 \qquad \text{(B-17b)}$$

where $v_1, v_2, \ldots$ are the across variables for branches $1, 2, \ldots$. We may eliminate the branch across variables in terms of through variables to obtain

$$-v_s + 0 + R_1 i_3 + \frac{1}{C} \int_{-\infty}^{t} i_6 \, d\lambda = 0 \qquad \text{(a)}$$

$$R_2 i_4 + L \frac{di_5}{dt} - \frac{1}{C} \int_{-\infty}^{t} i_6 \, d\lambda = 0 \qquad \text{(b)}$$

(B-18)

where it is assumed that the switch is closed. We note that

$$i_2 = i_A = i_3, \qquad i_4 = i_B = i_5$$

and

$$i_6 = i_A - i_B$$

When written in terms of the circuit currents, the Circuit Postulate equations are

$$R_1 \frac{di_A}{dt} + \frac{1}{C} i_A - \frac{1}{C} i_B = \frac{dv_s}{dt} \qquad \text{(a)}$$

$$-\frac{1}{C} i_A + L \frac{d^2 i_B}{dt^2} + R_2 \frac{di_B}{dt} + \frac{1}{C} i_B = 0 \qquad \text{(b)}$$

(B-19)

where each equation has been differentiated to eliminate the integrals. As with the mechanical system example, either i_A or i_B can be eliminated to obtain an equation involving one unknown variable. The order of the resulting equation would be second. The elimination of variables is left to the problems.

Topology of System Graphs

In order to select an independent set of node or circuit equations for a system, we introduce some basic concepts of graph topology. A subgraph has been defined as any subset of the branches of a graph. A subgraph that is of basic importance in determining a set of independent node or circuit equations for a system is a *tree*. A tree of a connected graph with N_v nodes has the following properties:

1. All nodes of the graph are included in the tree, and no nodes are left in isolated positions.
2. A tree contains $N_v - 1$ branches.
3. A tree includes no closed paths.

Those branches removed from a graph in forming a tree are called *chords* or *links*.

There are a number of different possible trees for a given graph. Several examples of trees for the graph of Figure B-4b are shown in Figure B-6.

The number of branches in a tree is obviously related to the number of nodes in the graph (and the tree), N_v, by

$$\text{number of branches in a tree} = N_v - 1 \qquad \text{(B-20)}$$

The number of chords, N_c, in a graph is the difference between the total number, N_b, of branches in the graph and the number of branches in a tree. Thus from (B-20) we obtain

$$N_c = N_b - (N_v - 1)$$
$$= N_b - N_v + 1 \qquad \text{(B-21)}$$

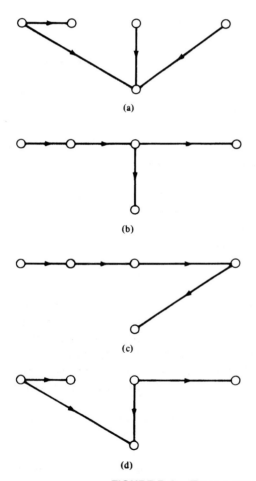

(a)

(b)

(c)

(d)

FIGURE B-6. Trees corresponding to the graph of Figure B-4b.

For a graph with P separate parts,

$$N_c = N_b - N_v + P \qquad \text{(B-22)}$$

Finally, the cotree consists of those branches not included in the tree.

EXAMPLE B-2

The number of branches in the graph of Figure B-4b is

$$N_b = 6$$

and the number of nodes is

$$N_v = 5$$

The number of chords is

$$N_c = 6 - 5 + 1 = 2$$

The number of branches in the tree is

$$N_{tb} = 6 - 2 = 4 = N_v - 1$$

which checks with (B-20).

We now use these ideas from the topology of graphs to establish the number of independent equations that can be written for a system.

B-4 Choosing Independent Equations for a System

We assume that each element of a schematic diagram of a system is represented by one branch in the system graph. Actually, there is no need to represent the switch of Figure B-4b by a branch if $t > 0$ (switch closed). Also, we assume that across-variable sources (voltage sources in the case of electrical systems) appear in series with other elements and that through-variable sources (current sources for electrical systems) appear in parallel with other elements.

The procedure we may follow is to use the Node Postulate (1) to write equations for $N_v - 1$ nodes after one has been chosen as a reference. We can then write, in addition, $N_b - (N_v - 1)$ equations using Circuit Postulate (2), giving a total of N_b equations. Since the Node Postulate equations involve through variables and the Circuit Postulate equations involve across variables, we will have $2N_b$ variables and only N_b equations. However, the through and across variables for each system element are related by a describing equation and we may reduce the number of variables from $2N_b$ to N_b. Hence, if the N_b equations obtained by applying Postulates 1 and 2 are independent, we may, in principle, solve for the N_b unknown through or across variables, depending on which ones remain after the describing equations have been applied.

It is perhaps a good thing at this point to review what is meant by independence of a set of equations. The equations

$$
\begin{aligned}
x + y &= 5 \qquad \text{(a)} \\
2x - 2y &= 10 \qquad \text{(b)}
\end{aligned}
\qquad \text{(B-23)}
$$

are independent, whereas the pair of equations

$$x + y = 5 \quad \text{(a)}$$
$$2x + 2y = 10 \quad \text{(b)}$$

(B-24)

are not. After some reflection, the reason for the latter set not being independent is clear. The second equation can be obtained from the first by multiplying through by two. Geometrically, both sets of equations represent lines in the x–y plane, the latter pair representing a pair of colinear lines, whereas the former set of equations represent a pair of lines that are perpendicular. Thus, for the first set, we may obtain a solution for x and y that corresponds to the crossing point; a unique solution is impossible to obtain for the second set.

We now turn to the problem of choosing a set of independent variables in terms of which to express the system equations. To keep the solution of these equations as simple as possible, we want the minimum number of independent variables that can be chosen. For simplicity, discussion will center on the choice of voltage and current variables for electrical systems, although it will apply equally well to the choice of independent through and across variables for mechanical systems.

To proceed, consider Figure B-7, which is the tree shown in Figure B-6d for the network of Figure B-4. Assume that node e is chosen as the reference, or datum, node. The voltages of nodes a, b, c, and d measured with respect to node e are defined as v_a, v_b, v_c, and v_d, respectively. Also shown in Figure B-7 are the branch voltages (across variables), with polarities indicated by plus and minus signs to emphasize the meaning of the reference arrows shown on the tree branches. Now each node voltage can be expressed in terms of tree branch voltages. For example, v_b is given by

$$v_b = -v_1 + v_2 \quad \text{(a)}$$

and v_c by

(B-25)

$$v_c = -v_3 + v_4 \quad \text{(b)}$$

Conversely, if we know the node voltages, we may obtain the tree branch voltages and, as a matter of fact, any branch voltage. Hence a KCL equation (Postulate 1) written for each of $N_v - 1$ nodes in terms of node voltages by using the describing equation of each element will provide a set of $N_v - 1$ independent equations in terms of the $N_v - 1$ node voltages relative to the reference node. Note that $N_v - 1 < N_b$. Other choices are also available for determining the branch voltages, but these will not be discussed here.

We now investigate the possibility for the choice of current variables fewer in number than the branch currents. From Figure B-7 we note that insertion of the link labeled 5 will allow the link current i_5 to flow. The loop current i_A is thereby established. Similarly, insertion of the link labeled 6 will allow link current i_6 to flow and the loop current i_B is thereby established. A little thought results in the conclusion that any of the branch currents can be expressed in terms of the loop currents i_A and i_B. For example,

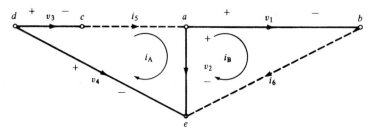

FIGURE B-7. Tree showing appropriate voltage and current variables to establish independent through and across variables.

$$i_2 = i_A - i_B \tag{B-26}$$

Thus the link currents and the loop currents defined thereby constitute an independent set of variables. We may write

$$N_c = N_b - N_v + 1 \tag{B-27}$$

KVL (Postulate 2) equations in terms of these N_c loop currents by employing the describing equations to eliminate the across variables for each element in favor of their through variables (branch currents in this discussion) and, in turn, express these branch currents in terms of loop currents defined by the link currents. Thus we will have $N_c = N_b - N_v + 1$ independent equations involving N_c loop current variables.

Usually, it will not be necessary to construct a tree to obtain an independent set of KCL or KVL equations. For example, we note that the circuit (KVL) equations written in terms of loop currents defined by the windows or meshes of a planar network will always give an independent set of equations. The node equations (KCL) written in terms of the $N_v - 1$ node voltages with one node used as a reference also result in a set of independent equations. Whether one chooses loop current equations or node voltage equations usually depends on whether $N_c = N_b - N_v + 1$ or $N_v - 1$ is smaller.

We close this section with an example that illustrates the use of the concepts introduced in this section.

EXAMPLE B-3

Consider the cubic array of 1-Ω resistors shown in Figure B-8a and its associated graph shown in Figure B-8b. Figure B-8c shows one possible tree. The links are the branches numbered 2, 8, 9, 11, 12, and 14. Insertion of these links establishes loop currents $i_A, i_B, \ldots, i_F$, respectively.

Suppose that we desire v_o with $v_I = 2$ V. To minimize the chance for error, we need a systematic procedure to solve for V_o. In order to conveniently relate branch currents and loop currents, we construct Table B-2, which is called a *tie-set schedule*. The method of construction is the following. If a loop current established by the insertion of a link includes the ith branch of the graph, a $+1$ is entered in column i of the row corresponding to that loop current if branch and loop currents are in the same direction; a -1 is entered if they are in opposite directions; a 0 is entered if the loop does not include the ith branch. The rows of the tie-set schedule give us the KVL equations around each loop in terms of branch voltages. For example, row A of Table B-2 tells us that

$$-v_1 + v_2 + v_3 = 0 \tag{B-28}$$

The columns of the tie-set schedule yield the branch currents in terms of the loop currents. For example, column 1 of Table B-2 tells us that

$$i_1 = -i_A \tag{B-29}$$

while column 5 gives us

$$i_5 = -i_B + i_E \tag{B-30}$$

The procedure in obtaining the KVL equations around each circuit defined by the link currents is then the following:

1. Write down the voltage-law equations for each loop in terms of the branch voltages.
2. Substitute for each branch voltage the describing equation for each branch with the branch current as the independent variable.
3. Substitute for the branch currents the expressions involving loop currents obtained from the columns of the tie-set schedule.

The result is $N_b - N_v + 1$ equations in terms of the $N_b - N_v + 1$ loop current variables.

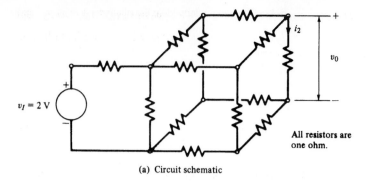

(a) Circuit schematic

$v_I = 2$ V

i_2

v_0

All resistors are one ohm.

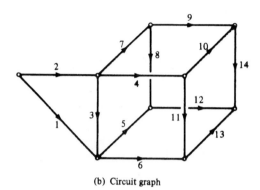

(b) Circuit graph

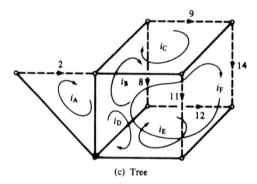

(c) Tree

FIGURE B-8. Circuit showing the selection of loop currents using a tree.

TABLE B-2
Tie-Set Schedule for Circuit of Figure B-8

Loops						Branch Number								
	1	2	3	4	5	6	7	8	9	10	11	12	13	14
A	−1	1	1	0	0	0	0	0	0	0	0	0	0	0
B	0	0	−1	0	−1	0	1	1	0	0	0	0	0	0
C	0	0	0	−1	0	0	1	0	1	−1	0	0	0	0
D	0	0	−1	1	0	−1	0	0	0	0	1	0	0	0
E	0	0	0	0	1	−1	0	0	0	0	0	1	−1	0
F	0	0	−1	1	0	−1	0	0	0	1	0	0	−1	1

560

In the present example, all components are 1-Ω resistors, so that these steps are particularly simple. Thus since $v_1 = 2$ V, the row A equation becomes

$$v_2 + v_3 = 2 \tag{B-31}$$

or, in terms of branch currents,

$$i_2 + i_3 = 2 \tag{B-32}$$

But columns 2 and 3 give us

$$i_2 = i_A \quad \text{and} \quad i_3 = i_A - i_B - i_D - i_F \tag{B-33}$$

Thus

$$i_A + (i_A - i_B - i_D - i_F) = 2 \tag{B-34}$$

or

$$2i_A - i_B - i_D - i_F = 2 \tag{B-35}$$

The student should obtain the remaining five equations.

B-5 Writing State Equations for Lumped Linear Systems Using Graph Theory

It is useful to have systematic procedures for writing state equations (see Chapter 7) for lumped linear systems. In this section, methods for obtaining the state equations, or normal form equations, for a lumped linear physical system from its graph are considered. One advantage of such equations is that they are ideally suited for computer solution. For a lumped linear system, the state equations have the form

$$\dot{\mathbf{x}}(t) = \mathbf{A}\mathbf{x}(t) + \mathbf{B}\mathbf{u}(t) \tag{B-36}$$

where **A** and **B** are matrices of constants which depend on the system parameters,

$$\mathbf{x}(t) = [x_1(t), x_2(t), \ldots, x_n(t)]^T \tag{B-37}$$

is the *state* vector, and

$$\mathbf{u}(t) = [u_1(t), u_2(t), \ldots, u_m(t)]^T \tag{B-38}$$

is a forcing function, or input, vector (see Appendix D for definitions and operations pertaining to matrices).

Before going through the procedure, we need to define the concept of a *normal tree*.

Normal Tree A tree that contains all *independent voltage (velocity) sources,* no *independent current (force) sources, the* maximum *possible number of capacitances (masses), and the* minimum *possible number of inductances (springs).*

If a normal tree can be found for a lumped, linear system graph, the equilibrium equations of the system can be written in the form (B-36) with the *minimal* number of state variables using the following procedure:

1. Choose a normal tree for the network under consideration.
2. Take either the voltages or the charges across the tree-branch capacitors and either the currents or the fluxes through the cotree-chord (link) inductors as the state variables.

3. Write the independent KCL equations, fundamental loop equations, and branch voltage current relations.
4. Eliminate all the network variables except the state variables chosen in Step 2 from the equation obtained in Step 3.
5. Rearrange the equations obtained in Step 4 and put them in the form of (B-36).

An example will illustrate the procedure.

EXAMPLE B-5

Consider the circuit shown in Fig. B-9a. A normal tree is shown in Fig. B-9b (this is one of the four possible normal trees). It is clear that the state variables are $x_1 = v_3$ and $x_2 = i_5$. Writing KCL equations at nodes n_1 and n_2 gives

$$i_4 - i_1 - x_2 = 0 \tag{B-39}$$

and

$$x_2 - i_2 - i_6 = 0 \tag{B-40}$$

respectively. Writing KVL equations by inserting links l_1, l_2, and l_3 in turn gives

$$V_1 - v_4 - v_1 = 0 \tag{B-41}$$

$$V_1 - v_4 - v_5 - V_2 - x_1 = 0 \tag{B-42}$$

$$v_2 - V_2 - x_1 = 0 \tag{B-43}$$

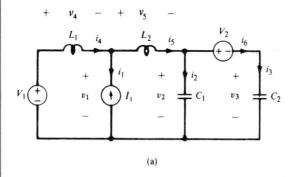

(a)

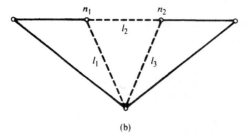

(b)

FIGURE B-9. The circuit for Example B-5 (a) and a normal tree for the circuit (b).

The voltage-current relationships for the various branches are

$$i_1 = -I_1 \tag{B-44}$$

$$v_4 = L_1 \frac{di_4}{dt} \tag{B-45}$$

$$v_5 = L_2 \frac{dx_2}{dt} \tag{B-46}$$

$$i_2 = C_1 \frac{dv_2}{dt} \tag{B-47}$$

$$i_6 = i_3 = C_2 \frac{dx_1}{dt} \tag{B-48}$$

Differentiating (B-43) yields

$$\frac{dv_2}{dt} - \frac{dV_2}{dt} - \frac{dx_1}{dt} = 0 \tag{B-49}$$

From (B-40) with (B-48) substituted, we get

$$x_2 - C_1 \frac{dv_2}{dt} - C_2 \frac{dx_1}{dt} = 0 \tag{B-50}$$

Eliminating dv_2/dt between (B-49) and (B-50), we obtain

$$x_2 - C_1 \frac{dV_2}{dt} - C_1 \frac{dx_1}{dt} - C_2 \frac{dx_1}{dt} = 0 \tag{B-51}$$

or

$$\frac{dx_1}{dt} = \frac{1}{C_1 + C_2} x_2 - \frac{C_1}{C_1 + C_2} \frac{dV_2}{dt} \tag{B-52}$$

which is the first state equation with input $u_3 \triangleq dV_2/dt$.

Substituting (B-45) and (B-46) into (B-42), we get

$$V_1 - L_1 \frac{di_4}{dt} - L_2 \frac{dx_2}{dt} - V_2 - x_1 = 0 \tag{B-53}$$

From (B-39) and (B-44) we have

$$i_4 = i_1 + x_2 = -I_1 + x_2 \tag{B-54}$$

When (B-54) is substituted into (B-53) and the result rearranged, we obtain

$$\frac{dx_2}{dt} = -\frac{1}{L_1 + L_2} x_1 + \frac{1}{L_1 + L_2} (V_1 - V_2) + \frac{L_1}{L_1 + L_2} \frac{dI_1}{dt} \tag{B-55}$$

which is the second state equation with inputs $u_1 = V_1$, $u_2 = V_2$, and $u_4 = dI_1/dt$.

Further Reading

W. A. BLACKWELL, *Mathematical Modeling of Physical Networks*. New York: Macmillan, 1968. Provides a unified and complete approach to the analysis of linear systems.

C. M. CLOSE AND D. K. FREDERICK, *Modeling and Analysis of Dynamic Systems*. Boston: Houghton Mifflin Company, 1978. Includes modeling of electrical, mechanical, electromechanical, thermal, and hydraulic systems.

S. C. GUPTA, J. W. BAYLESS, AND B. PEIKARI, *Circuit Analysis with Computer Applications to Problem Solving*. Scranton: Intext Educational Publishers, 1972. Provides a concise treatment of graph theory applications to solving network variables.

R. G. BUSACKER AND T. L. SAATY, *Finite Graphs and Networks: An Introduction with Applications*. New York: McGraw-Hill Book Company, 1965. A thorough treatment of graph theory with a coverage of applications to many areas, including the analysis of physical systems.

R. J. WILSON, *Introduction to Graph Theory*. New York: Academic Press, 1978. A concise introduction to graph theory with some applications included.

E. J. HENLEY AND R. A. WILLIAMS, *Graph Theory in Modern Engineering*. New York: Academic Press, 1973. Basic concepts of flow graphs with applications, including computer programs.

A. GIBBONS, *Algorithmic Graph Theory*, 1985, Cambridge: Cambridge University Press. As its title suggests, an algorithmic approach to graph theory.

Problems

B-1. Assume an applied sinusoidal force of the form $f(t) = A \cos \omega_0 t$ for the translational mechanical elements of Table B-1. The average power delivered to the device is

$$P_{av} = \frac{1}{T_0} \int_0^{T_0} f(t) \frac{d\Delta(t)}{dt} \, dt \qquad \text{where} \quad T = \frac{2\pi}{\omega_0}$$

(a) Show that the average power delivered to the translational damper is $A^2/2B$ watts.

(b) Show that the average power delivered to the spring and mass elements is zero.

B-2. Assume an applied sinusoidal torque of the form $T(t) = A \cos \omega_0 t$ for the rotational mechanical elements of Table B-1. The average power delivered to the device is

$$P_{av} = \frac{1}{T_0} \int_0^{T_0} T(t) \frac{d\theta(t)}{dt} \, dt \qquad \text{where} \quad \omega_0 = \frac{2\pi}{T_0}$$

(a) Show that the average power delivered to the rotational damper is $A^2/2B$ watts.

(b) Show that P_{av} for the rotational spring and mass is zero.

B-3. Eliminate $i_B(t)$ between (B-19a) and (B-19b) so that one equation in $i_A(t)$ is obtained.

B-4. For each of the system schematics in Figure PB-4, draw the system graph, choose a tree, and write down the Node and Circuit Postulate equations, (B-10) and (B-11).

B-5. For System 1 of Problem B-4, let C be the reference node with velocity zero. Write terminal equations for all elements in terms of V_A and V_B, the velocities of nodes A and B relative to node C. Express the Node Postulate equations in terms of V_A and V_B. How many independent equations are there?

B-6. For System 2 of Problem B-4, write terminal equations for all elements in terms of appropriately chosen loop currents. Substitute into the Circuit Postulate equations obtained in Problem B-4. How many independent Circuit Postulate equations are there?

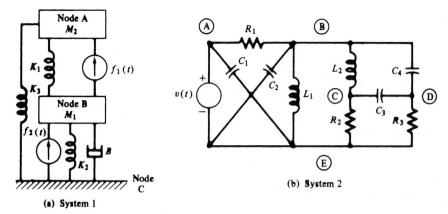

(a) System 1

(b) System 2

FIGURE PB-4

B-7. (a) Sketch trees for each of the system graphs in Figure PB-7.

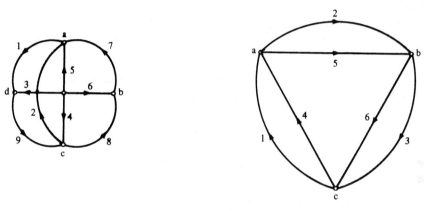

FIGURE PB-7

(b) Assuming that electrical networks are represented, compare the number of independent Postulate 1 (KCL) equations that can be written in terms of node voltages with the number of independent Postulate 2 (KVL) equations that can be written in terms of loop currents.

(c) Assume that link 1 for both parts (a) and (b) corresponds to a voltage source $v_s(t)$, and that all other links are 1-Ω resistors. Write down the KVL equations for each network in terms of loop currents determined by your choice of trees.

(d) Assume that link 1 for both parts (a) and (b) represents a current source $i_s(t)$, and that all other links are 1-Ω resistors. Write down the KCL equations in terms of node voltages with node C chosen as reference.

APPENDIX **C**

Functions of a Complex Variable— Summary of Important Definitions and Theorems

To cover complex variable theory in any detail in an appendix of the length of this one is impossible. Therefore, only the definitions and theorems important to the material in this text are summarized with few proofs given.

Let $s = \sigma + j\omega$ denote a complex variable. Another complex variable $W = U + jV$ is said to be a *function of the complex variable s* if, to each value of s in some set, there corresponds a value, or a set of values, of W. We denote the rule of correspondence between s and W by writing $W = F(s)$. If for each value of s there is only one value of W, $F(s)$ is said to be a *single-valued* function of s. If more than one value of W corresponds to some value or set of values of s, $F(s)$ is *multivalued*.

EXAMPLE C-1

(a) The function $W = F(s) = s^2$ is single-valued. We can express U and V in terms of σ and ω as

$$W = U + jV = (\sigma + j\omega)^2$$

$$= \sigma^2 - \omega^2 + j2\sigma\omega \qquad (C-1)$$

or

$$U = \sigma^2 - \omega^2 \qquad (C-2a)$$

and

$$V = 2\sigma\omega \qquad (C-2b)$$

For each value of $s = \sigma + j\omega$ there corresponds one, and only one, value for $W = U + jV$.

(b) The function $W = F(s) = \sqrt{s}$ is double-valued. To show this, we write s in polar form as

$$s = re^{j\phi} \qquad (C-3)$$

so that

$$W = (re^{j\phi})^{1/2} \qquad (C-4)$$

Now $e^{j\phi} = e^{j(\phi + 2k\pi)}$, where $k = 0, 1$. Therefore,

$$W = r^{1/2}e^{j(\phi/2 + k\pi)}, \qquad k = 0, 1 \qquad (C-5)$$

and for each value of s there correspond two values of W, which are

$$W_1 = r^{1/2}e^{j\phi/2} \qquad (C-6a)$$

and

$$W_2 = r^{1/2} e^{j(\phi/2 + \pi)} \tag{C-6b}$$

where $r^{1/2}$ is the positive square root of r. For example, with $s = -4$, these two values of W are

$$W_1 = 2e^{j\pi/2} = 2j \quad \text{and} \quad W_2 = 2e^{j(\pi/2 + \pi)} = -2j$$

(c) $W = s^*$, where the asterisk denotes the complex conjugate, is a single-valued function of s. If $s = \sigma + j\omega$, then $W = F(s) = \sigma - j\omega$.

We now state a few definitions pertaining to functions of a complex variable.

DEFINITION 1: *The* neighborhood *of a point s_0 in the complex plane is the open circular disk $|s - s_0| < \rho$ where $\rho > 0$ is a real arbitrary constant.*

DEFINITION 2: *A function $F(s)$ is said to have the* limit *L as s approaches s_0 if $F(s)$ is defined in a neighborhood of s_0 (except perhaps at s_0) and if, for every positive real number $\epsilon > 0$, we can find a real number δ such that*

$$|F(s) - L| < \epsilon \tag{C-7}$$

for all values of $s \neq s_0$ in the disk $|s - s_0| < \delta$ ($\delta > 0$). Note that s may approach s_0 from any direction in the complex plane.

EXERCISE

Show that the limit of $F(s) = s^2$ as $s \to 0$ is zero. Note that the value of $F(s)$ at $s = 0$ is immaterial.

DEFINITION 3: *The function $F(s)$ is said to be* continuous *at $s = s_0$ if $F(s_0)$ is defined and*

$$\lim_{s \to s_0} F(s) = F(s_0) \tag{C-8}$$

EXERCISE

Show that $F(s) = s^2$, all s, is continuous at $s = 0$, and in fact for all s.

DEFINITION 4: *A function $F(s)$ is said to be* differentiable *at a point $s = s_0$ if the limit*

$$F'(s_0) = \lim_{\Delta s \to 0} \frac{F(s_0 + \Delta s) - F(s_0)}{\Delta s} \tag{C-9}$$

exists. Letting $s = s_0 + \Delta s$, we may also write this as

$$F'(s_0) = \lim_{s \to s_0} \frac{F(s) - F(s_0)}{s - s_0} \tag{C-10}$$

Note that s may approach s_0 from any direction and the resulting limiting values of (C-10) must be the same regardless of the direction of approach.

EXAMPLE C-2 _____

Consider the derivative of the complex function $F(s) = s^2$. We find that

$$F(s) = \lim_{\Delta s \to 0} \frac{(s + \Delta s)^2 - s^2}{\Delta s}$$

$$= \lim_{\Delta s \to 0} \frac{2s\Delta s + (\Delta s)^2}{\Delta s}$$

$$= 2s \tag{C-11}$$

at any point s.

At this point, we simply state that all the familiar rules for differentiation of functions of a real variable continue to hold for functions of a complex variable. For example, the derivative of $\sin s$ is $\cos s$, etc. The following example gives a function that does not have a derivative anywhere.

EXAMPLE C-3 _____

Consider $F(s) = s^* = \sigma - j\omega$. For this function,

$$\frac{F(s + \Delta s) - F(s)}{\Delta s} = \frac{(\sigma + \Delta\sigma) - j(\omega + \Delta\omega) - (\sigma - j\omega)}{\Delta\sigma + j\Delta\omega}$$

$$= \frac{\Delta\sigma - j\Delta\omega}{\Delta\sigma + j\Delta\omega} \tag{C-12}$$

where $\Delta s = \Delta\sigma + j\Delta\omega$. The limit of this ratio depends on the way Δs approaches zero. For example, if $\Delta\sigma \to 0$ first and then $\Delta\omega \to 0$, we obtain as the limit -1. On the other hand, if $\Delta\omega \to 0$ first and then $\Delta\sigma \to 0$, the limit is $+1$. Hence this function is not differentiable anyplace, since s was arbitrarily chosen.

DEFINITION 5: *A function $F(s)$ is said to be* analytic *(regular) at a point $s = s_0$ if it is defined and has a derivative at every point in some neighborhood of s_0.*

Note that this definition is stronger than simply saying that a function is differentiable at a point, since the derivative must exist everywhere in a neighborhood of the point.

DEFINITION 6: *A function $F(s)$ that is analytic at every point in a domain D is said to be analytic in D. (Note: A domain is simply an open, connected set of points.)*

DEFINITION 7: *A function $F(s)$ that has at least one analytic point is called an analytic function.*

DEFINITION 8: *A point s_0 at which the analytic function $F(s)$ is not analytic is called a singular point.*

It can be shown that an analytic function $F(s)$ of a complex variable has derivatives of all orders, except at singular points, and can be represented as a Taylor series in a neighborhood of each point s_0 where $F(s)$ is analytic:

$$F(s) = \sum_{n=0}^{\infty} \frac{F^{(n)}(s_0)}{n!} (s - s_0)^n \tag{C-13}$$

where $F^{(n)}(s_0)$ denotes the nth derivative of $F(s)$ evaluated at $s = s_0$. For example,

$$e^s = \sum_{n=0}^{\infty} e^{s_0} \frac{(s - s_0)^n}{n!} \tag{C-14}$$

about any point s_0 in the complex plane. In fact, this is how the transcendental functions e^s, $\sin s$, $\cos s$, etc., of a complex variable are defined.

The function $F(s) = G(s)/(s - s_0)^m$, where $G(s)$ is analytic at $s = s_0$, is singular at $s = s_0$ because of the factor $(s - s_0)^{-m}$. Such a singularity is called a *pole* of order m. This type of singularity is said to be isolated and, in the neighborhood of an isolated singularity, the representation

$$F(s) = \sum_{n=-\infty}^{\infty} C_n(s - s_0)^n \tag{C-15}$$

can be used. More specifically:

DEFINITION 9: *If there is some neighborhood of a singular point s_0 of a function $F(s)$ throughout which F is analytic, except at the point itself, then s_0 is called an isolated singular point of F.*

The series (C-15) is known as the *Laurent series expansion* for the function at $s = s_0$. If $C_n = 0$, $n < -m$, $F(s)$ is said to have an mth-order pole at $s = s_0$ $(m > 0)$; if $m = 1$, $F(s)$ is said to have a simple pole at $s = s_0$; if $C_n \neq 0$ for infinitely many negative values of n, $F(s)$ has an essential singularity at $s = s_0$. Examples of functions with isolated singular points are $1/s$ and $e^s/(s - 1)^2$. The function $1/\sin(\pi/s)$ has isolated singular points at $s = \pm 1, \pm 1/2, \pm 1/3, \ldots$. At $s = 0$ this function has a non-isolated singular point since every neighborhood of $s = 0$ contains other singular points.

THEOREM 1: *A necessary condition that a function $F(s) = U + jV$ of a complex variable $s = \sigma + j\omega$ be analytic at a point s_0 is that its real and imaginary parts satisfy the Cauchy-Riemann equations*

$$\frac{\partial U}{\partial \sigma} = \frac{\partial V}{\partial \omega} \quad and \quad \frac{\partial U}{\partial \omega} = -\frac{\partial V}{\partial \sigma} \tag{C-16}$$

THEOREM 2: *Sufficient conditions for a function $F(s)$ of a complex variable s to be differentiable at a point $s = s_0$ is that $U(\sigma, \omega)$ and $V(\sigma, \omega)$ have continuous first partial derivatives that satisfy the Cauchy-Riemann equations at the point $s = s_0$.*

THEOREM 3 (CAUCHY'S INTEGRAL THEOREM): *Let a function of a complex variable $F(s)$ be analytic everywhere on and within a simple closed curve C (by simple, we mean that the curve does not cross itself). Then*

$$\oint_C F(s)\, ds = 0 \tag{C-17}$$

where $\oint_C (\cdot)\, ds$ denotes the line integral in a counterclockwise direction along C.

Figure C-1 illustrates the circumstances of the theorem. Suppose that $F(s)$ has an isolated singularity at $s = s_0$ which is inside C as shown in Figure C-2. Is $\oint_C F(s)\, ds = 0$ still? No, because $F(s)$ is not analytic everywhere within and on C. However, consider the simple closed curve shown in Figure C-3. Since $F(s)$ is analytic within and on the simple closed curve $C + C_1 + C' + C_2$, it follows by Cauchy's integral theorem that

$$\oint_{C+C_1+C'+C_2} F(s)\, ds = 0 \tag{C-18}$$

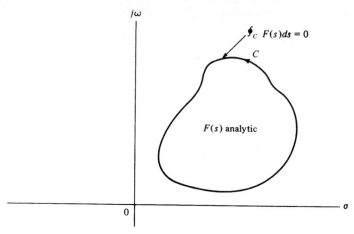

FIGURE C-1. Illustration of Cauchy's integral theorem.

Breaking the line integral up into the sum of four integrals over the separate curves C, C_1, C_2, and C', we obtain

$$\int_C F(s)\, ds + \int_{C_1} F(s)\, ds + \int_{C_2} F(s)\, ds + \int_{C'} F(s)\, ds = 0 \qquad (\text{C-19})$$

Now, as the gap between the curves C_1 and C_2 approaches zero, the second and third integrals will approach equal magnitude, opposite in sign values (both have the same integrand, are integrated over the same length, but are taken in opposite directions). It follows, therefore, that

$$\oint_C F(s)\, ds = -\oint_{C'} F(s)\, ds, \qquad \epsilon \to 0 \qquad (\text{C-20})$$

We now focus our attention on the integral over C' as $\epsilon \to 0$. For any point s on C', we may write

$$s = s_0 + \delta e^{j\theta} \qquad (s \text{ on } C') \qquad (\text{C-21})$$

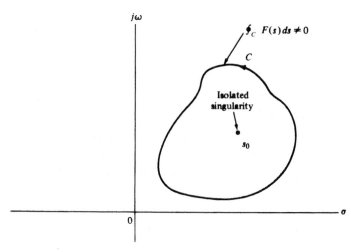

FIGURE C-2. Line integral around a closed contour that is not zero because of a singularity within the contour.

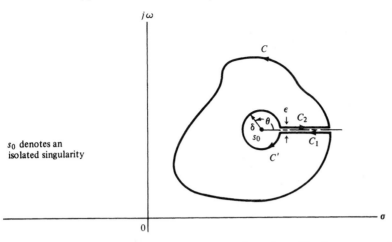

FIGURE C-3. Composite closed curve, $C + C_1 + C_2 + C'$, chosen so as to satisfy Cauchy's integral theorem.

Therefore, in the integral over C' of (C-20) we may substitute

$$ds = j\delta e^{j\theta}\, d\theta \qquad\qquad (C\text{-}22)$$

which results by taking the derivative of (C-21) (note that for any point on C', both s_0 and δ are fixed). Also, we may express $F(s)$ in the neighborhood of s_0 as a Laurent series as given by (C-15). In particular, on the curve C', s is given by (C-21), so that

$$s - s_0 = \delta e^{j\theta} \qquad\qquad (C\text{-}23)$$

and (C-15) may be expressed as

$$
\begin{aligned}
F(s) &= \sum_{n=-\infty}^{\infty} C_n (\delta e^{j\theta})^n \\
&= \sum_{n=-\infty}^{\infty} C_n \delta^n e^{jn\theta} \qquad (s \text{ on } C') \qquad\qquad (C\text{-}24)
\end{aligned}
$$

Substituting (C-22) and (C-24) into the integral over C', we obtain

$$\oint_{C'} F(s)\, ds = \int_{\theta=2\pi}^{0} \left[\sum_{n=-\infty}^{\infty} C_n \delta^n e^{jn\theta} \right] j\delta e^{j\theta}\, d\theta \qquad\qquad (C\text{-}25)$$

where the line integral can be written as an integral over the angle θ because of the circular path. Note that the integral over θ goes from $\theta = 2\pi$ to $\theta = 0$ because of the clockwise orientation of C'. Because the Laurent series for $F(s)$ is uniformly convergent to $F(s)$ in the neighborhood of s_0, the integral and sum in (C-25) can be interchanged to yield

$$\oint_{C'} F(s)\, ds = - \sum_{n=-\infty}^{\infty} jC_n \delta^{n+1} \int_{0}^{2\pi} e^{j(n+1)\theta}\, d\theta \qquad\qquad (C\text{-}26)$$

where the integral is now taken from $\theta = 0$ to $\theta = 2\pi$ with a minus sign inserted to account for the reversal of limits. Now, for $n + 1 \neq 0$, the integral of $\exp[j(n + 1)\theta]$ from 0 to 2π is zero (the student may show this by integration and substituting the limits). For $n + 1 = 0$, the integral in (C-26) evaluates to 2π. Thus all terms in the sum are zero except the term for $n = -1$, which gives

$$\oint F(s)\, ds = -2\pi j C_{-1} \tag{C-27}$$

Therefore, as δ and ϵ approach zero, (C-20) evaluates to

$$\oint_C F(s)\, ds = 2\pi j C_{-1} \tag{C-28}$$

where C_{-1} is the residue of $F(s)$ at $s = s_0$. If $F(s)$ has more than one isolated singular point within C, say at the points $s_1, s_2, \ldots, s_N$ with residues $K_1, K_2, \ldots, K_N$, then we may follow the procedure outlined above for each singular point and arrive at the *residue theorem:*

THEOREM 4 (Residue Theorem): *Let $F(s)$ be analytic everywhere within and on the simple closed curve C except at the isolated singular points $s_1, s_2, \ldots, s_N$. Then*

$$\oint_C F(s)\, ds = 2\pi j \sum_{n=1}^{N} K_n \tag{C-29}$$

where $\oint_C (\cdot)\, ds$ is the line integral in a counterclockwise sense of $F(s)$ around C and $K_1, K_2, \ldots, K_N$ are the residues of $F(s)$ at $s_2, s_2, \ldots, s_N$.

We complete this appendix with several theorems pertaining to the single-sided Laplace transform.

THEOREM 5 (Absolute Convergence of the Single-Sided Laplace Transform): *The integral*

$$X(s) = \int_0^\infty x(t) e^{-st}\, dt \tag{C-30}$$

converges absolutely for all $\mathrm{Re}(s) > c$ if

$$\int_0^L |x(t)|\, dt \le K < \infty \tag{C-31}$$

for $0 < L < \infty$ and $0 < K < \infty$, and

$$|x(t)| \le A e^{ct}, \, t > L \tag{C-32}$$

for some real A and c.

Proof. According to the definition on absolute convergence, we must show that

$$I = \int_0^\infty |x(t) e^{-st}|\, dt < \infty \qquad \text{for } \sigma = \mathrm{Re}(s) > c \tag{C-33}$$

We divide I up into two integrals as

$$I = \int_0^L |x(t)| e^{-\sigma t}\, dt + \int_L^\infty |x(t)| e^{-\sigma t}\, dt$$

$$= I_1 + I_2 \tag{C-34}$$

Now

$$I_1 \le e^{|\sigma|L} \int_0^L |x(t)|\, dt \le K e^{|\sigma|L} \tag{C-35}$$

and

$$I_2 \le A \int_L^\infty e^{-(\sigma-c)t} \, dt = \frac{Ae^{-(\sigma-c)L}}{\sigma - c} \qquad \text{since } \sigma > c \tag{C-36}$$

Hence,

$$I = I_1 + I_2 \le Ke^{|\sigma|L} + \frac{Ae^{-(\sigma-c)L}}{\sigma - c} < \infty \qquad \text{for } \sigma > c \tag{C-37}$$

DEFINITION 10: (*Uniform convergence of an integral*): *The integral*

$$I(s) = \int_0^\infty g(t, s) \, dt \tag{C-38}$$

is said to converge uniformly for s in a specific range (say 0 to ∞) if, for each $\epsilon > 0$, we can find a value d such that

$$\left| \int^b g(t, s) \, dt - I(s) \right| < \epsilon \qquad \text{for } b > d \tag{C-39}$$

THEOREM 6: *If the integral*

$$X(s) = \int_0^\infty x(t)e^{-st} \, dt \tag{C-40}$$

converges absolutely for $s_0 = \sigma_0 + j\omega_0$, it also converges absolutely and uniformly in the half plane $\text{Re}(s) \ge \sigma_0$.

THEOREM 7: *If the single-sided Laplace transform integral converges absolutely for $\sigma > \sigma_a$, then $X(s)$ is analytic for $\sigma > \sigma_a$.*

This theorem tells us that the singularities of $X(s)$ must lie to the left or on the line $\text{Re}(s) = \sigma_a$. By a process known as *analytic continuation*, $X(s)$ may be defined and shown to be analytic on the entire s-plane except, of course, at its singularities.

THEOREM 8: *If $x_1(t)$ and $x_2(t)$ both have the same single-sided Laplace transform, then*

$$x_1(t) = x_2(t) \tag{C-41}$$

except, possibly, at a set of points with "measure zero"; that is, the integral of $|x_1(t) - x_2(t)|$ over the range of integration is zero.

Further Reading

The books by Churchill and Kreyzig cited as references in Chapter 5 are recommended if the student wishes to have a more complete treatment of complex variable theory than that given here.

Matrix Algebra

The purpose of this appendix is to summarize the basic concepts of matrix theory.

An $n \times m$ matrix $\mathbf{A}$ is an array of numbers a_{ij} given by

$$\mathbf{A} = \begin{bmatrix} a_{11} & a_{12} & \cdots & a_{1m} \\ a_{21} & a_{22} & \cdots & a_{2m} \\ \vdots & & & \\ a_{n1} & a_{n2} & \cdots & a_{nm} \end{bmatrix} \tag{D-1}$$

It has n rows and m columns. If $m = 1$, there is only one column and $\mathbf{A}$ becomes a column matrix or vector $\mathbf{a}$ given by

$$\mathbf{a} = \begin{bmatrix} a_1 \\ a_2 \\ \vdots \\ a_n \end{bmatrix} \tag{D-2}$$

The transpose of an $n \times m$ matrix $\mathbf{A}$, denoted $\mathbf{A}'$, is an $m \times n$ matrix obtained from A by interchanging rows and columns; that is,

$$\mathbf{A}' = \begin{bmatrix} a_{11} & a_{21} & \cdots & a_{n1} \\ a_{12} & a_{22} & \cdots & a_{n2} \\ \vdots & & & \\ a_{1m} & a_{2m} & \cdots & a_{nm} \end{bmatrix} \tag{D-3}$$

A square matrix ($n = m$) is said to be *symmetric* if and only if $\mathbf{A}' = \mathbf{A}$. The transpose of a column vector is a row vector; that is,

$$\mathbf{a}' = [a_1 \quad a_2 \quad \ldots \quad a_n] \tag{D-4}$$

Two matrices (or vectors) which have the same dimensions may be added or subtracted. These operations are accomplished by adding or subtracting corresponding elements of the matrices. The operations may be defined as

$$\mathbf{A} \pm \mathbf{B} = \mathbf{C} \tag{D-5}$$

where the elements of C are obtained from those of A and B by the relation

$$c_{ij} = a_{ij} \pm b_{ij} \tag{D-6}$$

Matrix multiplication may be defined as follows. Let A be $n \times l$ with elements a_{ij} and B be $l \times m$ with elements b_{ij}. Then

$$AB = C \tag{D-7}$$

where C is $n \times m$ with elements c_{ij} given by

$$c_{ij} = \sum_{k=1}^{l} a_{ik}b_{kj} \tag{D-8}$$

Note that the number of columns of the first matrix must equal the number of rows of the second or multiplication cannot be accomplished. An an example, let

$$A = \begin{bmatrix} 1 & 2 & 1 \\ 3 & 0 & 1 \end{bmatrix} \qquad B = \begin{bmatrix} 2 & 1 \\ 1 & 0 \\ 3 & 2 \end{bmatrix} \tag{D-9}$$

Then

$$AB = C = \begin{bmatrix} 7 & 3 \\ 9 & 5 \end{bmatrix} \tag{D-10}$$

In general, matrix multiplication is not commutative; that is, in general

$$AB \neq BA \tag{D-11}$$

However, it is true that

$$A(BC) = (AB)C \qquad \text{and} \qquad A(B + C) = AB + AC \tag{D-12}$$

The identity matrix is a square matrix, I, having the property that for any square matrix A

$$AI = IA = A \tag{D-13}$$

The identity matrix is such that all principal diagonal elements are 1 and all other elements are zero. The identity matrix of dimension 2 is given by

$$I = \begin{bmatrix} 1 & 0 \\ 0 & 1 \end{bmatrix} \tag{D-14}$$

A square matrix A is said to be *nonsingular* if the determinant of A is nonzero. For a nonsingular matrix A there exists a matrix B with the property that

$$AB = BA = I \tag{D-15}$$

B is said to be the *inverse* of A, and is denoted as

$$B = A^{-1} \tag{D-16}$$

The inverse of a nonsingular matrix may be found from

$$A^{-1} = \frac{\text{adj } A}{|A|} \tag{D-17}$$

where adj $\mathbf{A}$ denotes the *adjoint* of $\mathbf{A}$, which is the transpose of the matrix of cofactors of $\mathbf{A}$.[†] $|\mathbf{A}|$ is the determinant of $\mathbf{A}$. As an example, let us find the inverse of

$$\mathbf{A} = \begin{bmatrix} 0 & 1 & 0 \\ 0 & 0 & 1 \\ -2 & -3 & -1 \end{bmatrix} \tag{D-18}$$

We have

$$|\mathbf{A}| = -2 \tag{D-19}$$

$$\text{adj } \mathbf{A} = \begin{bmatrix} 3 & -2 & 0 \\ 1 & 0 & -2 \\ 1 & 0 & 0 \end{bmatrix}^t = \begin{bmatrix} 3 & 1 & 1 \\ -2 & 0 & 0 \\ 0 & -2 & 0 \end{bmatrix} \tag{D-20}$$

Then

$$\mathbf{A}^{-1} = \begin{bmatrix} -\frac{3}{2} & -\frac{1}{2} & 0 \\ 1 & 0 & 0 \\ 0 & 1 & 0 \end{bmatrix} \tag{D-21}$$

As a check, the student may verify that $\mathbf{A}\mathbf{A}^{-1} = \mathbf{I}$.

The eigenvalues of an $n \times n$ matrix $\mathbf{A}$ are the roots λ_i, $i = 1, 2, \ldots, n$, of the characteristic equation

$$|\lambda\mathbf{I} - \mathbf{A}| = 0 \tag{D-22}$$

which is an nth-degree polynomial in λ. Eigenvalues play an important role in system analysis.

As an example, we shall find the eigenvalues of the matrix

$$\mathbf{A} = \begin{bmatrix} 0 & 1 \\ -4 & -5 \end{bmatrix} \tag{D-23}$$

We have

$$|\lambda\mathbf{I} - \mathbf{A}| = \begin{vmatrix} \lambda & -1 \\ 4 & \lambda + 5 \end{vmatrix} = \lambda(\lambda + 5) + 4 = \lambda^2 + 5\lambda + 4 = 0 \tag{D-24}$$

The polynomial may be factored and the roots found as

$$\lambda_1 = -1, \qquad \lambda_2 = -4 \tag{D-25}$$

These are the eigenvalues of the matrix given. The procedure is exactly the same, but considerably more tedious, for larger matrices.

The characteristic equation is an important quantity associated with a matrix, having other uses than merely the determination of the eigenvalues. Let us denote the characteristic equation of an $n \times n$ matrix $\mathbf{A}$ as

$$\lambda^n + a_{n-1}\lambda^{n-1} + \cdots + a_1\lambda + a_0 = 0 \tag{D-26}$$

The Cayley-Hamilton theorem states that a matrix satisfies its own characteristic equation. Therefore, $\mathbf{A}$ satisfies the *matrix* equation

$$\mathbf{A}^n + a_{n-1}\mathbf{A}^{n-1} + \cdots + a_1\mathbf{A} + a_0\mathbf{I} = 0 \tag{D-27}$$

[†]The i-jth cofactor of a matrix is obtained by striking out the ith row and jth column, evaluating the resulting determinant, and multiplying by $(-1)^{i+j}$.

from which

$$\mathbf{A}^n = -(a_{n-1}\mathbf{A}^{n-1} + a_{n-2}\mathbf{A}^{n-2} + \cdots + a_1\mathbf{A} + a_0\mathbf{I}) \tag{D-28}$$

If we multiply both sides of this equation by $\mathbf{A}$, we obtain

$$\mathbf{A}^{n+1} = -(a_{n-1}\mathbf{A}^n + a_{n-2}\mathbf{A}^{n-1} + \cdots + a_1\mathbf{A}^2 + a_0\mathbf{A}) \tag{D-29}$$

Clearly, we can substitute for $\mathbf{A}^n$ on the right to obtain an expression for $\mathbf{A}^{n+1}$ in terms of powers of $\mathbf{A}$ *no higher than* $n - 1$. The expression will be of the form

$$\mathbf{A}^{n+1} = b_0\mathbf{I} + b_1\mathbf{A} + \cdots + b_{n-1}\mathbf{A}^{n-1} \tag{D-30}$$

where the b_i can be determined from the a_i. By continuing this kind of operation and successively substituting, it is possible to write *any* power of $\mathbf{A}$ in terms of powers of $\mathbf{A}$ no higher than $n - 1$. This is the basis of one of the methods of finding the matrix exponential $\mathbf{e}^{\mathbf{A}t}$ in Chapter 7.

Differentiation or integration of a matrix with respect to time is accomplished by performing the appropriate operation on each element of the matrix. For example, suppose that

$$\mathbf{A} = \begin{bmatrix} e^{-t} & te^{-t} \\ 0 & e^{-t} \end{bmatrix} \tag{D-31}$$

Then

$$\frac{d\mathbf{A}}{dt} = \begin{bmatrix} -e^{-t} & e^{-t} - te^{-t} \\ 0 & -e^{-t} \end{bmatrix} \tag{D-32}$$

Since the Laplace transform is an integral operation, the Laplace transform of a matrix is obtained by Laplace transforming each of the elements. The Laplace transform of $\mathbf{A}$ above is

$$\mathscr{L}[\mathbf{A}] = \begin{bmatrix} \dfrac{1}{s+1} & \dfrac{1}{(s+1)^2} \\ 0 & \dfrac{1}{s+1} \end{bmatrix} \tag{D-33}$$

E

Analog Filters

In Chapter 9 a number of methods were studied for the design of digital filters. Many of these design methods assumed knowledge of an analog prototype from which the digital filter could be derived. The design process for the digital filter was then carried out in order to obtain a digital filter whose time-domain or frequency-domain characteristics approximated the corresponding characteristics of the assumed analog prototype. For example, the impulse invariance design technique results in a digital filter whose unit pulse response is the sampled unit impulse response of the analog filter from which it was derived. The bilinear z-transform design technique results in a digital filter whose frequency response approximates the frequency response of the analog filter from which it was derived.

The purpose of this appendix is to present the defining characteristics of several analog prototypes in order to better understand our starting point when we develop digital filters from an analog prototype. The subject of analog filter synthesis is an important area of study and this brief introduction is of necessity incomplete. The student interested in exploring this area in greater detail should consult one of the many textbooks available. Several are cited in the references.

The synthesis of analog filters is now a very mature subject area. Extensive sets of tables exist which give not only the frequency and phase responses of many analog prototypes, but also the element values necessary to realize those prototypes.[†] Many of the design procedures for digital filters have been developed in ways that allow this wide body of analog filter knowledge to be utilized effectively.

Generally, filter tables contain the amplitude response, phase response, pole-zero locations, and element values for the *normalized filter*, which is defined to be a low-pass filter having a reference frequency, which is often the bandwidth, of 1 rad/s. Then, using standard techniques, this filter can be transformed into a high-pass, bandpass, or notch filter having any desired bandwidth. We examine several of these transformation techniques later in this appendix.

The Signal and Systems MATLAB toolbox contains a number of routines for designing analog filters. Many of these are contained in *the Student Edition of MATLAB*. A wide variety of filter types are included along with routines for scaling filter bandwidths and for converting lowpass filters to bandpass, notch (bandstop) and highpass filters. For many applications, including all of those considered here, the availability of these routines has replaced the need for traditional filter tables.

E-1 Group Delay and Phase Delay

In Chapter 4 ideal filters were defined. Recall that the ideal filter resulted by investigating the conditions under which a linear system is distortionless. A distortionless system was defined in Chapter 3 as

---

[†]See, for example, Zverev (1967). See "Further Reading."

a system in which the system output is an amplitude-scaled and time-delayed version of the system input. In other words, if the input $x(t)$ is applied to a distortionless system, the output $y(t)$ is

$$y(t) = Gx(t - \tau) \tag{E-1}$$

in which G is the system gain and τ is the system time delay between system input and output. By Fourier transforming (E-1) we were able to show that the transfer function of an ideal system is

$$H(f) = Ge^{-j2\pi f\tau} \tag{E-2}$$

The amplitude response is the constant

$$A(f) = G \tag{E-3a}$$

and the phase response is the *linear* function of frequency

$$\phi(f) = -2\pi f\tau \tag{E-3b}$$

Often, instead of specifying the phase response of a filter, the group time delay is specified. The group time delay is defined as

$$\tilde{T}_g(\omega) = -\frac{d}{d\omega}\tilde{\phi}(\omega) \tag{E-4}$$

or, equivalently,

$$T_g(f) = -\frac{1}{2\pi}\frac{d}{df}\phi(f) \tag{E-5}$$

where tilde indicates a function of ω. Since an ideal filter has a phase response that is a linear function of frequency, the group delay of an ideal filter is constant for all frequencies. This is easily seen by substituting (E-3) into (E-5), which shows that the group delay of an ideal filter is

$$T_g(f) = -\frac{1}{2\pi}\frac{d}{df}(-2\pi f\tau) = \tau \tag{E-6}$$

Thus if an input to an ideal filter has a spectrum extending over a range of frequencies, the filter output will be distortionless if the filter gain and group delay are both constant over the nonzero range of the input spectrum.

Linear system theory tells us that if a sinusoidal input is applied to a linear time-invariant system, the output will also be sinusoidal, although the output may have a different amplitude and phase reference. Thus if a single spectral component is applied to a linear time-invariant system, the output is *always* distortionless. For this case we refer to another type of system time delay called *phase delay,* which is the actual time delay between input and output at the frequency of a single spectral component. Phase delay is easily illustrated by a simple example.

Assume that a system has input

$$x(t) = A \cos 2\pi f_0 t \tag{E-7a}$$

and output

$$y(t) = B \cos(2\pi f_0 t + \theta) \tag{E-7b}$$

The output can be written

$$y(t) = B \cos 2\pi f_0 \left(t + \frac{\theta}{2\pi f_0}\right) \tag{E-8}$$

Recognizing that since the output must be distortionless, $y(t)$ must be of the form

$$y(t) = B \cos 2\pi f_0(t - t_0) \tag{E-9}$$

Thus comparing (E-8) and (E-9) we see that the time delay between system input and system output is

$$t_0 = -\frac{\theta}{2\pi f_0} \tag{E-10}$$

which is the phase delay for this example. In general, the phase delay is written as

$$T_p(f) = -\frac{\phi(f)}{2\pi f} \tag{E-11}$$

in which $\phi(f)$ is the system phase shift, in radians, at the frequency f (hertz). Ideal filters have equal group and phase delays.

E-2 Approximation of Ideal Filters by Practical Filters

Since ideal filters are not physically realizable (recall that the unit impulse response for an ideal filter is noncausal), the filtering problem is to select, from among the many available filter prototypes, that prototype which satisfactorily approximates the desired filter characteristic with a minimum of hardware. For an ideal filter, each value of frequency either falls within the filter passband, which is characterized by constant gain, or falls within the filter stop band, which is characterized by infinite attenuation. Practical filters have a region between each passband and stop band known as a transition band.

A typical set of filter requirements is illustrated in Figure E-1. Within the passband, the filter amplitude response is not required to be constant; some variation about the nominal gain G_0 is allowed. In Figure E-1a, this variation has a peak-to-peak value of Δ_R. Within the stop band the amplitude response is required to be below the value A. In order for the transition band to be as small as possible, the amplitude response of the filter often passes through the two heavy dots shown in Figure E-1a. Typical applications also place requirements on the group delay of the filter, as illustrated in Figure E-1b. Some variation about the nominal value T_g is allowed. The peak-to-peak variation is denoted by Δ_T in Figure E-1b. The group delay in the transition band and in the stop band is often unimportant. Although Figure E-1 is drawn for a low-pass filter, the extension to other types should be obvious.

Many different trade-offs are possible among the parameters depicted in Figure E-1, and each of the standard prototypes makes a different trade-off. We now briefly review some of the most popular filter prototypes in order to become familiar with the various trade-offs possible.

A sketch of the amplitude response of each prototype to be discussed is illustrated in Figure E-2. Each filter is fourth order, and the highest passband frequency is 1 rad/s. Thus these are the amplitude responses of the normalized filters.

Butterworth Filter

The Butterworth filter is designed to have a maximally flat amplitude response in the passband. The edge of the passband is usually defined to be the frequency where the amplitude response is 0.707 times the maximum value. Thus ϵ in Figure E-1 is 0.707. The amplitude response of a Butterworth filter decreases monotonically in the transition band and in the stop band. Since the Butterworth filter is

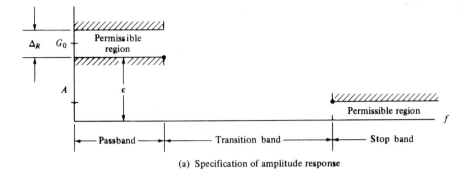

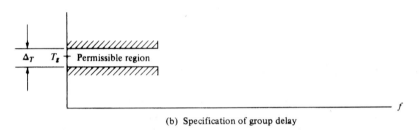

(a) Specification of amplitude response

(b) Specification of group delay

FIGURE E-1.   Typical set of filter requirements.

designed to have a maximally flat[†] passband amplitude response, no constraints are placed on the phase response or group delay. However, the group delay of a Butterworth filter is reasonably constant over most of the passband, as we shall observe later. The Butterworth filter is studied in more detail in the following section.

Bessel Filter

The Bessel filter is designed to exhibit a maximally linear phase response, which implies a maximally constant group delay. The Bessel filter has less stop band attenuation than a Butterworth filter of equal order and bandwidth. (See Problems E-3 and E-4 for more on Bessel filters.)

Chebyshev I Filter

The Chebyshev I filter allows the amplitude response to have ripple in the passband. This allows the amplitude response to have a steeper rolloff outside the passband for a fixed filter order. The ripples in the filter passband have equal amplitude and the amplitude response outside of the passband is monotonic. Where a given stopband attenuation is required, a Chebyshev I filter having lower order than a Butterworth filter can sometimes be used. This results in hardware simplifications assuming, of course, that the passband ripples can be tolerated in the given application. In Figure E-2c the passband ripple is 6 dB. This is an extremely large value but was done to illustrate this characteristic of the filter. We study the Chebyshev I filter in more detail in the following section.

[†]The nth order Butterworth filter amplitude response function has $(2n - 1)$ derivatives equal to zero at $\omega = 0$.

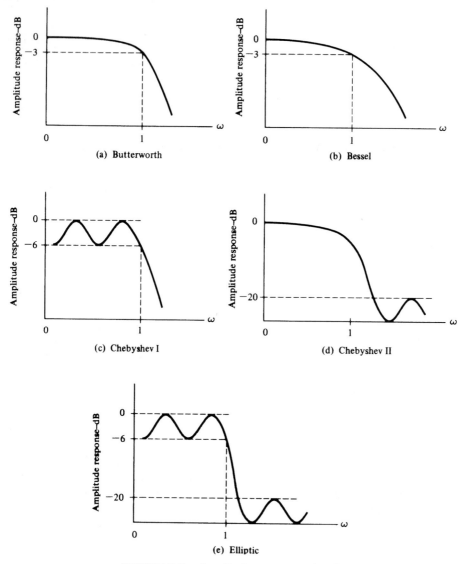

FIGURE E-2. Amplitude response of various prototypes.

Chebyshev II Filter

The Chebyshev II filter has a monotonic amplitude response and equal amplitude ripples in the stop band. For the filter illustrated in Figure E-2d, the minimum stop-band attenuation is 20 dB.

Elliptic Filter

The elliptic filter is a combination of the Chebyshev I and Chebyshev II filters in that the elliptic filter has amplitude response ripples in both the passband and stop band. The advantage of the elliptic filter is that for a given allowable ripple in the passband and a minimum attenuation in the stop band, the width of the transition band is minimized. In Figure E-2e, the passband ripple is 6 dB and the minimum attenuation in the stop band is 20 dB.

E-3 Butterworth and Chebyshev Filters

To gain a better understanding of filter characteristics, we now examine the Butterworth and Cheby-shev I prototypes. Only normalized filters will be considered. In the following section we shall see how to convert these normalized filters to other types of filters.

Butterworth Filters

The Butterworth filter having a 3-dB bandwidth of 1 rad/s has the steady-state amplitude response

$$|H_B(\omega)| = \frac{1}{\sqrt{1 + \omega^{2n}}} \tag{E-12}$$

where $\omega = 2\pi f$ and n is the order of the filter. As stated in the preceding section, the Butterworth filter is designed to have a maximally flat passband amplitude response. The system function corresponding to (E-12) is

$$H_B(s) = \frac{(-1)^n s_1 s_2 \cdots s_n}{(s - s_1)(s - s_2) \cdots (s - s_n)} \tag{E-13}$$

The poles of the Butterworth filter $(s_1, s_2, \ldots, s_n)$ all lie on a circle having a radius of 1 rad/s and centered on the origin of the s-plane. The poles are symmetrical with respect to the real axis. Figure E-3 illustrates the pole locations for $n = 2, 3, 4$, and 5. Note that the angle between the poles is $180°/n$ and that odd-order filters have one real pole at $s = -1$. Even-order Butterworth filters have no real poles.

The transfer function for an nth-order Butterworth filter is given by

$$H_B(s) = \frac{1}{B_n(s)} \tag{E-14}$$

where $B_n(s)$ represents the Butterworth polynomial of degree n. These polynomials are easily derived from knowledge of the pole locations (see Problem E-2). The first six Butterworth polynomials are tabulated in Table E-1. The amplitude responses, in decibels, of several Butterworth filters are illustrated in Figure E-4. All Butterworth filters have a gain at $\omega_c = 2\pi f_c = 1$ of

$$|H(\omega_c)| = \frac{1}{\sqrt{1 + 1^{2n}}} = 0.707 \tag{E-15}$$

which expressed in decibels is

$$20 \log|H(\omega_c)| = 20 \log 0.707 = -3 \tag{E-16}$$

The group delay of the nth-order Butterworth filter is given by[†]

$$\tilde{T}_g(\omega) = \frac{1}{1 + \omega^{2n}} \sum_{m=0}^{n-1} \frac{\omega^{2m}}{\sin(2m + 1)(\pi/2n)} \tag{E-17}$$

and is illustrated in Figure E-5 for several values of n. It can be seen that the group delay is reasonably constant over most of the filter passband.

[†]See Weinberg (1962), p. 497.

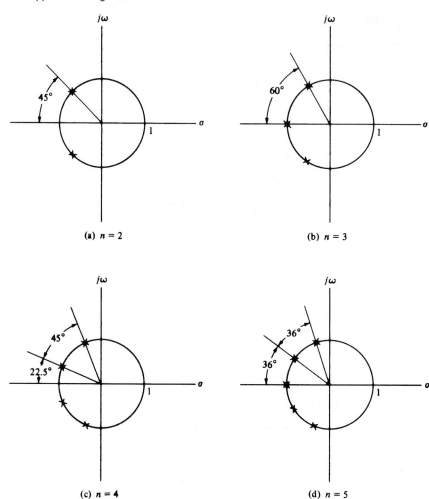

FIGURE E-3. Pole locations for normalized Butterworth filters.

TABLE E-1
Butterworth Polynomials $1 \leq n \leq 6$[a]

$B_1(s) = s + 1$
$B_2(s) = s^2 + 1.4142136s + 1$
$B_3(s) = s^3 + 2.0000000s^2 + 2.0000000s + 1$
$B_4(s) = s^4 + 2.6131259s^3 + 3.4142136s^2 + 2.6131259s + 1$
$B_5(s) = s^5 + 3.2360680s^4 + 5.2360680s^3 + 5.2360680s^2 + 3.2360680s + 1$
$B_6(s) = s^6 + 3.8637033s^5 + 7.4641016s^4 + 9.1416202s^3 + 7.4641016s^2 + 3.8637033s + 1$

[a]$H_n(s) = 1/B_n(s)$.

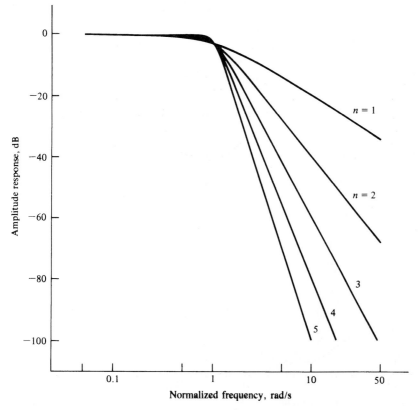

FIGURE E-4.   Amplitude response of Butterworth filters.

Chebyshev I Filter

The amplitude response of an nth-order Chebyshev I filter is defined by the equation

$$|H_c(\omega)| = \frac{1}{\sqrt{1 + \epsilon^2 C_n^2(\omega)}} \tag{E-18}$$

in which $C_n(\omega)$ is the Chebyshev polynomial and ϵ is a parameter to be discussed later. The Chebyshev polynomial is defined by

$$C_n(\omega) = \cos(n \cos^{-1} \omega) \tag{E-19}$$

Since the argument of the cosine function is only a real angle for ω in the range $-1 \leq \omega \leq 1$, it is convenient to define $C_n(\omega)$ in two ranges as

$$C_n(\omega) = \cos(n \cos^{-1} \omega), \qquad |\omega| \leq 1 \tag{E-20}$$

and

$$C_n(\omega) = \cosh(n \cosh^{-1} \omega), \qquad |\omega| > 1 \tag{E-21}$$

Equation (E-20) describes the filter in the passband and (E-21) describes the filter outside of the passband.

From (E-20), we see that at zero frequency

$$C_n^2(0) = \cos^2 \frac{n\pi}{2} = \begin{cases} 0, & n \text{ odd} \\ 1, & n \text{ even} \end{cases} \tag{E-22}$$

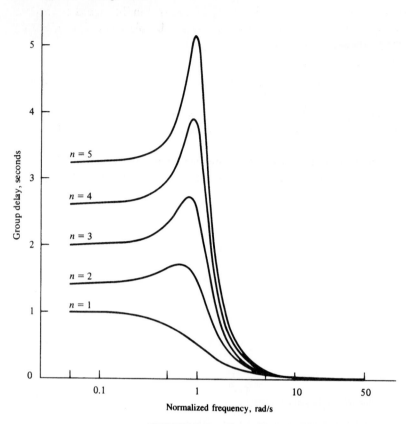

FIGURE E-5. Group delay of Butterworth filters.

so that

$$|H_c(0)| = 1, \qquad n \text{ odd} \tag{E-23}$$

and

$$|H_c(0)| = \frac{1}{\sqrt{1 + \epsilon^2}}, \qquad n \text{ even} \tag{E-24}$$

Since

$$C_n^2(1) = \cos^2 0 = 1, \qquad \text{all } n \tag{E-25}$$

both even-order and odd-order Chebyshev I filters have an amplitude response at band edge, $\omega = \omega_c = 2\pi f_c = 1$, equal to

$$|H(\omega_c)| = \frac{1}{\sqrt{1 + \epsilon^2}} \tag{E-26}$$

for all n. Thus, at band edge, the amplitude response of a Chebyshev I filter is

$$20 \log|H(f_c)| = -10 \log(1 + \epsilon^2) \tag{E-27}$$

The parameter ϵ establishes the amplitude of the passband ripple discussed in the preceding section. Since the maximum value of $|H_c(\omega)|$ within the filter passband is 1, and the minimum value of $|H_c(\omega)|$ within the passband is $1/\sqrt{1 + \epsilon^2}$, the peak-to-peak passband ripple is

$$\Delta_R = 1 - \frac{1}{\sqrt{1 + \epsilon^2}} \tag{E-28}$$

The total number of points within the passband (including $\omega = 0$) at which the derivative of $|H_c(\omega)|$ with respect to ω is zero is equal to the order of the filter. This observation makes it relatively easy to sketch the frequency response of a Chebyshev I filter.

It is relatively simple to compute the additional stop-band attenuation given by the Chebyshev I filter over the Butterworth filter by comparing the amplitude responses of both filters for $\omega \gg 1$. For the Butterworth filter

$$|H_B(\omega)| = \frac{1}{\sqrt{1 + \omega^{2n}}} \tag{E-29}$$

so that for $\omega \gg 1$,

$$|H_B(\omega)| \simeq \frac{1}{\omega^n} \tag{E-30}$$

which, expressed in decibels, is

$$20 \log|H_B(\omega)| = -20n \log \omega \tag{E-31}$$

For the Chebyshev I filter with $\omega \gg 1$, we have

$$|H_c(\omega)| = \frac{1}{\epsilon C_n(\omega)} \tag{E-32}$$

even for small ϵ (see Problem E-12).

Placing (E-32) in a form that allows comparison with (E-31) requires some effort. Since $\omega \gg 1$, we use (E-21)

$$C_n(\omega) = \cosh(n \cosh^{-1} \omega), \qquad |\omega| > 1 \tag{E-33}$$

For large x

$$\cosh^{-1} x \simeq \ln 2x, \qquad x \gg 1 \tag{E-34}$$

so that

$$C_n(\omega) \simeq \cosh(n \ln 2\omega) = \cosh[\ln(2\omega)^n], \qquad \omega \gg 1 \tag{E-35}$$

Equation (E-34) results from the fact that for large x

$$\cosh x \simeq \tfrac{1}{2}e^x, \qquad \omega \gg 1 \tag{E-36}$$

Also, for $\omega \gg 1$, we have

$$C_n(\omega) \simeq \tfrac{1}{2}\exp[\ln(2\omega)^n] = \tfrac{1}{2}(2\omega)^n, \qquad \omega \gg 1 \tag{E-37}$$

Substituting (E-37) into (E-32) yields

$$|H_c(\omega)| \simeq \frac{2}{\epsilon}(2\omega)^{-n}, \qquad \omega \gg 1 \tag{E-38}$$

which, expressed in decibels, is

$$20 \log|H_c(\omega)| \simeq 20 \log 2 - 20 \log \epsilon - 20n \log 2\omega \qquad \text{(E-39)}$$

or

$$20 \log|H_c(\omega)| \simeq -20(n-1) \log 2 - 20 \log \epsilon - 20n \log \omega \qquad \text{(E-40)}$$

The additional *attenuation* of the Chebyshev I filter over the Butterworth filter, expressed in decibels, is

$$\Delta_A = 20 \log|H_B(\omega)| - 20 \log|H_c(\omega)|$$
$$\simeq 20(n-1) \log 2 + 20 \log \epsilon \qquad \text{(E-41)}$$

For $\epsilon \gg 1$, the value of Δ_A can be negative, indicating less stop-band attenuation for the Chebyshev I filter. However, if $\epsilon < 1$, the bandwidth of the Chebyshev I filter is effectively larger than the bandwidth of the Butterworth to which it is being compared. Letting $\epsilon = 1$, which yields 3-dB attenuation at $\omega = 1$, results in

$$\Delta_A \simeq 20(n-1) \log 2 = 6(n-1) \text{ dB} \qquad \text{(E-42)}$$

indicating equal stop-band attenuation for $n = 1$. The value of Δ_A increases as n and ϵ increase above these values.

The preceding discussion illustrates that there is a fundamental difficulty in comparing the Butterworth and Chebyshev I filters. This difficulty results because, at bandedge ($\omega = 1$), the amplitude response of the Butterworth filter is $1/\sqrt{2}$ and the amplitude response of the Chebyshev I filter is $1/\sqrt{1 + \epsilon^2}$. In other words, $\omega = 1$ defines the half-power point for the Butterworth prototype and $\omega = 1$ defines the end of the ripple band in the Chebyshev prototype.

It is easy to determine the half-power frequency for the Chebyshev I filter. If the amplitude response is to be 0.707, then

$$\epsilon C_n(\omega_3) = 1 \qquad \text{(E-43)}$$

where ω_3 is the half-power frequency. The value of ω_3 will be greater than one unless $\epsilon > 1$. Since $\epsilon < 1$ for typical applications, we shall use (E-21) for $C_n(\omega)$. This yields

$$\epsilon \cosh(n \cosh^{-1} \omega_3) = 1 \qquad \text{(E-44)}$$

or

$$\omega_3 = \cosh\left(\frac{1}{n} \cosh^{-1} \frac{1}{\epsilon}\right) \qquad \text{(E-45)}$$

It is easily seen that for $\epsilon < 1$, the 3 dB frequency exceeds $\omega = 1$.

It is worth noting that for $n = 1$

$$C_1(\omega) = \omega \qquad \text{(E-46)}$$

so that

$$|H_c(\omega)| = \frac{1}{\sqrt{1 + (\omega\epsilon)^2}} \qquad \text{(E-47)}$$

We shall show in the following section that this is simply a first-order Butterworth filter with a scaled 3-dB frequency of $1/\epsilon$ radians per second.

It is also worth noting that since Δ_A is not a function of ω, the slopes of the stop-band amplitude responses of both the Butterworth and Chebyshev I filters for large ω are equal.

The amplitude responses of several Chebyshev I filters are illustrated in Figure E-6 for 1-dB pass-band ripple and in Figure E-7 for 3-dB passband ripple. A comparison of Figures E-6 and E-7 illustrates that, as the passband ripple increases, the stop-band attenuation also increases for fixed frequency and filter order. A comparison of Figures E-6 and E-7 with Figure E-4 illustrates the increased stop-band attenuation of the Chebyshev I filter compared with the Butterworth filter.

The passband responses of Chebyshev I filters are illustrated in greater detail in Figures E-8 and E-9. Figure E-8 shows the amplitude response assuming 1 dB passband ripple for the order $n = 1, 2, 3,$ and 4. In Figure E-9 the filter order is 4 and the amplitude response is shown for 0.1, 0.5, 1.0, and 2.0 dB passband ripple.

The pole locations of a normalized Chebyshev I filter are only slightly more difficult to specify than the pole locations of a Butterworth filter. The process of locating the poles is illustrated in Figure E-10 for the filter order $n = 4$ and 1 dB passband ripple. The first step is to draw two circles having radii a and b. The values of a and b are determined by the passband ripple factor, ϵ, according to

$$b, a = \tfrac{1}{2}[(\sqrt{1 + 1/\epsilon^2} + 1/\epsilon)^{1/n} \pm (\sqrt{1 - 1/\epsilon^2} + 1/\epsilon)^{-1/n}] \tag{E-48}$$

in which n is the order of the filter. The two circles define the major and minor axes of an ellipse, as shown by the dashed line in Figure E-10. The next step is to define the angles to the poles of a Butter-worth filter having order n and to mark these points on both of the circles. These points are indicated by the heavy dots on Figure E-10. Projecting these points horizontally from the b circle and vertically

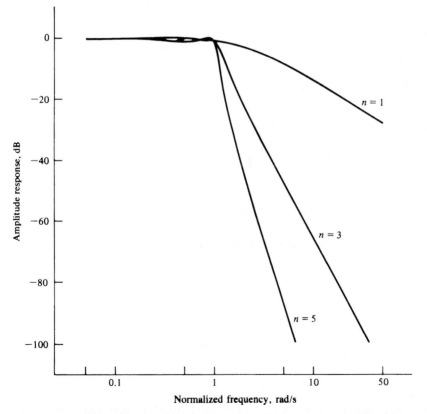

FIGURE E-6.   Amplitude response of Chebyshev I filter with 1-dB passband ripple.

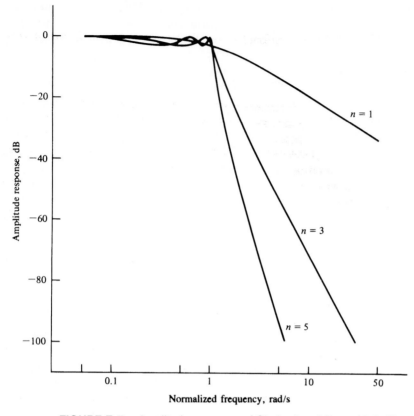

FIGURE E-7.   Amplitude response of Chebyshev I filter with 3-dB passband ripple.

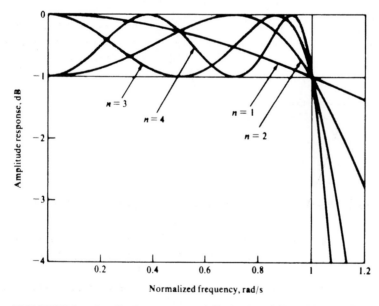

FIGURE E-8.   Amplitude response of Chebyshev I filters with 1-dB passband ripple and orders $n = 1$, 2, 3, and 4.

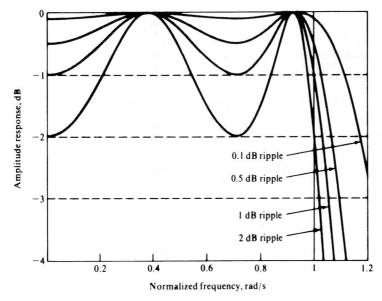

FIGURE E-9. Amplitude response of fourth-order Chebyshev I filters with passband ripple of 0.1, 0.5, 1, and 2 dB.

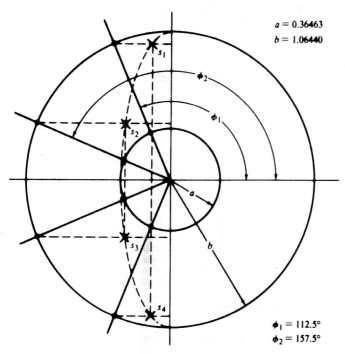

FIGURE E-10. Pole locations for a fourth-order normalized Chebyshev I filter with 1-dB passband ripple.

from the a circle defines the ellipse and the poles of the normalized Chebyshev I filter. It follows that the pole locations are

$$s_1 = a \cos \phi_1 + jb \sin \phi_1 \qquad \text{(E-49a)}$$

$$s_2 = a \cos \phi_2 + jb \sin \phi_2 \qquad \text{(E-49b)}$$

$$s_3 = a \cos \phi_2 - jb \sin \phi_2 \qquad \text{(E-49c)}$$

and

$$s_4 = a \cos \phi_1 - jb \sin \phi_1 \qquad \text{(E-49d)}$$

The numerical values are determined in the following example.

EXAMPLE E-1

In this example we determine the system function, $H(s)$, of a Chebyshev I filter in order to illustrate the concepts discussed in Section E-3. As an intermediate step the pole locations will be determined.
For a 1 dB passband ripple the value of ϵ is

$$\epsilon^2 = 10^{1/10} - 1 = 0.25893 \qquad \text{(E-50)}$$

from which

$$\epsilon = 0.50885 \qquad \text{(E-51)}$$

Using these values in (E-48) yields

$$b = 1.06440 \qquad \text{and} \qquad a = 0.36463 \qquad \text{(E-52)}$$

as shown in Figure E-10.
The angles are

$$\phi_1 = 90° + 22.5° = 112.5° \qquad \text{(E-53a)}$$

and

$$\phi_2 = \phi_1 + 45° = 157.5° \qquad \text{(E-53b)}$$

Therefore, from (E-49),

$$s_1 = -0.13954 + j0.98338 \qquad \text{(E-54a)}$$

$$s_2 = -0.33687 + j0.40733 \qquad \text{(E-54b)}$$

$$s_3 = -0.33687 - j0.40733 \qquad \text{(E-54c)}$$

and

$$s_4 = -0.13954 - j0.98338. \qquad \text{(E-54d)}$$

From these pole locations the transfer function can be determined. The result is

$$H(s) = \frac{A}{(s^2 + 0.27908s + 0.98651)(s^2 + 0.67374s + 0.24940)} \qquad \text{(E-55)}$$

The dc response of the filter is

$$H(0) = \frac{A}{(0.98651)(0.27940)} = \frac{A}{0.27563} \qquad \text{(E-56)}$$

If A is set equal to 0.27563, the product of the constant terms in the denominator, the dc gain will be one. Since this is an even-order Chebyshev I filter, we desire the dc gain to be $1/\sqrt{1 + \epsilon^2}$. Thus we must scale A by $1/\sqrt{1 + \epsilon^2}$. Therefore

$$A = \frac{0.27563}{\sqrt{1 + \epsilon^2}} = \frac{0.27563}{\sqrt{1.25893}} = 0.24566 \tag{E-57}$$

and $H(s)$ can be written

$$H(s) = \frac{0.24566}{s^4 + 0.95282s^3 + 1.45394s^2 + 0.74263s + 0.27563} \tag{E-58}$$

MATLAB *Application*

Using the routine `cheby1` from the MATLAB signals and Systems Toolbox allows the numerator and denominator polynomials to be easily determined and from the denominator polynomial the pole locations can be determined easily. The MATLAB code is

```
EDU» aeex1

n = 4;                          % Define filter order
Rp = 1;                         % Set passband ripple in dB
Wn = 1;                         % Set critical freq in rad/s
[b,a] = cheby1(n,Rp,Wn,'s');    % Design analog prototype
poles = roots(a);               % Poles
poles                           % Display poles

poles =

  -0.1395 + 0.9834i
  -0.1395 - 0.9834i
  -0.3369 + 0.4073i
  -0.3369 + 0.4073i

a                               % Display denominator coefficients

a =

     1.0000    0.9528    1.4539    0.7426    0.2756

b                               % Display numerator coefficients

b =

     0    0.0000    0.0000    0.0000    0.2457
```

We can see that the poles, numerator coefficients and the denominator coefficients are, to the precision displayed, in agreement with the previously obtained results.

E-4 Filter Transformations

In the previous sections emphasis has been placed on the normalized filter prototypes, that is, low-pass filters having a bandwidth of 1 rad/s. In this section we see how to convert the normalized filter to low-pass filters having arbitrary bandwidth and also how to convert the normalized filter to bandpass and notch (band reject) filters.

Frequency Scaling

The 1-rad/s bandwidth of a normalized prototype is converted to a bandwidth of ω_c radians per second by substituting s/ω_c for s in the normalized prototype system function $H(s)$. This is equivalent to substituting ω/ω_c or f/f_c in the sinusoidal steady-state transfer function.

EXAMPLE E-2 _____

This example illustrates the use of frequency scaling to convert the amplitude response of a proto-type filter to a lowpass filter having an arbitrary bandwidth.

The amplitude response of an nth-order Butterworth filter having a 3-dB bandwidth of 150 Hz is, from (E-29),

$$|H_B(\omega)| = \frac{1}{\sqrt{1 + [\omega/2\pi(150)]^{2n}}} = \frac{1}{\sqrt{1 + (f/150)^{2n}}} \tag{E-59}$$

EXAMPLE E-3 _____

This example illustrates the development of the transfer function for a second-order Butterworth filter having an arbitrary 3-dB bandwidth. The result of this example is used in Example 9-7 in which the bilinear z-transform digital filter, based on a second-order Butterworth analog filter, is developed.

The transfer function of a second-order Butterworth filter with a 3-dB bandwidth of ω_c radians per second is found by using $B_2(s)$ in Table E-1 with s replaced by s/ω_c. The result is

$$H(s) = \frac{1}{(s/\omega_c)^2 + \sqrt{2}\,(s/\omega_c) + 1} \tag{E-60}$$

which yields

$$H(s) = \frac{\omega_c^2}{s^2 + \sqrt{2}\,\omega_c s + \omega^{2c}} \tag{E-61}$$

EXAMPLE E-4 _____

In this example frequency scaling is illustrated using a third-order Butterworth filter prototype. The system function $H(s)$ for a third-order Butterworth filter having a 3-dB bandwidth of 150 Hz is determined by first substituting

$$\frac{s}{2\pi(150)} \tag{E-62}$$

for s in $B_3(s)$ in Table E-1. The required system function is then

$$H(s) = 1/B_3\left(\frac{s}{2\pi(150)}\right) \tag{E-63}$$

Since

$$B_3\left(\frac{s}{2\pi(150)}\right) = \frac{s^3 + 2[2\pi(150)]s^2 + 2\,[2\pi(150)]^2\,s + [2\pi(150)]^3}{[2\pi(150)]^3} \tag{E-64}$$

which is

$$B_3\left(\frac{s}{2\pi(150)}\right) = \frac{s^3 + 1.88496(10^3)s^2 + 1.77653(10^6)s + 8.37169(10^8)}{8.37169(10^8)} \tag{E-65}$$

the required system function is

$$H(s) = \frac{8.37169(10^8)}{s^3 + 1.88496(10^3)s^2 + 1.77653(10^6)s + 8.37169(10^8)} \tag{E-66}$$

MATLAB *Application*

In the following MATLAB program, we first explicitly perform frequency scaling using the MATLAB command `lp2lp` from the Signals and Systems Toolbox. As a check, the scaled filter is developed directly by using the scaled frequency in the argument of the `butter` function. The results, of course, agree. Note the roundoff error present in the calculated numerator coefficients. Note that the two design methods result in different roundoff errors.

```
EDU» aeex4
format short e              % Set display format
n =3;                      % Define order
Wn = 1;                    % Prototype bandwidth
Wo = 2*pi*150;             % Specify new bandwidth
[b,a] = butter(n,Wn,'s');  % Prototype coefficients
[b1,a1] = lp2lp(b,a,Wo);   % Convert bandwidth
b1 % Numerator coefficients

b1 =

            0   1.8190e-012   4.6566e-009   8.3717e+008

a1                         % Denominator coefficients

a1 =

1.0000e+000 1.8850e+003 1.7765e+006 8.3717e+008

%
% Check on previous results
%
[b2,a2] = butter(n,Wo,'s');  % Determine coefficients
b2                           % Numerator coefficients

b2 =

            0 6.8212e-013 -4.6566e-010   8.3717e+008
a2                           % Denominator coefficients
```

```
a2 =

  1.0000e+000   1.8850e+003   1.7765e+006   8.3717e+008
```

Low-Pass-to-Bandpass Transformation

Assume that $H(s)$ represents the system function of a low-pass filter having a bandwidth of 1 rad/s (i.e., a normalized prototype). This filter is easily converted into a bandpass filter having a *geometric center frequency* of ω_c radians per second and a bandwidth of ω_b radians per second by substituting for s in $H(s)$ the quantity

$$\frac{s^2 + \omega_c^2}{s\omega_b}$$

In other words, the bandpass transfer function is defined by

$$H_{bp}(s) = H\left(\frac{s^2 + \omega_c^2}{s\omega_b}\right) \tag{E-67}$$

This transformation illustrates that the order of the bandpass filter is twice the order of the low-pass filter from which it is derived.

The bandwidth of the low-pass prototype is $\omega = 1$. Thus the corresponding critical frequencies[†]of the bandpass filter are given by the two solutions of the quadratic equation

$$\left.\frac{s^2 + \omega_c^2}{s\omega_b}\right|_{s=j\omega} = \frac{-\omega^2 + \omega_c^2}{j\omega\omega_b} = j1 \tag{E-68}$$

This yields

$$\omega^2 - \omega\omega_b - \omega_c^2 = 0 \tag{E-69}$$

whose solutions are

$$\omega = \tfrac{1}{2}\omega_b \pm \tfrac{1}{2}\sqrt{\omega_b^2 + 4\omega_c^2} \tag{E-70}$$

The upper critical frequency ω_u is obtained by using the plus sign. Thus

$$\omega_u = \tfrac{1}{2}\omega_b + \tfrac{1}{2}\sqrt{\omega_b^2 + 4\omega_c^2} \tag{E-71}$$

Using the minus sign in (E-70) clearly results in a negative quantity. This is reconciled by realizing that, since the amplitude response is an even function of frequency, the lower critical frequency ω_l can be written

$$\omega_l = -\tfrac{1}{2}\omega_b + \tfrac{1}{2}\sqrt{\omega_b^2 + 4\omega_c^2} \tag{E-72}$$

Taking the difference between (E-71) and (E-72) illustrates that

$$\omega_u - \omega_l = \omega_b \tag{E-73}$$

which is the standard definition of bandwidth.

The relationship between the two critical frequencies and ω_c is also easily derived. The product of the two critical frequencies is, from (E-71) and (E-72),

$$\omega_u\omega_l = \tfrac{1}{4}(\omega_b + \sqrt{\omega_b^2 + 4\omega_c^2})(-\omega_b + \sqrt{\omega_b^2 + 4\omega_c^2})$$

[†]Usually we speak in terms of 3-dB break frequencies. However, for some filter prototypes the attenuation at $\omega = 1$ is not 3 dB (consider, for example, the Chebyshev filter). Thus we must be careful in our terminology.

which yields

$$\omega_u \omega_l = \omega_c^2 \tag{E-74}$$

Thus

$$\omega_c = \sqrt{\omega_u \omega_l} \tag{E-75}$$

which states that the center frequency of a bandpass filter, derived by use of the transformation (E-67), is the geometric mean of the upper and lower critical frequencies. Thus we refer to ω_c as the geometric center frequency.

It should be noted that if $\omega_c \gg \omega_b$, which is typically the case, (E-71) and (E-72) become

$$\omega_u \simeq \omega_c + \tfrac{1}{2}\omega_b \tag{E-76a}$$

and

$$\omega_l \simeq \omega_c - \tfrac{1}{2}\omega_b \tag{E-76b}$$

respectively. Adding these two equations yields

$$\omega_c \simeq \tfrac{1}{2}(\omega_u + \omega_l) \tag{E-77}$$

An example will illustrate the application of the low pass-to-bandpass transformation.

EXAMPLE E-5

The purpose of this problem is to illustrate the lowpass bandpass conversion. In this example the transfer function of a fourth-order Butterworth bandpass filter having a geometric center frequency of 1,000 Hz and a bandwidth of 100 Hz will be determined. Equation (E-67) illustrates that a second-order low-pass filter will yield a fourth-order bandpass filter upon application of the low-pass-to-bandpass transformation. Thus the required transfer function is

$$H(s) = \frac{1}{B_2\left[(s^2 + \omega_c^2)/s\omega_b\right]} \tag{E-78}$$

Using $B_2(s)$ from Table E-1 yields

$$H(s) = \frac{1}{\left[(s^2 + \omega_c^2)/s\omega_b\right]^2 + 1.41421\left[(s^2 + \omega_c^2)/s\omega_b\right] + 1} \tag{E-79}$$

which can be written as

$$H(s) = \frac{\omega_b^2 s^2}{s^4 + 1.41421\omega_b s^3 + (2\omega_c^2 + \omega_b^2)s^2 + 1.41421\omega_c^2\omega_b s + \omega_c^4} \tag{E-80}$$

The required specifications are realized by letting

$$\omega_c = 2\pi(1000) = 2000\pi \tag{E-81}$$

and

$$\omega_b = 2\pi(100) = 200\pi \tag{E-82}$$

Substituting these values into (E-80) yields

$$H(s) = \frac{3.94784(10^5)s^2}{s^4 + 8.88577(10^2)s^3 + 7.93516(10^7)s^2 + 3.50796(10^{10})s + 1.55855(10^{15})} \tag{E-83}$$

MATLAB Application

Lowpass to bandpass conversion is carried out easily using the routine 1p2bp in the MATLAB Signals and Systems Toolbox. The following code illustrates the process:

```
EDU» aeex5

n = 2;                          % Define order
Wn = 1;                         % Define 3-dB freq in rad/s
Wo = 2*pi*1000;                 % fc in rad/s (Wc)
Wb = 2*pi*100;                  % fb in rad/s (Wb)
[b,a] = butter(n,Wn,'s');       % Establish analog prototype
[bt,at] = 1p2bp(b,a,Wo,Wb);     % Convert to bandpass filter
bt                              % Display numerator coefficients


bt =

          0   2.2737e-013   3.9478e+005   1.5259e-005   2.5000e-001

at                              % Display denominator coeffs.

at =

  1.0000e+000   8.8858e+002   7.9352e+007   3.5080e+010   1.5585e+015

% The rest of this file is to do the plot.

w = 2*pi*logspace(2,4,500);     % Compute freq. vector in rad/s
h = freqs(bt,at,w);             % Compute frequency response
mag = 20*log10(abs(h));         % Determine amplitude response
f=w/(2*pi);                     % Convert to Hz for plot
semilogx(f,mag)                 % Execute plot
xlabel('Frequency - Hz')        % Label x axis
ylabel('Amplitude Response - dB') % Label y axis
```

Note that the denominator polynimial appears correct (and it is). The numerator polynomial indicates small values as coefficients to terms other than the s^2 term. The small values shown result from roundoff errors in the MATLAB program.

Low-Pass-to-Band-Reject (Notch) Transformation

A band-reject filter can be obtained from a standard low-pass prototype by using the transformation

$$H_{br}(s) = H\left(\frac{s\omega_b}{s^2 + \omega_c^2}\right) \qquad \text{(E-84)}$$

which is the reciprocal of the low-pass-to-bandpass transformation. The analysis of the resulting parameters (bandwidth and center frequency) is exactly parallel to the analysis of the bandpass filter; thus it will not be pursued here. However, it can be easily shown that the bandwidth of the notch filter is ω_b radians per second and that ω_c is the geometric center frequency of the notch in radians per second.

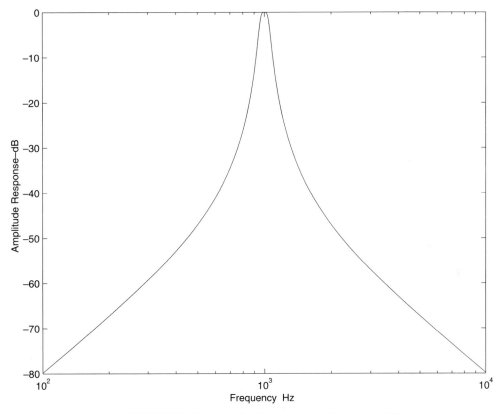

FIGURE E-11.   Amplitude response of bandpass filter.

EXAMPLE E-6

The purpose of this example is to design a notch filter, suitable for removing 60 Hz power line interference from signals within a control system, based on the lowpass to notch transformation. The analytical development is left as an end-of-chapter problem. We, therefore, go immediately to the MATLAB Application.

MATLAB Application

Except for the fact that we use the MATLAB routine `lp2bs` to realize the lowpass to notch filter transformation and that we change the center frequency and bandwidth to appropriate values to remove the 60 Hz power line interference, this problem is identical to the previous problem. We set the center frequency to 60 Hz and the bandwidth to 2 Hz. The corresponding MATLAB code is as follows:

```
EDU» aeex6

n = 2;                          % Define order
Wn = 1;                         % Define 3-dB freq in rad/s
Wo = 2*pi*60;                   % fc in rad/s (Wc)
Wb = 2*pi*2;                    % fb in rad/s (Wb)
[b,a] = butter(n,Wn,'s');       % Establish analog prototype
[bt,at] = lp2bs(b,a,Wo,Wb);     % Convert to bandstop filter
bt                              % Display numerator coefficients
```

```
bt =

1.0000e+000          0 2.8424e+005          0  2.0199e+010

at                              % Display denominator coeffs.

at =

1.0000e+000  1.7772e+001  2.8440e+005  2.5257e+006  2.0199e+010
```

% The rest of this file is to do the plot.

```
w = 2*pi*logspace(1,2,500);      % Compute freq. vector in rad/s
h = freqs(bt,at,w);              % Compute frequency response
mag = 20*log10(abs(h));          % Determine amplitude response
f=w/(2*pi);                      % Convert to Hz for plot
semilogx(f,mag)                  % Execute plot
xlabel('Frequency - Hz')         % Label x axis
ylabel('Amplitude Response - dB') % Label y axis
```

Note that we have a deep notch at 60 Hz as required to remove power line noise.

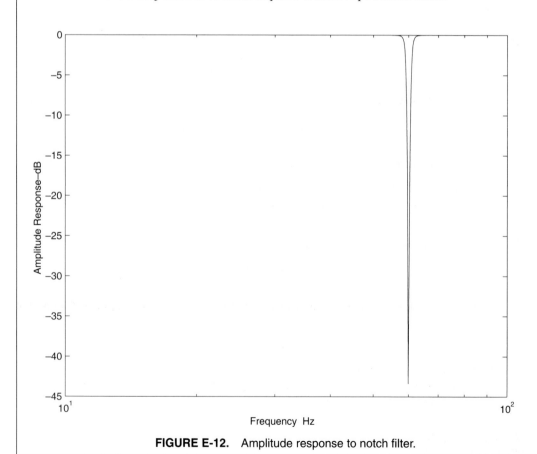

FIGURE E-12. Amplitude response to notch filter.

Further Reading

L. WEINBERG, *Network Analysis and Synthesis,* New York: McGraw-Hill, 1962. This is the classic text on filter synthesis. It is written at the senior or beginning graduate level and is extremely complete. The transfer functions of all the filter prototypes discussed in this chapter are derived in detail. Implementations using lumped-parameter, passive circuit elements are discussed.

A. I. ZVEREV, *Handbook of Filter Synthesis,* New York: Wiley, 1967. This volume contains extensive collections of amplitude response and group delay characteristics that are valuable in learning the characteristics of the various prototypes. It also contains extensive sets of tables giving the required element values for various implementations.

L. P. HUELSMAN and P. E. ALLEN, *Introduction to the Theory and Design of Active Filters,* New York: McGraw-Hill, 1980. This book is a standard text on active filter theory, which is an important area not covered in this book. The student who has an interest in the implementation of analog filters should be aware of the concept of the active filter and should appreciate the techniques used in implementing active filters.

M. E. VAN VALKENBURG, *Analog Filter Design.* New York: Holt, Rinehart and Winston, 1982. This book contains much information on classical filter theory as well as information on the synthesis of active analog filters based on a large number of prototypes.

Problems

Section E-3

E-1. It can be shown that the coefficients of the nth-order Butterworth polynomial

$$B_n(s) = \sum_{k=0}^{n} a_k s^k$$

are given by the recursion formula

$$a_{k+1} = \frac{\cos(k\pi/2n)}{\sin[(k+1)(\pi/2n)]} a_k$$

with $a_0 = 1$.[†] By using this recursion formula, derive $B_n(s)$ in Table E-1 for $n = 2, 3,$ and 4.

E-2. Use the fact that the poles of the Butterworth filter are distributed on the unit circle to find the Butterworth polynomials $B_n(s)$ for $n = 1, 2, 3, 4, 5,$ and 6. Compare the results with those given in Table E-1.

E-3. It can be shown that the system function for a low-pass Bessel filter is given by

$$H(s) = \frac{k}{Y_n(s)}$$

where $Y_n(s)$ is the Bessel polynomial of degree n and k is a suitably chosen constant. Assume that $Y_2(s)$ is given by[‡]

$$Y_2(s) = s^2 + 3s + 3$$

(a) Determine k so that the dc gain is 1.

(b) Plot the amplitude response of both the second-order Bessel and Butterworth filters together using a single coordinate system. Comment on your results.

[†]Weinberg (1962), p. 494.

[‡]Weinberg (1962), p. 500.

(c) Plot the phase response of both the Bessel and Butterworth filters together using a single co-ordinate system and linear scales. Comment on your results.

(d) The bandwidth of a low-pass Bessel filter is defined to be

$$f_c = \frac{\omega_c}{2\pi} = \frac{1}{2\pi t_0}$$

where t_0 is the nominal passband group delay of the filter. Using this definition, compare the bandwidths of the Bessel and Butterworth filters.

E-4. Repeat Problem E-3 for third-order Bessel and Butterworth filters assuming that

$$Y_3(s) = s^3 + 6s^2 + 15s + 15$$

E-5. Determine the values of ϵ required to give a Chebyshev I filter having a passband ripple of (a) 0.2 dB, (b) 0.5 dB, (c) 2 dB, (d) 2.5 dB, (e) 3 dB, and (f) 5 dB.

E-6. Determine the 3-dB bandwidth for a normalized fourth-order Chebyshev I filter assuming a pass-band ripple of (a) 0.05 dB, (b) 0.2 dB, (c) 1 dB, (d) 2 dB, (e) 3 dB, and (f) 5 dB.

E-7. Repeat Problem E-6 assuming that the Chebyshev I filter is fifth order.

E-8. Determine the values of a and b for a normalized Chebyshev I filter assuming that the pass-band ripple is 0.75 dB and that the filter order, n, is three. Determine the pole locations and de-termine $H(s)$.

E-9. Repeat Problem E-8 assuming that the passband ripple is 2 dB and that the filter order, n, is four.

E-10. Repeat Problem E-8 assuming that the passband ripple is 2 dB and that the filter order, n, is five.

E-11. Sketch, approximately to scale, the amplitude response of a Chebyshev I filter having a 3-dB passband ripple for $n = 2, 3, 4,$ and 5. Pay particular attention to the response at $\omega = 0$ and at $\omega = 1$. Also pay particular attention to the number of passband ripples.

E-12. Show by plotting $C_n(\omega)$ as a function of ω in the range $0 \le \omega \le \omega_x$, where $\omega_x > 1$, that $C_n(\omega)$ oscillates in the filter passband and increases rapidly for $\omega > 1$. Let $n = 2, 3,$ and 4.

E-13. A low-pass filter is to have a dc ($f = 0$) gain of 1 and is to have a gain of 0.707 (-3dB) at 100 Hz. The attenuation is to be at least 45 dB at 400 Hz. Determine the necessary filter order as-suming that a Butterworth filter is used.

E-14. Repeat the preceeding problem assuming that a Chebyshev I filter is used and that the allowable passband ripple is 0.5 dB. How does the answer change if we allow 2 dB passband ripple? Re-peat for 3 dB passband ripple.

Section E-4

E-15. Derive the transfer function for a second-order Butterworth low-pass filter having a 3-dB band-width of 300 Hz. Plot accurately the amplitude and phase responses of the low-pass filter.

E-16. Derive the transfer function for a second-order Butterworth bandpass filter having a geometric center frequency of 300 Hz and a bandwidth of 600 Hz. Verify your result by plotting the am-plitude and phase responses.

E-17. Determine the transfer function for a fourth-order Butterworth bandpass filter with a geometric center frequency of 100 Hz and a bandwidth of 50 Hz. Determine the upper and lower 3-dB frequencies.

E-18. Repeat Problem E-17 assuming that the bandpass filter is to be sixth order.

E-19. Determine the transfer function of a fourth-order Butterworth notch filter with a notch geometric center frequency of 1,000 Hz and a notch bandwidth of 300 Hz. Plot accurately the amplitude and phase responses of the notch filter.

E-20. A bandpass filter has a center frequency of 900 Hz and a bandwidth of 600 Hz. Determine f_l and f_u, the lower and upper critical frequencies.

E-21. Determine the transfer function of a fourth-order Butterworth notch filter for which the geometric center frequency of the notch is 1,000 Hz and the bandwidth of the notch is 100 Hz. Compare the denominator polynomial of the transfer function for the notch filter with the corresponding denominator polynomial found for the bandpass filter found in Example E-5. What do you conclude?

Computer Projects

E-1. By writing an appropriate MATLAB program, verify the results given in Table E-1.

E-2. Develop a MATLAB M-file that will yield amplitude response plots for a set of Butterworth filters as shown in Figure E-4. The input to the program is to be a vector containing the filter orders for which the amplitude response is to be determined. Include other features in the program that you deem useful.

E-3. Repeat the previous exercise for a Chebyshev I filter. For this case the input to the program is to be a 2 by n matrix **X**, in which n is total number of curves to be displayed on the plot (n different filter orders). The first row of **X** defines the filter orders for which an amplitude response is to be generated and the second row of **X** corresponds to the passband passband ripple in dB. In other words, each column vector defines a filter with the first element defining the order and the second element defining the passband ripple of the filter.

E-4. Develop a MATLAB M-file for plotting the pole-zero locations of a system function $H(s)$. The inputs to the function are to be the minimum and maximum values of the real parts of the pole-zero locations, the minimum and maximum values of the imaginary parts of the pole-zero locations and two vectors containing the coefficients of the denominator and numerator polynomials of $H(s)$. The M-file plots the poles and zeros and labels the plot accordingly. The resulting M-file is to be used for investigating the relationship between the poles and zeros of a prototype filter and the pole and zero locations of the bandpass and notch filters derived from the prototype.

Using the M-file plot the poles and zeros of a tenth-order Butterworth prototype filter (lowpass with a bandwidth of 1 radian/second). From the prototype filter develop a bandpass Butterworth filter with a center frequency of 10 Hz and a bandwidth of 2 Hz and plot the poles and zeros of the bandpass filter. From the prototype filter develop a notch filter with a notch center frequency of 10 Hz and a notch bandwidth of 2 Hz. Plot the poles and zeros of the notch filter using the M-file developed in the first part of his MATLAB project.

E-5. Figure E-5 illustrates the group delay of Butterworth prototype filters. The calculations used to produce the group delay characteristics shown in Figure E-5 were based on (E-17). Develop a MATLAB M-file (function) that evaluates (E-17) and plots, and appropriately labels, the group delay. The inputs to the function file are to be the 3-dB frequency of the filter and a vector containing the filter orders for which the group delay is to be determined. Use the resulting M-file to verify the group delay characteristics shown in Figure E-5.

As shown in Appendix E, the group delay can be calculated by differentiating the phase response of the filter. Develop a second M-file that determines the group delay of a filter by numerical differentiation of the phase response of a filter. Write a MATLAB program to generate the phase response of a Butterworth filter and use the M-file to determine the group delays of prototype Butterworth filters having orders 1-5. Compare the results with those shown in Figure E-5. Experiment with the sampling rate applied to the phase response. What do you conclude?

After you have confidence in the M-file you developed for the first part of this problem, use the M-file to determine the group delay of a fourth-order Butterworth bandpass filter having a center frequency of 1,000 Hz and a bandwidth of 200 Hz.

F

Mathematical Tables

The Sinc Function

z	sinc z	sinc^2 z	z	sinc z	sinc^2 z
0.0	1.0	1.0	1.6	−0.18921	0.03580
0.1	0.98363	0.96753	1.7	−0.15148	0.02295
0.2	0.93549	0.87514	1.8	−0.10394	0.01080
0.3	0.85839	0.73684	1.9	−0.05177	0.00268
0.4	0.75683	0.57279	2.0	0	0
0.5	0.63662	0.40528	2.1	0.04684	0.00219
0.6	0.50455	0.25457	2.2	0.08504	0.00723
0.7	0.36788	0.13534	2.3	0.11196	0.01254
0.8	0.23387	0.05470	2.4	0.12614	0.01591
0.9	0.10929	0.01194	2.5	0.12732	0.01621
1.0	0	0	2.6	0.11643	0.01356
1.1	−0.08942	0.00800	2.7	0.09538	0.00910
1.2	−0.15591	0.02431	2.8	0.06682	0.00447
1.3	−0.19809	0.03924	2.9	0.03392	0.00115
1.4	−0.21624	0.04676	3.0	0	0
1.5	−0.21221	0.04503			

Trigonometric Identities

Euler's theorem: $e^{\pm ju} = \cos u \pm j \sin u$

$\cos u = \frac{1}{2}(e^{ju} + e^{-ju})$

$\sin u = (e^{ju} - e^{-ju})/2j$

$\sin^2 u + \cos^2 u = 1$

$\cos^2 u - \sin^2 u = \cos 2u$

$2 \sin u \cos u = \sin 2u$

$\cos^2 u = \frac{1}{2}(1 + \cos 2u)$

$\sin^2 u = \frac{1}{2}(1 - \cos 2u)$

$\sin(u \pm v) = \sin u \cos v \pm \cos u \sin v$

$\cos(u \pm v) = \cos u \cos v \mp \sin u \sin v$

$\sin u \sin v = \frac{1}{2}[\cos(u - v) - \cos(u + v)]$

$\cos u \cos v = \frac{1}{2}[\cos(u - v) + \cos(u + v)]$

$\sin u \cos v = \frac{1}{2}[\sin(u - v) + \sin(u + v)]$

Indefinite Integrals

$$\int \sin(ax)\, dx = -\frac{1}{a}\cos(ax)$$

$$\int \cos(ax)\, dx = \frac{1}{a}\sin(ax)$$

$$\int \sin^2(ax)\, dx = x/2 - \sin(2ax)/4a$$

$$\int \cos^2(ax)\, dx = x/2 + \sin(2ax)/4a$$

$$\int x \sin(ax)\, dx = [\sin(ax) - ax\cos(ax)]/a^2$$

$$\int x \cos(ax)\, dx = [\cos(ax) + ax\sin(ax)]/a^2$$

$$\int x^m \sin(x)\, dx = -x^m \cos(x) + m \int x^{m-1} \cos(x)\, dx$$

$$\int x^m \cos(x)\, dx = x^m \sin(x) - m \int x^{m-1} \sin(x)\, dx$$

$$\int \sin(ax) \sin(bx)\, dx = \frac{\sin(a-b)x}{2(a-b)} - \frac{\sin(a+b)x}{2(a+b)} \qquad a^2 \neq b^2$$

$$\int \sin(ax) \cos(bx)\, dx = -\left[\frac{\cos(a-b)x}{2(a-b)} + \frac{\cos(a+b)x}{2(a+b)}\right], \qquad a^2 \neq b^2$$

$$\int \cos(ax) \cos(bx)\, dx = \frac{\sin(a-b)x}{2(a-b)} + \frac{\sin(a+b)x}{2(a+b)} \qquad a^2 \neq b^2$$

$$\int e^{ax}\, dx = e^{ax}/a$$

$$\int x^m e^{ax}\, dx = \frac{x^m e^{ax}}{a} - \frac{m}{a} \int x^{m-1} e^{ax}\, dx$$

$$\int e^{ax} \sin(bx)\, dx = \frac{e^{ax}}{a^2 + b^2}[a \sin(bx) - b \cos(bx)]$$

$$\int e^{ax} \cos(bx)\, dx = \frac{e^{ax}}{a^2 + b^2}[a \cos(bx) + b \sin(bx)]$$

Definite Integrals

$$\int_0^\infty \frac{a\,dx}{a^2 + x^2} = \pi/2, \qquad a > 0$$

$$\int_0^{\pi/2} \sin^n(x)\, dx = \int_0^{\pi/2} \cos^n(x)\, dx = \begin{cases} \dfrac{1 \cdot 3 \cdot 5 \cdots (n-1)}{2 \cdot 4 \cdot 6 \cdots (n)} \dfrac{\pi}{2}, & n \text{ even, } n \text{ an integer} \\[2mm] \dfrac{2 \cdot 4 \cdot 6 \cdots (n-1)}{1 \cdot 3 \cdot 5 \cdots (n)}, & n \text{ odd} \end{cases}$$

$$\int_0^\pi \sin^2(nx)\, dx = \int_0^\pi \cos^2(mx)\, dx = \pi/2, \qquad n \text{ an integer}$$

$$\int_0^\pi \sin(mx) \sin(nx)\, dx = \int_0^\pi \cos(mx) \cos(nx)\, dx = 0, \qquad m \neq n, m \text{ and } n \text{ integer}$$

$$\int_0^\pi \sin(mx) \cos(nx)\, dx = \begin{cases} 2m/(m^2 - n^2), & m + n \text{ odd} \\ 0, & m + n \text{ even} \end{cases}$$

$$\int_0^\infty \frac{\sin(ax)}{x}\, dx = \frac{\pi}{2}, \qquad a > 0$$

Definite Integrals (*continued*)

$$\int_0^\infty \frac{\sin^2 x}{x^2}\, dx = \frac{\pi}{2}$$

$$\int_0^\infty e^{-a^2 x^2}\, dx = \sqrt{\pi}/2a, \qquad a > 0$$

$$\int_0^\infty x^n e^{-ax}\, dx = n!/a^{n+1}, \qquad n \text{ an integer and } a > 0$$

$$\int_0^\infty x^{2n} e^{-ax^2}\, dx = \frac{1 \cdot 3 \cdot 5 \cdots (2n-1)}{2^{n+1} a^n} \sqrt{\frac{\pi}{a}}$$

$$\int_0^\infty e^{-ax} \cos(bx)\, dx = \frac{a}{a^2 + b^2}, \qquad a > 0$$

$$\int_0^\infty e^{-ax} \sin(bx)\, dx = \frac{b}{a^2 + b^2}, \qquad a > 0$$

$$\int_0^\infty e^{-a^2 x^2} \cos(bx)\, dx = \frac{\sqrt{\pi}}{2a} e^{-b^2/4a^2}, \qquad a > 0$$

Index of MATLAB Functions

This appendix contains a list of all MATLAB commands used in the text together with the page numbers on which the command can be found. The list is subdivided into three groups, a list of the non-graphical commands, a list of the commands controlling graphics (graphical commands) and a list of user-defined special functions that are specified early in the book and used in later chapters.

A. Non-graphical Commands

The following commands are part of the MATLAB language. Each use is identified in the following list to give the student insight into the many ways that the various commands can be used.

B. Graphical Commands

The following commands are used in many examples throughout the book for plotting functions and for annotating plots. Only the first few uses are listed.

C. User-defined Special Functions

The following special functions are defined in Chapter 1 (Section 1.6) and are used later in the book, especially in Chapter 2. Since they are not part of the MATLAB language, they must be defined in the MATLAB workspace prior to use.

Function	Usage	Page on which defined
impls_fn	impulse function	32
pls_fn	unit pulse function	32
rmp_fn	ramp function	32
stp_fn	step function	32

H

Answers to Selected Problems

1-1. (a) $x(t) = (5/54)t^3$ km, $t \le 72$ seconds; $x(72 \text{ s}) = 35.56$ km.

1-3. (a) $K = 4$ kg/s^2; (b) $\Delta a_{min} = 1$ m/s^2.

1-7. (a) $x_{imp.samp.} = \sum_{n=-\infty}^{\infty} \cos(0.2\pi n)\delta(t - 0.1n)$.

1-9. (a) $T_0 = 0.04$ s; (c) $T_0 = 0.0286$ s; (e) $T_0 = 0.2$ s.

1-10. (b) $A + B = 8 + j11.6603$; (d) $A \cdot B = -10.9808 + j40.9808$.

1-11. (b) $T_0 = 0.1176$ s; (d) $T_0 = 2$ s; (f) $T_0 = 2$ s.

1-12. (a) $x_c(t) = \text{Re}\{3 \exp[j(10\pi t - 5\pi/6)]\}$;

  (b) $x_c(t) = 1.5 \exp[j(10\pi t - 5\pi/6)] + 1.5 \exp[-j(10\pi t - 5\pi/6)]$.

1-14. (a) $x(t) = u(t) + u(t - 3) - u(t - 5) - u(t - 6)$.

1-17. In general, $u_{-n}(t) = t^{n-1}/(n - 1)!$, $t \ge 0$ and 0 otherwise.

1-20. $x_b(t) = \sum_{n=0}^{\infty} u(t - 4n)u(2 - t - 4n)$.

1-22. $x_1(t) = u(t) + r(t - 1) - 2r(t - 2) + r(t - 3) - u(t - 4)$ (one possible representation);

  $x_3(t) = r(t) - r(t - 1) - r(t - 3) + r(t - 4)$ (one possible representation).

1-26. (b) 1; (d) 0.

1-27. (a) 9 exp(6); (c) 3.

1-28. (b) $C_1 = -3$ and all other C's are 0.

1-31. (b) -3.

1-33. (a) $E = 1/20$ J; (c) $E = 1$ J.

1-34. (b) $E_2 = 3$ J; (d) $E_4 = 2/3$ J.

1-37. (a) $P = 2^2/2 = 2$ W; (c) $P = 3^2/2 = 4.5$ W; (e) $P = 2^2/2 + 3^2/2 = 6.5$ W

  (permissible to add powers because separate sinusoids have harmonically related frequencies).

1-38. (b) Energy, with $E = 37$ J; (d) Power, with $P = \frac{1}{2}$W.

1-40. (a) Energy, with $E = 1$ J; (d) Power, with $P = 2$W.

1-41. (c) Only (3) is an energy signal. Signal (4) is neither energy nor power.

1-45. (b) 26 W.

2-2. (a) First; (c) Zero.

2-5. Hint: consider a value of $t < 1$.

2-11. (b) Causal, fixed, dynamic, order 3; (d) causal, dynamic, order 2; (f) dynamic, order 2.

2-17. (b) $y(t) = 0$, $t < 10$; $y(t) = t - 10$, $10 \le t < 12$; $y(t) = 2$, $t \ge 12$.

  (d) $y(t) = \{0.5[1 - \exp(-4)]\exp[-2(t - 2)] - 0.1[1 - \exp(-20)]$

    $\times \exp[-10(t - 2)]\}u(t - 2) + \{0.5[1 - \exp(-2t)] - 0.1[1 - \exp(-10t)]\}u(t)u(2 - t)$.

2-20. $h(t) = \delta(t)/3$.

2-23. $h(t) = \dfrac{R_2}{R_1 + R_2}\left[\delta(t) - \dfrac{1}{(R_1 + R_2)C}\exp\left(-\dfrac{t}{(R_1 + R_2)C}\right)u(t)\right]$.

2-25. (a) $a(t) = -r(t)/RC$.

2-26. $y(t) = [1 - e^{-Rt/L}]u(t)u(1 - t) + [1 - e^{-R/L}]e^{-R(t-1)/L}u(t - 1)$.

2-29. (b) $a_r(t) = \dfrac{L}{R}[1 - e^{-Rt/L}]u(t)$.

2-31. (b) $y(t) = a_s(t) - 2 a_s(t - 1) + a_s(t - 2)$ where $a_s(t) = [1 - \exp(-t/RC)]u(t)$.

2-33. (b) $A(f) = \dfrac{|f|/f_3}{\sqrt{1 + (f/f_3)^2}}$ where $f_3 = \dfrac{R}{2\pi L}$; **(c)** $\theta(f) = \dfrac{\pi}{2} - \tan^{-1}(f/f_3)$.

2-36. The system is BIBO stable.

2-37. The system is not BIBO stable.

3-3. (b) $x_2(t) = \cos(200\,\pi t) + j \sin(200\,\pi t)$;

(d) $x_4(t) = \dfrac{5}{8}\cos(20\pi t) + \dfrac{5}{16}\cos(60\pi t) + \dfrac{1}{16}\cos(100\pi t)$.

3-5. The Fourier series for the string is

$$y(x) = \frac{4}{\pi^2}\left[\cos\left(\frac{\pi x}{18}\right) + \frac{1}{9}\cos\left(\frac{3\pi x}{18}\right) + \frac{1}{25}\cos\left(\frac{5\pi x}{18}\right) + \ldots\right].$$

3-8. (a) 3/4 W; **(b)** 5/8 W at the output.

3-9. (b) 90.3%.

3-11. (a) $f_0 = \frac{3}{4}$ Hz; **(c)** $2/(1 + 9\,\pi^2)^{1/2}$.

3-13. (b) $x_b(t) = -\dfrac{j}{8}e^{-j90\pi t} + \dfrac{j3}{8}e^{-j30\pi t} + e^{-j25\pi t} - \dfrac{j3}{8}e^{j30\pi t} + \dfrac{j}{8}e^{j90\pi t}$.

3-14. (a) $Y_n = X_n \exp(-jn\,\omega_0\tau)$; **(b)** $Y_n = X_{n-1}$.

3-19. $A(t) = 2|\cos(\Delta\omega t/2)|$ and $\theta(t) = \tan^{-1}\left(\dfrac{\sin \Delta\omega t}{1 + \cos \Delta\omega t}\right)$.

3-24. $y(t) = \displaystyle\sum_{n=-\infty,\,n\ \text{odd}}^{\infty}\left(\dfrac{8A}{(\pi n)^2\sqrt{1 + n^2}}\right)\cos[n\omega_0 t - \tan^{-1}(n)]$.

3-25. (b) Phase or delay distortion only; **(d)** amplitude distortion only.

3-26. (a) Mean-square error $= 4A^2\varepsilon/T_0$; **(c)** find the Fourier series of the distorted square wave and find the ratio of the second harmonic power to fundamental power at the filter output to get

$$\dfrac{P_{2,\,\text{out}}}{P_{1,\,\text{out}}} = \left[\dfrac{\sin[2\pi(\varepsilon/T_0 + 0.5)]}{2 \sin[\pi(\varepsilon/T_0 + 0.5)]}\right]^2 \dfrac{1}{1 + 9Q^2/4}$$ with a similar expression for the ratio of third

harmonic to fundamental power. In the second harmonic case setting the ratio $= 0.001$ gives $Q = 3.23$.

3-28. $1/n^2$.

3-30. (b) $\phi_2(t) = (6e^{-2t} - 4e^{-t})u(t)$.

3-31. (b) ISE $= A^2 T_0(1 - 80/9\pi^2)$.

4-2. $|X_a(f)| = |X_b(f)| = \dfrac{A}{\sqrt{\alpha^2 + (2\pi f^2)}}$; $\theta_a(f) = -\theta_b(f) = -\tan^{-1}(2\pi f/\alpha)$.

4-6. (b) $X_b(f) = \dfrac{1}{j2\pi f}\left[\dfrac{1}{j\pi f} - 1\right]e^{-j2\pi f} + \dfrac{2}{(j2\pi f)^3}[1 - e^{-j2\pi f}]$.

4-7. The spectrum can be plotted from $X(f) = A\tau \operatorname{sinc}(f\tau)\dfrac{\cos(\pi f\varepsilon)}{1 - (2\varepsilon f)^2}$.

4-8. **(b)** $X_2(f) = \dfrac{1}{2}\Pi(f) - \dfrac{j}{2\pi}\ln\left|\dfrac{2f-1}{2f+1}\right|$ (use the Fourier transforms of the sinc-function and the unit step along with the multiplication theorem of Fourier transforms).

4-10. **(b)** The inverse Fourier transform of this signal is $\dfrac{4\pi t}{\alpha^2 + (2\pi t)^2}$.

4-11. **(c)** $G_c(f) = \dfrac{(2A\alpha)^2}{[\alpha^2 + (2\pi f)^2]^2}$.

4-12. **(a)** 70.5%; **(b)** 50%.

4-16. **(a)** $X_a(f) = 2\,\mathrm{sinc}(f)\cos(3\pi f)$; $X_c(f) = 4\,\mathrm{sinc}(4f) + 2\,\mathrm{sinc}(2f)$.

4-18. Duality gives $x(t) = X_c(-t) = 4\,\mathrm{sinc}(4t) + 2\,\mathrm{sinc}(2t)$.

4-21. **(b)** $X_2(f) = -\dfrac{\mathrm{sinc}(f)}{j\pi f}\cos(\pi f)$; **(c)** $X_3(f) = 2\,\mathrm{sinc}(2f) - \mathrm{sinc}^2(f)$.

4-23. **(b)** $x(t) = 45\,\mathrm{sinc}^2(3t)$.

4-24. **(b)** For the raised cosine pulse:
$$X_{RC}(f) = \sqrt{\dfrac{2}{3}}\,A\,\tau\left[\mathrm{sinc}(f\tau) + \dfrac{1}{2}\mathrm{sinc}(f\tau - 1) + \dfrac{1}{2}\mathrm{sinc}(f\tau + 1)\right].$$

4-25. **(a)** $x_1(t) * h_1(t) = \dfrac{1}{\beta - \alpha}\left[\dfrac{1}{\beta - \alpha}e^{-\beta t} - \left(\dfrac{1}{\beta - \alpha} - t\right)e^{-\alpha t}\right]u(t);$

(c) $x_3(t) * h_3(t) = \dfrac{1}{\beta - \alpha}[e^{-\alpha t}u(t) + e^{\beta t}u(-t)].$

4-29. **(b)** Phase distortion only.

4-31. $H(j\omega) = \dfrac{1}{1 + 2\left(\dfrac{j\omega}{2\pi \times 10^4}\right) + 2\left(\dfrac{j\omega}{2\pi \times 10^4}\right)^2 + \left(\dfrac{j\omega}{2\pi \times 10^4}\right)^3}.$

4-32. **(b)** $y(t) = [\mathrm{sinc}(t - 0.5) + \mathrm{sinc}(t + 0.5)]\cos(2\pi f_0 t).$

4-33. **(c)** $\dfrac{P_{fund.}}{P_{nth\ harm.}} = \dfrac{\left[1 - \left(\dfrac{nf_0}{f_b}\right)^2\right]^2 + \left(\dfrac{nf_0}{f_b}\right)^2}{\left[1 - \left(\dfrac{f_0}{f_b}\right)^2\right]^2 + \left(\dfrac{f_0}{f_b}\right)^2}$; **(d)** $f_0/f_b = 1.92$.

4-37. The step response is $a(t) = \left[1 - e^{-\omega_c t} - \dfrac{2}{\sqrt{3}}e^{-\omega_c t/2}\sin\left(\dfrac{\sqrt{3}}{2}\omega_c t\right)\right]u(t)$. From this, the 10% and 90% times can be solved for and used to compute the rise time ($\omega_c = 20{,}000\pi$).

4-39. **(b)** $X_b(f) = 2\sum_{n=-\infty, n\neq 0}^{\infty}\mathrm{sinc}^2(n/2)\delta(f - n/12).$

5-1. **(a)** $X(s) = \dfrac{2}{s(s + 2)}$; **(d)** $X(s) = 1 - e^{-10s}.$

5-3. **(a)** $x_1(t) = (1 + e^{-10t})u(t)$; **(c)** $x_3(t) = (e^{-5t} - e^{-10t})u(t).$

5-6. $X(s) = \dfrac{1}{s^2}(1 - e^{-s})^2.$

5-7. **(b)** $y(t) = \left[\dfrac{1}{2}(1 + e^{-2t}) - e^{-t}\right]u(t).$

5-9. $i(t) = \left(e^{-2t} - \dfrac{1}{2}e^{-t}\right)u(t).$

5-10. **(b)** $x_2(t) = [\cos(t) + \sin(t)]\exp(-2t)u(t)$; **(d)** $x_4(t) = \dfrac{10}{3}\sin(3t)\exp(-5t)u(t).$

5-11. (b) $X_\Delta(s) = \dfrac{1}{\tau s^2} \dfrac{[1 - \exp(-\tau s)]^2}{1 - \exp(-T_0 s)}$.

5-13. (a) Initial value = 1; final value = 0; **(c)** initial value doesn't exist; final value = 0.

5-14. (b) $Y_2(s) = \dfrac{s^2 + 4s + 3}{(s^2 + 4s + 5)^2}$; **(d)** $Y_4(s) = \dfrac{(s + 2)^2}{(s^2 + 4s + 5)^2}$;

  (f) $Y_6(s) = 2\dfrac{s + 1/2}{s^2 + s + 5/16} + 3\dfrac{s + 10}{s^2 + 20s + 125}$.

5-15. (b) $y(t) = \dfrac{3}{16}[-\cos(5t) + 3\cos(3t)]u(t)$; **(d)** $y(t) = \dfrac{1}{3}\{1 - \sin[3(t - 5)]\}u(t - 5)$.

5-16. (a) $x_1(t) = [9\exp(-t) - 8\exp(-2t)]u(t)$;
  (c) $x_3(t) = \delta(t) + [3\exp(-t) - \exp(-2t)]u(t)$

5-18. (a) $x(t) = \left[5\exp(-2t) + 2\exp(-3t) + \dfrac{1}{2}\sin(2t)\right]u(t)$.

5-20. $x(t = \left[\exp(-t) - \dfrac{27}{2}t^2\exp(-3t)\right]u(t)$.

5-21. (b) $x(t) = \dfrac{1}{4}\left\{1 - \left[\cos(2t) + \dfrac{1}{2}\sin(2t) - t\cos(2t)\right]\right\}\exp(-t)u(t)$.

5-26. (b) $x_2(t) = \left\{2 - \sin(3t) + \dfrac{1}{54}[\sin(3t) - 3t\cos(3t)]\right\}u(t)$;
  (d) $x_4(t) = \exp(-3t)u(t) - \exp[-3(t - 2)]u(t - 2) + \cdots$.

5-28. (b) $i(t) = \dfrac{V}{R}\cos\left(\dfrac{t}{\sqrt{LC}}\right)u(t)$.

5-30. (b) $x_2(t) = \exp(-4t)u(t) - \exp[-4(t - 2)]u(t - 2)$.

5-31. (a) Initial value = 1, final value = 0.

5-32. (b) $x_2(t) = \left[1 - \left(1 - t - \dfrac{t^2}{2}\right)\exp(-t)\right]u(t)$.

5-33. (b) $X_2(s) = -\dfrac{1}{s(s + 1)}$, $-1 < \mathrm{Re}(s) < 0$.

5-34. (b) $x_2(t) = [\cos(t) - \sin(t)]u(-t) + \exp(-t)u(t)$.

6-5. $V_{oc}(s) = \dfrac{V_s(s)}{1 + sRC_1}$ and $Z_{eq}(s) = \dfrac{s^2LC_1R + sL + R}{sC_1R + 1}$.

6-6. $v_o(t) = \dfrac{5}{6}[\exp(-t/3) - \exp(-4t/3)]u(t)$.

6-7. (b) $\begin{bmatrix} sL + R_2 & -sL \\ -sL & sL + R_1 + 1/sC \end{bmatrix}\begin{bmatrix} I_1(s) \\ I_2(s) \end{bmatrix} = \begin{bmatrix} V_s(s) + LI_0 \\ -LI_0 - V_0/s \end{bmatrix}$.

6-10. (a) $V_{oc}(s) = 3(9s + 1)/s^2$ and $Z_{thev}(s) = 6(9s - 1)$;
  (c) $V_{oc}(s) = -\dfrac{1}{2s^2}$ and $Z_{thev}(s) = 0$.

6-11. $I_R(s) = \dfrac{5s^3 + 2s^2 + 10s + 2}{s(9s^3 + 5s^2 + 18s + 6)}$.

  6-14. $v_o(t) = A\ \exp(-Rt/L)u(t) - RI_0\ \exp(-Rt/L)u(t)$
  (identify the ZSR and ZIR terms)

6-16. (d) The zero is placed at 60 hertz by setting $2\pi(60) = 1/RC$.

6-17. $H(s) = \dfrac{1}{R^3 C_1 C_2 C_3 s^3 + 2R^2 C_2 (C_1 + C_3)s^2 + R(C_1 + 3C_2)s + 1}$.

6-19. Number of RHP roots: **(a)** none; **(c)** 1.

6-20. **(b)** $0 < K < \frac{3}{4}$.

6-22. Construct the Routh array from the denominator polynomial of
$$H(s) = \frac{A_0 \omega_{p_1} \omega_{p_2}^2}{s^3 + (\omega_{p_1} + 2\omega_{p_2})s^2 + (2\omega_{p_1}\omega_{p_2} + \omega_{p_2}^2)s + \omega_{p_1}\omega_{p_2}^2(1 + A_0)}.$$
With the parameter values given it is found that the 748 configuration is unstable.

6-23. The transfer function is the same as that for Problem 6-22 except that A_0 is divided by 10. This results in stability for both amplifier types.

6-30. $H(s) = \dfrac{(G_1 + G_4)G_2 G_3}{1 + G_1 G_2 H_1 + G_1 G_2 G_3 H_2}$.

6-32. **(a)** $a = 8$ and $b = 12$; **(b)** $b > -3$.

6-33. **(a)** $\varepsilon(t) = [1 - \exp(-t)]u(t)$ which approaches 0 as $t \to \infty$; **(b)** $\varepsilon(t) = \sin(t)u(t)$ which does not approach 0 as $t \to \infty$; **(c)** choose $A(s)$ to produce two LHP real poles in $E(s)$.

7-5. $e^{\mathbf{A}t} = \begin{bmatrix} e^{-t} & 0 \\ 0 & e^{-4t} \end{bmatrix}$

7-6. $e^{\mathbf{A}t} = \begin{bmatrix} e^{-2t} & 0 \\ 0 & e^{-5t} \end{bmatrix}$

7-8. $\begin{bmatrix} x_1(t) \\ x_2(t) \end{bmatrix} = \begin{bmatrix} (1 + 2e^{-2t})u(t) \\ (2 + e^{-4t})u(t) \end{bmatrix}$

7-14. $\mathbf{H}(s) = \left(\dfrac{\dfrac{2}{(s+1)(s+3)}}{\dfrac{s}{(s+1)(s+3)}} \right)$

7-20. **(a)** $\begin{bmatrix} \dot{x}_1 \\ \dot{x}_2 \end{bmatrix} = \begin{bmatrix} -1 & 0 \\ 0 & -3 \end{bmatrix}\begin{bmatrix} x_1 \\ x_2 \end{bmatrix} + \begin{bmatrix} 1 \\ 1 \end{bmatrix}u$

$y = \begin{bmatrix} \frac{3}{2} & -\frac{3}{2} \end{bmatrix}\begin{bmatrix} x_1 \\ x_2 \end{bmatrix}$

(b) $\begin{bmatrix} \dot{x}_1 \\ \dot{x}_2 \\ \dot{x}_3 \end{bmatrix} = \begin{bmatrix} -1 & 0 & 0 \\ 0 & -5 & 0 \\ 0 & 0 & -6 \end{bmatrix}\begin{bmatrix} x_1 \\ x_2 \\ x_3 \end{bmatrix} + \begin{bmatrix} 1 \\ 1 \\ 1 \end{bmatrix}u$

$y = \begin{bmatrix} 0 & 1 & -1 \end{bmatrix}\begin{bmatrix} x_1 \\ x_2 \\ x_3 \end{bmatrix}$

7-25. **(a)** $\mathbf{F} = \begin{bmatrix} 0.7788 & 0 \\ 0 & 0.6065 \end{bmatrix}$, $\mathbf{G} = \begin{bmatrix} 0.4424 \\ -0.1968 \end{bmatrix}$

(b) $\mathbf{F} = \begin{bmatrix} 0.9328 & 0.1920 \\ -0.3840 & 0.3568 \end{bmatrix}$, $\mathbf{G} = \begin{bmatrix} 0.0672 \\ 0.3848 \end{bmatrix}$

8-5. $(SNR)_{dB} = -1.20 + 6.02n$

8-18. $x(nT) = \begin{bmatrix} 0 & 2.29389 & -2.7553 & 6.7553 & -0.9389 & 0 & 2.9389 & -4.7553 & 4.7553 \end{bmatrix}$

8-19. (a) $X(z) = \dfrac{1}{1 - \dfrac{1}{5} z^{-1}}$

(b) $X(z) = \dfrac{1}{1 + \dfrac{1}{5} z^{-1}}$

(c) $X(z) = \dfrac{1}{1 - z^{-1}} + \dfrac{\left(\dfrac{3}{4}\right)^4 z^{-4}}{1 - \dfrac{3}{4} z^{-1}}$

(d) $X(z) = \dfrac{2 - 2z^{-8}}{1 - z^{-1}}$

8-23. $x(5T) = \dfrac{a^5}{120}$

8-24. $X(z) = \dfrac{az^{-1}}{1 + a^2 z^{-2}}$

8-28. (a) $x(nT) = [0 \quad 0 \quad 1 \quad 0.3 \quad -0.7 \quad 1 \quad 0.3 \quad -0.7]$
(b) $x(nT) = [0 \quad 0 \quad 1 \quad 0.6 \quad -1.31 \quad 0.58 \quad 1.09 \quad -1.31 \quad -0.42 \quad 0.49]$

8-36. $x(nT) = \cos(0.9n)$

8-41. $\dfrac{10}{9} + \dfrac{32}{117}\left(-\dfrac{4}{5}\right)^n - \dfrac{5}{13}\left(\dfrac{1}{2}\right)^n$

8-45. (a) $H(z) = \dfrac{1}{1 - z^{-1}}$ (b) $H(z) = \dfrac{1 + 3z^{-2}}{1 - 2z^{-1} + z^{-2}}$

(c) $H(z) = \dfrac{1}{1 - z^{-1} + 0.16z^{-2}}$ (d) $H(z) = \dfrac{1 + 4z^{-1}}{1 + 0.6z^{-1} - 0.16z^{-2}}$

(e) $H(z) = \dfrac{1 + z^{-2}}{1 - 0.707z^{-1} + 0.25z^{-2}}$

8-51. $y = [0 \quad 0 \quad 0 \quad 0 \quad 0 \quad 0 \quad 4 \quad 8 \quad 12 \quad 16 \quad 16 \quad 12 \quad 8 \quad 4 \quad 0 \quad 0 \quad 0 \quad 0 \quad 0 \quad 0 \quad 0]$
$y = [0 \quad 1 \quad 3 \quad 6 \quad 9 \quad 10 \quad 7 \quad 1 \quad -6 \quad -9 \quad -8 \quad -4 \quad 0 \quad 2 \quad 2 \quad 1 \quad 0 \quad 0 \quad 0 \quad 0 \quad 0]$

8-60. $y(nT) = \left(\dfrac{1}{4}\right)^n (n - 7)u(n - 8)$

8-66. (a) $H(1) = 2; \quad H(-1) = \dfrac{2}{3}$ (b) $H(1) = 3; \quad H(-1) = \dfrac{1}{3}$

(c) $H(1) = \dfrac{2}{3}; \quad H(-1) = -2$ (d) $H(1) = \dfrac{1}{2}; \quad H(-1) = -\dfrac{1}{3}$

8-71. $a_1 = -4.5; \quad a_2 = 4.5$

9-11. $H(z) = \dfrac{8 + 2.2z^{-1}}{1 + 0.3z^{-1}}$

9-14. $H(z) = \dfrac{1}{1 - \dfrac{1}{4} z^{-2}}$

9-16. $H(z) = \dfrac{2 - \dfrac{3}{8}z^{-1} - \dfrac{1}{16}z^{-2}}{1 - \dfrac{1}{4}z^{-2}}$

9-23. $H(z) = \dfrac{1}{N}\dfrac{1 - z^{-N}}{1 - z^{-1}}$; $A(r) = \dfrac{1}{N}\left|\dfrac{\sin(N\pi r)}{\sin(\pi r)}\right|$

9-30.   **(a)** $H(z) = 4T\left(\dfrac{1}{1 - e^{-2T}z^{-1}} - \dfrac{1}{1 - e^{-4T}z^{-1}}\right)$

   (b) $H(z) = T\left(\dfrac{1}{1 - z^{-1}} - \dfrac{2}{1 - e^{-2T}z^{-1}} + \dfrac{1}{1 - e^{-4T}z^{-1}}\right)$

   (c) $H(z) = \dfrac{2T}{7}\left(\dfrac{1}{1 - e^{-T/2}z^{-1}} - \dfrac{1}{1 - e^{-4T}z^{-1}}\right)$

9-40. $H(z) = \dfrac{0.2201z^{-1}}{1 - 0.7788z^{-1}}$

9-48. $H(z) = \dfrac{0.00149 + 0.00297z^{-1} + 0.00149z^{-2}}{1 - 1.7760z^{-1} + 0.7819z^{-2}}$

9-53. $H(z) = \dfrac{(10^{-4})(1.1338 + 3.4015z^{-1} + 3.4015z^{-2} + 1.1338z^{-3})}{1 - 2.8001z^{-1} + 2.6197z^{-2} - 0.8187z^{-3}}$

9-56. $\dfrac{\omega_r}{0.5\omega_s} = 0.538$

9-63. $A(r) = 10\sin(2\pi r)$; $\phi(r) = \dfrac{\pi}{2} - 2\pi r$

10-1. $X_k = 0$ for $k \neq 0$ and $X_k = N$ for $k = 0$.

10-2.   **(a)** $X_k = \dfrac{1 - \exp[-j2\pi kn(K + 1)/N]}{1 - \exp(-j2\pi K/N)}$, $0 < k \le N - 1$ and $X_0 = K + 1$.

10-7.   For the 8-point case, $X_k = 4\delta_{k,0} + 2\delta_{k,2} + 2\delta_{k,-2}$.

10-10. $X_0 = 7$; $X_1 = X_3 = X_5 = X_7 = 1$; $X_2 = X_4 = X_6 = -1$.

10-14.   Both require exactly the same number of both.

10-22.   **(b)** Take the square root of the sum of the squares of the real and imaginary parts of each FFT output point.

10-24.   The FFTs of the real and imaginary parts are $\{X(k)\} = \{0,\ 1.5,\ 2,\ 1.5\}$ and $\{Y(k)\} = \{-1,\ 1 + j0.5,\ -1,\ 1 - j0.5\}$.

10-25. $N = 2^7$ is the smallest power-of-2 DFT that will work.

10-27.   The result for the DFT of the window can be put into the form:

$$W(k) = A\dfrac{1 - \exp(-j2\pi kM/N)}{1 - \exp(-j2\pi k/N)} - \dfrac{B}{2}\dfrac{1 - \exp[j2\pi M(1/M - k/N)]}{1 - \exp[j2\pi(1/M - k/N)]}$$
$$- \dfrac{B}{2}\dfrac{1 - \exp[-j2\pi M(1/M - k/N)]}{1 - \exp[-j2\pi(1/M - k/N)]}$$

In the case of the Hanning window, $A = B = \frac{1}{2}$.

10-29.   The smallest power-of-2 DFT that will work is 2048.

E-5.   **(a)** 0.2171; **(b)** 0.3493; **(c)** 0.7648; **(d)** 0.8822; **(e)** 0.9976; **(f)** 1.4705

E-15. $H(s) = \dfrac{3.5531(10^6)}{s^2 + 2.6657(10^3)s + 3.5531(10^6)}$

Index